Lectures on Electromagnetism

Second Edition

T0345328

Lectures on Electromagnetism

Second Edition

Ashok Das

University of Rochester, USA

HINDUSTAN
BOOK AGENCY

World Scientific

NEW JERSEY · LONDON · SINGAPORE · BEIJING · SHANGHAI · HONG KONG · TAIPEI · CHENNAI

Published by

World Scientific Publishing Co. Pte. Ltd.

5 Toh Tuck Link, Singapore 596224

USA office: 27 Warren Street, Suite 401-402, Hackensack, NJ 07601

UK office: 57 Shelton Street, Covent Garden, London WC2H 9HE

British Library Cataloguing-in-Publication Data
A catalogue record for this book is available from the British Library.

LECTURES ON ELECTROMAGNETISM
Second Edition

ISBN 978-981-4508-26-1

Printed in India, bookbinding made in Singapore.

To
Jayanthi, Judy
and
Silvana

Contents

Preface

Over the years, I have taught a two semester graduate course on electromagnetism at the University of Rochester. The present book follows that material almost word-for-word. I have not attempted to polish the writing, and these lecture notes, therefore, reflect the informality of the class room. In fact, I even considered presenting the material in the original format, but lectures have a way of ending and starting in the middle of a topic, which is neither very appropriate nor expected of a book. Nonetheless, the subject is presented exactly in the order it was taught in class.

Some of the material is repeated in places, but this was deemed important for clarifying the lectures. The book is self-contained, in the sense that most of the steps in the development of the subject are derived in detail, and integrals are either evaluated or listed when needed. A motivated student should be able to work through the notes independently and without difficulty. Throughout the book, I followed the convention of representing three dimensional vectors by bold-faced symbols, and I use CGS units because of their relevance in special relativity.

In preparing lectures for the course, I relied, at least partially, on the material contained in the following texts:

1. P. C. Clemmow, "Introduction to Electromagnetic Theory", Cambridge University Press (1973).

2. L. Eyges, "The Classical Electromagnetic Field", Dover Publication (1972).

3. J. Frenkel, "Princípios de Eletrodinâmica Clássica", Editora da Universidade de São Paulo (1996).

4. W. Hauser, "Introduction to the Principles of Electromagnetism", Addison-Wesley Publishing (1971).

Several of my colleagues at Rochester and at other universities, as well as many of my students, have influenced the development of these lectures. Most important were, of course, the excellent questions raised by students in class and during private discussions. I sincerely appreciate everyone's input.

Part of the lecture notes were converted to LaTeX by Diane Pickersgill. The present format of the book in LaTeX is largely due to the meticulous work of Dr. Alex Constandache, who succeeded in giving it a more "user friendly" appearance. Most of the figures were drawn using PSTricks, while a few were done using Gnuplot and Xfig. I would like to thank Drs A. Constandache and F. T. Brandt for help with several figures.

It is also a pleasure to thank the editors of the TRiPS series, as well as the publisher, for being so accommodating to all my requests in connection with the book.

Finally, I thank the members of my family, in particular for their patient support and understanding during the completion of this book.

Ashok Das

Rochester

Preface to the second edition

The modifications in this second edition of the book arose mainly from the requests made by various readers. Several typos in the earlier version have been fixed and the presentation made clearer at many places, sometimes with longer derivations of the results. The figures now carry captions with references to them in the text. In addition to the numerous examples and exercises that were already present in the text, I have now included a few selected problems at the end of (almost) every chapter in the present edition.

Ashok Das

Rochester

CHAPTER 1

Electrostatics

Electrostatics is the study of the properties of electric charge distributions at rest. This is the first step in the understanding of electromagnetic phenomena. In the next few lectures, we will discuss various properties of time independent charge distributions.

1.1 Coulomb's law

It was observed quite early that when particles carrying electric charge are brought closer, they experience a force, and this force was called the electric force. The main question that one studies in electrostatics is the analysis of the electric force experienced by a given charge due to a complicated distribution of static electric charges in space. (As a side remark, let us simply note here that although at the moment it may seem like the course is only concerned with developing techniques for solving problems in electromagnetism, the techniques are quite general and are so powerful that they may be used in any other field of research as well. In that sense, we can think of the material of the course as developing powerful techniques for solving theoretical problems through examples of electromagnetism.) Fundamental to this study, therefore, is the understanding of the force between a pair of static charges separated by a given distance. This question was, in fact, studied by Coulomb in a series of impressive experiments and he found that the electric force between a pair of static particles carrying electric charge

1. is linearly proportional to the individual charges on each of the particles,

2. varies inversely as the square of the distance between the two particles,

1

3. and is a vector along the line joining the positions of the two particles carrying charge. It is attractive if the charges of the two particles have opposite sign and is repulsive otherwise.

Quantitatively, we can, therefore, write that the force experienced by a particle with charge q at the coordinate $\mathbf{r}$ because of the presence of a static particle carrying charge q_1 at the coordinate $\mathbf{r}_1$ is given by

$$\mathbf{F}(\mathbf{r} - \mathbf{r}_1) = k\,\frac{qq_1}{|\mathbf{r} - \mathbf{r}_1|^3}\,(\mathbf{r} - \mathbf{r}_1) = k\,\frac{qq_1}{|\mathbf{R}|^3}\,\mathbf{R} = k\,\frac{qq_1}{|\mathbf{R}|^2}\,\hat{\mathbf{R}}. \qquad (1.1)$$

Here, we have defined $\mathbf{R} = \mathbf{r} - \mathbf{r}_1$ (see Fig. 1.1) and have used the notation (which we will use throughout the lectures) that a boldface quantity represents a vector while a boldface quantity with a "hat" simply stands for a unit vector and, in the above equation, k represents the constant of proportionality.

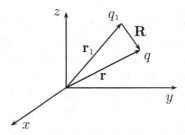

Figure 1.1: Direction of the electrostatic force experienced by a charge q at $\mathbf{r}$ due to a charge q_1 of the same sign at coordinate $\mathbf{r}_1$.

There are several things to note about the force from the expression in (1.1). First, it is translationally invariant, namely, under a translation of the coordinate system, the expression is unchanged. Sometimes, one takes advantage of this to translate the coordinate system such that q_1 is at the origin, in which case, the force has the simple form

$$\mathbf{F}(\mathbf{r}) = k\,\frac{qq_1}{|\mathbf{r}|^3}\,\mathbf{r} = k\,\frac{qq_1}{|\mathbf{r}|^2}\,\hat{\mathbf{r}}, \qquad (1.2)$$

where $\mathbf{r}$ denotes the coordinate of the charge q in the new frame (with respect to the new origin). Second, the value of the constant

k depends on the system of units used. In the CGS (Gaussian) units (which we will work with throughout these lectures), where distances are measured in centimeters and the force in dynes, $k = 1$ while in the MKS (SI) units (which is used mostly in engineering applications), where distances are measured in meters and the force in Newtons, $k = 10^{-7}c^2 \ \frac{\text{N}-\text{m}^2}{\text{C}^2}$. Here "N" stands for Newton, "C" for Coulomb and "m" for meter and c is the speed of light (just the value $\approx 3 \times 10^8$ in the MKS system without the dimensions). The rationale behind the choice of different units is as follows. In applications (SI units), currents are measured in units of Ampere which is defined to be Coulomb/sec. Therefore, one can define the unit of electric charge, Coulomb, from Ampere's law which involves the magnetic force between two currents. Once the unit of charge, Coulomb, is defined in this way, consistency determines the constant k in the Coulomb's law. (Basically, once the unit of charge has been defined from the magnetic force, it has to be consistent with the definition of force in Coulomb's law. The peculiar value of the constant arises because it involves the vacuum magnetic permeability since it originates from the definition of the magnetic force and the constant can be identified with $k = \frac{1}{4\pi\epsilon_0}$ where ϵ_0 is the dielectric permittivity of the vacuum through usual relations.) On the other hand, in the CGS (Gaussian) units, the desire is more to be consistent with the requirements of relativistic invariance. We know from the studies in relativity that electric and magnetic fields can be mapped to each other under a Lorentz transformation. The simplest way to see this is to note that if we have a static charge, it only produces an electric field. However, in a different Lorentz frame, the charge would be moving giving rise to a current and, therefore, to a magnetic field as well. Thus, relativistic invariance treats electric and magnetic fields on the same footing. From the form of Lorentz transformations, we can also see that they have to have the same dimension. The choice $k = 1$ enforces this and we see that if we choose $k = 1$, then, Coulomb's law can be thought of as defining the unit of electric charge in the CGS units, which is an "esu". (The electric force between two charges of magnitude 1 esu each, at a separation of 1 cm, is defined to be 1 dyne.) It is quite easy to see from this that (remember that $1\text{N} = 10^5 \text{dynes}$)

$$1 \text{ esu} = 1 \text{ statCoulomb} = (10\,c)^{-1}\,\text{C} = (3 \times 10^9)^{-1}\,\text{C}. \qquad (1.3)$$

In fact, let us suppose, $1\,\text{esu} = \alpha\,\text{C}$. Then,

$$k = 10^{-7} c^2 \, \frac{N - m^2}{C^2}$$

$$= 10^{-7} c^2 \times \frac{10^5 \times 10^4}{\alpha^{-2}} \, \frac{dyne - cm^2}{esu^2}$$

$$= 10^2 c^2 \alpha^2 = 1, \tag{1.4}$$

which determines $\alpha = (10\,c)^{-1}$ and, therefore, $1\,esu = (10c)^{-1}C$. (We note here that the magnitude of electric charge on an electron or on a proton is known to be $1.6 \times 10^{-19}C = 4.8 \times 10^{-10}esu$.)

Experimentally, it is also observed that when more charges are present, the force on a given charge adds vectorially as if each of the source charges interacts (pairwise) with the given charge independent of the presence of the others. Namely, if we have a distribution of static source charges $q_i, i = 1, 2, \ldots, n$, at the coordinates $\mathbf{r}_i$, then, the force experienced by a test charge q at the coordinate $\mathbf{r}$ is given by (in the CGS system that we have chosen)

$$\mathbf{F}(\mathbf{r}) = \sum_{i=1}^{n} \frac{qq_i}{|\mathbf{r} - \mathbf{r}_i|^3} (\mathbf{r} - \mathbf{r}_i). \tag{1.5}$$

This is known as the law of superposition of electric forces.

As a digression, let us talk briefly about the experimental validity of Coulomb's law. Of course, there are various uncertainties associated with any experiment and it is impossible to say experimentally that the electric force is given exactly by Coulomb's law. One can only put limits on its validity. There are two meaningful ways one puts limits on the experimental validity of Coulomb's law. The first is to assume that the force, instead of varying exactly as the inverse square of the distance, varies as $|\mathbf{r} - \mathbf{r}_1|^{-2-\delta}$ and find an experimental bound for δ. Experiments, of course, have become much more sensitive since the time of Coulomb (although the philosophy of the experiments remains essentially the same) and the present day experiments give us a bound of $\delta \sim 10^{-16}$ and, therefore, we can think of the electric force as varying inversely as the square of the distance for all practical purposes.

A second way of putting an experimental limit on the form of Coulomb's law is to parameterize the potential (from which the force is obtained) as $\frac{e^{-\mu r}}{r}$ where r represents the magnitude of the separation between the two charges (in the earlier notation $|\mathbf{r} - \mathbf{r}_1|$) and μ a mass parameter so that the exponential will be dimensionless (in some units). If Coulomb's law is exact, we should have $\mu = 0$ giving the Coulomb potential. However, if $\mu \neq 0$, the potential is called the Yukawa potential and the present experimental bound on the mass is given by $\mu < 10^{-20} m_e$, where m_e represents the mass of the electron. The reason for such a parameterization of the potential is of a quantum nature. Quantum mechanically, if we think of a force as arising due to the exchange of a particle (quantum), then μ would represent the mass of the exchanged particle which, in the case of electromagnetic forces, is called the photon. The experimental bound suggests that the photon is massless which is what we would expect from gauge invariance of the Maxwell's equations which we will discuss later. Furthermore, all these results are obtained from laboratory as well as satellite experiments. Thus, to summarize, Coulomb's law holds quite well over a wide range of distances – from very small to very large – and experimentally, we find that the mass of the exchanged particle (photon) associated with the Coulomb force is consistent with zero.

1.2 Electric field

From the form of the force in (1.5) it is clear that, even in the presence of a distribution of static charges, the force is linearly proportional to the charge of the test particle. Therefore, by dividing out the charge of the test particle, we can define an auxiliary quantity which we call the electric field. For a single source charge q_1 at $\mathbf{r}_1$, the electric field at the coordinate $\mathbf{r}$ is given by ($k = 1$ in our units)

$$\mathbf{E}(\mathbf{r}) = \frac{q_1}{|\mathbf{r} - \mathbf{r}_1|^3} (\mathbf{r} - \mathbf{r}_1), \tag{1.6}$$

while, for a distribution of static charges, the electric field at $\mathbf{r}$ is given by (the electric field also depends on the locations of the source charges which we are suppressing for simplicity)

$$\mathbf{E}(\mathbf{r}) = \sum_{i=1}^{n} \frac{q_i}{|\mathbf{r} - \mathbf{r}_i|^3} (\mathbf{r} - \mathbf{r}_i). \tag{1.7}$$

This quantity is inherently a characteristic of a given distribution of static charges and has no reference to the test charge. The electric field is not directly measurable, rather the electric force on a particle is and, from the definition of the electric field, we can think of the electric field as the force per unit charge. Namely, when an electrically charged particle with charge q is brought to the coordinate $\mathbf{r}$ in the presence of a distribution of charges, it would experience an electric force

$$\mathbf{F}(\mathbf{r}) = q\mathbf{E}(\mathbf{r}). \qquad (1.8)$$

However, the assumption here is that in bringing the test charge to the coordinate $\mathbf{r}$, we should not alter the electric field of the charge distribution appreciably (namely, the magnitude of the test charge should not be large). The concept of an electric field is not as crucial for the study of problems in electrostatics as it is for the study of time-dependent phenomena which we will deal with later. Finally, let us simply note here that the standard unit of the electric field in MKS (SI) system is $\frac{\text{Volt}}{\text{m}}$, while it is $\frac{\text{statVolt}}{\text{cm}}$ in the CGS system and they are related as

$$1\,\frac{\text{Volt}}{\text{m}} = 10^4\,c^{-1}\,\frac{\text{statVolt}}{\text{cm}} = \frac{1}{3} \times 10^{-4}\,\frac{\text{statVolt}}{\text{cm}}. \qquad (1.9)$$

So far, we have talked about a distribution of discrete charges. However, in many physical examples, we may have a continuous distribution of charges. Of course, as we know, electric charge is quantized in units of the charge of the electron. However, the magnitude of the charge carried by an electron is quite small,

$$e = 4.8 \times 10^{-10}\,\text{esu} = 1.6 \times 10^{-19}\,\text{C}. \qquad (1.10)$$

Consequently, it is quite meaningful to talk about a continuous distribution of charges. Furthermore, as we know, in quantum mechanics a particle has a probabilistic description and that the probability density associated with a particle is a continuous function (of the coordinates) leading, in the case of a charged particle, to a continuous distribution of charge. For all of these reasons, it is meaningful for us to develop various concepts for a continuous distribution of charges which is really quite simple. For example, if $\rho(\mathbf{r})$ represents the volume density of a continuous charge distribution, then, the electric field produced by this charge distribution at $\mathbf{r}$ has the form

$$\mathbf{E}(\mathbf{r}) = \int_V \mathrm{d}^3 r'\,\frac{\rho(\mathbf{r}')}{|\mathbf{r} - \mathbf{r}'|^3}\,(\mathbf{r} - \mathbf{r}'), \qquad (1.11)$$

where V denotes the volume containing the charge. This can be seen simply as follows. The total charge contained in a small volume $\mathrm{d}^3 r'$ around $\mathbf{r}'$ is given by $\Delta q = \mathrm{d}^3 r'\,\rho(\mathbf{r}')$ which will produce an electric field at the coordinate $\mathbf{r}$,

$$\Delta\mathbf{E}(\mathbf{r}) = \frac{\Delta q}{|\mathbf{r} - \mathbf{r}'|^3}\,(\mathbf{r} - \mathbf{r}') = \mathrm{d}^3 r'\,\frac{\rho(\mathbf{r}')}{|\mathbf{r} - \mathbf{r}'|^3}\,(\mathbf{r} - \mathbf{r}'). \qquad (1.12)$$

Integrating this over the volume containing the charge, we obtain
the electric field due to a continuous distribution of charges as given
in (1.11). Similarly, if we have a surface density $\sigma(\mathbf{r})$ or a linear
density $\lambda(\mathbf{r})$ of charges, the corresponding expressions for the electric
field would involve a surface integral or a line integral respectively.
However, to make a connection between a continuous and a discrete
charge distribution, we need the concept of the Dirac delta function
which we will discuss next.

The Dirac delta function is one of the most fundamental concepts in the study of microscopic systems and will probably be cov-
ered in greater detail in your study of quantum mechanics. However,
let us discuss here only what we need from the point of view of the
present discussions. The Dirac delta function is, in some sense, a generalization of the Kronecker delta to the case of continuous indices.
It is denoted by $\delta^3(\mathbf{r} - \mathbf{r}')$ (this is true in three dimensions and, in
general, in n dimensions it is denoted by $\delta^n(\mathbf{r} - \mathbf{r}')$) and is defined
such that

$$\int d^3r'\, f(\mathbf{r}')\, \delta^3(\mathbf{r} - \mathbf{r}') = f(\mathbf{r}),$$

$$\int d^3r'\, \delta^3(\mathbf{r} - \mathbf{r}') = 1, \tag{1.13}$$

for any well behaved function $f(\mathbf{r})$. The second relation in (1.13), in
fact, follows from the first if we choose $f(\mathbf{r}) = 1$. The two relations
imply that the delta function must vanish at points where its argument does not vanish and that at points where its argument vanishes,
it must diverge (see Fig. 1.2) in such a way that its integral is unity
(namely, the area under the curve is normalized).

This does not correspond to the behavior of any simple function
that we know of. In fact, the Dirac delta function is truly not a function, rather it can be thought of as a limit of a sequence of functions.
Without going into detail, let us note some explicit representations
for the delta function. In one dimension, for example, we can write

$$\delta(x - x') = \lim_{g \to \infty} \frac{1}{\pi} \frac{\sin g(x - x')}{(x - x')}$$

$$= \lim_{g \to \infty} \frac{1}{2\pi} \int_{-g}^{g} dk\, e^{ik(x - x')} = \int_{-\infty}^{\infty} \frac{dk}{2\pi}\, e^{ik(x - x')},$$

$$\delta(x - x') = \lim_{\alpha \to \infty} \sqrt{\frac{\alpha}{\pi}}\, e^{-\alpha(x - x')^2},$$

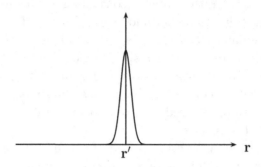

Figure 1.2: A graphical representation of the Dirac delta function $\delta^3(\mathbf{r} - \mathbf{r}')$.

$$\delta(x - x') = \frac{\mathrm{d}\theta(x - x')}{\mathrm{d}x}, \qquad (1.14)$$

where $\theta(x - x')$ is the step function defined to be

$$\theta(x - x') = \begin{cases} 1, & \text{for} \quad x - x' > 0, \\ 0, & \text{for} \quad x - x' < 0. \end{cases} \qquad (1.15)$$

The one dimensional representations in (1.14) can be easily generalized to higher dimensions. For example, in Cartesian coordinates, $\delta^3(\mathbf{r} - \mathbf{r}') = \delta(x - x')\delta(y - y')\delta(z - z')$.

The delta function has some important and useful properties which can be easily derived from its definition in (1.13). Let us simply note some of them here (in one dimension)

$$(\delta(x - x'))^* = \delta(x - x'),$$

$$\delta(x' - x) = \delta(x - x'),$$

$$\frac{\mathrm{d}}{\mathrm{d}x}\delta(x - x') = -\frac{\mathrm{d}}{\mathrm{d}x'}\delta(x - x'),$$

$$(x - x')\delta(x - x') = 0,$$

$$(x - x')\frac{\mathrm{d}}{\mathrm{d}x}\delta(x - x') = -\delta(x - x'),$$

$$\delta(a(x - x')) = \frac{1}{|a|}\delta(x - x'), \qquad (1.16)$$

and so on. It is important to note from the defining relation for the delta function in (1.13) that it has the dimension of a density (inverse volume of the space on which it is defined). Consequently, it is clear that we can define the charge density for a discrete charge q_1 at the point $\mathbf{r}_1$ as

$$\rho(\mathbf{r}) = q_1\,\delta^3(\mathbf{r} - \mathbf{r}_1). \qquad (1.17)$$

It follows then, from the defining relation in (1.11), that the electric field produced at the coordinate $\mathbf{r}$ by a point charge q_1 located at the point $\mathbf{r}_1$ is given by

$$\begin{aligned}
\mathbf{E}(\mathbf{r}) &= \int d^3r'\, \frac{\rho(\mathbf{r}')}{|\mathbf{r} - \mathbf{r}'|^3}\,(\mathbf{r} - \mathbf{r}') \\
&= \int d^3r'\, \delta^3(\mathbf{r}' - \mathbf{r}_1)\,\frac{q_1}{|\mathbf{r} - \mathbf{r}'|^3}\,(\mathbf{r} - \mathbf{r}') \\
&= \frac{q_1}{|\mathbf{r} - \mathbf{r}_1|^3}\,(\mathbf{r} - \mathbf{r}_1), \qquad (1.18)
\end{aligned}$$

as it should be (see (1.6)). A similar derivation follows as well for the electric field produced by a distribution of point charges.

1.3 Gauss' law

Let us consider a continuous charge distribution given by the volume charge density $\rho(\mathbf{r})$ contained in a finite region bounded by a surface as shown in Fig. 1.3. It is a simple matter to calculate the flux of electric field out of this surface in the following manner.

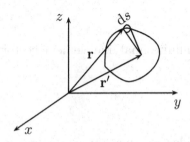

Figure 1.3: A finite volume containing a continuous distribution of charge.

From the defining relation in (1.11), we obtain the flux of the electric field to be

$$\int_S d\mathbf{s} \cdot \mathbf{E}(\mathbf{r}) = \int d^3r' \int_S d\mathbf{s} \cdot (\mathbf{r} - \mathbf{r}') \frac{\rho(\mathbf{r}')}{|\mathbf{r} - \mathbf{r}'|^3}, \tag{1.19}$$

where we have interchanged the orders of integration. The surface integral, on the right hand side, can be simplified by calculating it around $\mathbf{r}'$. The infinitesimal surface (area) element at $\mathbf{r}$ would, then, be a vector pointing radially outwards along $(\mathbf{r} - \mathbf{r}')$ and would have the form

$$d\mathbf{s} = d\Omega \, |\mathbf{r} - \mathbf{r}'|(\mathbf{r} - \mathbf{r}'), \tag{1.20}$$

where $d\Omega$ represents the solid angle subtended at $\mathbf{r}'$ by the infinitesimal surface area so that we have

$$d\mathbf{s} \cdot (\mathbf{r} - \mathbf{r}') = d\Omega \, |\mathbf{r} - \mathbf{r}'|^3. \tag{1.21}$$

Let us add here some clarification on the expression for the surface element in (1.20). First, consider an infinitesimal line element vector in polar coordinates (see Fig. 1.4) which has the form

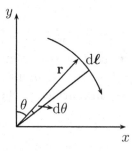

Figure 1.4: An infinitesimal line element vector in polar coordinates.

$$d\boldsymbol{\ell} = r \, d\theta \, \hat{\boldsymbol{\theta}}. \tag{1.22}$$

Now, let us consider a surface (area) element in spherical coordinates shown in Fig. 1.5. For an infinitesimal change $d\theta$ and $d\phi$ of the angular coordinates (and no change in r), we can write the surface (area) element vector to be (note that $\hat{\boldsymbol{\theta}} \times \hat{\boldsymbol{\phi}} = \hat{\mathbf{r}}$)

$$d\mathbf{s} = r \, d\theta \, \hat{\boldsymbol{\theta}} \times r \sin\theta \, d\phi \, \hat{\boldsymbol{\phi}} = r^2 \sin\theta \, d\theta \, d\phi \, \hat{\mathbf{r}} = r^2 \, d\Omega \, \hat{\mathbf{r}}. \tag{1.23}$$

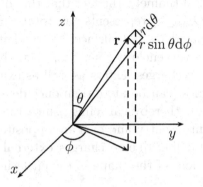

Figure 1.5: An infinitesimal surface (area) element vector in spherical coordinates.

Here $d\Omega$ is the solid angle subtended by the surface element at the origin and we have

$$\int d\Omega = \int_0^\pi d\theta \, \sin\theta \int_0^{2\pi} d\phi = 4\pi, \tag{1.24}$$

showing that the total solid angle around a point is 4π. An alternative way to understand the surface element is to note that the line element vector in spherical coordinates has the form

$$d\boldsymbol{\ell} = dr \, \hat{\mathbf{r}} + r \, d\theta \, \hat{\boldsymbol{\theta}} + r \, \sin\theta \, d\phi \, \hat{\boldsymbol{\phi}}, \tag{1.25}$$

so that the area along the direction of $\hat{\mathbf{r}}$ is simply given by (here r is fixed and $\hat{\boldsymbol{\theta}} \times \hat{\boldsymbol{\phi}} = \hat{\mathbf{r}}$)

$$d\mathbf{s} = d\boldsymbol{\ell}_\theta \times d\boldsymbol{\ell}_\phi = r \, d\theta \, \hat{\boldsymbol{\theta}} \times r \, \sin\theta \, d\phi \, \hat{\boldsymbol{\phi}}$$
$$= r^2 \, \sin\theta \, d\theta \, d\phi \, \hat{\mathbf{r}} = r^2 \, d\Omega \, \hat{\mathbf{r}}, \tag{1.26}$$

which is the result obtained earlier in (1.23). Using (1.21), the surface integral on the right hand side of (1.19) simplifies and we have

$$\int_S d\mathbf{s} \cdot \mathbf{E}(\mathbf{r}) = \int d^3r' \, \rho(\mathbf{r}') \int d\Omega$$

$$= 4\pi \int d^3r' \, \rho(\mathbf{r}') = 4\pi \, Q_{\text{enclosed}}, \tag{1.27}$$

where we have used (1.24) (namely, the fact that the total solid angle around $\mathbf{r}'$ is 4π) and Q_{enclosed} represents the total charge enclosed inside the volume bounded by the surface. Note that if there are electric charges outside the enclosing surface, then the flux of the electric field, due to such charges, enters as well as exits the surface the same number of times (can be more than once depending on the topology of the volume), thereby canceling (since the directions of the surfaces at the point of entry and exit are opposite) any further contribution to the total flux due to charges external to the given volume. Thus, independent of the shape of the enclosing surface, we have the general result

$$\int_S \mathbf{ds} \cdot \mathbf{E}(\mathbf{r}) = 4\pi\, Q_{\text{enclosed}}. \tag{1.28}$$

This is known as Gauss' law which says that the total electric flux out of a closed surface equals 4π times the total electric charge enclosed in the volume bounded by the surface. In fact, (1.28) represents the integral form of Gauss' law. We can also write a differential form for Gauss' law by appealing to Gauss' theorem which says that, for any vector function $\mathbf{A}(\mathbf{r})$,

$$\int_V d^3r\, \boldsymbol{\nabla} \cdot \mathbf{A}(\mathbf{r}) = \int_S \mathbf{ds} \cdot \mathbf{A}(\mathbf{r}), \tag{1.29}$$

where S represents the surface enclosing the volume V. Using this in (1.28), we obtain

$$\int_S \mathbf{ds} \cdot \mathbf{E}(\mathbf{r}) = 4\pi\, Q_{\text{enclosed}},$$

$$\text{or,} \quad \int_V d^3r\, \boldsymbol{\nabla} \cdot \mathbf{E}(\mathbf{r}) = 4\pi \int_V d^3r\, \rho(\mathbf{r}),$$

$$\text{or,} \quad \boldsymbol{\nabla} \cdot \mathbf{E}(\mathbf{r}) = 4\pi\, \rho(\mathbf{r}). \tag{1.30}$$

This is the differential form of Gauss' law. It is worth commenting here that Gauss' law is a consequence of the $\frac{1}{r^2}$ form of the Coulomb force (1.1). (Otherwise the coordinate dependent factors from the surface element and the force law would not cancel.) Such laws also arise for other forces which vary as inverse square of the distance and, in particular, one can write down a Gauss' law for gravitation as well. Furthermore, an experimental test of $\frac{1}{r^2}$ behavior of the Coulomb force corresponds to testing Gauss' law as well.

▶ **Example (Modified Coulomb's law).** Let us consider the hypothetical example of a point charge q at the origin producing an electric field at a point $\mathbf{r}$ of the form

$$\mathbf{E}(\mathbf{r}) = \frac{q}{r^{2+\delta}} \,\hat{\mathbf{r}}. \tag{1.31}$$

For $\delta = 0$, we have the Coulomb behavior, but if $\delta \neq 0$ (δ can be positive or negative), the charge will carry a nontrivial dimensionality.

In this case, the electric flux out of a large sphere of radius R enclosing the point charge (origin) will be given by

$$\int d\mathbf{s} \cdot \mathbf{E}(\mathbf{R}) = \int R^2 d\Omega \, \frac{q}{R^{2+\delta}} = \frac{4\pi q}{R^\delta}. \tag{1.32}$$

Namely, we observe that a modification of the Coulomb's law will lead to a modified Gauss' law where the electric flux out of a sphere will not be $4\pi Q_{\text{enclosed}}$, rather it will depend on the radius of the spherical surface. Consequently, the electric flux out of a spherical surface will change as the radius of the sphere changes even though the charge contained inside is the same.

Similarly, recalling that the divergence of a vector $\mathbf{A}$ in spherical coordinates has the form

$$\boldsymbol{\nabla} \cdot \mathbf{A} = \frac{1}{r^2} \frac{\partial(r^2 A_r)}{\partial r} + \frac{1}{r\sin\theta} \frac{\partial(\sin\theta A_\theta)}{\partial\theta} + \frac{1}{r\sin\theta} \frac{\partial A_\phi}{\partial\phi}, \tag{1.33}$$

we obtain the modified differential form of Gauss' law on the surface of a large sphere of radius R to be

$$\boldsymbol{\nabla} \cdot \mathbf{E}|_R = \frac{1}{r^2} \frac{\partial}{\partial r} \left(r^2 \frac{q}{r^{2+\delta}} \right) \bigg|_R = -\frac{q\delta}{R^{3+\delta}}, \tag{1.34}$$

which also differs from the usual differential form of Gauss' law.   ◀

Gauss' law is quite useful in determining the electric field when we know the charge distribution. However, the integral form is meaningful (easier to solve) only when there is a symmetry in the problem. As examples, let us solve some problems with symmetry.

▶ **Example (Infinitely long charged wire).** Let us consider a thin wire which is infinitely long along the z-axis and which carries a constant linear charge density of λ. We would like to determine the electric field produced by such a charge distribution.

This problem has a cylindrical symmetry and hence we expect the electric field to point radially perpendicular to the wire and that its magnitude will be the same at any point on the surface of a cylinder whose axis lies along the wire. (Any non-radial component of the electric field would vanish by the up-down symmetry of the system. In other words, any non-radial component of the electric field due to charges in the upper half of the wire would be canceled exactly by that due to the charges in the lower half of the wire. The magnitude of the electric field will be a constant on a cylindrical surface because of the radial symmetry.) Thus, drawing a cylindrical Gaussian surface of radius r (normally the radius in cylindrical coordinates is denoted by ρ, but we use r instead to avoid any confusion with the volume charge density) and height h around the wire as shown in Fig. 1.6, we obtain from Gauss' law (1.28) (applied to this surface and note that there

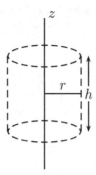

Figure 1.6: The dashed surface represents a cylindrical Gaussian surface enclosing a part of the infinitely long thin wire along the z-axis carrying a constant linear charge density λ.

is no flux through the top or the bottom of the cylinder because of the radial nature of the electric field),

$$\int_S d\mathbf{s} \cdot \mathbf{E}(\mathbf{r}) = 4\pi \, Q,$$

or, $|\mathbf{E}(\mathbf{r})| \, 2\pi r h = 4\pi \, h\lambda,$

or, $|\mathbf{E}(\mathbf{r})| = \dfrac{2\lambda}{r},$ (1.35)

and the field points radially perpendicular to the wire. (Normally, one writes this as $\mathbf{E}(\mathbf{r}) = \frac{2\lambda}{\rho}\, \hat{\rho}$, where ρ represents the radial vector on the plane in cylindrical coordinates. The cylindrical coordinates are conventionally denoted by (ρ, ϕ, z).) We see that the strength of the electric field in this case decreases inversely with the perpendicular distance from the wire. We also note here that $|\mathbf{E}(\mathbf{r})|$ denotes the magnitude of the electric field, namely, $\mathbf{E}(\mathbf{r}) = |\mathbf{E}(\mathbf{r})|\hat{\mathbf{r}}$.   ◄

Exercise. Compare this with the behavior of the electric field due to a point charge.

► **Example (Charged spherical shell).** Let us next consider a spherical shell of radius R, carrying a uniform distribution of charge characterized by the constant surface charge density σ. We would like to calculate the electric field due to such a charge distribution.

Choosing the center of the shell to be the origin, let us note that the problem has spherical symmetry and, consequently, the only direction that is physically meaningful is the radial direction from the origin (center of the shell). The electric field, being a vector, can only point along this direction. Furthermore, because of the spherical symmetry, the magnitude of the electric field at any point on the surface of a spherical shell of radius r around the origin (concentric spherical shell) must be the same. Therefore, let us imagine a spherical Gaussian surface of radius $r > R$ as shown in Fig. 1.7 and apply Gauss' law to determine the electric flux through this surface. This gives,

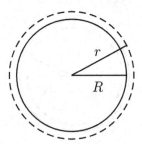

Figure 1.7: The dashed surface represents a spherical Gaussian surface enclosing the spherical shell of radius R carrying a constant surface charge density σ.

$$\int_S d\mathbf{s} \cdot \mathbf{E}(\mathbf{r}) = 4\pi\, Q,$$

or, $\quad \int r^2 d\Omega\, |\mathbf{E}(\mathbf{r})| = 4\pi\, Q,$

or, $\quad |\mathbf{E}(\mathbf{r})| = \dfrac{Q}{r^2},$ \hfill (1.36)

where, $Q = 4\pi R^2 \sigma$ is the total charge carried by the spherical shell. Furthermore, recalling that the electric field can only point radially, for $r > R$, we have

$$\mathbf{E}(\mathbf{r}) = \frac{Q}{r^2}\,\hat{\mathbf{r}} = \frac{4\pi R^2 \sigma}{r^2}\,\hat{\mathbf{r}}. \tag{1.37}$$

Namely, outside the spherical shell, the electric field behaves as if the entire charge on the surface of the shell were located at the center of the shell.

On the other hand, for points inside the shell, if we apply Gauss' law and calculate the electric flux through a spherical shell of radius $r < R$ (see Fig. 1.8), we obtain,

$$\int_S d\mathbf{s} \cdot \mathbf{E}(\mathbf{r}) = 0, \tag{1.38}$$

since there is no charge inside the Gaussian surface and this leads to the fact that inside the shell the electric field vanishes. Therefore, we determine the electric field to have the general form

$$\mathbf{E}(\mathbf{r}) = \theta(r - R)\,\frac{Q}{r^2}\,\hat{\mathbf{r}} = \theta(r - R)\,\frac{4\pi R^2 \sigma}{r^2}\,\hat{\mathbf{r}}, \tag{1.39}$$

which shows explicitly that the electric field is discontinuous across the charged surface. This is, of course, reminiscent of the behavior of the gravitational field which satisfies the inverse square law as well. Let us note from (1.39) that the

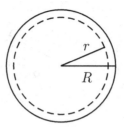

Figure 1.8: The dashed surface represents a spherical Gaussian surface inside the spherical shell of radius R carrying a constant surface charge density σ.

discontinuity in the normal component of the electric field across the surface carrying charge is given by (R and L refer respectively to right and left of the shell, or even more appropriate outside and inside the shell)

$$\hat{\mathbf{r}} \cdot (\mathbf{E}_{\mathrm{R}} - \mathbf{E}_{\mathrm{L}})\Big|_R = E_{\mathrm{n,R}} - E_{\mathrm{n,L}}\Big|_R = \frac{Q}{R^2} = 4\pi\sigma. \tag{1.40}$$

◀

▶ **Example (Charged infinite plane).** As another example of a system with symmetry, let us consider a thin rectangular plane of infinite extent carrying a constant surface charge density of σ and we would like to calculate the electric field that it produces.

From the symmetry of the problem, we realize that the direction of the electric field at any point would be perpendicular to the plane and its magnitude will be the same at points whose perpendicular distance from the plane is the same. Thus, drawing a rectangular Gaussian surface (see Fig. 1.9) whose end surfaces are equidistant from the plane and have area A, we obtain from Gauss' law that

$$\int_S d\mathbf{s} \cdot \mathbf{E} = 4\pi Q,$$

or, $2|\mathbf{E}|A = 4\pi\sigma A,$

or, $|\mathbf{E}| = 2\pi\sigma.$ \tag{1.41}

Thus, we can write

$$\mathbf{E}(\mathbf{r}) = 2\pi\sigma\,\hat{\mathbf{n}}, \tag{1.42}$$

where $\hat{\mathbf{n}}$ represents the outward unit vector normal to the plane. We see that, in this case, the magnitude of the electric field is constant at every point in space (although the direction changes on either side of the plane). We note again that

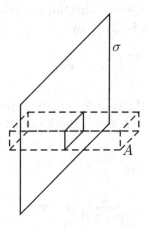

Figure 1.9: The dashed surface represents a rectangular Gaussian surface enclosing a part of the infinite plane with a constant surface charge density σ.

the normal component of the electric field is discontinuous across the surface carrying charge with the discontinuity given by

$$\hat{\mathbf{n}} \cdot (\mathbf{E}_{\mathrm{R}} - \mathbf{E}_{\mathrm{L}})\Big| = E_{\mathrm{n,R}} - E_{\mathrm{n,L}}\Big| = 4\pi\sigma, \qquad (1.43)$$

where the restriction refers to the location of the plane and (1.43) can be compared with (1.40). ◄

This demonstrates how the integral form of Gauss' law (1.28) is useful in determining the electric field when there is enough symmetry in the problem. However, we note that if there is no symmetry in the problem, the electric field on the Gaussian surface need not be constant and, consequently, the surface integration becomes more involved and the method is less useful.

1.4 Potential

We have seen that the electric field produced by a static charge distribution is a vector given by (1.11). Let us discuss briefly the nature of this vector for a static distribution of charges. The analysis of this section obviously may not hold when the charge distribution is non-static and we will come back to this question later. For the time being, however, we are interested in solving problems only in electrostatics and, therefore, this is quite meaningful as we will see shortly.

We have already seen that the electric field for an arbitrary distribution of charges is given by

$$\mathbf{E}(\mathbf{r}) = \int d^3 r' \, \frac{\rho(\mathbf{r}')}{|\mathbf{r} - \mathbf{r}'|^3} \, (\mathbf{r} - \mathbf{r}'), \tag{1.44}$$

where $\rho(\mathbf{r})$ represents the charge density of the distribution of charges in a given volume. Let us recall the identities

$$\nabla \left(\frac{1}{|\mathbf{r} - \mathbf{r}'|} \right) = -\frac{1}{|\mathbf{r} - \mathbf{r}'|^3} \, (\mathbf{r} - \mathbf{r}'),$$

$$\nabla \left(\frac{1}{|\mathbf{r} - \mathbf{r}'|^2} \right) = -\frac{2}{|\mathbf{r} - \mathbf{r}'|^4} \, (\mathbf{r} - \mathbf{r}'), \tag{1.45}$$

$$\nabla \left(\frac{1}{|\mathbf{r} - \mathbf{r}'|^3} \right) = -\frac{3}{|\mathbf{r} - \mathbf{r}'|^5} \, (\mathbf{r} - \mathbf{r}'),$$

and so on, which are readily verified in the Cartesian coordinates. For example, we note that

$$\nabla \left(\frac{1}{|\mathbf{r} - \mathbf{r}'|} \right)$$

$$= \left(\hat{\mathbf{x}} \frac{\partial}{\partial x} + \hat{\mathbf{y}} \frac{\partial}{\partial y} + \hat{\mathbf{z}} \frac{\partial}{\partial z} \right) \frac{1}{\sqrt{(x - x')^2 + (y - y')^2 + (z - z')^2}}$$

$$= -\frac{\hat{\mathbf{x}}(x - x') + \hat{\mathbf{y}}(y - y') + \hat{\mathbf{z}}(z - z')}{((x - x')^2 + ((y - y')^2 + (z - z')^2)^{\frac{3}{2}}}$$

$$= -\frac{1}{|\mathbf{r} - \mathbf{r}'|^3} \, (\mathbf{r} - \mathbf{r}'), \tag{1.46}$$

and so on.

Using the last relation in (1.45), it is now clear from the definition of the electric field in (1.44) that

$$\nabla \times \mathbf{E}(\mathbf{r}) = -3 \int d^3 r' \, \frac{\rho(\mathbf{r}')}{|\mathbf{r} - \mathbf{r}'|^5} \, (\mathbf{r} - \mathbf{r}') \times (\mathbf{r} - \mathbf{r}')$$

$$= 0, \tag{1.47}$$

where we have also used the familiar vector identity

$$\nabla \times \mathbf{r} = 0. \tag{1.48}$$

We could also have seen the identity (1.47) alternatively as follows. Note that upon using the first relation in (1.45), we can also write the electric field (1.44) as

$$\mathbf{E}(\mathbf{r}) = -\boldsymbol{\nabla} \int d^3 r' \, \frac{\rho(\mathbf{r}')}{|\mathbf{r} - \mathbf{r}'|}, \tag{1.49}$$

from which it follows that

$$\boldsymbol{\nabla} \times \mathbf{E}(\mathbf{r}) = -\boldsymbol{\nabla} \times \boldsymbol{\nabla} \int d^3 r' \, \frac{\rho(\mathbf{r}')}{|\mathbf{r} - \mathbf{r}'|} = 0. \tag{1.50}$$

Namely, we see from the definition of a static electric field that its curl vanishes and, consequently, the electric field, in this case, is a conservative field. (This is not necessarily true if the source charges are not static as we will see later.) We know from Stokes' theorem that, for any vector function $\mathbf{A}(\mathbf{r})$,

$$\int_S d\mathbf{s} \cdot (\boldsymbol{\nabla} \times \mathbf{A}) = \oint_C d\boldsymbol{\ell} \cdot \mathbf{A}, \tag{1.51}$$

where C represents a closed contour which bounds the surface S. Since the curl of the electric field vanishes, it follows from Stokes' theorem that along any closed curve

$$\oint_C d\boldsymbol{\ell} \cdot \mathbf{E} = 0, \tag{1.52}$$

so that the value of the line integral of the electric field between any two points depends only on the end points independent of the path of integration.

It is well known that any vector, which falls off rapidly at infinite, can be decomposed uniquely into the sum of two vectors, one of which is divergence free while the other has vanishing curl. This is commonly known as the Helmholtz theorem. Explicitly, the Helmholtz theorem says that, if $\mathbf{A}$ is a vector function, we can write it as

$$\mathbf{A} = \mathbf{B} + \mathbf{C}, \tag{1.53}$$

where

$$\boldsymbol{\nabla} \cdot \mathbf{B} = 0 = \boldsymbol{\nabla} \times \mathbf{C}. \tag{1.54}$$

Furthermore, we can easily express $\mathbf{B}$ and $\mathbf{C}$ in terms of $\mathbf{A}$ to write

$$\mathbf{A}(\mathbf{r}) = \frac{1}{4\pi} \boldsymbol{\nabla} \times \int d^3 r' \, \frac{\boldsymbol{\nabla}' \times \mathbf{A}(\mathbf{r}')}{|\mathbf{r} - \mathbf{r}'|}$$

$$- \frac{1}{4\pi} \boldsymbol{\nabla} \int d^3 r' \, \frac{\boldsymbol{\nabla}' \cdot \mathbf{A}(\mathbf{r}')}{|\mathbf{r} - \mathbf{r}'|}, \tag{1.55}$$

which can be readily verified by using the identity

$$\nabla^2 \left(\frac{1}{|\mathbf{r} - \mathbf{r}'|} \right) = -4\pi \, \delta^3 (\mathbf{r} - \mathbf{r}'). \tag{1.56}$$

A more physical way to understand the Helmholtz theorem is to recognize that any vector can always be uniquely decomposed into a longitudinal and a transverse component with respect to a given direction. The gradient operation (∇) indeed provides a direction and any vector can be decomposed into a parallel and a perpendicular component with respect to this direction. In modern parlance, one says that a vector can be projected parallel to the direction of ∇ and perpendicular to it and writes formally (repeated indices are summed)

$$A_i = \left(\frac{\nabla_i \nabla_j}{\nabla^2} + \left(\delta_{ij} - \frac{\nabla_i \nabla_j}{\nabla^2} \right) \right) A_j, \tag{1.57}$$

where it is understood that $\nabla^2 = \nabla_i \nabla_i$ with all the components (indices) summed. Furthermore, written this way, it is quite clear that

$$P_{ij}^{(L)} = \frac{\nabla_i \nabla_j}{\nabla^2},$$

$$P_{ij}^{(T)} = \delta_{ij} - \frac{\nabla_i \nabla_j}{\nabla^2}, \tag{1.58}$$

denote respectively the longitudinal (curl free) and the transverse (divergence free) projection operators with respect to the gradient.

The Helmholtz theorem is particularly interesting in the case of a static electric field because, in this case, the electric field has vanishing curl. Consequently, we see that for a static distribution of charges we can write the electric field as the gradient of a scalar function (the negative sign is a convention whose rationale would become clear when we deal with relativistic systems)

$$\mathbf{E}(\mathbf{r}) = -\nabla \Phi(\mathbf{r}), \tag{1.59}$$

and the form of the scalar function follows from the general form of an arbitrary vector in (1.55) (from the Helmholtz theorem) to be

$$\Phi(\mathbf{r}) = \frac{1}{4\pi} \int d^3 r' \frac{\nabla' \cdot \mathbf{E}(\mathbf{r}')}{|\mathbf{r} - \mathbf{r}'|} = \int d^3 r' \frac{\rho(\mathbf{r}')}{|\mathbf{r} - \mathbf{r}'|}. \tag{1.60}$$

Here, we have used the differential form of Gauss' law (1.30) in the last step. Let us note that we could have derived the result in (1.60) also directly from the definitions of the electric field in (1.49) and (1.59), namely,

$$\mathbf{E}(\mathbf{r}) = -\boldsymbol{\nabla} \int d^3r' \, \frac{\rho(\mathbf{r}')}{|\mathbf{r} - \mathbf{r}'|} = -\boldsymbol{\nabla}\Phi(\mathbf{r}),$$

$$\text{or,} \quad \Phi(\mathbf{r}) = \int d^3r' \, \frac{\rho(\mathbf{r}')}{|\mathbf{r} - \mathbf{r}'|}, \tag{1.61}$$

where we have ignored a constant of integration (which is related to the choice of a reference point, as we will see). Alternatively, from (1.59) we can also write

$$\Phi(\mathbf{r}) = - \int_{\infty}^{\mathbf{r}} d\boldsymbol{\ell} \cdot \mathbf{E}. \tag{1.62}$$

The scalar function $\Phi(\mathbf{r})$ is called the potential or the scalar potential or the potential function (not to be confused with the potential energy) and it is clear that if we know $\Phi(\mathbf{r})$, we can determine the electric field simply by the gradient operation. The surfaces on which $\Phi(\mathbf{r})$ is a constant are known as equipotential surfaces. Such surfaces are important in the study of electrostatics because the electric field lines are normal to them (because of the gradient nature of the electric field). The importance of $\Phi(\mathbf{r})$ lies in the fact that it is a scalar function and, consequently, is much easier to handle than the vector field $\mathbf{E}(\mathbf{r})$. Going back to the differential form of Gauss' law in (1.30), we see that with the identification in (1.59) the potential for a given distribution of charges satisfies the partial differential equation

$$\boldsymbol{\nabla} \cdot \mathbf{E}(\mathbf{r}) = 4\pi\rho(\mathbf{r}),$$

$$\text{or,} \quad \boldsymbol{\nabla}^2\Phi(\mathbf{r}) = -4\pi\rho(\mathbf{r}), \tag{1.63}$$

which is known as the Poisson equation. In regions where there are no source charges, the Poisson equation reduces to the form

$$\boldsymbol{\nabla}^2\Phi(\mathbf{r}) = 0, \tag{1.64}$$

which is known as the Laplace equation. ($\boldsymbol{\nabla}^2$ is conventionally called the Laplacian.) Solving problems in electrostatics, therefore, corresponds to solving the Poisson equation or the Laplace equation subject to appropriate boundary conditions. (Once the potential is determined, the electric field can be determined by taking the gradient.) We will develop general methods for solving these equations later when the meaning of the particular integral representation of $\Phi(\mathbf{r})$ obtained in (1.60) will also become clear.

▶ **Example.** Let us consider a charge distribution that produces an electric field of the form

$$\mathbf{E}(\mathbf{r}) = \frac{q\mathbf{r}}{r^a}, \tag{1.65}$$

where a is a real parameter and $a \neq 3$. (For $a = 3$, we have the electric field of a point charge that we have already studied in (1.6).) We would like to calculate the charge density that produces such an electric field as well as the potential associated with it.

Using (1.45) and (1.48) in (1.65) we note that

$$\boldsymbol{\nabla} \times \mathbf{E}(\mathbf{r}) = 0, \tag{1.66}$$

namely, the electric field (1.65) is curl free. Second, we obtain the divergence of the electric field to be

$$\boldsymbol{\nabla} \cdot \mathbf{E}(\mathbf{r}) = q\boldsymbol{\nabla} \cdot \frac{\mathbf{r}}{r^a}$$

$$= q\left[\boldsymbol{\nabla}\left(\frac{1}{r^a}\right) \cdot \mathbf{r} + \frac{1}{r^a}\boldsymbol{\nabla} \cdot \mathbf{r}\right]$$

$$= q\left[-\frac{a}{r^{a+2}}\mathbf{r} \cdot \mathbf{r} + \frac{3}{r^a}\right] = \frac{(3-a)q}{r^a}. \tag{1.67}$$

Comparing this with the differential form of Gauss' law (1.30), we identify the charge density that produces such an electric field to be

$$\rho(\mathbf{r}) = \frac{(3-a)q}{4\pi r^a}. \tag{1.68}$$

We can use (1.62) to calculate the potential as

$$\Phi(\mathbf{r}) = -\int_\infty^{\mathbf{r}} d\boldsymbol{\ell} \cdot \mathbf{E}$$

$$= -q\int_\infty^r \frac{dr'}{(r')^{a-1}} = -\frac{qr^{2-a}}{2-a}, \tag{1.69}$$

where we have used the fact that the electric field is conservative (curl free) and correspondingly chosen a radial path to do the line integral. (Recall that the electric field is along the radial direction. We have also thrown away a divergent constant for the case $a < 2$.) Note that when $a = 2$, the above expression needs to be calculated in a limiting manner and gives

$$\Phi_{a=2}(\mathbf{r}) = -q\ln r, \tag{1.70}$$

where we have thrown away an infinite constant.

This problem can also be studied in an alternative manner as follows. Note from (1.65) that the electric field can be rewritten as

$$\mathbf{E}(\mathbf{r}) = \frac{q\mathbf{r}}{r^a} = -\frac{q}{a-2}\boldsymbol{\nabla}\left(\frac{1}{r^{a-2}}\right) = \frac{q}{2-a}\boldsymbol{\nabla}\left(r^{2-a}\right). \tag{1.71}$$

It follows from this that the electric field is curl free

$$\boldsymbol{\nabla} \times \mathbf{E}(\mathbf{r}) = 0, \tag{1.72}$$

and the potential can be identified with

$$\Phi(\mathbf{r}) = -\frac{qr^{2-a}}{2-a}, \tag{1.73}$$

as derived earlier. Furthermore, the divergence of the electric field is given by

$$\nabla \cdot \mathbf{E}(\mathbf{r}) = -\nabla^2\Phi(\mathbf{r}) = \frac{q}{2-a}\,\nabla^2(r^{2-a}) = \frac{q}{2-a}\frac{1}{r^2}\frac{\partial}{\partial r}\left(r^2\frac{\partial}{\partial r}\right)r^{2-a}$$

$$= \frac{q}{2-a}\frac{1}{r^2}\frac{\partial}{\partial r}\left((2-a)r^{3-a}\right) = \frac{(3-a)q}{r^a}, \tag{1.74}$$

which coincides with the earlier result and leads to the charge density through Gauss' law. ◀

▶ **Example.** As another example, let us consider the Yukawa potential that we have discussed earlier. Namely, let us assume that the potential due to a charge distribution has the form

$$\Phi(\mathbf{r}) = \frac{qe^{-\mu r}}{r}, \tag{1.75}$$

where μ represents a mass scale.

In this case, the electric field can be easily calculated to have the form

$$\mathbf{E}(\mathbf{r}) = -\nabla\Phi(\mathbf{r}) = -q\hat{\mathbf{r}}\,\frac{\partial}{\partial r}\frac{e^{-\mu r}}{r}$$

$$= \frac{q\hat{\mathbf{r}}}{r^2}\left(1+\mu r\right)e^{-\mu r}. \tag{1.76}$$

We note that when $\mu = 0$, this reduces to the Coulomb field (1.6) for a point charge at the origin.

We can also calculate the charge density that produces this potential by using the differential form of Gauss' law. Namely, we note that

$$\nabla \cdot \mathbf{E}(\mathbf{r}) = q\nabla \cdot \frac{\mathbf{r}}{r^3}\left(1+\mu r\right)e^{-\mu r}$$

$$= q\left[\left(\nabla \cdot \frac{\mathbf{r}}{r^3}\right)(1+\mu r)\,e^{-\mu r} + \frac{\mathbf{r}}{r^3}\cdot\nabla\left((1+\mu r)\,e^{-\mu r}\right)\right]$$

$$= q\left[-\left(\nabla^2\left(\frac{1}{r}\right)\right)(1+\mu r)\,e^{-\mu r} + \frac{\mathbf{r}}{r^3}\cdot\left(-\mu^2\mathbf{r}\right)e^{-\mu r}\right]$$

$$= q\left[4\pi\delta^3(\mathbf{r})\,(1+\mu r)\,e^{-\mu r} - \frac{\mu^2}{r}\,e^{-\mu r}\right]$$

$$= q\left[4\pi\delta^3(\mathbf{r}) - \frac{\mu^2}{r}\,e^{-\mu r}\right] = 4\pi\rho(\mathbf{r}). \tag{1.77}$$

Here we have used (1.56) in the intermediate step. This determines the charge density associated with the Yukawa potential to be

$$\rho(\mathbf{r}) = q\,\delta^3(\mathbf{r}) - \frac{q\mu^2}{4\pi r}\,e^{-\mu r}. \tag{1.78}$$

We note that for $\mu = 0$, this reduces to (1.17). ◀

Meaning of the potential. To get a feeling for the meaning of the potential function, let us consider a test charge q being brought in from a reference point $\mathbf{r}_A$ to the point $\mathbf{r}$ in the presence of an electric field $\mathbf{E}$. Since the electric field exerts a force on the charged particle, work has to be done to move the electric charge and the amount of work needed to bring it to $\mathbf{r}$ is given by

$$W = -\int_{\mathbf{r}_A}^{\mathbf{r}} d\boldsymbol{\ell} \cdot \mathbf{F} = -q\int_{\mathbf{r}_A}^{\mathbf{r}} d\boldsymbol{\ell} \cdot \mathbf{E} = q\int_{\mathbf{r}_A}^{\mathbf{r}} d\boldsymbol{\ell} \cdot \boldsymbol{\nabla}\Phi$$

$$= q(\Phi(\mathbf{r}) - \Phi(\mathbf{r}_A)). \tag{1.79}$$

Namely, the work done is given by the difference of the potential at the two coordinates up to the multiplicative factor of the charge of the test particle. This is indeed a consequence of the conservative nature of the electric field and the result is independent of the path along which the test charge is brought to the final point. Furthermore, if we choose the reference point to be at infinity, where the potential for most physical systems vanishes (namely, if we assume that the electric field vanishes at infinity as is done in the Helmholtz theorem), then, we can write

$$W = q\Phi(\mathbf{r}). \tag{1.80}$$

Namely, we can think of the potential as the work done in bringing a unit charge from spatial infinity to a given coordinate point in the presence of an electric field. The work done, of course, can be thought of as the potential energy of the charge. The potential is measured in Volts in the MKS (SI) system whereas the unit in the CGS system is statVolt. From the definition of the potential above, it is quite clear that

$$1\,\text{Volt} = 1\,\frac{\text{Joule}}{\text{C}} = \frac{10^7\,\text{erg}}{10\,c\,\text{esu}}$$

$$= 10^6\,c^{-1}\,\text{statVolt}$$

$$= \frac{1}{3} \times 10^{-2}\,\text{statVolt}, \tag{1.81}$$

where we have used (1.3). Eq. (1.81) is consistent with (1.9).

1.5 Electrostatic energy

Let us next calculate the electrostatic potential energy for a given distribution of charges. As we will see, the form of the result is different depending on whether we are calculating the energy for a discrete distribution of charges or a continuous distribution of charges. So, let us analyze this feature in some detail. First, let us assume that we have a discrete distribution of point charges. Let us not worry about how the charges were produced and assume that the potential energy for such a system of charges, when they are infinitely separated from one another, is zero. We want to calculate the electrostatic energy associated with such a distribution of charges. To this end, we note that, given a charge q_1 at $\mathbf{r}_1$, if we bring in a second charge q_2 to the coordinate $\mathbf{r}_2$, the work done (and hence the potential energy) would be given by

$$W_{12} = -\int_{\infty}^{\mathbf{r}_2} d\boldsymbol{\ell} \cdot \mathbf{F}_{12}$$

$$= -q_1 q_2 \int_{\infty}^{\mathbf{r}_2} d\mathbf{r} \cdot \frac{(\mathbf{r} - \mathbf{r}_1)}{|\mathbf{r} - \mathbf{r}_1|^3}$$

$$= -q_1 q_2 \int_{\infty}^{\mathbf{r}_2 - \mathbf{r}_1} d\mathbf{r} \cdot \frac{\hat{\mathbf{r}}}{|\mathbf{r}|^2}$$

$$= -q_1 q_2 \int_{\infty}^{|\mathbf{r}_1 - \mathbf{r}_2|} \frac{dr}{r^2}$$

$$= \frac{q_1 q_2}{|\mathbf{r}_1 - \mathbf{r}_2|}, \tag{1.82}$$

where $|\mathbf{r}| = r$ and, since the integral is independent of the path, owing to the conservative nature of the electric field, we have chosen a radial path in evaluating the integral. We can now keep adding more and more charges and since the electric force is additive, the calculation simplifies. For example, to bring in a third charge q_3 to the point $\mathbf{r}_3$, the total work done is given by

$$W_{123} = W_{12} - \int_{\infty}^{\mathbf{r}_3} d\boldsymbol{\ell} \cdot (\mathbf{F}_{13} + \mathbf{F}_{23})$$

$$= \frac{q_1 q_2}{|\mathbf{r}_1 - \mathbf{r}_2|} + \frac{q_1 q_3}{|\mathbf{r}_1 - \mathbf{r}_3|} + \frac{q_2 q_3}{|\mathbf{r}_2 - \mathbf{r}_3|}. \tag{1.83}$$

It is obvious that this expression is completely symmetric in the permutation of any pair of charged particles (charges and coordinates) and hence, the order in which the charges are brought in does not matter. Carrying this out for n charged particles, it is easily obtained that the total work required is

$$W = \frac{1}{2} \sum_{i,j,i \neq j} \frac{q_i q_j}{|\mathbf{r}_i - \mathbf{r}_j|}, \tag{1.84}$$

where the factor of $\frac{1}{2}$ is there to avoid double counting. This is, therefore, the electrostatic energy for a distribution of point charges.

Let us next calculate the electrostatic energy for a continuous distribution of charges. Once again, generalizing (1.84) to a continuous distribution of charges, we can write

$$W = \frac{1}{2} \int d^3r \, d^3r' \, \frac{\rho(\mathbf{r})\rho(\mathbf{r}')}{|\mathbf{r} - \mathbf{r}'|}$$

$$= \frac{1}{2} \int d^3r \, \rho(\mathbf{r}) \int d^3r' \, \frac{\rho(\mathbf{r}')}{|\mathbf{r} - \mathbf{r}'|} = \frac{1}{2} \int d^3r \, \rho(\mathbf{r})\Phi(\mathbf{r})$$

$$= -\frac{1}{8\pi} \int d^3r \, \left(\nabla^2 \Phi(\mathbf{r})\right) \Phi(\mathbf{r})$$

$$= \frac{1}{8\pi} \int d^3r \, (\nabla\Phi(\mathbf{r})) \cdot (\nabla\Phi(\mathbf{r}))$$

$$= \frac{1}{8\pi} \int d^3r \, \mathbf{E}^2(\mathbf{r}), \tag{1.85}$$

where we have used the differential form of Gauss' law in (1.63) (Poisson equation) as well as the relation between the electric field and the potential. We have also neglected surface terms in the integration by parts with the assumption that the electric field falls off rapidly at infinity.

The difference between the two cases needs to be discussed. First, we note that the energy for a continuous distribution of charges is completely given in terms of the electric field and is non-negative. On the other hand, for a distribution of point charges the sign of the energy in (1.84) clearly depends on the signs of various charges and can, in fact, be negative. Let us recall that the electrostatic energy for a pair of similarly charged particles is positive while it is negative if the charges of the two particles are opposite in sign. This difference in the behavior of the electrostatic energy between a discrete and a continuous distribution of charges arises mainly because of our choice of the reference system in the case of point charges. We assumed the zero of the energy to correspond to the system of point charges separated by an infinite distance thereby discarding the self-energy associated with the point charges. (Experimentally, one measures only the differences in energy and, therefore, this is quite acceptable.) The reason for such a choice is that the self-energy for a point charge is known to diverge and hence, it is meaningful to throw away such a divergent constant in the calculation. In contrast, the expression for the electrostatic energy of a continuous distribution of charges is more complete in that it contains the self-energy.

To summarize what we have learnt so far, all of electrostatics can be described by the two equations

$$\boldsymbol{\nabla} \times \mathbf{E} = 0 \implies \mathbf{E}(\mathbf{r}) = -\boldsymbol{\nabla}\,\Phi(\mathbf{r}), \tag{1.86}$$

and

$$\boldsymbol{\nabla} \cdot \mathbf{E}(\mathbf{r}) = 4\pi\,\rho(\mathbf{r}) \implies \nabla^2\Phi(\mathbf{r}) = -4\pi\,\rho(\mathbf{r}). \tag{1.87}$$

These equations can also be written in the respective integral forms as

$$\oint_C d\boldsymbol{\ell} \cdot \mathbf{E} = 0,$$

$$\int_S d\mathbf{s} \cdot \mathbf{E} = 4\pi\,Q. \tag{1.88}$$

Furthermore, the solution of the Poisson equation has the general form

$$\Phi(\mathbf{r}) = \int d^3r'\,\frac{\rho(\mathbf{r}')}{|\mathbf{r} - \mathbf{r}'|}. \tag{1.89}$$

It is worth remarking here that the true dynamical equations of electromagnetism (which we will study later) are coupled equations, involving both electric and magnetic fields, as one would expect from relativistic invariance. However, in the static limit these equations decouple, making it possible and meaningful to study electrostatics and magnetostatics separately.

1.6 Selected problems

1. Prove the following relations from vector analysis ($\mathbf{A}$ and $\mathbf{B}$ are three dimensional vectors)

$$\boldsymbol{\nabla} \times (\boldsymbol{\nabla} \times \mathbf{A}) = \boldsymbol{\nabla}(\boldsymbol{\nabla} \cdot \mathbf{A}) - \nabla^2\mathbf{A},$$

$$\nabla \cdot (\mathbf{A} \times \mathbf{B}) = \mathbf{B} \cdot (\nabla \times \mathbf{A}) - \mathbf{A} \cdot (\nabla \times \mathbf{B}),$$

$$\nabla \cdot \mathbf{r} = 3, \qquad \nabla \times \mathbf{r} = 0,$$

$$\nabla^2 \left(\frac{1}{|\mathbf{r}|} \right) = -4\pi \delta^3(\mathbf{r}).$$

2. Show that if $T_{ij}, i, j = 1, 2, 3$, denotes a second rank three dimensional tensor, then a generalization of Gauss' theorem would lead to

$$\int_V d^3 r \, \partial_j T_{ij} = \int_S ds_j \, T_{ij},$$

where S denotes the surface bounding the volume V and repeated indices are assumed to be summed.

3. Using the result from the last problem show that for an arbitrary vector $\mathbf{A}$ we have

$$\int_V d^3 r \, \nabla \times \mathbf{A} = \int_S ds \times \mathbf{A},$$

where S denotes the surface bounding the volume V.

4. Given the following two vectors, $\mathbf{E}$, which do you think would describe a true static electric field?

   (i) $\mathbf{E} = k[xy\hat{\mathbf{x}} + 2yz\hat{\mathbf{y}} + 3xz\hat{\mathbf{z}}]$,

   (ii) $\mathbf{E} = k[y^2\hat{\mathbf{x}} + (2xy + z^2)\hat{\mathbf{y}} + 2yz\hat{\mathbf{z}}]$.

   Here, k is a constant. For the true electric field, determine the potential with the origin as the reference point.

5. Consider the hypothetical case that the Yukawa potential given by

$$\Phi(\mathbf{r}) = q \, \frac{e^{-\mu r}}{r},$$

where $r = |\mathbf{r}|$ and μ is a mass parameter (in units of $c, \hbar$) is produced by a point charge q at the origin. Would Gauss' law be valid for such a case? Show that, for $\mathbf{r} \neq 0$, this potential satisfies the equation

$$\nabla^2 \Phi(\mathbf{r}) = \mu^2 \Phi(\mathbf{r}).$$

6. Calculate the electric field inside and outside of a solid sphere of radius R carrying a constant volume charge density ρ.

7. Consider a spherical region V of radius R without any charge and a point charge q outside the spherical region at a distance r from the center. Evaluate explicitly the net electric flux out of the surface S which bounds the spherical region V.

8. Consider a spherical shell of radius R with a uniform surface charge density σ. What is the electrostatic energy stored in such a system. What happens as the radius of the sphere decreases? What is the electrostatic energy for a solid sphere of radius R carrying a uniform volume charge density ρ?

9. Consider a spherical distribution of charge for which the volume charge density is nonzero only for $0 \le r \le R$ and has the form

$$\rho(\mathbf{r}) = k\,r^{-n},$$

where both k and n are positive constants. Calculate the electrostatic energy associated with such a distribution of charges. For what values of n is the energy finite?

CHAPTER 2
Potential for simple systems

Let us next calculate the potential function for some simple systems with known charge distributions, to get a feeling for its properties. The simplest example, of course, is the potential for a point charge. Let us assume that a charge q_1 is located at the coordinate $\mathbf{r}_1$. In such a case, we can write the charge density as (see (1.17))

$$\rho(\mathbf{r}) = q_1 \, \delta^3(\mathbf{r} - \mathbf{r}_1), \tag{2.1}$$

so that the potential at the coordinate $\mathbf{r}$ is easily obtained to be

$$\Phi(\mathbf{r}) = \int \mathrm{d}^3 r' \, \frac{\rho(\mathbf{r}')}{|\mathbf{r} - \mathbf{r}'|}$$

$$= \int \mathrm{d}^3 r' \, \delta^3(\mathbf{r}' - \mathbf{r}_1) \, \frac{q_1}{|\mathbf{r} - \mathbf{r}'|}$$

$$= \frac{q_1}{|\mathbf{r} - \mathbf{r}_1|}, \tag{2.2}$$

as we would expect. Similarly, we can also calculate the potential for a system of point charges. One thing to note from this calculation is the fact that the potential for a single charge is a continuous function of $\mathbf{r}$. Furthermore, it vanishes as $|\mathbf{r}| \to \infty$ for a fixed $\mathbf{r}_1$. This is consistent with our choice of the reference point for the potential, namely, that the potential vanishes for infinite separation.

▶ **Example (Hydrogen atom).** The electric charge distribution (due to the electron) in the ground state of the Hydrogen atom is given by

$$\rho(\mathbf{r}) = \frac{q}{\pi a^3} \, e^{-\frac{2r}{a}}, \tag{2.3}$$

where q represents the charge of the electron and a, the Bohr radius. We would like to calculate the potential as well as the electric field due to such a charge distribution. We note that the total charge of the distribution is given by

$$\int \mathrm{d}^3 r \, \rho(\mathbf{r}) = \frac{q}{\pi a^3} \, (4\pi) \int_0^\infty \mathrm{d}r \, r^2 \, e^{-\frac{2r}{a}} = \frac{q}{2} \int_0^\infty \mathrm{d}r \, r^2 e^{-r} = \frac{q}{2} \Gamma(3) = q, \tag{2.4}$$

31

which is to be expected since the charge density in (2.3) is simply q times the (quantum mechanical) probability density in the ground state.

The calculation of the potential and the electric field can be carried out in one of two ways. First, let us note that the charge distribution is spherically symmetric. Consequently, we expect the electric field as well as the potential to reflect this. Using Gauss' law to integrate the electric field over a spherical (Gaussian) surface of radius r we obtain

$$\int d\mathbf{s} \cdot \mathbf{E} = 4\pi \int d^3 r' \, \rho(\mathbf{r}')$$

or, $$4\pi r^2 |\mathbf{E}(\mathbf{r})| = 4\pi \int d\Omega \int_0^r r'^2 dr' \, \frac{q}{\pi a^3} e^{-\frac{2r'}{a}}$$

or, $$|\mathbf{E}(\mathbf{r})| = \frac{q}{\pi a^3 r^2} (4\pi) \int_0^r dr' \, r'^2 \, e^{-\frac{2r'}{a}}$$

$$= \frac{q}{2r^2} \int_0^{\frac{2r}{a}} dr' \, r'^2 \, e^{-r'}$$

$$= \frac{q}{2r^2} \left[(-r'^2 - 2r' - 2)e^{-r'} \right]_0^{\frac{2r}{a}}$$

$$= \frac{q}{r^2} \left[1 - \left(1 + \frac{2r}{a} + \frac{2r^2}{a^2} \right) e^{-\frac{2r}{a}} \right]. \tag{2.5}$$

Therefore, we conclude that the radial electric field is given by

$$\mathbf{E}(\mathbf{r}) = \frac{q\hat{\mathbf{r}}}{r^2} \left[1 - \left(1 + \frac{2r}{a} + \frac{2r^2}{a^2} \right) e^{-\frac{2r}{a}} \right]. \tag{2.6}$$

The potential is then easily determined from

$$\Phi(\mathbf{r}) = - \int_\infty^r d\boldsymbol{\ell} \cdot \mathbf{E}(\mathbf{r}')$$

$$= - \int_\infty^r dr' \, \frac{q}{r'^2} \left[1 - \left(1 + \frac{2r'}{a} + \frac{2r'^2}{a^2} \right) e^{-\frac{2r'}{a}} \right]$$

$$= q \int_\infty^r dr' \, \frac{d}{dr'} \left[\frac{1}{r'} \left(1 - \left(1 + \frac{r'}{a} \right) e^{-\frac{2r'}{a}} \right) \right]$$

$$= \frac{q}{r} \left[1 - \left(1 + \frac{r}{a} \right) e^{-\frac{2r}{a}} \right]. \tag{2.7}$$

The second way of solving the problem is to note that given the charge density, we can obtain the potential simply as

$$\Phi(\mathbf{r}) = \int d^3 r' \, \frac{\rho(\mathbf{r}')}{|\mathbf{r} - \mathbf{r}'|} = \frac{q}{\pi a^3} \int d^3 r' \, \frac{e^{-\frac{2r'}{a}}}{|\mathbf{r} - \mathbf{r}'|}$$

$$= \frac{q}{\pi a^3} (2\pi) \int\limits_{0}^{\infty} dr' \, r'^2 e^{-\frac{2r'}{a}} \int\limits_{-1}^{1} \frac{d\cos\theta}{(r^2 + r'^2 - 2rr'\cos\theta)^{\frac{1}{2}}}$$

$$= \frac{q}{\pi a^3} \frac{(-2\pi)}{r} \int\limits_{0}^{\infty} dr' \, r' e^{-\frac{2r'}{a}} \left(|r - r'| - (r + r') \right)$$

$$= \frac{4q}{ra^3} \left[\int\limits_{0}^{r} dr' \, r'^2 e^{-\frac{2r'}{a}} + r \int\limits_{r}^{\infty} dr' \, r' e^{-\frac{2r'}{a}} \right]$$

$$= \frac{q}{r} \left[1 - \left(1 + \frac{r}{a} \right) e^{-\frac{2r}{a}} \right]. \tag{2.8}$$

This is the same result as obtained in the earlier method in (2.7). The electric field now follows from the definition

$$\mathbf{E}(\mathbf{r}) = -\boldsymbol{\nabla}\Phi(\mathbf{r}) = -\hat{\mathbf{r}} \frac{\partial}{\partial r} \frac{q}{r} \left[1 - \left(1 + \frac{r}{a} \right) e^{-\frac{2r}{a}} \right]$$

$$= \frac{q\hat{\mathbf{r}}}{r^2} \left[1 - \left(1 + \frac{2r}{a} + \frac{2r^2}{a^2} \right) e^{-\frac{2r}{a}} \right], \tag{2.9}$$

which is what we had obtained earlier in (2.6). ◀

Before we proceed further with calculations, let us recall some of the results from our earlier analyses in the last chapter. We have noted from the calculations of the electric fields in the presence of surface charges that the electric field is discontinuous across a surface carrying charge. In fact, the general result is (see (1.40))

$$\hat{\mathbf{n}} \cdot (\mathbf{E}_R - \mathbf{E}_L) = 4\pi\sigma, \tag{2.10}$$

where $\mathbf{E}_R$ and $\mathbf{E}_L$ represent the electric fields infinitesimally close to the surface to the right and to the left respectively and $\hat{\mathbf{n}}$ denotes the unit vector normal to the surface on the right. In contrast, the tangential components of the electric fields are continuous across the surface. Let us next examine the behavior of the electric potential across a surface carrying a charge density.

2.1 Potential for a thin spherical shell

Let us consider a thin spherical shell of radius R with a constant surface charge density σ as shown in Fig. 2.1. When we have a surface charge density, as opposed to a volume charge density, the expression for the potential can be rewritten as a surface integral

$$\Phi(\mathbf{r}) = \int ds' \, \frac{\sigma(\mathbf{r}')}{|\mathbf{r} - \mathbf{r}'|}. \tag{2.11}$$

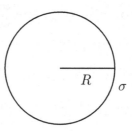

Figure 2.1: A thin spherical shell of radius R with a constant surface charge density σ.

Alternatively, we note that we can express the surface charge density as a volume charge density through the use of the Dirac delta function, which in the present case takes the form (remember that the delta function has the inverse dimension of its argument)

$$\rho(\mathbf{r}) = \sigma\delta(r - R). \tag{2.12}$$

Using this, then, we obtain from (2.1),

$$
\begin{aligned}
\Phi(\mathbf{r}) &= \int d^3r' \, \frac{\rho(\mathbf{r}')}{|\mathbf{r} - \mathbf{r}'|} \\
&= \int r'^2 dr' \, \sin\theta' \, d\theta' \, d\phi' \, \frac{\sigma\delta(r' - R)}{(r^2 + r'^2 - 2rr'\cos\theta')^{\frac{1}{2}}} \\
&= 2\pi R^2 \sigma \int_{-1}^{1} \frac{d\alpha}{(r^2 + R^2 - 2rR\alpha)^{\frac{1}{2}}} \\
&= 2\pi R^2 \sigma \left(-\frac{1}{rR}\right) (r^2 + R^2 - 2rR\alpha)^{\frac{1}{2}} \Big|_{-1}^{1} \\
&= 2\pi R^2 \sigma \left(-\frac{1}{rR}\right) \left(\sqrt{(r - R)^2} - \sqrt{(r + R)^2}\right). \tag{2.13}
\end{aligned}
$$

Here, we have simplified our calculation by choosing the z-axis to be parallel to $\mathbf{r}$ and have defined $\alpha = \cos\theta'$ in the intermediate step. In evaluating the final form of the potential, we have to be careful in choosing the positive square root, particularly in the first factor.

Recognizing that $Q = 4\pi R^2 \sigma$, we obtain from (2.13)

$$\Phi(\mathbf{r}) = \frac{2\pi R\sigma}{r} (r + R - |r - R|)$$

$$= \begin{cases} \frac{4\pi R^2 \sigma}{r} = \frac{Q}{r} & \text{for } r > R, \\ \frac{4\pi R^2 \sigma}{R} = \frac{Q}{R} & \text{for } r < R. \end{cases} \tag{2.14}$$

We can also rewrite this as

$$\Phi(\mathbf{r}) = \theta(r - R)\frac{Q}{r} + \theta(R - r)\frac{Q}{R}. \tag{2.15}$$

There are several things to observe from the result in (2.15) (or (2.14)). First, the potential is spherically symmetric, as it should be because of the symmetry in the problem. Second, it is a continuous function across $r = R$, namely, across the surface carrying the charge. This has to be contrasted with the behavior of the electric field. Outside the shell, the potential behaves as if all the charge were located at the origin. We also note that the potential is a constant inside the shell. In fact, the value of the potential at the origin is readily seen to be the average of the potential over any closed surface within the shell. This, as we will see, is a general feature of the solutions of Laplace equation. Finally, given the potential, we can determine the electric field by taking the gradient (see (1.59)). Recalling from (2.15) that the potential only depends on the radial coordinate, we obtain

$$\mathbf{E}(\mathbf{r}) = -\boldsymbol{\nabla}\Phi(\mathbf{r})$$

$$= -\hat{\mathbf{r}}\frac{\partial}{\partial r}\left(\theta(r - R)\frac{Q}{r} + \theta(R - r)\frac{Q}{R}\right)$$

$$= \theta(r - R)\frac{Q}{r^2}\hat{\mathbf{r}}. \tag{2.16}$$

The important thing to note in this derivation is that the derivatives of the two theta functions give delta functions of opposite sign which cancel each other. Eq. (2.16) is, of course, our previous result obtained in (1.37) and (1.38), namely, the electric field is non-vanishing only outside the shell and, at such points, it behaves as if the total charge were concentrated at the origin. Furthermore, expressed as in (2.16), the discontinuity in the electric field across the surface carrying charge is manifest.

2.2 Potential for an infinitely long wire

Although this problem is quite simple, it is worth going through the derivation which brings out some particular property of the choice of the reference point. Let us consider an infinitely long and thin wire carrying a constant linear charge density λ. For simplicity, we assume the wire to lie along the z-axis as shown in Fig. 2.2.

Figure 2.2: An infinitely long wire along the z-axis carrying a constant linear charge density λ.

In this case the problem has cylindrical symmetry and, consequently, it is meaningful to use cylindrical coordinates (see Fig. 2.3) given by

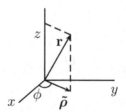

Figure 2.3: Cylindrical coordinates $(\tilde{\rho}, \phi, z)$.

$$x = \tilde{\rho} \cos \phi,$$

$$y = \tilde{\rho} \sin \phi,$$

$$z = z, \tag{2.17}$$

where $\tilde{\rho}$ represents the radial coordinate on the plane $z = 0$ ($z =$ constant). (Normally, it is denoted as ρ. Here, we have added a tilde to distinguish it from the volume charge density.)

Once again, since we have a problem with a linear charge density along the z-axis, we can write the potential as a line integral

$$\Phi(\mathbf{r}) = \int dz' \frac{\lambda(\mathbf{r}')}{|\mathbf{r} - \mathbf{r}'|}. \qquad (2.18)$$

Alternatively, we can express the linear density as a volume density, which, for the present problem, can be done in the following manner. We note that the wire carrying charge lies along the z-axis and, therefore, the charge density is nonzero only for $x = 0 = y$. Thus, we can write a volume charge density for the system as

$$\rho(\mathbf{r}) = \lambda \delta(x)\delta(y) = \frac{\lambda}{2\pi\tilde{\rho}} \delta(\tilde{\rho}), \qquad (2.19)$$

where the normalization factor of $\frac{1}{2\pi\tilde{\rho}}$ arises from the observation that if, $\delta(x)\delta(y) = c\delta(\tilde{\rho})$ (c has to be independent of the angle ϕ by rotational symmetry), then,

$$\int dx\, dy\, \delta(x)\delta(y) = 1,$$

$$\text{or,} \quad \int \tilde{\rho}\, d\tilde{\rho}\, d\phi\, c\delta(\tilde{\rho}) = 1,$$

$$\text{or,} \quad c = \frac{1}{2\pi\tilde{\rho}}. \qquad (2.20)$$

With (2.19), we can now calculate the potential due to a thin wire from the definition in (2.1).

$$\Phi(\mathbf{r}) = \int d^3r' \frac{\rho(\mathbf{r}')}{|\mathbf{r} - \mathbf{r}'|}$$

$$= \int \tilde{\rho}'\, d\tilde{\rho}'\, d\phi'\, dz' \frac{\lambda}{2\pi\tilde{\rho}'} \frac{\delta(\tilde{\rho}')}{((\tilde{\boldsymbol{\rho}} - \tilde{\boldsymbol{\rho}}')^2 + (z - z')^2)^{\frac{1}{2}}}$$

$$= \lambda \int\limits_{-\infty}^{\infty} \frac{dz'}{(\tilde{\rho}^2 + (z - z')^2)^{\frac{1}{2}}}$$

$$= \lambda \int\limits_{-\infty}^{\infty} \frac{d\bar{z}}{(\tilde{\rho}^2 + \bar{z}^2)^{\frac{1}{2}}}$$

$$= \lambda \ln\left(\bar{z} + \sqrt{\tilde{\rho}^2 + \bar{z}^2}\right)\Big|_{-\infty}^{\infty}. \qquad (2.21)$$

It is clear that the right hand side of (2.21) diverges and the reason
for this is not hard to see. In writing an expression for the potential, we had chosen the potential to vanish at infinity which we had
taken as a reference point. However, in the present problem, such
a choice is not consistent simply because the charge density extends
to spatial infinity. (We have an infinitely long wire as is clear from
the integration limits.) The proper way to analyze this problem is
to recognize that we must choose a different reference point for this
problem (or equivalently allow for a constant potential at infinity).
In particular, let us note that a different choice of the reference point
simply corresponds to adding a constant term to the potential, be it
infinite. Thus, we can extract the finite meaningful potential from
(2.21) by writing

$$
\begin{aligned}
\Phi(\mathbf{r}) &= \lim_{\Lambda \to \infty} \lambda \, \ln \left(\tilde{z} + \sqrt{\tilde{\rho}^2 + \tilde{z}^2} \right) \Big|_{-\Lambda}^{\Lambda} \\
&= \lim_{\Lambda \to \infty} \lambda \, \ln \frac{(\Lambda + \sqrt{\tilde{\rho}^2 + \Lambda^2})}{(-\Lambda + \sqrt{\tilde{\rho}^2 + \Lambda^2})} \\
&= \lim_{\Lambda \to \infty} \lambda \, \ln \frac{(\Lambda + \sqrt{\tilde{\rho}^2 + \Lambda^2})^2}{\tilde{\rho}^2} \\
&\approx \lim_{\Lambda \to \infty} \lambda \, \ln \frac{4\Lambda^2 + 2\tilde{\rho}^2 + O(\frac{1}{\Lambda^2})}{\tilde{\rho}^2} \\
&= -2\lambda \, \ln \tilde{\rho} + \text{ constant},
\end{aligned}
\tag{2.22}
$$

where the constant, on the right hand side, is a divergent constant.
(The important thing to remember is that the potential is not observable, but the electric field is through the electric force. The electric
field is obtained from the potential by the gradient operation so that
a constant term in the potential does not contribute to the electric
field.) Thus, discarding the constant we determine the potential for
the infinitely long charged wire to be

$$
\Phi(\mathbf{r}) = -2\lambda \, \ln \tilde{\rho}.
\tag{2.23}
$$

We see that the potential has cylindrical symmetry and it is continuous. The electric field can be obtained from (2.23) (or (2.22))
by taking the gradient and since the potential only depends on the
coordinate $\tilde{\rho}$, we obtain

$$
\mathbf{E}(\mathbf{r}) = -\boldsymbol{\nabla}\Phi(\mathbf{r}) = -\hat{\tilde{\rho}} \, \frac{\partial}{\partial \tilde{\rho}} \left(-2\lambda \, \ln \tilde{\rho} \right) = \frac{2\lambda}{\tilde{\rho}} \, \hat{\tilde{\rho}},
\tag{2.24}
$$

which is the same result that we had obtained earlier in (1.35).

2.3 Potential for a circular charged disc

Let us next consider a thin circular disc of radius R which carries a uniform surface charge density σ. For simplicity, we choose the disc to lie in the $x - y$ plane (at $z = 0$) with the center at the origin of the coordinate system as shown in Fig. 2.4.

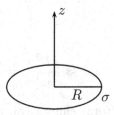

Figure 2.4: A thin circular disc of radius R carrying a uniform surface charge density σ.

Let us calculate the potential due to the disc at points along the z-axis, namely, at $(x = 0, y = 0, z)$. Clearly, this is a problem with cylindrical symmetry and it is meaningful to use the cylindrical coordinates $\mathbf{r} = (r, \phi, z)$ (namely, we are simply going to write r for what we called $\tilde{\rho}$ before). Then, as before, we can write the surface density of charge as a volume density of the form

$$\rho(\mathbf{r}) = \sigma\,\delta(z)\,\theta(R - r). \tag{2.25}$$

Note that the theta function implements the finite extension of the disc (and, therefore, the charge distribution).

We can now calculate the potential at a point on the z-axis $(x = 0 = y)$ simply as

$$\Phi(x = 0, y = 0, z) = \int d^3 r'\, \frac{\rho(\mathbf{r}')}{|\mathbf{r} - \mathbf{r}'|}$$

$$= \int r'\,dr'd\phi'\,dz'\, \frac{\sigma\delta(z')\theta(R - r')}{(r'^2 + (z - z')^2)^{\frac{1}{2}}}$$

$$= 2\pi\sigma \int\limits_0^R \frac{r'dr'}{(r'^2 + z^2)^{\frac{1}{2}}}$$

$$= 2\pi\sigma \left. \left(r'^2 + z^2\right)^{\frac{1}{2}} \right|_0^R$$

$$= 2\pi\sigma \left(\sqrt{R^2 + z^2} - |z|\right). \tag{2.26}$$

This shows that the potential is a continuous function. Note that, at very far off distances, namely, when $|z| \gg R$, we can expand the potential in (2.26) in a power series to obtain

$$\Phi(x = 0, y = 0, |z| \gg R) = 2\pi\sigma \left(|z| \left(1 + \frac{R^2}{z^2}\right)^{\frac{1}{2}} - |z|\right)$$

$$\approx \frac{\pi R^2 \sigma}{|z|} = \frac{Q}{|z|}, \tag{2.27}$$

where $Q = \pi R^2 \sigma$ is the total charge contained on the disc. Thus, we see that very far away from the disc, the potential along the z-axis behaves as if all the charge were concentrated at the origin (like a point charge). We also note that, at the center of the disc $(x = 0, y = 0, z = 0)$, the potential has the value

$$\Phi(x = 0, y = 0, z = 0) = 2\pi\sigma R. \tag{2.28}$$

Furthermore, the electric field along the z-axis can be obtained from the potential in (2.26) by taking the gradient which gives

$$\mathbf{E}(x = 0, y = 0, z) = -\boldsymbol{\nabla}\Phi(x = 0, y = 0, z)$$

$$= -\hat{\mathbf{z}} \frac{\partial}{\partial z} 2\pi\sigma \left(\sqrt{R^2 + z^2} - |z|\right)$$

$$= -2\pi\sigma \left(\frac{z}{\sqrt{R^2 + z^2}} - \text{sgn}(z)\right) \hat{\mathbf{z}}. \tag{2.29}$$

Here, sgn (z) stands for the sign of z which can also be represented as an alternating step function. It is obvious from (2.29) that very close to the disc, namely, for $z \approx 0$, the electric field has the leading behavior

$$\mathbf{E} \approx \text{sgn}(z)\, 2\pi\sigma\, \hat{\mathbf{z}}. \tag{2.30}$$

First of all, this explicitly shows that the electric field is discontinuous across the surface. But, more interestingly, the form of the electric field, at such points, is like the electric field for the infinite plane which we have calculated earlier (see (1.42)). This can be understood qualitatively as a consequence of the fact that very close to the disc,

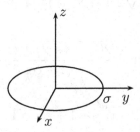

Figure 2.5: A thin circular disc of radius R carrying a uniform surface charge density σ.

the disc appears to be of infinite extent even though it has a finite size.

For this system (see Fig. 2.5), the value of the potential (or the electric field) at a general point is not easy to obtain in closed form, nor does it provide any meaningful insight. However, we can calculate the value of the potential at the rim of the disc quite easily. By rotational invariance on the plane, the value of the potential will be the same at any point on the rim and, for simplicity, we choose the point $(x = R, y = 0, z = 0)$ to calculate the potential. Then, the potential has the form

$$\Phi(x = R, y = 0, z = 0)$$

$$= \int d^3 r' \, \frac{\rho(\mathbf{r}')}{|\mathbf{r} - \mathbf{r}'|}$$

$$= \int r' dr' d\phi' dz' \, \frac{\sigma \delta(z') \theta(R - r')}{((R - r' \cos \phi')^2 + r'^2 \sin^2 \phi' + z'^2)^{\frac{1}{2}}}$$

$$= \sigma \int_0^R \int_0^{2\pi} \frac{r' dr' d\phi'}{(r'^2 - 2Rr' \cos \phi' + R^2)^{\frac{1}{2}}}$$

$$= \sigma \int_0^{2\pi} d\phi' \left[\sqrt{r'^2 - 2Rr' \cos \phi' + R^2} \right.$$

$$\left. + R \cos \phi' \ln \left(2\sqrt{r'^2 - 2Rr' \cos \phi' + R^2} + 2r' - 2R \cos \phi' \right) \right]_0^R$$

$$= \sigma R \int_0^{2\pi} d\phi' \left[\sqrt{2(1 - \cos \phi')} - 1 \right.$$

$$\left. + \cos \phi' \ln \frac{\sqrt{2(1 - \cos \phi')} + (1 - \cos \phi')}{(1 - \cos \phi')} \right]$$

$$= \sigma R \int_0^{2\pi} d\phi' \left[2 \sin \frac{\phi'}{2} - 1 + \cos \phi' \ln \left(1 + \frac{1}{\sin \frac{\phi'}{2}} \right) \right]. \tag{2.31}$$

Here, in the intermediate steps, we have used some standard integrals from the tables (for example, see Gradshteyn and Ryzhik, 2.261 and 2.264). Finally, integrating by parts the last term inside the bracket in (2.31), we have (the first term in the integration by parts vanishes at the limits)

$$\int_0^{2\pi} d\phi' \cos \phi' \ln \left(1 + \frac{1}{\sin \frac{\phi'}{2}} \right) = \int_0^{2\pi} d\phi' \frac{\left(\cos \frac{\phi'}{2} \right)^2}{\left(1 + \sin \frac{\phi'}{2} \right)}$$

$$= \int_0^{2\pi} d\phi' \left(1 - \sin \frac{\phi'}{2} \right). \tag{2.32}$$

Using this in (2.31), we obtain

$$\Phi(x = R, y = 0, z = 0) = \sigma R \int_0^{2\pi} d\phi' \left[2 \sin \frac{\phi'}{2} - 1 + 1 - \sin \frac{\phi'}{2} \right]$$

$$= \sigma R \int_0^{2\pi} d\phi' \sin \frac{\phi'}{2}$$

$$= 4 \sigma R. \tag{2.33}$$

It is interesting to compare this with the value of the potential at the center of the disc (2.28), which makes it clear that the potential decreases as we move out of the center of the disc. Consequently, there must exist a radial component of the electric field on the disc itself.

2.4 Potential for a charge displaced along the z-axis

The next example that we consider is really not that different from what we have studied earlier and yet has many features which will be useful later. Let us consider a point charge q located at $(x = 0, y = 0, z = R)$ as shown in Fig. 2.6.

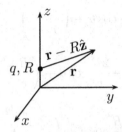

Figure 2.6: A point charge q displaced from the origin along the z-axis by a distance R.

In this case, we can write the charge density as

$$\rho(\mathbf{r}) = q\delta(x)\delta(y)\delta(z - R) = \frac{q}{2\pi r^2 \sin\theta} \delta(r - R)\delta(\theta), \qquad (2.34)$$

where, in the second step, we have rewritten the charge density in spherical coordinates. Note that the multiplicative factor arises from the normalization of the delta function and that, even though the factor appears to be singular (for example, $\sin\theta$ in the denominator), it is, in fact, well behaved inside an integral.

With this, we are now ready to calculate the potential at $\mathbf{r}$ due to this charge located on the z-axis,

$$
\begin{aligned}
\Phi(\mathbf{r}) &= \int d^3 r' \frac{\rho(\mathbf{r}')}{|\mathbf{r} - \mathbf{r}'|} \\
&= q \int r'^2 dr' \sin\theta' d\theta' d\phi' \frac{1}{2\pi r'^2 \sin\theta'} \\
&\quad \times \frac{\delta(r' - R)\delta(\theta')}{(r^2 + r'^2 - 2rr'\cos(\theta - \theta'))^{\frac{1}{2}}} \\
&= \frac{q}{(r^2 + R^2 - 2Rr\cos\theta)^{\frac{1}{2}}}. \qquad (2.35)
\end{aligned}
$$

At this point, let us assume that we are interested in the potential very far away from the point charge. Namely, let us assume that $r \gg R$. Then, we can expand the denominator in (2.35) in a power series as

$$\frac{1}{(r^2 + R^2 - 2Rr\cos\theta)^{\frac{1}{2}}}$$

$$= \frac{1}{r}\left(1 - \frac{2R}{r}\cos\theta + \left(\frac{R}{r}\right)^2\right)^{-\frac{1}{2}}$$

$$= \frac{1}{r}\left[1 - \frac{1}{2}\left(-\frac{2R}{r}\cos\theta + \left(\frac{R}{r}\right)^2\right)\right.$$

$$\left. + \frac{3}{8}\left(-\frac{2R}{r}\cos\theta + \left(\frac{R}{r}\right)^2\right)^2 + \cdots\right]$$

$$= \frac{1}{r}\left[1 + \frac{R\cos\theta}{r} + \frac{R^2(3\cos^2\theta - 1)}{2r^2} + \cdots\right]$$

$$= \left[\frac{1}{r} + \frac{R\cos\theta}{r^2} + \frac{R^2(3\cos^2\theta - 1)}{2r^3} + \cdots\right]. \tag{2.36}$$

Substituting this back into the potential in (2.35), we obtain (for $r \gg R$),

$$\Phi(\mathbf{r}) = \frac{q}{r} + \frac{qR\cos\theta}{r^2} + \frac{qR^2(3\cos^2\theta - 1)}{2r^3} + \cdots. \tag{2.37}$$

As we will see later, the angular coefficients of the expansion of the denominator in (2.36) can be identified with the Legendre polynomials, namely,

$$P_0(\cos\theta) = 1,$$

$$P_1(\cos\theta) = \cos\theta,$$

$$P_2(\cos\theta) = \frac{(3\cos^2\theta - 1)}{2}, \tag{2.38}$$

and so on. Thus, very far away from the point charge, we can write the potential due to a charge displaced along the z-axis, as

$$\Phi(\mathbf{r}) = \sum_{n=0}^{\infty} \frac{qR^n}{r^{n+1}} P_n(\cos\theta). \tag{2.39}$$

(Incidentally, by a similar expansion we can also show that when $R \gg r$ the potential has the form

$$\Phi(\mathbf{r}) = \sum_{n} \frac{qr^n}{R^{n+1}} P_n(\cos\theta). \tag{2.40}$$

This also follows from the fact that the expression for the potential in (2.35) is symmetric under $r \leftrightarrow R$.)

Such an expansion of the potential is known as the multipole expansion. We note that, very far away from the charge, the dominant term is the first term which we recognize to be the potential due to a point charge at the origin (also called a monopole term). However, if for some reason, the first term is absent (namely, if we have a charge neutral system), then the dominant term will be the second term which is the potential due to a dipole. Furthermore, if we have a system for which the first two terms vanish, then the leading term would be the third term which is the potential due to a quadrupole and so on.

As a parenthetical discussion, let us analyze the expansion of the denominator in (2.35) a bit more in detail. Let us recall that a translation by an amount a in one dimension is implemented by $e^{a\frac{\mathrm{d}}{\mathrm{d}x}}$ so that we can write

$$f(x + a) = \left(e^{a\frac{\mathrm{d}}{\mathrm{d}x}} f(x) \right). \tag{2.41}$$

The exponential operator simply generates the Taylor series for the expansion of the function. In higher dimensions, this generalizes so that we can write

$$f(\mathbf{r} + \mathbf{a}) = \left(e^{\mathbf{a}\cdot\boldsymbol{\nabla}} f(\mathbf{r}) \right). \tag{2.42}$$

If we apply this to the denominator in (2.35) (say, for $r \gg R$), we obtain (in spherical coordinates $\hat{\mathbf{z}} = \hat{\mathbf{r}}\cos\theta - \hat{\boldsymbol{\theta}}\sin\theta$)

$$\frac{1}{(r^2 + R^2 - 2rR\cos\theta)^{\frac{1}{2}}} = \frac{1}{|\mathbf{r} - \mathbf{R}|}$$

$$= \left(e^{-\mathbf{R}\cdot\boldsymbol{\nabla}} \frac{1}{|\mathbf{r}|} \right) = \left(e^{-\mathbf{R}\cdot\boldsymbol{\nabla}} \frac{1}{r} \right)$$

$$= \left(e^{-R\left(\cos\theta\frac{\partial}{\partial r} - \frac{\sin\theta}{r}\frac{\partial}{\partial\theta}\right)} \frac{1}{r} \right), \tag{2.43}$$

which, in fact, generates the series in (2.36). The other thing to observe is that, for $r \gg R$, we can write the denominator in (2.35) as

$$\frac{1}{(r^2 + R^2 - 2Rr\cos\theta)^{\frac{1}{2}}} = \frac{1}{r}\frac{1}{(1 - 2z\cos\theta + z^2)^{\frac{1}{2}}}, \tag{2.44}$$

where we have defined

$$z = \frac{R}{r}. \tag{2.45}$$

Furthermore, it is well known that the second fraction on the right hand side of (2.44) is the generator of Legendre polynomials

$$\frac{1}{(1 - 2z \cos \theta + z^2)^{\frac{1}{2}}} = \sum_{n=0}^{\infty} z^n P_n(\cos \theta), \tag{2.46}$$

which explains the structure of the series in (2.39).

2.5 Dipole

Let us consider next a charge configuration as shown in Fig. 2.7 where we are assuming that a charge q is located at $z = R$ while a second charge $(-q)$ is located at $z = -R$.

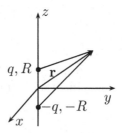

Figure 2.7: A dipole system with two point charges where charge q is at $z = R$ while charge $(-q)$ is at $z = -R$.

The potential for this system can be simply obtained from what we have already calculated (see (2.39) or (2.40)), namely, for $r \gg R$, we have (the potential due to the second charge, $(-q)$ at $z = -R$, is obtained by letting $q \to -q$ and $\theta \to \pi - \theta$)

$$\Phi(\mathbf{r}) = \left[\frac{q}{r} + \frac{qR \cos \theta}{r^2} + \frac{qR^2(3 \cos^2 \theta - 1)}{2r^3} + \cdots \right]$$

$$+ \left[-\frac{q}{r} + \frac{qR \cos \theta}{r^2} - \frac{qR^2(3 \cos^2 \theta - 1)}{2r^3} + \cdots \right]$$

$$= \frac{2qR \cos \theta}{r^2} + \cdots . \tag{2.47}$$

Here, we note that the potential at very large distances does not behave like that of a point charge. In fact, the total charge of the system is zero as seen from large distances. Therefore, it is the second term in the expansion that gives the leading contribution. Such a configuration of (equal and opposite) charges, when the separation between them is small (or, when we are interested in the large distance behavior), is called a dipole (centered at the origin) and the dipole moment associated with the system is defined to be

$$\mathbf{p} = q\mathbf{d} = 2qR\,\hat{\mathbf{z}}, \tag{2.48}$$

where $\mathbf{d}$ represents the vector from the negative charge to the positive charge. Incidentally, a more complete definition of the electric dipole moment for a continuous distribution of charges is given by

$$\mathbf{p} = \int d^3r\,\mathbf{r}\,\rho(\mathbf{r}), \tag{2.49}$$

which can be seen to reduce to the earlier definition in the case of point charges. In terms of the dipole moment in (2.48), we can rewrite the potential for the dipole (in (2.47)) to be

$$\Phi_{\text{dipole}}(\mathbf{r}) = \frac{\mathbf{p}\cdot\hat{\mathbf{r}}}{r^2} = \frac{\mathbf{p}\cdot\mathbf{r}}{r^3}$$

$$= -\mathbf{p}\cdot\nabla\left(\frac{1}{|\mathbf{r}|}\right) = -\nabla\cdot\left(\frac{\mathbf{p}}{|\mathbf{r}|}\right), \tag{2.50}$$

which shows that the potential for the dipole can, in fact, be written as a divergence ($\mathbf{p}$ is constant and hence can be taken inside the gradient operation). Thus, comparing with the potential for a point charge, we realize that the potential for the dipole behaves more like the electric field of a charge (just the gradient nature or the dependence on the distance and not the vector aspect). We can also calculate the electric field associated with the dipole system by recalling that in spherical coordinates,

$$\nabla = \hat{\mathbf{r}}\frac{\partial}{\partial r} + \frac{\hat{\boldsymbol{\theta}}}{r}\frac{\partial}{\partial\theta} + \frac{\hat{\boldsymbol{\phi}}}{r\sin\theta}\frac{\partial}{\partial\phi}, \tag{2.51}$$

so that the electric field for a dipole has the form

$$\mathbf{E}_{\text{dipole}}(\mathbf{r}) = -\nabla\Phi_{\text{dipole}}(\mathbf{r})$$

$$= -\hat{\mathbf{r}}\frac{\partial}{\partial r}\left(\frac{2qR\cos\theta}{r^2}\right) - \frac{\hat{\boldsymbol{\theta}}}{r}\frac{\partial}{\partial\theta}\left(\frac{2qR\cos\theta}{r^2}\right)$$

$$= \frac{4qR\cos\theta}{r^3}\,\hat{\mathbf{r}} + \frac{2qR\sin\theta}{r^3}\,\hat{\boldsymbol{\theta}}$$

$$= \frac{6qR\cos\theta}{r^3}\,\hat{\mathbf{r}} - \frac{2qR}{r^3}(\cos\theta\,\hat{\mathbf{r}} - \sin\theta\,\hat{\boldsymbol{\theta}})$$

$$= \frac{3(\mathbf{p}\cdot\hat{\mathbf{r}})\,\hat{\mathbf{r}} - \mathbf{p}}{r^3}, \tag{2.52}$$

where we have used the relations between the unit vectors in the Cartesian and the spherical coordinates in the last step, namely, $\hat{\mathbf{z}} = \hat{\mathbf{r}}\cos\theta - \hat{\boldsymbol{\theta}}\sin\theta$ (see Fig. 2.8), to rewrite the expression in terms of the dipole moment and the direction of observation.

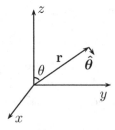

Figure 2.8: Unit vectors $\hat{\mathbf{r}}, \hat{\boldsymbol{\theta}}, \hat{\mathbf{z}}$.

Alternatively, the electric field in (2.52) can also be derived in a simpler manner in the following way.

$$\mathbf{E}_{\text{dipole}}(\mathbf{r}) = -\boldsymbol{\nabla}\Phi_{\text{dipole}}(\mathbf{r})$$

$$= -\boldsymbol{\nabla}\left(\frac{\mathbf{p}\cdot\mathbf{r}}{|\mathbf{r}|^3}\right)$$

$$= -(\mathbf{p}\cdot\mathbf{r})\boldsymbol{\nabla}\left(\frac{1}{|\mathbf{r}|^3}\right) - \left(\frac{\boldsymbol{\nabla}(\mathbf{p}\cdot\mathbf{r})}{|\mathbf{r}|^3}\right)$$

$$= \frac{3(\mathbf{p}\cdot\mathbf{r})\,\mathbf{r}}{|\mathbf{r}|^5} - \frac{\mathbf{p}}{|\mathbf{r}|^3}$$

$$= \frac{3(\mathbf{p}\cdot\hat{\mathbf{r}})\,\hat{\mathbf{r}} - \mathbf{p}}{|\mathbf{r}|^3}, \tag{2.53}$$

where $|\mathbf{r}| = r$. We see that both the potential and the electric field for a dipole decrease faster than the corresponding quantities for a point charge (monopole). Note also that the electric field for the dipole continues to be curl free ($\boldsymbol{\nabla}\times\mathbf{p} = 0$ since $\mathbf{p}$ is constant and $\boldsymbol{\nabla}\times\hat{\mathbf{r}} = 0$). The curl free nature also follows from the fact that $\mathbf{E} = -\boldsymbol{\nabla}\Phi_{\text{dipole}}$.

► **Example (Force on a dipole).** Let us consider a dipole of length ℓ where the vector from the origin of the coordinate system to the negative charge $(-q)$ is $\mathbf{r}$ and $\ell \ll r$ (see Fig. 2.9). Let us further assume that the dipole is placed in an electric field $\mathbf{E}(\mathbf{r})$ which is not necessarily uniform.

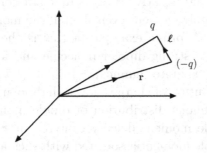

Figure 2.9: Dipole in an electric field.

In this case, the negative and the positive charges of the dipole will experience an electric force given respectively by

$$\mathbf{F}_{(-q)} = -q\mathbf{E}(\mathbf{r}), \qquad \mathbf{F}_{(q)} = q\mathbf{E}(\mathbf{r} + \ell), \tag{2.54}$$

where ℓ denotes the vector from the negative charge to the positive charge. As a result, the total force acting on the dipole can be written as

$$\begin{aligned}
\mathbf{F} = \mathbf{F}_{(-q)} + \mathbf{F}_{(q)} &= -q\mathbf{E}(\mathbf{r}) + q\mathbf{E}(\mathbf{r} + \ell) \\
&\approx -q\mathbf{E}(\mathbf{r}) + q\left(\mathbf{E}(\mathbf{r}) + (\ell \cdot \nabla)\mathbf{E}(\mathbf{r})\right) \\
&= q\left(\ell \cdot \nabla\right)\mathbf{E}(\mathbf{r}) = (\mathbf{p} \cdot \nabla)\,\mathbf{E}(\mathbf{r}). \tag{2.55}
\end{aligned}$$

Here we have used the definition of the dipole moment in (2.48) to identify $\mathbf{p} = q\ell$. We note that the dipole experiences an electrostatic force in the presence of a nonuniform electric field. On the other hand, if the electric field is uniform (constant), then it is clear that the net force on the dipole vanishes, namely, the positive and the negative charges experience equal and opposite forces.

We can also calculate the torque exerted on the dipole due to such a force given in (2.55). The torque around the origin, for example, will have the form

$$\begin{aligned}
\boldsymbol{\tau} = \mathbf{r} \times \mathbf{F}_{(-q)} + (\mathbf{r} + \ell) \times \mathbf{F}_{(q)} \\
\approx -q\mathbf{r} \times \mathbf{E}(\mathbf{r}) + q(\mathbf{r} + \ell) \times (\mathbf{E}(\mathbf{r}) + (\ell \cdot \nabla)\mathbf{E}(\mathbf{r})) \\
\approx q\ell \times \mathbf{E}(\mathbf{r}) + q\mathbf{r} \times (\ell \cdot \nabla)\,\mathbf{E}(\mathbf{r}) \\
= \mathbf{p} \times \mathbf{E}(\mathbf{r}) + \mathbf{r} \times (\mathbf{p} \cdot \nabla)\,\mathbf{E}(\mathbf{r}). \tag{2.56}
\end{aligned}$$

We note that when the electric field is uniform, the second term on the right hand side vanishes, but the first term is nonzero. Therefore, the dipole experiences a torque even though the net force acting on the dipole is zero in this case. ◄

2.6 Continuous distribution of dipoles

It is worth pointing out here that there are many physical systems in nature that behave like a dipole. In many molecules, for example, the centers of positive and negative charges may not coincide giving rise to an associated dipole moment even though the molecule as a whole is charge neutral. A prime example of this is the water molecule which behaves like a strong dipolar molecule and for these reasons, the study of dipoles is quite significant. In fact, just as we can define a continuous distribution of charges, for such dipolar material we can also define a continuous distribution of dipole moments. Let $\mathbf{P}(\mathbf{r})$ represent the dipole moment density centered at $\mathbf{r}$ so that we can write the total dipole moment associated with such a physical system as

$$\mathcal{P} = \int_V d^3r \, \mathbf{P}(\mathbf{r}), \tag{2.57}$$

where V denotes the volume containing the distribution. From the form of the potential for a single dipole in (2.50),

$$\Phi_{\text{dipole}}(\mathbf{r}) = \frac{\mathbf{p} \cdot (\mathbf{r} - \mathbf{r}')}{|\mathbf{r} - \mathbf{r}'|^3}, \tag{2.58}$$

where we are assuming that the dipole is centered at $\mathbf{r}'$, we see that for a continuous distribution of dipole moments we can write the potential as

$$\Phi_{\text{dipole}}(\mathbf{r}) = \int_V d^3r' \, \frac{\mathbf{P}(\mathbf{r}') \cdot (\mathbf{r} - \mathbf{r}')}{|\mathbf{r} - \mathbf{r}'|^3}$$

$$= -\int_V d^3r' \, \mathbf{P}(\mathbf{r}') \cdot \nabla \left(\frac{1}{|\mathbf{r} - \mathbf{r}'|} \right)$$

$$= -\nabla \cdot \int_V d^3r' \, \frac{\mathbf{P}(\mathbf{r}')}{|\mathbf{r} - \mathbf{r}'|}. \tag{2.59}$$

Here, in the last step we have used the fact that $\mathbf{P}(\mathbf{r}')$ does not depend on the coordinate $\mathbf{r}$ and hence the gradient operator does not act on it and can be taken outside the integral. Thus, we see that the potential for a continuous distribution of dipole moments can be written as a divergence as well.

Given a potential we can always relate it to a charge distribution through the Poisson equation. Therefore, it is meaningful to ask what kind of a continuous charge distribution would give rise to the potential (2.59) for a given continuous dipole distribution. The answer is surprisingly not very difficult. Let us note from (2.59) that we can write

$$
\Phi_{\text{dipole}}(\mathbf{r}) = -\int_V d^3r' \, \mathbf{P}(\mathbf{r}') \cdot \boldsymbol{\nabla} \left(\frac{1}{|\mathbf{r} - \mathbf{r}'|} \right)
$$

$$
= \int_V d^3r' \, \mathbf{P}(\mathbf{r}') \cdot \boldsymbol{\nabla}' \left(\frac{1}{|\mathbf{r} - \mathbf{r}'|} \right)
$$

$$
= \int_V d^3r' \, \boldsymbol{\nabla}' \cdot \left(\frac{\mathbf{P}(\mathbf{r}')}{|\mathbf{r} - \mathbf{r}'|} \right) - \int_V d^3r' \, \frac{\boldsymbol{\nabla}' \cdot \mathbf{P}(\mathbf{r}')}{|\mathbf{r} - \mathbf{r}'|}
$$

$$
= \int_S \frac{d\mathbf{s}' \cdot \mathbf{P}(\mathbf{r}')}{|\mathbf{r} - \mathbf{r}'|} - \int_V d^3r' \, \frac{\boldsymbol{\nabla}' \cdot \mathbf{P}(\mathbf{r}')}{|\mathbf{r} - \mathbf{r}'|}. \tag{2.60}
$$

This shows that a dipole potential can equivalently be thought of as being produced by a combination of a surface charge density as well as a volume charge density given respectively by

$$
\sigma(\mathbf{r}) = \hat{\mathbf{n}} \cdot \mathbf{P}(\mathbf{r}),
$$

$$
\rho(\mathbf{r}) = -\boldsymbol{\nabla} \cdot \mathbf{P}(\mathbf{r}), \tag{2.61}
$$

where $\hat{\mathbf{n}}$ represents a unit vector normal to the given surface. (Here the volume and the surface integrals refer respectively to the volume where the dipole moments are localized and the surface bounding such a volume.)

▶ **Example.** As an example of such a system, let us consider a sphere of radius R with a uniform distribution of dipole moments given by the density $\mathbf{P}$ along the z-axis as shown in Fig. 2.10. (A vector field is said to be uniform when its magnitude as well as its direction are the same at every point.)

To calculate the potential due to such a distribution of dipoles we note that the total charge of the system is zero. Therefore, there is no monopole contribution and the leading term in the expansion of the potential is the potential due to the dipoles. We choose the origin of the coordinate system to coincide with the center of the sphere. Then, the calculation of the potential is straightforward,

$$
\Phi_{\text{dipole}}(\mathbf{r}) = -\boldsymbol{\nabla} \cdot \int d^3r' \, \frac{\mathbf{P} \, \theta(R - r')}{|\mathbf{r} - \mathbf{r}'|}
$$

$$
= -\mathbf{P} \cdot \boldsymbol{\nabla} \int d^3r' \, \frac{\theta(R - r')}{|\mathbf{r} - \mathbf{r}'|}. \tag{2.62}
$$

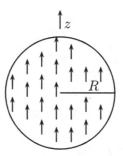

Figure 2.10: A sphere of radius R with a uniform distribution of dipole moments along the z-axis.

The integral can be evaluated in the following way. (We use the standard trick to simplify the evaluation of this integral, namely, let us assume that $\mathbf{r}$ lies along the z-axis, or alternatively that the angle θ' is measured from $\mathbf{r}$.)

$$\int d^3r' \, \frac{\theta(R-r')}{|\mathbf{r}-\mathbf{r}'|}$$

$$= \int r'^2 \, dr' \, \sin\theta' \, d\theta' \, d\phi' \, \frac{\theta(R-r')}{(r^2 + r'^2 - 2rr'\cos\theta')^{\frac{1}{2}}}$$

$$= 2\pi \int_0^R dr' \, r'^2 \left(\frac{1}{rr'}\right) (r^2 + r'^2 - 2rr'\cos\theta')^{\frac{1}{2}} \Big|_0^\pi$$

$$= \frac{2\pi}{r} \int_0^R dr' \, r' \left[(r+r') - |r-r'|\right]. \tag{2.63}$$

The value of this integral clearly depends on whether $r > R$ or $r < R$ because of the second term. For $r > R$, we have

$$\int d^3r' \, \frac{\theta(R-r')}{|\mathbf{r}-\mathbf{r}'|} = \frac{2\pi}{r} \int_0^R dr' \, r' \, (r+r'-r+r')$$

$$= \frac{4\pi}{r} \int_0^R dr' \, r'^2 = \frac{4\pi R^3}{3r}, \tag{2.64}$$

while, for $r < R$, we have

$$\int d^3r' \frac{\theta(R - r')}{|\mathbf{r} - \mathbf{r}'|}$$

$$= \frac{2\pi}{r} \left[\int_0^r dr' \, r'(r + r' - r + r') + \int_r^R dr' \, r'(r + r' - r' + r) \right]$$

$$= \frac{4\pi}{r} \left[\int_0^r dr' \, r'^2 + \int_r^R dr' \, rr' \right] = \frac{4\pi}{r} \left[\frac{r^3}{3} + \frac{r(R^2 - r^2)}{2} \right]$$

$$= \frac{4\pi(3R^2 - r^2)}{6} = \frac{4\pi R^3}{6R} \left(3 - \left(\frac{r}{R} \right)^2 \right). \tag{2.65}$$

Therefore, we can write the potential (2.62) for the dipole distribution to be

$$\Phi_{\text{dipole}}(\mathbf{r}) = -\boldsymbol{P} \cdot \nabla \left(\theta(r - R) \frac{1}{r} + \theta(R - r) \frac{1}{2R} \left(3 - \left(\frac{r}{R} \right)^2 \right) \right), \tag{2.66}$$

where we have defined the total dipole moment of the sphere (V is the volume of the sphere) as

$$\boldsymbol{P} = \mathbf{P}V = \frac{4\pi R^3 \mathbf{P}}{3}. \tag{2.67}$$

Furthermore, since the polarization is along the z-axis, we obtain

$$\Phi_{\text{dipole}}(\mathbf{r})$$

$$= -|\boldsymbol{P}| \frac{\partial}{\partial z} \left(\theta(r - R) \frac{1}{r} + \theta(R - r) \frac{1}{2R} \left(3 - \left(\frac{r}{R} \right)^2 \right) \right)$$

$$= |\boldsymbol{P}| \left[\frac{z}{r^3} \theta(r - R) - \frac{z}{R^2} \delta(r - R) + \frac{z}{R^2} \delta(r - R) + \frac{z}{R^3} \theta(R - r) \right]$$

$$= \frac{\boldsymbol{P} \cdot \mathbf{r}}{r^3} \theta(r - R) + \frac{\boldsymbol{P} \cdot \mathbf{r}}{R^3} \theta(R - r). \tag{2.68}$$

It is clear from the above calculation that the potential is continuous. Outside the sphere the potential behaves as if we have a single dipole with moment $\boldsymbol{P}$ centered at the origin while the potential inside the sphere is that of a uniform electric field $(-\frac{\boldsymbol{P}}{R^3})$ (the negative sign is from the definition of the potential $(-\mathbf{E} \cdot \mathbf{r})$ for a uniform field, or alternatively, from the definition $\mathbf{E} = -\nabla\Phi$). ◀

2.7 Quadrupole

Let us consider a configuration of four charges as shown in Fig. 2.11. Once again, we can obtain the potential for this system from what we have already calculated in (2.39). The potential for the charge at the origin is obtained by setting $R = 0$ and $q \to -2q$, while the potential

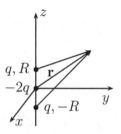

Figure 2.11: A quadrupole configuration with two point charges q at $z = \pm R$ and a point charge $(-2q)$ at the origin.

for the charge on the lower z-axis is obtained by letting $\theta \to \pi - \theta$. The complete potential is given by

$$\Phi(\mathbf{r}) = \left[\frac{q}{r} + \frac{qR\cos\theta}{r^2} + \frac{qR^2(3\cos^2\theta - 1)}{2r^3} + \cdots \right] - \frac{2q}{r}$$

$$+ \left[\frac{q}{r} - \frac{qR\cos\theta}{r^2} + \frac{qR^2(3\cos^2\theta - 1)}{2r^3} + \cdots \right]$$

$$= \frac{qR^2(3\cos^2\theta - 1)}{r^3} + \cdots . \tag{2.69}$$

This shows that for this system of charges both the monopole and the dipole terms in the potential vanish. The total charge of the system is zero which is why the monopole term in (2.69) vanishes, but we can also think of the system of four charges in Fig. 2.11 as two dipoles with opposite dipole moments which makes the dipole term vanish as well (total dipole moment is zero). As a result it is the third term in the multipole expansion which gives the leading behavior of the potential for large distances. Such a configuration of charges is known as a quadrupole. We note that unlike the case of dipoles, other configurations of quadrupoles are possible and this is not the unique quadrupole configuration. Furthermore, while the dipole moment is a vector, the quadrupole moment, in general, is a second rank symmetric traceless tensor. In fact, the n-th term in the multipole expansion (see, for example, (2.39)) leads to the $2n$-th pole moment which is an n-th rank tensor. The monopole gives a 0-th rank tensor $q = \int \mathrm{d}^3r\, \rho(\mathbf{r})$ which is a scalar, the dipole gives a rank one tensor which is a vector and so on. In Cartesian coordinates the quadrupole moment tensor is defined to be

$$Q_{ij} = \int d^3x \left(3x_i x_j - \delta_{ij} \mathbf{x}^2\right) \rho(\mathbf{x}), \qquad (2.70)$$

which is manifestly symmetric and traceless. For the charge configuration in Fig. 2.11, which can be thought of as two dipoles aligned back to back along the z-axis, we have (note that $\rho(\mathbf{x}) = q\left((\delta(x_3 - R) + \delta(x_3 + R)) - 2\delta(x_3)\right)\delta(x_1)\delta(x_2)$)

$$Q_{11} = -2qR^2 = Q_{22},$$

$$Q_{33} = 4qR^2 = -(Q_{11} + Q_{22}) = -2Q_{11}, \qquad (2.71)$$

with all other components vanishing. (Note that knowing Q_{11} and Q_{22}, we could have predicted the value for Q_{33} from the tracelessness condition of the quadrupole moment tensor.) In general, the potential for the quadrupole can be written in terms of the quadrupole moment as (for large values of $r = |\mathbf{x}|$)

$$\Phi_{\text{quadrupole}}(\mathbf{r}) = \frac{1}{6} \sum_{i,j} Q_{ij} \frac{\left(3x_i x_j - \delta_{ij} \mathbf{x}^2\right)}{r^5}, \qquad (2.72)$$

which can be readily checked to give the leading term (the quadrupole potential) in (2.69) for the system under consideration. We can again obtain the electric fields associated with the quadrupole system by taking the gradient. But, we will not get into the details of this except for noting that the potential as well as the electric field for the quadrupole decrease even faster than those for the dipole.

2.8 Potential due to a double layer of charges

Let us consider a single plane of infinite extent carrying a constant surface charge density σ. We have already obtained the electric field for such a configuration in the last chapter using Gauss' law (see (1.42)). Let us now calculate the potential associated with such a configuration as shown in Fig. 2.12. Let us assume that the plane is at $z = 0$ and because of the symmetry in the problem, we can always choose the z-axis to lie along the perpendicular to the plane from the coordinate where we are interested in evaluating the potential.

The calculation of the potential is straightforward. We can write the volume charge density for the system to be

$$\rho(\mathbf{r}) = \sigma\delta(z), \qquad (2.73)$$

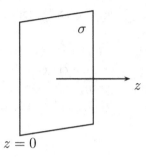

Figure 2.12: An infinite plane located at $z = 0$ and carrying a constant surface charge density σ.

where σ denotes the constant surface charge density. Consequently, using cylindrical coordinates, we obtain,

$$\Phi(\mathbf{r}) = \int d^3 r' \, \frac{\rho(\mathbf{r}')}{|\mathbf{r} - \mathbf{r}'|}$$

$$= \int r' dr' \, d\phi' \, dz' \, \frac{\sigma \delta(z')}{(r'^2 + (z - z')^2)^{\frac{1}{2}}} = \sigma \int \frac{r' dr' \, d\phi'}{(r'^2 + z^2)^{\frac{1}{2}}}$$

$$= \lim_{\Lambda \to \infty} 2\pi\sigma \int_0^\Lambda \frac{r' dr'}{(r'^2 + z^2)^{\frac{1}{2}}} = \lim_{\Lambda \to \infty} 2\pi\sigma \left. (r'^2 + z^2)^{\frac{1}{2}} \right|_0^\Lambda$$

$$= \lim_{\Lambda \to \infty} 2\pi\sigma \left[(\Lambda^2 + z^2)^{\frac{1}{2}} - |z| \right]$$

$$= -2\pi\sigma \, |z| + \text{constant}, \tag{2.74}$$

where we recognize that the constant on the right hand side of (2.74) is a divergent constant. We also understand the origin of this divergent constant, namely, we have a charge distribution which extends to infinity. Consequently, the reference point for the potential has to be chosen differently or one has to allow for a (possibly) divergent constant in the potential. However, an additive constant is not meaningful if we are interested in physical quantities such as the electric field and, therefore, ignoring the constant we can write the potential for such a system of charge distribution to be

$$\Phi(\mathbf{r}) = -2\pi\sigma |z|. \tag{2.75}$$

Taking the gradient of (2.75) we obtain the electric field to be

$$\mathbf{E}(\mathbf{r}) = -\boldsymbol{\nabla}\Phi(\mathbf{r}) = -\hat{z}\frac{\partial}{\partial z}\left(-2\pi\sigma|z|\right) = 2\pi\sigma\,\mathrm{sgn}(z)\,\hat{z}, \qquad (2.76)$$

which is the result we had obtained earlier in (1.42).

Let us next consider two thin planes of infinite extent separated by a small distance d. Namely, let us assume that the two planes are located at $z = \frac{d}{2}$ and $z = -\frac{d}{2}$ and carry a uniform surface charge density of σ and $-\sigma$ respectively as shown in Fig. 2.13. Thus, in some sense we have an infinite distribution of dipoles. Since the electric potential is additive, the potential for this distribution of charges can be easily obtained from the calculation for a single layer of infinite extent in (2.75) which leads to (remember that the locations of the planes are displaced from $z = 0$)

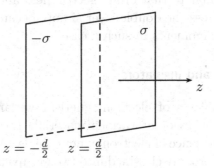

Figure 2.13: Two infinite planes carrying constant surface charge densities σ and $(-\sigma)$.

$$\Phi(\mathbf{r}) = \Phi_1(\mathbf{r}) + \Phi_2(\mathbf{r}) = -2\pi\sigma\left(\left|z - \frac{d}{2}\right| - \left|z + \frac{d}{2}\right|\right). \qquad (2.77)$$

This can be simplified and written as

$$\Phi(\mathbf{r}) = \begin{cases} 2\pi\sigma d & \text{for } z > \frac{d}{2}, \\ 4\pi\sigma z & \text{for } -\frac{d}{2} < z < \frac{d}{2}, \\ -2\pi\sigma d & \text{for } z < -\frac{d}{2}. \end{cases} \qquad (2.78)$$

Taking the gradient of the potential it is now easy to obtain that

$$\mathbf{E}(\mathbf{r}) = -\boldsymbol{\nabla}\Phi(\mathbf{r}) = \begin{cases} 0 & \text{for } z > \frac{d}{2}, \\ -4\pi\sigma\,\hat{\mathbf{z}} & \text{for } -\frac{d}{2} < z < \frac{d}{2}, \\ 0 & \text{for } z < -\frac{d}{2}. \end{cases} \tag{2.79}$$

Let us note that for this system the electric field is nontrivial only between the two planes. Outside the two planes the electric field vanishes because the contributions from the two planes (which are oppositely charged) cancel exactly. Furthermore, there is one aspect of the results in (2.78) and (2.79) which is worth emphasizing. Namely, when d is very small, the potential is discontinuous across the double layer, while the normal component of the electric field is continuous. The continuity of the normal component of the electric can be easily understood from the fact that across the double layer, the net surface charge density is zero. On the other hand, the discontinuity in the dipole potential is understood from the fact that it has the behavior similar to that of an electric field and, consequently, is discontinuous across the double layer simply because there is surface density of dipole moments in such a case.

2.9 Conductors and insulators

In studying problems of electromagnetism we are quite often interested in the behavior of various materials (not simply of point charges) in the presence of electromagnetic fields. All materials found in nature can be broadly classified into two groups depending on their response to an external electric field. One class of materials is known as conductors and metals are prime examples of such systems which contain a large number of free electrons. These electrons are free to move around inside the material (they cannot, of course, leave the material under normal circumstances) and they respond almost instantaneously (in an idealized situation) to any external applied field. When an electric field is set up within the material due to some external source the free electrons conduct electricity.

In contrast, the second class of materials, known as insulators, are non-conductors. In such materials, the electrons are tightly bound to the atomic nuclei and are not free to move. When an external electric field is applied, the electrons may be displaced slightly from their normal position (the atomic nuclei are heavier, so they do not move appreciably), but they are still bound to the nuclei. As a result, insulators do not conduct electricity. Dielectrics are insulators which can be polarized in the presence of an external electric field (namely, the negative and the positive charge centers can be separated). In some dielectrics permanent atomic (molecular) dipole moments may be present independent of any applied external field. The water molecule that we discussed earlier is an example of such a substance. In such materials, even though polarization (dipole moment) is non-vanishing at smaller scales (for individual molecules), macroscopically the dielectric may be unpolarized. However, a small external electric field would be able to orient the dipoles giving rise to a net polarization for the dielectric. (Let us note here that, for all practical purposes, "dielectric" and "insulator" mean the same thing. Vacuum is the only insulator which cannot be polarized and hence is not a dielectric.)

Although an ideal dielectric has zero conductivity (it does not conduct electricity at all), in reality, dielectrics may have a small conductivity. However, a typical conductor has a conductivity which is about 10^{20} times larger than that of a typical dielectric and so, in our discussions, we can safely assume the conductivity of dielectrics to be zero. Similarly, even though conductors have a finite conductivity (collision of the electrons during motion gives rise to a resistivity to the flow of currents), for the purposes of our discussions we will assume that the conductors have an infinite conductivity.

Besides conductors and insulators (dielectrics), there are also semi-conductors and electrolytes with intermediate properties as far as conductivity is concerned, but they have many other properties which make them interesting independently. However, we would limit ourselves only to conductors and dielectrics for the purposes of our discussions.

Let us summarize here the properties of conductors. First, it is clear that if a conductor is in static equilibrium in the presence of an external electric field, then, the electric field inside the conductor must be zero. This must be so, because if the electric field is nonzero inside the conductor, the free electrons will experience an electric force and would move, violating the assumption of static equilibrium. The way conductors achieve static equilibrium is really quite simple. In response to an external electric field, the electrons move to one edge of the surface of the conductor leaving the opposite edge positively charged (see Fig. 2.14) and set up an internal electric field which exactly cancels the external field within the conductor.

Since $\mathbf{E} = 0$ inside a conductor, it follows from Gauss' law that the net charge density must vanish inside a conductor. This simply means that there is an equal number of positive and negative charges

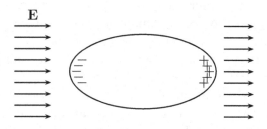

Figure 2.14: A conductor in an external electric field.

in any small volume inside a conductor. The external electric field simply leads to a redistribution of the free electrons on the surface of the conductor giving rise to a nonzero surface density of charges. (It also follows from this that any charge that one puts inside a conductor must necessarily move to the surface for static equilibrium.) Furthermore, the potential must be a constant inside the conductor ($\mathbf{E} = 0$) all the way up to the surface of the conductor. Thus, the surface of a conductor (in fact, any surface inside the conductor) defines an equipotential surface. There cannot be any tangential component of the electric field on the surface of the conductor (otherwise, static equilibrium will not hold). Outside the conductor, of course, the electric field will not be zero and, in fact, from our earlier discussions, we can conclude that immediately outside the conductor the electric field would be normal to the surface with the value

$$E_{\mathrm{n,R}} = 4\pi\,\sigma, \tag{2.80}$$

where σ denotes the surface density of the redistributed charges.

The conductors need not always have a simple configuration. Sometimes, a conductor may contain a cavity inside. If such a conductor is placed in an external electric field with no charge in the cavity (see Fig. 2.15), once again there will be a surface distribution of charges (now there are two surfaces, one exterior and one interior) such that there would be no electric field inside the conductor as well as inside the cavity in the static equilibrium. The vanishing of the electric field inside the conductor can, of course, be understood along the lines of reasoning given above. The vanishing of the electric field inside the cavity follows simply from the fact that along any closed

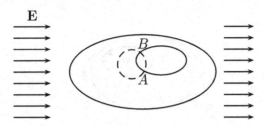

Figure 2.15: A conductor with an internal cavity in an external electric field.

curve cutting the inner surface (see Fig. 2.15), we must have

$$\oint_C d\boldsymbol{\ell} \cdot \mathbf{E} = \int_A^B d\boldsymbol{\ell} \cdot \mathbf{E} = \int_A^B d\ell \, |\mathbf{E}| = 0, \tag{2.81}$$

implying that $\mathbf{E} = 0$ inside the cavity. In deriving this result, we have chosen the path of integration inside the cavity to lie along the electric field (recall that $\mathbf{E} = 0$ inside the conductor and because of the conservative nature of the electric field, the choice of path is irrelevant). This result is quite interesting because this shows that the electric field within the conductor as well as inside the cavity vanishes. Consequently, there is no discontinuity of the electric field across the inner surface. This is possible only if the inner surface does not carry any surface charge density. Therefore, we conclude that in such a case, the free electrons redistribute themselves in a way so that only the outer surface carries a surface charge density. This also shows that any external field cannot penetrate inside a cavity within a conductor. This is the principle behind electrical shielding which is used to shield electrical equipment by putting them inside a metal container (commonly known as the Faraday's cage) and this is also the reason we do not get electrocuted inside a car during lightning. (One may think that this result can also be derived using a Gaussian surface inside the cavity. It is true that there is no charge inside and hence there will be no net flux. However, that does not mean that there is no electric field in the cavity.)

Thus, the mechanism by which a conductor maintains a vanish-

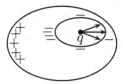

Figure 2.16: A cavity carrying a charge q inside a conductor.

ing **E** field within itself is that in the presence of external fields or charges, induced charges appear on the surface of the conductor to precisely cancel the external field. The actual distribution of the surface charges is, of course, a hard problem to calculate, but the charges rearrange themselves precisely in a way so as to give **E** = 0 within the conductor. If there is a cavity inside a conductor and the cavity carries a charge q (see Fig. 2.16), it would induce a charge density on the inner as well as the outer surfaces of the conductor so as to have a vanishing electric field within the conductor. On the other hand, the electric field would be non-vanishing within the cavity as well as outside the conductor.

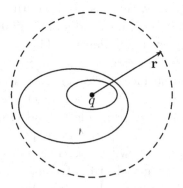

Figure 2.17: A spherical Gaussian surface of radius r surrounding the conductor.

The electric field outside the conductor can be calculated simply by using Gauss' law. Drawing a spherical Gaussian surface (centered at the charge q as in Fig. 2.17) of radius r which is much bigger than any dimension of the conductor, we see that the field at such points must be radial with the value

$$\mathbf{E}(\mathbf{r}) = \frac{q}{|\mathbf{r}|^2}\,\hat{\mathbf{r}}, \tag{2.82}$$

independent of the shape of the cavity or the conductor.

This discussion shows that in the presence of external electric fields the behavior of conductors is quite complex and interesting and needs to be analyzed carefully which we will do.

2.10 Capacitor

When we have two conductors carrying equal and opposite charge, they are said to form a capacitor system. The surfaces of the conductors are equipotential surfaces and let us denote the potentials on the surfaces of the two conductors as Φ_1 and Φ_2 respectively with $\Phi_1 > \Phi_2$. Consequently, the potential difference between the two surfaces can be written as

$$V = \Phi_1 - \Phi_2, \tag{2.83}$$

and is conventionally known as the voltage between the two surfaces. From our discussion so far we know that the potential and, therefore, the voltage depend linearly on the charge Q of the conductors. Consequently, we can write

$$Q = CV, \tag{2.84}$$

where the constant of proportionality C is known as the capacitance of the system which can be thought of as the charge necessary to maintain a unit voltage across the two surfaces. In the CGS system, it is clear that the unit of capacitance is given by

$$\frac{\text{esu}}{\text{statVolt}} = \frac{(\text{esu})^2}{\text{esu}-\text{statVolt}} = \frac{(\text{esu})^2/\text{cm}}{\text{esu}-\text{statVolt}}\,\text{cm} = \text{cm}, \tag{2.85}$$

where we have used the fact that both "$\frac{(\text{esu})^2}{\text{cm}}$" and "esu $-$ statVolt" correspond to units of work, namely, "erg". Since the capacitance has the dimension of a length, it is intuitively clear that it must be a geometrical property of the system. We also note here that the MKS unit of capacitance is a farad (F) which is defined as

$$1 \text{ farad} = 1 \, \frac{\text{Coulomb}}{\text{Volt}} = \frac{3 \times 10^9 \text{esu}}{\frac{1}{3} \times 10^{-2} \text{statVolt}} = 9 \times 10^{11} \text{ cm}. \quad (2.86)$$

We see that a farad is quite large and, in fact, the capacitance of a typical capacitor is of the order of a microfarad (μF) or even smaller, a picofarad (pF or sometimes also written as $\mu\mu$F). Let us also note here that sometimes one even talks about the capacitance of a single conductor. It is understood, in such a case, that the second conducting surface lies at infinity.

▶ **Example (Parallel plate capacitor).** Let us calculate the capacitance of some typical capacitors. The simplest is to consider a pair of parallel, rectangular conducting plates with area A separated by a distance d and carrying charges Q and $(-Q)$ respectively, distributed uniformly over the two surfaces as in Fig. 2.18. If the separation d is small compared to the area A of the plates, we can think of them as two infinite plates carrying an equal and opposite surface charge density. This is a problem we have worked out earlier in detail and we conclude that the electric field will be nonzero only between the two plates and would be uniform given by (see (2.79))

$$|\mathbf{E}| = 4\pi\sigma, \quad (2.87)$$

pointing from the upper to the lower plate. Here $\sigma = \frac{Q}{A}$ represents the magnitude of the surface charge density.

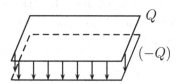

Figure 2.18: Two parallel plates separated by a distance d and carrying charges Q and $(-Q)$.

It follows, therefore, that the voltage across the plates is given by

$$V = |\mathbf{E}|d = 4\pi\sigma d = \frac{4\pi d}{A} Q. \quad (2.88)$$

As a result, we immediately identify the capacitance for the parallel plate system to be

$$C = \frac{A}{4\pi d}, \quad (2.89)$$

and we see explicitly that it is determined by the geometry of the system.

We note here that in deriving this result, we have pretended as if the two plates are of infinite extension. This is, of course, not true and consequently the electric fields are not uniform all over. In particular, at the edges of the plates, the electric fields are not uniform. Thus, our determination of the capacitance is not quite correct. If one determines the correction due to the edge effect, it turns out that the capacitance increases slightly, but to a first approximation, the idealized answer obtained above is quite good when d is very small.

Let us now calculate the energy stored in this capacitor system. First, let us note that if the capacitance is C and the magnitude of the charges on the two plates is $\widetilde{Q}$ with a voltage $\widetilde{V}$, then, to increase the charge on the upper plate infinitesimally by an amount $d\widetilde{Q}$ (and, therefore, to decrease the charge on the lower plate by $d\widetilde{Q}$), that is to take an amount of charge $d\widetilde{Q}$ from the lower plate and move it to the upper plate, we must do work against the electric force and the amount of work is given by

$$dW = d\widetilde{Q}\,(\widetilde{\Phi}_1 - \widetilde{\Phi}_2) = \widetilde{V}\,d\widetilde{Q} = \frac{\widetilde{Q}\,d\widetilde{Q}}{C}. \tag{2.90}$$

Integrating this we can obtain the total work necessary to charge the capacitor plates starting from the uncharged state and the result is

$$W = \int_0^Q \frac{\widetilde{Q}\,d\widetilde{Q}}{C} = \frac{Q^2}{2C} = \frac{(CV)^2}{2C} = \frac{1}{2}CV^2. \tag{2.91}$$

This work must, of course, be stored in the capacitor as electrostatic energy. Now, let us recall that for the parallel plate system

$$V = |\mathbf{E}|d,$$

$$C = \frac{A}{4\pi d}. \tag{2.92}$$

Using these we can also write

$$W = \frac{1}{2} \times \frac{A}{4\pi d}\,(|\mathbf{E}|d)^2 = \frac{1}{8\pi}\,\mathbf{E}^2 \times Ad = \frac{1}{8\pi}\,\mathbf{E}^2 \times \text{volume}, \tag{2.93}$$

which is exactly the result we had obtained earlier for the electrostatic energy for a continuous distribution of charges in the last chapter (see (1.85)). ◄

▶ **Example (Spherical capacitor).** As a second example, let us consider two spherical conducting shells of radii R_1 and R_2 carrying charges Q_1 and Q_2 respectively. We assume that $R_1 > R_2$ (see Fig. 2.19).

The spherical symmetry of the problem determines that the charges would distribute uniformly over the two surfaces giving rise to the surface densities σ_1 and σ_2 respectively. Thus, we can write

$$\rho(\mathbf{r}) = \sigma_1\delta(r - R_1) + \sigma_2\delta(r - R_2), \tag{2.94}$$

where

$$\sigma_1 = \frac{Q_1}{4\pi R_1^2}, \quad \sigma_2 = \frac{Q_2}{4\pi R_2^2}. \tag{2.95}$$

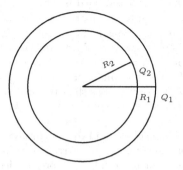

Figure 2.19: Two spherical shells of radius R_1 and R_2 carrying charges Q_1 and Q_2 respectively.

The potential for this system can now be easily calculated.

$$\Phi(\mathbf{r}) = \int d^3r' \, \frac{\rho(\mathbf{r}')}{|\mathbf{r} - \mathbf{r}'|}$$

$$= \int r'^2 dr' \, \sin\theta' \, d\theta' \, d\phi' \, \frac{(\sigma_1\delta(r' - R_1) + \sigma_2\delta(r' - R_2))}{(r^2 + r'^2 - 2rr'\cos\theta')^{\frac{1}{2}}}$$

$$= 2\pi \left[R_1^2\sigma_1 \left(-\frac{1}{rR_1} \right) (|r - R_1| - (r + R_1)) \right.$$

$$\left. + R_2^2\sigma_2 \left(-\frac{1}{rR_2} \right) (|r - R_2| - (r + R_2)) \right]$$

$$= \theta(r - R_1)\frac{Q_1}{r} + \theta(R_1 - r)\frac{Q_1}{R_1}$$

$$+ \theta(r - R_2)\frac{Q_2}{r} + \theta(R_2 - r)\frac{Q_2}{R_2}. \tag{2.96}$$

This shows that the surface at $r = R_1$ is an equipotential surface with the potential given by

$$\Phi(R_1) = \frac{(Q_1 + Q_2)}{R_1}, \tag{2.97}$$

while the surface $r = R_2$ has a constant potential given by

$$\Phi(R_2) = \frac{Q_1}{R_1} + \frac{Q_2}{R_2}. \tag{2.98}$$

This can, of course, be qualitatively understood in terms of the properties of electric systems as well as the calculations we have already done. Namely, at $r = R_1$, the system behaves as if the entire charge $(Q_1 + Q_2)$ were concentrated at the origin. On the other hand, at $r = R_2$, the charge Q_2 behaves like it is at the origin while the charge Q_1 gives a constant potential characteristic of the potential inside a shell of radius R_1.

This is quite general so far. Let us next assume that we have a spherical capacitor system in which case

$$Q_2 = -Q_1 = Q, \quad Q > 0. \tag{2.99}$$

(The other possibility is $Q_2 = -Q_1 = -Q$ which can also be worked out in a parallel manner and leads to the same result for the capacitance.) In this case, the voltage across the two surfaces of the spherical capacitor system is given by

$$V = \Phi(R_2) - \Phi(R_1) = Q \left(\frac{1}{R_2} - \frac{1}{R_1} \right)$$

$$= \frac{(R_1 - R_2)}{R_1 R_2} Q. \tag{2.100}$$

Therefore, it follows that the capacitance of this capacitor is again a geometrical quantity given by

$$C = \frac{R_1 R_2}{(R_1 - R_2)}. \tag{2.101}$$

Furthermore, if we let $R_1 \to \infty$, then, we have a single spherical shell of radius R_2 with a capacitance given by

$$C = R_2. \tag{2.102}$$

Let us note here that capacitors are widely used to store charge.   ◀

2.11 Selected problems

1. Consider a particle as a sphere of radius R carrying a volume charge density

$$\rho(\mathbf{r}) = \frac{3Q}{\pi R^4}(R - r), \quad r \le R,$$

where the coordinate origin is assumed to coincide with the center of the sphere. What is the total charge carried by the sphere? Calculate the electric fields both inside and outside the sphere.

2. Consider a dipole centered at the origin in a uniform electric field. Calculate the torque on the dipole due to the electric field. If the dipole is initially parallel to the electric field, how much work would be needed to rotate it by an angle θ?

3. a) Calculate the dipole moment for the following configuration of charges: three, each of value q, located at $(x = 0, y = 0, z = d)$, $(x = R, y = 0, z = 0)$, $(x = -R, y = 0, z = 0)$, as well as a charge of $-3q$ located at $(x = 0, y = 0, z = -d)$.

b) Calculate the quadrupole moment for the charge configuration where two charges of value q are at $(x = 0, y = 0, z = R)$ and $(x = 0, y = 0, z = -R)$, while two other charges of value $-q$ are at $(x = R, y = 0, z = 0)$ and $(x = -R, y = 0, z = 0)$.

4. Consider a localized distribution of charge given by a density

$$\rho(\mathbf{r}) = r^2 e^{-r} \sin^2 \theta.$$

a) What is the potential due to this charge distribution at very far away distances?

b) What are the nontrivial multipole moments present in this potential?

5. Two long cylindrical conductors (wires) of radii a_1 and a_2 respectively are separated by a distance $d \gg a_1, a_2$. Find the capacitance *per unit length* for the system. If $d = 0.5$cm and $a_1 = a_2 = a$, what must be the diameter of the wires to give a capacitance per unit length of 9×10^{-3}?

6. Consider a pair of coaxial, conducting, hollow cylinders of infinite length and with radii R_1 and R_2, where $R_2 > R_1$. The outer and the inner cylinders carry charges Q and $-Q$ respectively. What is the capacitance per unit length for such a system?

CHAPTER 3

Boundary value problems

3.1 Method of images

When we have a system of conductors, the physical problem of interest may be, in general, much more complex than calculating the potential for a given distribution of point charges. For example, we may have a system of conductors in an electric field with the surfaces of the conductors held fixed at some given value of the potential. (A conductor may be grounded meaning that the surface is maintained at zero potential.) Such problems where boundary values of the potential may be specified are commonly known as boundary value problems. The proper way to solve such a problem is, of course, by determining the solution of the Laplace or the Poisson equation subject to the appropriate boundary conditions, which we will discuss later in this chapter. However, sometimes the physical system may be simple enough that one can determine the potential in a simple manner without having to derive the explicit solution of the Laplace or the Poisson equation. One such method is known as the method of images. Here, the idea is very simple. If we are given a physical system with a set of boundary conditions, we try to reproduce these conditions by a simple choice of fictitious "image" charge distributions. If we can do this, then, solving the original problem subject to the boundary conditions is equivalent to solving the problem with these additional fictitious charges without any boundary condition. It is clear that if the boundary conditions are extremely complicated, finding the "image" charge distribution may prove difficult and, consequently, the method will not be very useful. However, for relatively simple boundary conditions this method works quite well as we will see from the following examples.

▶ **Example (Point charge above a conducting plane).** Let us consider a system consisting of a conducting plane of infinite extent which is maintained at zero potential (grounded), and a point charge q which is above the plane at a height

d as shown in Fig. 3.1. Let us assume for concreteness that the charge is positive although this is not essential for our discussion.

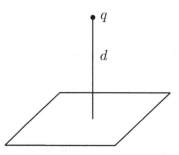

Figure 3.1: A point charge q above an infinite conducting plane (grounded) at a height d.

It is clear that the electric field is meaningful and nonzero only above the plane. This is because, for an insulated conductor, the point charge would induce charges of opposite sign on the two sides of the conductor in such a way as to cancel the field within the conductor. The charges on the lower surface of the conductor, positive charges in this case, would then give rise to an electric field on the lower half of the plane. Here, however, we have a conductor that is grounded and the ground has an infinite supply of negative charges which would move on to the conductor to annihilate all the positive charges. As a result of this, the surface would have a net negative charge and all the field lines originating from the point charge q would terminate on the plane and there will be no field lines below the plane. Namely, the electric field cannot penetrate a grounded conducting plane of infinite extent. Another way to see this is to note that, by grounding, the surface of the conductor is maintained at zero potential as is the surface of the plane infinitely below the conducting plane. Consequently, the potential difference (voltage) across these two planes is zero and there cannot be an electric field in this region. It is only at points above the plane that the electric field will be nonzero.

Therefore, to determine the potential and the electric field above the plane for this physical system, we need to find a fictitious charge distribution which can reproduce the boundary condition. Without loss of generality, we can assume that the conducting plane lies in the $x - y$ plane and that the charge q is located at a height d on the positive z-axis. Clearly the problem has cylindrical symmetry and, consequently, it is meaningful to use cylindrical coordinates to analyze this problem. Let us next assume that a second point charge q' is located at a height d' below the plane on the negative z-axis (see Fig. 3.2). In the presence of these two charges, the potential at any point can be easily calculated and in particular on the plane at $z = 0$ it is given by

$$\Phi(r, \phi, z = 0) = \frac{q}{\sqrt{r^2 + d^2}} + \frac{q'}{\sqrt{r^2 + d'^2}}, \tag{3.1}$$

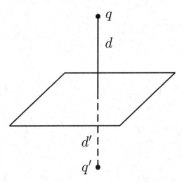

Figure 3.2: The system in Fig. 3.1 with an image charge q' at a distance d' on the opposite side of the plane directly below the charge q.

where r represents the radial coordinate on the plane (in cylindrical coordinates). Requiring the potential to vanish on the plane (which is our boundary condition), we obtain

$$q^2 \left(r^2 + d'^2\right) - q'^2 \left(r^2 + d^2\right) = 0,$$

$$\text{or,} \quad \left(q^2 - q'^2\right) r^2 + \left(q^2 d'^2 - q'^2 d^2\right) = 0. \tag{3.2}$$

Requiring this to be true for any value of r (namely, at any point on the plane), we determine

$$q' = \pm q, \qquad d' = \pm d. \tag{3.3}$$

Putting this back into the expression for Φ in (3.1) and noting that both d and d' are positive, we determine that the potential vanishes on the plane only for

$$q' = -q, \qquad d' = d. \tag{3.4}$$

We see that if we were to introduce a fictitious charge $(-q)$ located at $z = -d$ into the problem, then we can reproduce in a natural manner the boundary condition of the problem (namely, the plane is grounded).

Thus, as far as the calculations of the potential and the electric field for positive z are concerned, we can forget about the plane and the boundary condition and work with only these two charges. The charge above the plane is, of course, the physical charge. The second charge which is not real but which we can use in lieu of the boundary condition is known as the "image" charge borrowing the terminology from optics. At any point $\mathbf{r}$ above the plane $(z > 0)$, the potential is given by (see Fig. 3.3)

$$\Phi(\mathbf{r}) = \frac{q}{\sqrt{r^2 + (z - d)^2}} - \frac{q}{\sqrt{r^2 + (z + d)^2}}. \tag{3.5}$$

Consequently, the electric field at any point above the plane $(z > 0)$ is obtained to be

$$\mathbf{E}(\mathbf{r}) = -\boldsymbol{\nabla}\Phi(\mathbf{r})$$

$$= -\left(\hat{\mathbf{r}}\frac{\partial}{\partial r} + \hat{\mathbf{z}}\frac{\partial}{\partial z}\right)\left(\frac{q}{\sqrt{r^2 + (z-d)^2}} - \frac{q}{\sqrt{r^2 + (z+d)^2}}\right)$$

$$= q\hat{\mathbf{r}}\left(\frac{r}{(r^2 + (z-d)^2)^{\frac{3}{2}}} - \frac{r}{(r^2 + (z+d)^2)^{\frac{3}{2}}}\right)$$

$$+ q\hat{\mathbf{z}}\left(\frac{(z-d)}{(r^2 + (z-d)^2)^{\frac{3}{2}}} - \frac{(z+d)}{(r^2 + (z+d)^2)^{\frac{3}{2}}}\right), \tag{3.6}$$

where we have used the fact that the potential is independent of the azimuthal angle ϕ and correspondingly have dropped the angular derivative in the gradient.

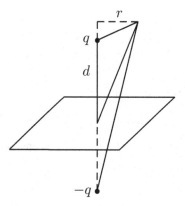

Figure 3.3: The potential at a point above the plane due to the charge q and the "image" charge $(-q)$ at a distance d above and below the plane respectively.

Let us emphasize that the expressions for both the potential and the electric field are meaningful only for $z > 0$. As is obvious, they give the wrong result for $z < 0$ as they should because we do not have a real charge for $z < 0$. It is worth noting here that for $z \to 0^+$, the electric field (3.6) takes the form

$$\mathbf{E}(\mathbf{r})|_{z\to 0^+} = -\frac{2qd}{(r^2 + d^2)^{\frac{3}{2}}}\,\hat{\mathbf{z}}. \tag{3.7}$$

That is, there is only a normal component of the electric field on the surface along the z-axis as we would expect. It is not a uniform electric field on the plane, but has cylindrical symmetry and, in fact, the field becomes weaker as we move away from the origin. (By the way, remember that r is the radial coordinate on the plane.) On the lower surface ($z < 0$), of course, there is no electric field and, consequently, the normal component of the electric field is discontinuous across the surface. And as we have seen (see, for example, (2.10)), the discontinuity is proportional to the surface charge density

$$\sigma(r) = \frac{1}{4\pi} \hat{z} \cdot (\mathbf{E}_R - \mathbf{E}_L)\Big|_{z=0} = -\frac{qd}{2\pi \left(r^2 + d^2\right)^{\frac{3}{2}}}. \tag{3.8}$$

Namely, the presence of the point charge q induces a charge density of opposite sign on the surface of the grounded conductor. (For our case, with the choice $q > 0$, there would be a negative induced surface charge.) The surface charge density is invariant under rotations on the plane, but is not uniform. In fact, like the electric field, it falls off rapidly as we move away from the center. We can, of course, obtain the total induced surface charge by integrating over the entire area and we have

$$Q_{\text{induced}} = \int ds\, \sigma(r)$$

$$= -\frac{qd}{2\pi} \int \frac{r\,dr\,d\phi}{\left(r^2 + d^2\right)^{\frac{3}{2}}}$$

$$= -qd \int_0^\infty \frac{r\,dr}{\left(r^2 + d^2\right)^{\frac{3}{2}}}$$

$$= qd\, \frac{1}{\sqrt{r^2 + d^2}}\Big|_0^\infty = -q. \tag{3.9}$$

Namely, the total induced charge on the plane is equal to the "image" charge (which, in this case, is equal in magnitude to the physical charge, but opposite in sign).

The induced charge, of course, would lead to an attractive force between the point charge and the plane and this can be calculated in the following way. Let us note that at any point on the z-axis, the electric field produced by the induced surface charge density on the plane would be along the z-axis by symmetry. We can calculate it in a simple manner. The potential at any point along the z-axis due to the surface charges is given by ($z > 0$)

$$\Phi(x = 0, y = 0, z) = \int r'\,dr'\,d\phi'\, \frac{\sigma(r')}{\left(r'^2 + z^2\right)^{\frac{1}{2}}}$$

$$= -\frac{qd}{2\pi}\,(2\pi) \int \frac{r'\,dr'}{\left(r'^2 + d^2\right)^{\frac{3}{2}} \left(r'^2 + z^2\right)^{\frac{1}{2}}}$$

$$= \frac{q}{2}\, \frac{\partial}{\partial d} \int_0^\infty \frac{dx}{\left(x + d^2\right)^{\frac{1}{2}} \left(x + z^2\right)^{\frac{1}{2}}}$$

$$= \frac{q}{2}\, \frac{\partial}{\partial d} \int_0^\infty \frac{dx}{\sqrt{x^2 + (d^2 + z^2)x + d^2 z^2}}$$

$$= \frac{q}{2}\, \frac{\partial}{\partial d} \ln\left[2\sqrt{x^2 + (d^2 + z^2)\,x + d^2 z^2} + 2x + \left(d^2 + z^2\right)\right]\Big|_0^\infty$$

$$= \frac{q}{2}\, \frac{\partial}{\partial d}(-)\ln(z + d)^2 = -\frac{q}{z + d}. \tag{3.10}$$

(It can be seen in two different, but equivalent, ways that the upper limit does not contribute. If we take the derivative $\frac{\partial}{\partial d}$ first and then the limit, it is obvious.

Alternatively, we note that the upper limit gives rise to an infinite constant which vanishes upon taking the derivative $\frac{\partial}{\partial d}$.) We recognize this to be the potential at a point along the z-axis in the upper half plane due to the fictitious "image" charge (see (3.5)). The electric field due to the induced surface charge can now be calculated and gives the value for $z = d$ to be (actually, this can be obtained from the general expression for the $\mathbf{E}$ field derived in (3.6) as well)

$$\mathbf{E}(x = 0, y = 0, z = d) = -\hat{\mathbf{z}} \frac{\partial}{\partial z} \left(-\frac{q}{z + d} \right)_{z=d} = -\frac{q}{4d^2} \, \hat{\mathbf{z}}. \tag{3.11}$$

Therefore, the force of attraction experienced by the point charge has the value (this is the force of attraction between the conducting plane and the point charge)

$$\mathbf{F} = q\mathbf{E}(x = 0, y = 0, z = d) = -\frac{q^2}{4d^2} \, \hat{\mathbf{z}}. \tag{3.12}$$

Once again, we see that this is exactly the attractive force between the point charge and the "image" charge and, consequently, on the positive z-axis the "image" charge truly reproduces the effect of the induced surface charge on the plane.

Finally, let us note that we can calculate the electrostatic energy of the system in a simple way. Let us recall that if the point charge is at a distance z from the plane on the positive z-axis, then, the force between the conducting plate and the charge is given by (see (3.12))

$$\mathbf{F}(z) = -\frac{q^2}{4z^2} \, \hat{\mathbf{z}}. \tag{3.13}$$

Using this, we can calculate the work that must be done to bring the point charge from infinity to a distance d above the plane on the z-axis. From the defining relation, we have

$$W = -\int_{\infty}^{d\hat{\mathbf{z}}} \mathrm{d}\boldsymbol{\ell} \cdot \mathbf{F} = \frac{q^2}{4} \int_{\infty}^{d} \frac{\mathrm{d}z}{z^2} = -\frac{q^2}{4z} \Big|_{\infty}^{d} = -\frac{q^2}{4d}. \tag{3.14}$$

In deriving this result, we have used the fact that the work is independent of the choice of the path and, consequently, we have chosen a path along the z-axis for simplicity. It is worth noting here that this energy is half the energy between the point charge and its "image". This can be understood in the following way. The "image" charge is not real. But, if we had calculated the work done to bring the point charge and its "image" from infinity to the final position, we would have done twice the work because we have to move the "image" charge as well. This is, of course, wrong because the "image" charge is not real. So, had we calculated the electrostatic energy using the "image" charge, we would have obtained an erroneous result. (An alternative way to see why one would get twice the result using the "image" charge is to note that the electrostatic energy is related to the square of the electric field integrated over the entire volume, as discussed in chapter 1. With the "image" charge, of course, there is an associated electric field even in the lower half of the plane and by symmetry, it contributes an exact amount as the physical electric field above the plane. That is why the electrostatic energy calculated with the "image" charge would give twice the actual value. In reality, of course, there is no electric field in the lower half plane and that is how the error creeps in.)

Although we have found a solution to the problem of a point charge above a grounded conducting plane of infinite extent by the method of images, it is not clear whether this solution is unique. The uniqueness of the solution can be seen only from an analysis of the Laplace or the Poisson equation which we will discuss later in this chapter.   ◀

▶ **Example (Point charge between two grounded intersecting planes).** As a second example we study a physical system consisting of a point charge q in front of two intersecting orthogonal conducting planes which are of infinite extent and are grounded as shown in Fig. 3.4. We will use the method of images to solve this problem (without going into too much detail). We will see that the solution for this system can be determined provided we have three "image" charges.

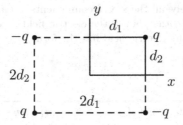

Figure 3.4: The point charge q at $(x = d_1, y = d_2, z = 0)$ and the three image charges.

Let the two infinite, orthogonal and intersecting conducting planes be described by $x = 0, y \geq 0$ and $y = 0, x \geq 0$ respectively. If we assume that the point charge is on the plane $z = 0$, then, it is easy to conclude that all the "image" charges would also lie on the same plane. In fact, it is easy to check that with the choice of the "image" charges shown in the figure, the potential at any point on the plane at $x = 0$ is

$$\Phi(x = 0, y, z) = \frac{q}{(d_1^2 + (d_2 - y)^2 + z^2)^{\frac{1}{2}}} - \frac{q}{(d_1^2 + (d_2 - y)^2 + z^2)^{\frac{1}{2}}}$$
$$+ \frac{q}{(d_1^2 + (d_2 + y)^2 + z^2)^{\frac{1}{2}}} - \frac{q}{(d_1^2 + (d_2 + y)^2 + z^2)^{\frac{1}{2}}}$$
$$= 0. \tag{3.15}$$

Similarly, the potential at any point on the plane at $y = 0$ also vanishes.

$$\Phi(x, y = 0, z) = \frac{q}{((d_1 - x)^2 + d_2^2 + z^2)^{\frac{1}{2}}} - \frac{q}{((d_1 + x)^2 + d_2^2 + z^2)^{\frac{1}{2}}}$$
$$+ \frac{q}{((d_1 + x)^2 + d_2^2 + z^2)^{\frac{1}{2}}} - \frac{q}{((d_1 - x)^2 + d_2^2 + z^2)^{\frac{1}{2}}}$$
$$= 0. \tag{3.16}$$

Thus, these image charges indeed reproduce the boundary condition of vanishing potential on the two infinite planes.

Once we have determined the "image" charges, we can forget about the conducting planes and determine the potential in the region $x > 0$ and $y > 0$ simply to be

$$\Phi(x, y, z) = \frac{q}{((x - d_1)^2 + (y - d_2)^2 + z^2)^{\frac{1}{2}}}$$

$$- \frac{q}{((x + d_1)^2 + (y - d_2)^2 + z^2)^{\frac{1}{2}}} - \frac{q}{((x - d_1)^2 + (y + d_2)^2 + z^2)^{\frac{1}{2}}}$$

$$+ \frac{q}{((x + d_1)^2 + (y + d_2)^2 + z^2)^{\frac{1}{2}}}. \tag{3.17}$$

The electric field can be obtained from (3.17) by taking the gradient. It has a general structure (namely, all the $\hat{\mathbf{x}}, \hat{\mathbf{y}}, \hat{\mathbf{z}}$ components are nonzero) at an arbitrary point. However, close to the planes, the electric field takes a simpler form. For example, when $x = 0$, the electric field has the form

$$\mathbf{E}(x = 0, y > 0, z)$$

$$= 2qd_1 \, \hat{\mathbf{x}} \left[\frac{1}{(d_1^2 + (y + d_2)^2 + z^2)^{\frac{3}{2}}} - \frac{1}{(d_1^2 + (y - d_2)^2 + z^2)^{\frac{3}{2}}} \right], \tag{3.18}$$

while, for $y = 0$, it has the form

$$\mathbf{E}(x > 0, y = 0, z)$$

$$= 2qd_2 \, \hat{\mathbf{y}} \left[\frac{1}{((x + d_1)^2 + d_2^2 + z^2)^{\frac{3}{2}}} - \frac{1}{((x - d_1)^2 + d_2^2 + z^2)^{\frac{3}{2}}} \right]. \tag{3.19}$$

The surface charge densities now follow from the discontinuities of the electric field across the two planes. Namely,

$$\sigma(x = 0, y > 0, z) = \frac{1}{4\pi} \, \hat{\mathbf{x}} \cdot \mathbf{E}(x = 0, y > 0, z)$$

$$= \frac{qd_1}{2\pi} \left[\frac{1}{(d_1^2 + (y + d_2)^2 + z^2)^{\frac{3}{2}}} - \frac{1}{(d_1^2 + (y - d_2)^2 + z^2)^{\frac{3}{2}}} \right],$$

$$\sigma(x > 0, y = 0, z) = \frac{1}{4\pi} \, \hat{\mathbf{y}} \cdot \mathbf{E}(x > 0, y = 0, z)$$

$$= \frac{qd_2}{2\pi} \left[\frac{1}{((x + d_1)^2 + d_2^2 + z^2)^{\frac{3}{2}}} - \frac{1}{((x - d_1)^2 + d_2^2 + z^2)^{\frac{3}{2}}} \right]. \tag{3.20}$$

Using (3.20), the total induced charge on the conducting planes can now be determined as follows.

$$Q_{\text{induced}} = \int_0^\infty dy \int_{-\infty}^\infty dz \, \sigma(x = 0, y > 0, z)$$

$$+ \int_0^\infty dx \int_{-\infty}^\infty dz \, \sigma(x > 0, y = 0, z)$$

$$= \frac{qd_1}{\pi} \iint_0^\infty dydz \left[\frac{1}{(d_1^2 + (y + d_2)^2 + z^2)^{\frac{3}{2}}} - \frac{1}{(d_1^2 + (y - d_2)^2 + z^2)^{\frac{3}{2}}} \right]$$

$$+ \frac{qd_2}{\pi} \iint_0^\infty dxdz \left[\frac{1}{((x + d_1)^2 + d_2^2 + z^2)^{\frac{3}{2}}} - \frac{1}{((x - d_1)^2 + d_2^2 + z^2)^{\frac{3}{2}}} \right]$$

$$= -\frac{2qd_1}{\pi} \int_0^\infty dy \left[\frac{1}{y^2 + 2d_2y + d_1^2 + d_2^2} - \frac{1}{y^2 - 2d_2y + d_1^2 + d_2^2} \right]$$

$$- \frac{2qd_2}{\pi} \int_0^\infty dx \left[\frac{1}{x^2 + 2d_1x + d_1^2 + d_2^2} - \frac{1}{x^2 - 2d_1x + d_1^2 + d_2^2} \right]$$

$$= -\frac{2qd_1}{\pi} \frac{1}{d_1} \tan^{-1}\left(\frac{d_2}{d_1}\right) - \frac{2qd_2}{\pi} \frac{1}{d_2} \tan^{-1}\left(\frac{d_1}{d_2}\right)$$

$$= -\frac{2q}{\pi} \left(\tan^{-1}\left(\frac{d_2}{d_1}\right) + \tan^{-1}\left(\frac{d_1}{d_2}\right) \right)$$

$$= -\frac{2q}{\pi} \times \frac{\pi}{2} = -q. \tag{3.21}$$

In deriving this result, we have used some standard integrals from the tables (Gradshteyn and Ryzhik, 2.172 and 2.271) as well as the trigonometric relation that, for $x > 0$,

$$\tan^{-1} x + \tan^{-1}\left(\frac{1}{x}\right) = \frac{\pi}{2}. \tag{3.22}$$

Once again, we see that the total induced charge (on the two plates) is equal to the sum of all the "image" charges (which is equal in magnitude to the physical point charge, but opposite in sign). However, the amount of charges on the two plates depends on the ratio of the perpendicular distances d_1 and d_2 of the point charge from the two planes. ◀

▶ **Example (Point charge outside a grounded conducting sphere).** Another example of the method of images is the system of a point charge q at a distance d from the center of a conducting sphere of radius R where we will assume that $d > R$. The sphere is grounded, namely, the surface of the sphere is maintained at zero potential. Clearly, the potential inside the conducting sphere will be zero (since every surface inside the conducting sphere defines an equipotential surface with the same potential as the surface of the sphere) and the electric field within the sphere will also be zero. Thus, the region where the potential and the electric field would be nontrivial lies outside the sphere.

Let us assume the origin of our coordinate system to coincide with the center of the sphere. Without any loss of generality, we can choose the charge to lie along the z-axis. Then, let us consider the following system of charges, charge q at $z = d$ and another charge q' at $z = d'$. Namely, we are considering the effect of an additional charge located on the line connecting the point charge and the center of the sphere as shown in Fig. 3.5. It is clear now that the potential for this combined system of charges at any point $\mathbf{r}$ with $r \geq R$ is given by

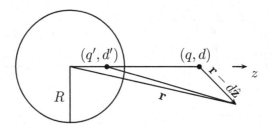

Figure 3.5: A point charge q outside a grounded conducting sphere and its image charge.

$$\Phi(\mathbf{r}) = \frac{q}{|\mathbf{r} - d\hat{\mathbf{z}}|} + \frac{q'}{|\mathbf{r} - d'\hat{\mathbf{z}}|}$$

$$= \frac{q}{(r^2 + d^2 - 2rd\cos\theta)^{\frac{1}{2}}} + \frac{q'}{(r^2 + d'^2 - 2rd'\cos\theta)^{\frac{1}{2}}}. \qquad (3.23)$$

If we require that this potential vanishes when $r = R$ for any angle θ and ϕ (namely, on the surface of the sphere), we obtain,

$$q^2(R^2 + d'^2 - 2Rd'\cos\theta) - q'^2(R^2 + d^2 - 2Rd\cos\theta) = 0,$$

$$\text{or,} \quad R^2\left(1 - \left(\frac{q'}{q}\right)^2\right) + d'^2\left(1 - \left(\frac{q'd}{qd'}\right)^2\right) - 2Rd'\cos\theta\left(1 - \left(\frac{q'}{q}\right)^2\frac{d}{d'}\right) = 0. \qquad (3.24)$$

Since this must hold for any θ, the coefficient of the last term in (3.24) must vanish leading to

$$q'^2 = q^2\frac{d'}{d}. \qquad (3.25)$$

Substituting this back into (3.24), the other two terms lead to

$$d' = \frac{R^2}{d}, \quad q' = -q\left(\frac{R}{d}\right). \qquad (3.26)$$

Actually, there are two solutions for the charge (as is clear from (3.25)), but this is the one which gives a vanishing potential (see (3.23)) on the surface of the sphere $r = R$.

Thus, we see that, studying the problem of a point charge outside a conducting sphere which is grounded, is equivalent to studying the point charge in the presence of an "image" charge inside the sphere. The potential at an arbitrary point outside the sphere is now easily determined to be

$$\Phi(\mathbf{r}) = \frac{q}{(r^2 + d^2 - 2rd\cos\theta)^{\frac{1}{2}}} + \frac{q'}{(r^2 + d'^2 - 2rd'\cos\theta)^{\frac{1}{2}}}$$

$$= q\left(\frac{1}{(r^2 + d^2 - 2rd\cos\theta)^{\frac{1}{2}}} - \frac{R}{(r^2d^2 + R^4 - 2rdR^2\cos\theta)^{\frac{1}{2}}}\right). \quad (3.27)$$

The electric field can, of course, be calculated from this by taking the gradient and it is clear that it would, in general, have both a radial and an angular component. However, near the surface of the sphere, namely, when $r = R$, the angular components cancel out (The simplest way to see this is to note that when taking derivative with respect to θ, we can set $r = R$, but then the potential vanishes and so does the θ component of the gradient at such points. Physically, of course, this means that there is no tangential component of $\mathbf{E}$ on the surface of the conductor as we would expect.) and we have only a radial component

$$\mathbf{E}(\mathbf{r})|_{r=R} = -\boldsymbol{\nabla}\Phi(\mathbf{r})|_{r=R} = -\hat{\mathbf{r}}\frac{\partial}{\partial r}\Phi(\mathbf{r})\bigg|_{r=R}$$

$$= q\hat{\mathbf{r}}\left(\frac{(R - d\cos\theta)}{(d^2 + R^2 - 2dR\cos\theta)^{\frac{3}{2}}} - \frac{R(Rd^2 - R^2d\cos\theta)}{R^3(d^2 + R^2 - 2dR\cos\theta)^{\frac{3}{2}}}\right)$$

$$= -\frac{q(d^2 - R^2)\,\hat{\mathbf{r}}}{R\,(d^2 + R^2 - 2dR\cos\theta)^{\frac{3}{2}}}. \quad (3.28)$$

Consequently, we can determine the surface distribution of the charges from the discontinuity of the electric field, namely,

$$\sigma(r = R, \theta, \phi) = \frac{1}{4\pi}\,\hat{\mathbf{r}}\cdot\mathbf{E}(\mathbf{r})|_{r=R}$$

$$= -\frac{q(d^2 - R^2)}{4\pi R}\frac{1}{(d^2 + R^2 - 2dR\cos\theta)^{\frac{3}{2}}}. \quad (3.29)$$

The total induced charge on the surface of the sphere can now be obtained from (3.29) through a simple integration and we obtain,

$$Q_{\text{induced}} = \int R^2\sin\theta\,\mathrm{d}\theta\,\mathrm{d}\phi\,\sigma(r = R, \theta, \phi)$$

$$= -\frac{q(d^2 - R^2)}{4\pi R}\times 2\pi R^2\int_{-1}^{1}\frac{\mathrm{d}x}{(d^2 + R^2 - 2dRx)^{\frac{3}{2}}}$$

$$= -\frac{q(d^2 - R^2)R}{2}\left(\frac{1}{dR}\right)\frac{1}{(d^2 + R^2 - 2dRx)^{\frac{1}{2}}}\bigg|_{-1}^{1}$$

$$= -\frac{q(d^2 - R^2)}{2d}\left[\frac{1}{d - R} - \frac{1}{d + R}\right]$$

$$= -q\left(\frac{R}{d}\right) = q'. \quad (3.30)$$

This again shows that the total charge induced on the sphere is identical to the "image" charge (which is not equal in magnitude to the point charge in this case).

The force of attraction between the point charge and the sphere can again be calculated directly or from the "image" charge.

$$\mathbf{F}(x=0, y=0, z=d) = \frac{qq'}{(d-d')^2}\, \hat{\mathbf{z}} = -q \times \frac{q\left(\frac{R}{d}\right)}{\left(d - \frac{R^2}{d}\right)^2}\, \hat{\mathbf{z}}$$

$$= -\frac{q^2 dR}{(d^2 - R^2)^2}\, \hat{\mathbf{z}}. \tag{3.31}$$

It is clear from this that, at any point on the z-axis, the force experienced by the point charge is

$$\mathbf{F}(x=0, y=0, z) = -\frac{q^2 zR}{(z^2 - R^2)^2}\, \hat{\mathbf{z}}. \tag{3.32}$$

Therefore, the work done in bringing the charge from infinity is easily obtained to be

$$W = -\int_{\infty}^{d\hat{\mathbf{z}}} d\boldsymbol{\ell} \cdot \mathbf{F} = -\int_{\infty}^{d} dz \left(-\frac{q^2 zR}{(z^2 - R^2)^2} \right)$$

$$= \frac{q^2 R}{2} \times \left[-\frac{1}{z^2 - R^2} \right]_{\infty}^{d} = -\frac{q^2 R}{2(d^2 - R^2)}. \tag{3.33}$$

Once again, this is half of the energy that we would have found from a calculation using the "image" charge for reasons which we have discussed earlier. The method of images works well for all quantities except for the energy of the system. ◀

Although we have discussed the method of images only within the context of point charges, it works well for other systems such as line charges etc. But, it is clear from our discussions that only when there is some symmetry in the problem, it may be easier to determine the "image" charges, otherwise, the method may not be very useful.

3.2 Boundary conditions for differential equations

In solving dynamical equations of second order such as Newton's equation, we normally require two initial conditions (namely, the initial position and the initial velocity) to solve the equation uniquely. The Laplace equation as well as the Poisson equation are also second order equations and yet, we saw that given just one condition such as the potential on the surface of a conductor, we could solve the problem completely. Therefore, this raises the question of how one determines what kind of boundary conditions (or initial conditions) are necessary for solving a differential equation uniquely.

To examine this, let us start with an ordinary differential equation of order m of the form

$$a_0(x)\frac{d^m f(x)}{dx^m} + a_1(x)\frac{d^{m-1}f(x)}{dx^{m-1}} + \cdots + a_m(x)f(x) = F, \tag{3.34}$$

where $f(x)$ represents the unknown variable to be determined. In general, the function F may depend on x as well as $f(x)$. When it does depend on $f(x)$ in a nonlinear manner, the equation is said to be nonlinear. However, in most of our discussions we will be interested in linear equations where F can at the most depend on x (if it depends linearly on $f(x)$, the linear terms in F can always be combined with the terms on the left). Of course, an equation can have an infinity of solutions in general and the one appropriate for a particular physical situation is uniquely selected by the given boundary conditions. Thus, for example, we know from the study of the harmonic oscillator equation (which is second order in the time variable) that, in general, there is an infinity of solutions given by an arbitrary linear superposition of $e^{\pm i\omega t}$ where ω is the natural frequency of the oscillator. However, if we further specify the initial position as well as the initial velocity of the oscillator, then the solution is uniquely determined. Thus, the boundary conditions as well as the surface on which they are prescribed (in the case of Newton's equation, they are known as initial conditions because they are prescribed on the initial surface $t = 0$) are quite crucial in determining the unique solution of a given physical problem.

For the case of the mth order equation in (3.34), it is clear from our experience with Newton's equation that a unique solution may be possible if we know the function $f(x)$ as well as its first $(m - 1)$ derivatives at some point x_0. (It is assumed that $F(x)$ is a known function whose value at x_0 is known.) This is because at any point in space, the function $f(x)$ (which we assume to be continuous as is the case for the potential) has a Taylor expansion of the form

$$f(x) = f(x_0) + (x - x_0) \left. \frac{\mathrm{d}f}{\mathrm{d}x} \right|_{x_0} + \cdots$$
$$+ \frac{(x - x_0)^n}{n!} \left. \frac{\mathrm{d}^n f}{\mathrm{d}x^n} \right|_{x_0} + \cdots . \tag{3.35}$$

We see from this that determining $f(x)$ involves a knowledge of the derivatives of the function at x_0 to all orders. However, we note that if we know the function and its first $(m - 1)$ derivatives at x_0, we can determine all the higher order derivatives from the differential equation itself. For example, from (3.34) we have

$$\frac{\mathrm{d}^m f}{\mathrm{d}x^m}\bigg|_{x_0} = -\frac{1}{a_0(x_0)}\left[-F(x_0) + a_m(x_0)f(x_0) + a_{m-1}(x_0)\frac{\mathrm{d}f}{\mathrm{d}x}\bigg|_{x_0}\right.$$

$$\left. + \cdots + a_1(x_0)\frac{\mathrm{d}^{m-1}f}{\mathrm{d}x^{m-1}}\bigg|_{x_0}\right], \qquad (3.36)$$

and so on for the higher order derivatives. Thus, we see that a unique solution of an mth order ordinary differential equation needs m boundary conditions, namely, the values of the function as well as its first $(m-1)$ derivatives at a coordinate, say x_0. Note that the mth derivative (as well as the higher order ones) can no longer be specified independently, but is (are) determined from the differential equation itself. Specifying the mth derivative as well will only over-specify the solution. In contrast, specifying a fewer number of derivatives will not determine the solution uniquely.

Let us also note that if we specify the boundary conditions at a coordinate where $a_0(x_0) = 0$, then it is clear from (3.36) that $\frac{\mathrm{d}^m f}{\mathrm{d}x^m}|_{x_0}$ as well as the higher derivatives cannot be determined and hence a solution cannot be uniquely determined. In fact, in such a case, either the boundary conditions are consistent with the equation itself in which case infinitely many solutions are possible, or the boundary conditions are inconsistent with the equation implying that no solution satisfying the given boundary conditions is possible. The point x_0 where the coefficient of the highest derivative term of the differential equation vanishes is known as the characteristic point of the equation and it is clear that to have a unique solution, we need to specify the right number of boundary conditions at points which are not characteristic points of the equation. Boundary conditions where one specifies the function as well as its first derivative on a boundary (for a second order equation) are known as Cauchy conditions and finding a solution subject to such boundary conditions is known as the Cauchy problem (Cauchy initial value problem if time is involved). However, as we will see for some equations other boundary conditions are more appropriate to obtain a unique solution.

3.2.1 Partial differential equations.

Let us next analyze the boundary conditions for partial differential equations. For simplicity of discussion, let us start with a general two dimensional second order partial differential equation of the form

$$A \frac{\partial^2 f}{\partial x^2} + 2B \frac{\partial^2 f}{\partial x \partial y} + C \frac{\partial^2 f}{\partial y^2} = F\left(x, y, \frac{\partial f}{\partial x}, \frac{\partial f}{\partial y}\right). \tag{3.37}$$

Here, we are going to assume that the equation is linear and that the coefficients A, B and C are, in general, functions of x, y. Furthermore, the two independent variables x, y can both be space coordinates or one space and one time coordinate.

Unlike the one dimensional ordinary differential equation that we discussed earlier, here specifying the function and its (two) first derivatives at a point will not be enough to determine the solution uniquely. Rather, we need to specify appropriate boundary conditions on a curve. (In general, the solution of a partial differential equation in n variables needs boundary conditions specified on a $(n-1)$ dimensional hypersurface.) Furthermore, as we will see Cauchy boundary conditions may not always work in these cases because they may over-specify the solution.

To understand the nature of boundary conditions and the curve on which they must be specified, let us represent the boundary curve parametrically by $\xi = x(s)$ and $\eta = y(s)$ where s is the distance of a point on the curve from some reference point. At any point on the curve, there are two orthogonal directions – one along the tangent to the curve and the other normal to it (see Fig. 3.6). The unit vectors along these directions are easily determined to be

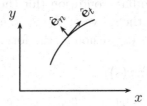

Figure 3.6: Unit vectors $\hat{e}_t$ and $\hat{e}_n$ which are respectively tangential and normal to the curve at a given point.

$$\hat{e}_t = \frac{d\xi}{ds}\hat{x} + \frac{d\eta}{ds}\hat{y},$$

$$\hat{e}_n = \hat{z} \times \hat{e}_t = -\frac{d\eta}{ds}\hat{x} + \frac{d\xi}{ds}\hat{y}. \tag{3.38}$$

These are orthogonal by construction and are easily checked to be unit vectors from the relation that

$$\left(\frac{\mathrm{d}\xi}{\mathrm{d}s}\right)^2 + \left(\frac{\mathrm{d}\eta}{\mathrm{d}s}\right)^2 = 1, \tag{3.39}$$

which follows from the fact that the infinitesimal distance between two points on the curve can be written as

$$\mathrm{d}s^2 = \mathrm{d}\xi^2 + \mathrm{d}\eta^2. \tag{3.40}$$

The value of the function on the boundary curve can be represented as $f(\xi, \eta) = f(s)$. It is clear then that we can define the derivative of the function along the curve as well as along the normal to it as

$$f_t = \hat{\mathbf{e}}_t \cdot \boldsymbol{\nabla} f|_{x(s),y(s)} = \frac{\mathrm{d}\xi}{\mathrm{d}s}\left(\frac{\partial f}{\partial x}\right)_{x(s),y(s)} + \frac{\mathrm{d}\eta}{\mathrm{d}s}\left(\frac{\partial f}{\partial y}\right)_{x(s),y(s)}$$

$$\equiv \frac{\mathrm{d}f}{\mathrm{d}s}, \tag{3.41}$$

$$f_n = \hat{\mathbf{e}}_n \cdot \boldsymbol{\nabla} f|_{x(s),y(s)} = -\frac{\mathrm{d}\eta}{\mathrm{d}s}\left(\frac{\partial f}{\partial x}\right)_{x(s),y(s)} + \frac{\mathrm{d}\xi}{\mathrm{d}s}\left(\frac{\partial f}{\partial y}\right)_{x(s),y(s)}.$$

This analysis makes it clear that once we know $f(s)$ along the curve, we also know its derivative along the curve, since $f_t = \frac{\mathrm{d}f}{\mathrm{d}s}$. Therefore, the first order derivative which needs to be specified as an independent boundary condition (for the Cauchy problem) is the derivative normal to the curve or f_n. A solution at any point will, of course, have a Taylor expansion of the form

$$f(x, y) = f(x(s), y(s))$$

$$+ \sum_{n=1}^{\infty} \sum_{m=0}^{n} \frac{(x - x(s))^{n-m}(y - y(s))^m}{n!} \left.\frac{\partial^n f}{\partial x^{n-m}\partial y^m}\right|_{x(s),y(s)}, \tag{3.42}$$

and if we know all the partial derivatives in the expansion, then the solution can be uniquely determined. The question, therefore, is whether from a knowledge of the values of $f(s)$ and $f_n(s)$, we can determine all the partial derivatives and, therefore, the solution uniquely.

To start with, let us note that using $(\frac{\mathrm{d}\xi}{\mathrm{d}s})^2 + (\frac{\mathrm{d}\eta}{\mathrm{d}s})^2 = 1$ in (3.41), we obtain (this basically corresponds to inverting (3.41))

$$\left(\frac{\partial f}{\partial x}\right)_{x(s),y(s)} = \frac{d\xi}{ds} f_t - \frac{d\eta}{ds} f_n = a(s),$$

$$\left(\frac{\partial f}{\partial y}\right)_{x(s),y(s)} = \frac{d\eta}{ds} f_t + \frac{d\xi}{ds} f_n = b(s). \tag{3.43}$$

Thus, we see that given $f(s)$ and f_n, we can determine the (two) first derivatives directly from the data (remember $f_t = \frac{df}{ds}$). Furthermore, since $a(s)$ and $b(s)$ are known functions, by taking their derivative as well as using the differential equation, we have

$$\frac{d\xi}{ds} \left(\frac{\partial^2 f}{\partial x^2}\right)_{x(s),y(s)} + \frac{d\eta}{ds} \left(\frac{\partial^2 f}{\partial x \partial y}\right)_{x(s),y(s)} = \frac{da(s)}{ds},$$

$$\frac{d\xi}{ds} \left(\frac{\partial^2 f}{\partial x \partial y}\right)_{x(s),y(s)} + \frac{d\eta}{ds} \left(\frac{\partial^2 f}{\partial y^2}\right)_{x(s),y(s)} = \frac{db(s)}{ds},$$

$$\left[A(s)\frac{\partial^2 f}{\partial x^2} + 2B(s)\frac{\partial^2 f}{\partial x \partial y} + C(s)\frac{\partial^2 f}{\partial y^2}\right]_{x(s),y(s)} = F(s). \tag{3.44}$$

This is a set of three coupled inhomogeneous equations in the three unknown second order derivatives and has a unique solution only if the determinant of the coefficient matrix does not vanish. Thus, as long as

$$\Delta = \begin{vmatrix} \frac{d\xi}{ds} & \frac{d\eta}{ds} & 0 \\ 0 & \frac{d\xi}{ds} & \frac{d\eta}{ds} \\ A & 2B & C \end{vmatrix} \neq 0, \tag{3.45}$$

we can determine the second derivatives uniquely. Furthermore, once these are known, by successive differentiation, the higher derivatives can also be determined and, consequently, it would appear that the Cauchy problem can be uniquely solved, but as we will see shortly it is not that simple.

On the other hand, if the characteristic determinant vanishes, we have (upon expanding the determinant)

$$A\left(\frac{d\eta}{ds}\right)^2 - 2B\frac{d\xi}{ds}\frac{d\eta}{ds} + C\left(\frac{d\xi}{ds}\right)^2 = 0, \tag{3.46}$$

which is a quadratic equation with solutions

$$\frac{d\eta}{ds} = \frac{B \pm \sqrt{B^2 - AC}}{A} \frac{d\xi}{ds},$$

$$\text{or,} \quad d\eta = \frac{B \pm \sqrt{B^2 - AC}}{A} d\xi. \tag{3.47}$$

If this holds, then the Cauchy problem cannot be uniquely solved. These two equations are equations for two curves which in the present case are known as the characteristic curves and we see that the Cauchy problem cannot be solved if the Cauchy data are specified on any of the characteristics. This is reminiscent of the behavior in the case of ordinary differential equations.

For a given second order equation, $A, B,$ and C are known functions and depending on the behavior of the radical in (3.47), partial differential equations can be classified into three different groups. If $B^2 > AC$, then we see that there are two real characteristic curves of the equation. Such equations are known as hyperbolic equations. The most familiar of the hyperbolic equations is the wave equation of the form (in $1 + 1$ dimensions)

$$\frac{\partial^2 f}{\partial x^2} - \frac{1}{v^2} \frac{\partial^2 f}{\partial t^2} = 0, \tag{3.48}$$

where v represents the speed of propagation of the wave.

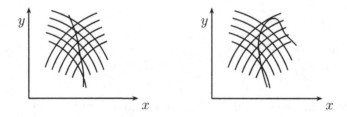

Figure 3.7: A curve intersects the characteristic curves (on the left) while it is tangential to a characteristic (on the right).

For a hyperbolic equation, there are two families of characteristic curves. Of course, if the Cauchy data are specified along any one of the two characteristics, the solution cannot be uniquely obtained. On the other hand, if the Cauchy data are given on a curve which intersects all the characteristics exactly once, then, it is easy to see that the Cauchy problem can be solved uniquely. But, if the boundary is a curve which is tangent to any of the characteristic curves at any point (this, therefore, includes closed curves, see, for example, the second plot in Fig. 3.7), specifying the Cauchy data along the entire curve will over-specify the solution. In such a case, one needs to specify the Cauchy data only on part of the curve and either the function or the normal derivative alone on the other parts of the curve. Specifying only the value of the function on the boundary is known as the Dirichlet boundary condition while specifying the normal derivative on a boundary is known as the Neumann boundary condition.

If the differential equation is such that $B^2 < AC$, then, it is known as an elliptic equation and clearly, in this case, the characteristic curves are complex, namely, we do not have any real characteristics. A familiar example of such an equation which shows up in many physical problems is the Laplace equation (in two dimensions),

$$\nabla^2 f = \frac{\partial^2 f}{\partial x^2} + \frac{\partial^2 f}{\partial y^2} = 0, \tag{3.49}$$

or its higher dimensional generalizations. (In electrostatics, for example, we can identify $f = \Phi$ in a region free of charge.) In such a case, since no real characteristic curves are present, it would seem that the Cauchy boundary conditions would be sufficient to solve the problem. However, further analysis shows that, in this case, Cauchy boundary conditions over-specify the solution and the problem can be solved uniquely only if the function or the normal derivative or a linear combination of the two is specified on a closed curve. Thus, either Dirichlet or Neumann boundary condition or a linear combination of the two, specified along a closed boundary, solves an elliptic equation such as the Laplace equation uniquely. This is familiar from the examples we have already worked out involving the conductors. We have seen that we can solve the Laplace equation (in three dimensions) when the potential or the charge density is specified on a given surface. (Remember that the surface charge density is related to the normal component of the electric field which is the normal derivative of the potential.) When only the Dirichlet or Neumann boundary condition is specified, the corresponding problem is known as a Dirichlet/Neumann problem. On the other hand, if a linear combination of the two is specified, then, the problem is referred to as a mixed boundary value problem. Mixed boundary value problems are harder to solve analytically and are, in general, solved numerically.

Finally, if the partial differential equation is such that $B^2 = AC$, then, it is known as a parabolic equation. The diffusion equation,

$$\frac{\partial^2 f}{\partial x^2} = \alpha \frac{\partial f}{\partial t}, \tag{3.50}$$

is an example of a parabolic equation. In the case of a parabolic equation, there is only one characteristic curve and a unique solution can be obtained only for either Dirichlet or Neumann (or a mixed) boundary condition on an open curve.

Although our discussion so far has been within the context of two dimensions, it can be generalized to higher dimensions as well. For a general second order partial differential equation in n dimensions, we can always find a suitable coordinate transformation to diagonalize the equation. When, only one of the coefficients of the diagonalized equation is negative, the equation is known as a hyperbolic equation. If none of the coefficients of the diagonalized equation is negative, then, it is known as an elliptic equation and if any one of the coefficients vanishes, then, the equation is called a parabolic equation. In general, hyperbolic equations are solved uniquely by specifying Cauchy boundary conditions on an open hypersurface, elliptic equations by specifying Dirichlet/Neumann (or mixed) boundary conditions on a closed hypersurface and parabolic equations by specifying Dirichlet/Neumann (or mixed) boundary conditions on an open hypersurface.

3.2.2 Uniqueness theorem. Let us now go back to the Laplace equation in three dimensions involving the electrostatic potential. Let us assume that

$$\nabla^2 \Phi = 0, \tag{3.51}$$

holds in a region of space denoted by the volume V whose bounding surface is S. This is an elliptic equation and, as we have seen, for a unique solution, we must specify either the value of the potential Φ or $\frac{\partial \Phi}{\partial n}$ (which is the negative of the normal component of the electric field) on S. Suppose Φ_1 and Φ_2 represent two solutions of the Laplace equation satisfying the same boundary condition on S, then we will show that they can at the most differ by a constant.

To prove this, let us first note that the Laplace equation is a linear partial differential equation and hence a superposition of distinct solutions also defines a solution. Therefore, let us define

$$\Phi = \Phi_1 - \Phi_2, \tag{3.52}$$

which also satisfies the Laplace equation, namely,

$$\nabla^2 \Phi = \nabla^2 \Phi_1 - \nabla^2 \Phi_2 = 0. \tag{3.53}$$

Using Gauss' theorem, let us next note that we can write

$$\int_V d^3r\, \nabla \cdot (\Phi(\nabla\Phi)) = \int_S ds \cdot (\nabla\Phi)\Phi,$$

$$\text{or,} \quad \int_V d^3r\, ((\nabla\Phi)^2 + \Phi(\nabla^2\Phi)) = \int_S ds \left(\frac{\partial\Phi}{\partial n}\right)\Phi,$$

$$\text{or,} \quad \int_V d^3r\, (\nabla\Phi)^2 = 0. \tag{3.54}$$

Here, we have used the fact that Φ satisfies the Laplace equation (3.53) so that the second term on the left hand side does not contribute. Furthermore, for either Dirichlet or Neumann boundary conditions on S, either Φ or $\frac{\partial\Phi}{\partial n}$ vanishes on S so that the right hand side identically vanishes. The final result, as it stands, shows that the integral of a positive quantity vanishes and, consequently, the integrand on the left in (3.54) must vanish, namely,

$$\nabla\Phi = 0. \tag{3.55}$$

This would seem to say that $\Phi = \Phi_1 - \Phi_2$ is at most a constant. For Dirichlet boundary condition this constant must vanish since it vanishes on the boundary S. On the other hand, for Neumann boundary condition for which

$$\hat{\mathbf{n}} \cdot \nabla\Phi = 0, \tag{3.56}$$

on S, we can only say that Φ_1 and Φ_2 can at most differ by a constant. However, since a constant is not relevant for calculations of physical quantities like the electric field, we can say that for a given set of boundary conditions, either Dirichlet or Neumann, the solution of the Laplace equation is unique. Namely, if we find a solution to a given problem, it must be the unique solution. This shows that the solutions we found earlier using the method of images must be the unique solutions for those particular physical problems.

Incidentally, the uniqueness theorem can also be extended to the solutions of the Poisson equation quite easily. For, suppose Φ_1 and Φ_2 satisfy the Poisson equation

$$\nabla^2\Phi_1 = -4\pi\rho(\mathbf{r}), \qquad \nabla^2\Phi_2 = -4\pi\rho(\mathbf{r}), \tag{3.57}$$

with the same Dirichlet/Neumann boundary condition on S, then, $\Phi = \Phi_1 - \Phi_2$ would satisfy the Laplace equation. This would, therefore, imply as before in the derivation of (3.55) that

$$\boldsymbol{\nabla}\Phi = 0. \tag{3.58}$$

For solutions satisfying the Dirichlet boundary condition on S, we have

$$\Phi = 0, \tag{3.59}$$

and, consequently, the two solutions are the same. However, for Neumann boundary conditions, $\Phi = $ constant which for the purposes of calculating electric field etc. still implies that the solution is unique.

3.3 Solutions of the Laplace equation

Let us now analyze the solutions of the Laplace equation in three dimensions, subject to specific boundary conditions. In Cartesian coordinates the Laplace equation has the form

$$\boldsymbol{\nabla}^2\Phi(\mathbf{r}) = \left(\frac{\partial^2}{\partial x^2} + \frac{\partial^2}{\partial y^2} + \frac{\partial^2}{\partial z^2}\right)\Phi(x,y,z) = 0, \tag{3.60}$$

which, in spherical coordinates, takes the form

$$\left(\frac{1}{r^2}\frac{\partial}{\partial r}r^2\frac{\partial}{\partial r} + \frac{1}{r^2\sin\theta}\frac{\partial}{\partial\theta}\sin\theta\frac{\partial}{\partial\theta} + \frac{1}{r^2\sin^2\theta}\frac{\partial^2}{\partial\phi^2}\right)\Phi(r,\theta,\phi) = 0, \tag{3.61}$$

while, in cylindrical coordinates, it has the form

$$\left(\frac{1}{r}\frac{\partial}{\partial r}r\frac{\partial}{\partial r} + \frac{1}{r^2}\frac{\partial^2}{\partial\phi^2} + \frac{\partial^2}{\partial z^2}\right)\Phi(r,\phi,z) = 0. \tag{3.62}$$

Depending on the symmetry in the problem, it is meaningful to solve the equation in the appropriate coordinates. The Laplace equation appears in many branches of physics and its solutions are known as harmonic functions. To begin with, let us summarize some of the general properties of harmonic functions.

3.3.1 General properties of harmonic functions. The harmonic functions possess several interesting properties which can be derived without explicitly solving the Laplace equation.

1. Let us assume that Φ represents a solution of the Laplace equation in a given region V, namely,

$$\nabla^2 \Phi = 0, \tag{3.63}$$

in V. Then, it follows from Gauss' theorem that

$$\int_S ds \cdot (\nabla \Phi) = \int_V d^3 r \, \nabla^2 \Phi = 0. \tag{3.64}$$

Here S is assumed to be the surface which bounds the volume V. (In electrostatics this would correspond to the fact that the integral of the electric field over a closed surface or the electric flux out of the region must vanish.) Furthermore, using Stokes' theorem, it follows that for any function Φ (this is independent of whether Φ satisfies the Laplace equation or not and in electrostatics corresponds to the electric field being conservative)

$$\oint_C d\boldsymbol{\ell} \cdot (\nabla \Phi) = \int_S ds \cdot (\nabla \times \nabla \Phi) = 0. \tag{3.65}$$

2. Let us note that, given any two functions A and B, we have the identity

$$\nabla \cdot (A \nabla B - B \nabla A) = (A \nabla^2 B - B \nabla^2 A), \tag{3.66}$$

so that using Gauss' theorem, we can write

$$\int_V d^3 r \, (A \nabla^2 B - B \nabla^2 A) = \int_S ds \cdot (A \nabla B - B \nabla A). \tag{3.67}$$

This is known as Green's identity. If we now choose $A = \frac{1}{|\mathbf{r}|}$ and $B = \Phi$, namely, if B represents a solution of the Laplace equation (3.63) and A satisfies (this is known as the Green's function for the Laplacian)

$$\nabla^2 A = \nabla^2 \frac{1}{|\mathbf{r}|} = -4\pi \, \delta^3(\mathbf{r}), \tag{3.68}$$

then, from (3.67) (we are going to assume that S is the surface of a sphere of radius R), we obtain

$$4\pi \int_V d^3 r \, \Phi(\mathbf{r}) \delta^3(\mathbf{r}) = \int_S ds \cdot \left(\frac{1}{R}(\nabla \Phi) + \frac{\hat{\mathbf{r}}}{R^2} \Phi \right),$$

$$\text{or,} \quad \Phi(\mathbf{r} = 0) = \frac{1}{4\pi R^2} \int_S ds \, \Phi(R, \theta, \phi). \tag{3.69}$$

Here we have used Eq. (3.64). Relation (3.69) leads to the result that if Φ satisfies the Laplace equation (3.63) inside a sphere of radius R, then the value of Φ at the origin of the sphere is the average of the value of Φ over the surface of the sphere (recall that the surface area of a sphere of radius R is $4\pi R^2$). This is something which we have seen earlier in specific calculations, but this is, in fact, a general property of harmonic functions. This is a very important result, for it has many interesting consequences.

3. One of the most interesting consequences of (3.69) is the fact that a harmonic function without any singularities in a given region (namely, satisfying the Laplace equation in the entire region) cannot have a maximum or a minimum value in that region. This can be proved easily in the following manner. Let us suppose that Φ (satisfying the Laplace equation in a given region) has a maximum at a point $\mathbf{r}_0$ in that region. If we draw a small sphere around $\mathbf{r}_0$, then, it is clear that $d\mathbf{s} \cdot (\boldsymbol{\nabla}\Phi)$ which is proportional to the normal derivative of Φ must be negative at every point on the surface of the sphere since $\mathbf{r}_0$ represents a maximum of Φ. It follows, then, that

$$\int_S d\mathbf{s} \cdot (\boldsymbol{\nabla}\Phi) < 0. \tag{3.70}$$

This is, however, inconsistent with the general property (3.64) of the solution of the Laplace equation, namely,

$$\int_S d\mathbf{s} \cdot (\boldsymbol{\nabla}\Phi) = 0. \tag{3.71}$$

Consequently, there cannot be a maximum of Φ at any point in the region. By a similar argument, it is straightforward to show that Φ cannot have a minimum at any point in the region either. (Incidentally, this is the reason why there can be no stable equilibrium in pure electrostatics. To have electrostatic equilibrium, one must apply some other form of force as well.)

4. It follows from the above property that if Φ satisfies the Laplace equation in a given region and if its value is a constant on the bounding surface of the region, then, it must have the same constant value in the entire region. The proof of this is really quite simple. As we have seen, Φ cannot have any maximum or minimum inside the region. Therefore, the maximum and the

minimum can occur at best on the bounding surface. However, Φ is a constant on the surface and, consequently, its maximum and minimum are the same and equal to its value on the surface (also equal to its value in the interior). This implies that Φ is a constant inside the region and all the way up to the surface. This also has the implication that if the solution of the Laplace equation is valid in the entire space, it must vanish if it vanishes asymptotically.

3.3.2 Solution in Cartesian coordinates. The solutions of the Laplace equation are not hard to work out because of the very special structure of the Laplacian operator which leads to separable solutions in a number of coordinate systems. In this section, we will work out the solutions in the Cartesian coordinates subject to appropriate boundary conditions.

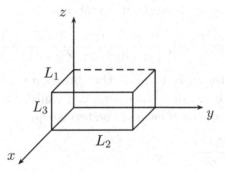

Figure 3.8: A conducting rectangular box with the faces maintained at fixed potentials.

Let us consider the physical problem of a conducting rectangular box of sides L_1, L_2 and L_3 respectively as shown in Fig. 3.8. All the faces of the box are grounded, except for the top face which is maintained at the given potential $f(x, y)$. We would like to determine the potential everywhere in the interior of the box. It is clear that the potential can be easily obtained by solving the Laplace equation in Cartesian coordinates, namely,

$$\left(\frac{\partial^2}{\partial x^2} + \frac{\partial^2}{\partial y^2} + \frac{\partial^2}{\partial z^2} \right) \Phi(x, y, z) = 0, \tag{3.72}$$

in the region $0 \leq x \leq L_1$, $0 \leq y \leq L_2$ and $0 \leq z \leq L_3$, subject to the boundary conditions

$$\Phi(0, y, z) = \Phi(x, 0, z) = \Phi(x, y, 0) = 0,$$

$$\Phi(L_1, y, z) = \Phi(x, L_2, z) = 0,$$

$$\Phi(x, y, L_3) = f(x, y), \tag{3.73}$$

in this region.

In trying to solve the Laplace equation for this system, let us note that the Laplacian in Cartesian coordinates (see (3.72)) is really a sum of three terms which commute with one another. In such a case, it is a general result that the solution can be written as a product of three terms, each depending on only one coordinate, namely, in such a case, we expect the solution to have the factorized form

$$\Phi(x, y, z) = X(x)Y(y)Z(z). \tag{3.74}$$

Substituting this form of the solution into the Laplace equation (3.72) and dividing by Φ throughout, we obtain

$$\frac{1}{X(x)} \frac{d^2 X(x)}{dx^2} + \frac{1}{Y(y)} \frac{d^2 Y(y)}{dy^2} + \frac{1}{Z(z)} \frac{d^2 Z(z)}{dz^2} = 0. \tag{3.75}$$

Since each of the three terms in the above expression depends on only one coordinate x, y or z, their sum can vanish for arbitrary values of the coordinates only if each of the terms equals a constant such that

$$\frac{1}{X(x)} \frac{d^2 X(x)}{dx^2} = \alpha_1,$$

$$\frac{1}{Y(y)} \frac{d^2 Y(y)}{dy^2} = \alpha_2,$$

$$\frac{1}{Z(z)} \frac{d^2 Z(z)}{dz^2} = \alpha_3, \qquad \alpha_1 + \alpha_2 + \alpha_3 = 0. \tag{3.76}$$

The boundary conditions (3.73) for Φ can now be translated to conditions on the individual component functions as

$$X(0) = X(L_1) = Y(0) = Y(L_2) = Z(0) = 0,$$

$$Z(L_3) = \text{constant}. \tag{3.77}$$

The three ordinary differential equations in (3.76) can now be solved subject to the boundary conditions (3.77) and they lead to

$$X_m(x) = \sin\left(\frac{m\pi x}{L_1}\right), \qquad \alpha_1^{(m)} = -\left(\frac{m\pi}{L_1}\right)^2,$$

$$Y_n(y) = \sin\left(\frac{n\pi y}{L_2}\right), \qquad \alpha_2^{(n)} = -\left(\frac{n\pi}{L_2}\right)^2, \tag{3.78}$$

$$Z_{mn}(z) = \frac{\sinh(\alpha_{mn}z)}{\sinh(\alpha_{mn}L_3)}, \qquad \alpha_3^{(m,n)} = \alpha_{mn}^2 = \left(\frac{m\pi}{L_1}\right)^2 + \left(\frac{n\pi}{L_2}\right)^2,$$

with $m, n = 1, 2, \ldots$ and we have normalized the z solution for later convenience. The determination of these solutions uses only the five homogeneous boundary conditions on $\Phi(x, y, z)$ with the condition for the surface $z = L_3$ in (3.73) yet to be implemented.

A general solution of the problem can now be written as a linear superposition of the form (it is not unique yet)

$$\Phi(x, y, z) = \sum_{m,n=1}^{\infty} A_{mn} X_m(x) Y_n(y) Z_{mn}(z)$$

$$= \sum_{m,n=1}^{\infty} A_{mn} \sin\left(\frac{m\pi x}{L_1}\right) \sin\left(\frac{n\pi y}{L_2}\right) \frac{\sinh(\alpha_{mn}z)}{\sinh(\alpha_{mn}L_3)}, \tag{3.79}$$

where A_{mn} are constants which can be determined by imposing the last of the boundary conditions in (3.73), namely,

$$\Phi(x, y, L_3) = \sum_{m,n=1}^{\infty} A_{mn} \sin\left(\frac{m\pi x}{L_1}\right) \sin\left(\frac{n\pi y}{L_2}\right) = f(x, y). \tag{3.80}$$

Since $f(x, y)$ is a given function, this relation can be inverted to determine

$$A_{mn} = \frac{4}{L_1 L_2} \int_0^{L_1} dx \int_0^{L_2} dy \, f(x, y) \sin\left(\frac{m\pi x}{L_1}\right) \sin\left(\frac{n\pi y}{L_2}\right), \tag{3.81}$$

where we have used the standard orthonormality relation for trigonometric functions, namely,

$$\int_0^L dx \sin\left(\frac{n\pi x}{L}\right) \sin\left(\frac{n'\pi x}{L}\right) = \frac{L}{2} \delta_{nn'}. \tag{3.82}$$

With this, the determination of the (unique) solution of the Laplace equation (namely, the potential) in the interior of the box is complete. Physical quantities such as the electric field can now be obtained from the potential. This method of solving the Laplace equation is known as the method of separation of variables.

3.3.3 Solution in spherical coordinates.

Let us next solve the Laplace equation in spherical coordinates given in (3.61)

$$\left(\frac{1}{r^2} \frac{\partial}{\partial r} r^2 \frac{\partial}{\partial r} + \frac{1}{r^2 \sin\theta} \frac{\partial}{\partial\theta} \sin\theta \frac{\partial}{\partial\theta} + \frac{1}{r^2 \sin^2\theta} \frac{\partial^2}{\partial\phi^2} \right) \Phi(r, \theta, \phi) = 0. \tag{3.83}$$

Once again, the Laplace equation is separable in spherical coordinates and, therefore, let us try a solution in the product form

$$\Phi(r, \theta, \phi) = R(r)\Theta(\theta)Q(\phi). \tag{3.84}$$

Substituting this back into the Laplace equation and dividing through-out by $\frac{\Phi}{r^2}$, we obtain

$$\frac{1}{R} \frac{d}{dr} \left(r^2 \frac{dR}{dr} \right) + \frac{1}{\Theta \sin\theta} \frac{d}{d\theta} \left(\sin\theta \frac{d\Theta}{d\theta} \right) + \frac{1}{\sin^2\theta} \left(\frac{1}{Q} \frac{d^2 Q}{d\phi^2} \right) = 0. \tag{3.85}$$

Since the expression within the parenthesis in the last term of (3.85) is the only term which depends on ϕ, this equation cannot be satisfied (for arbitrary values of ϕ) unless this term equals a constant, namely, we must have

$$\frac{1}{Q} \frac{d^2 Q}{d\phi^2} = -m^2, \tag{3.86}$$

where the choice of the sign of the constant is for convenience. Equation (3.86) can be readily integrated to give

$$Q_m(\phi) = e^{\pm im\phi}. \tag{3.87}$$

Since we expect the solutions to be single valued the constant m is required to be an integer (so that ϕ and $\phi + 2\pi$ lead to the same solution). Allowing for both positive as well as negative integers (including zero) for m, we note that we can write the solution depending on the azimuthal angle as

$$Q_m(\phi) = e^{im\phi}, \qquad m = 0, \pm 1, \pm 2, \ldots. \tag{3.88}$$

Substituting the form of the solution (3.88) into (3.85), we obtain

$$\frac{1}{R} \frac{d}{dr} \left(r^2 \frac{dR}{dr} \right) = -\frac{1}{\Theta \sin\theta} \frac{d}{d\theta} \left(\sin\theta \frac{d\Theta}{d\theta} \right) + \frac{m^2}{\sin^2\theta} = k. \tag{3.89}$$

Here we have used the fact that since the left hand side and the right hand side of (3.89) depend on independent variables, this relation will hold only if each side equals a constant which we denote by k. We see from (3.89) that the θ equation takes the simple form

$$\frac{1}{\sin\theta}\frac{d}{d\theta}\left(\sin\theta\frac{d\Theta}{d\theta}\right)+\left(k-\frac{m^2}{\sin^2\theta}\right)\Theta=0. \tag{3.90}$$

Introducing the variable $x=\cos\theta$, this equation can be written in the simple form

$$\frac{d}{dx}\left((1-x^2)\frac{d\Theta}{dx}\right)+\left(k-\frac{m^2}{(1-x^2)}\right)\Theta=0. \tag{3.91}$$

A systematic analysis of this equation reveals that its solutions become unphysical at $x=\pm1$ unless $k=\ell(\ell+1)$ where $\ell=0,1,2,\ldots$ and $m=-\ell,-\ell+1,\ldots,\ell-1,\ell$ for any given value of ℓ. (Incidentally, if the physical problem excludes the regions $x=\pm1$ or $\theta=0,\pi$, then these restrictions would not apply. However, in most applications, we will have $0\leq\theta\leq\pi$, so that we will consider only these integer values for the separation constants k,m.)

For these integer values of the separation constants, the θ equation takes the form

$$\frac{d}{dx}\left((1-x^2)\frac{dP_{\ell,m}(x)}{dx}\right)+\left(\ell(\ell+1)-\frac{m^2}{(1-x^2)}\right)P_{\ell,m}(x)=0. \tag{3.92}$$

This is known as the associated Legendre equation and the solutions, $P_{\ell,m}(x)$, are known as the associated Legendre polynomials. In particular, when $m=0$ the equation $(P_{\ell,m=0}(x)=P_\ell(x))$

$$\frac{d}{dx}\left((1-x^2)\frac{dP_\ell(x)}{dx}\right)+\ell(\ell+1)P_\ell(x)=0, \tag{3.93}$$

is known as the Legendre equation and the solutions, $P_\ell(x)$, are polynomials of order ℓ known as the Legendre polynomials. It can be easily checked that the associated Legendre polynomials are related to the Legendre polynomials through the relation (associated Legendre polynomials are not really polynomials for odd values of m as is clear from (3.94))

$$P_{\ell,m}(x)=(1-x^2)^{\frac{|m|}{2}}\frac{d^{|m|}P_\ell(x)}{dx^{|m|}},\quad\ell\geq|m|. \tag{3.94}$$

The Legendre polynomials have a closed form expression given by the Rodrigues' formula

$$P_\ell(x) = \frac{1}{2^\ell \, \ell!} \frac{\mathrm{d}^\ell}{\mathrm{d}x^\ell}(x^2 - 1)^\ell, \tag{3.95}$$

from which the explicit forms of the first few Legendre polynomials can be easily determined to be

$$P_0(x) = P_0(\cos\theta) = 1,$$

$$P_1(x) = P_1(\cos\theta) = x = \cos\theta,$$

$$P_2(x) = P_2(\cos\theta) = \frac{1}{2}(3x^2 - 1) = \frac{1}{2}(3\cos^2\theta - 1), \tag{3.96}$$

and so on. These are precisely the functions that we encountered in the last chapter in connection with the expansion of the potential for a point charge displaced along the z-axis (see (2.38)). It is also clear that we can write a closed form expression for the associated Legendre polynomials from the Rodrigues' formula as well.

It is also possible to write a generating function for the Legendre polynomials in a simple manner. Consider a function of two variables

$$T(x, s) = \frac{1}{(1 - 2sx + s^2)^{\frac{1}{2}}}. \tag{3.97}$$

The Taylor expansion of this function around $s = 0$ gives the Legendre polynomials as the coefficients of the expansion, namely,

$$T(x, s) = \frac{1}{(1 - 2sx + s^2)^{\frac{1}{2}}} = \sum_{\ell=0}^{\infty} s^\ell P_\ell(x). \tag{3.98}$$

In fact, this generating function leads to recursion relations satisfied by the Legendre polynomials and using these it is straightforward to show that the $P_\ell(x)$'s in the Taylor expansion in (3.98) do satisfy the Legendre equation. This relation is particularly interesting because it immediately gives

$$\frac{1}{|\mathbf{r} - \mathbf{r}'|} = \frac{1}{(r^2 - 2rr'\cos\theta + r'^2)^{\frac{1}{2}}}$$

$$= \frac{1}{r} \frac{1}{\left(1 - \frac{2r'}{r}\cos\theta + \left(\frac{r'}{r}\right)^2\right)^{\frac{1}{2}}}$$

$$= \frac{1}{r} \sum_{\ell=0}^{\infty} \left(\frac{r'}{r}\right)^{\ell} P_{\ell}(\cos\theta), \tag{3.99}$$

where we have assumed that $r > r'$. This is, of course, the expansion of the potential that we have discussed earlier in (2.39) and (2.46). A similar expansion is trivially obtained when $r < r'$, simply by interchanging r and r' in the previous expression (which is symmetric in r, r').

The Legendre polynomials can be shown to satisfy the orthonormality relation

$$\int_{-1}^{1} dx\, P_{\ell}(x)\, P_{\ell'}(x) = \frac{2}{2\ell+1}\, \delta_{\ell\ell'}. \tag{3.100}$$

In general, when $m \neq 0$, the total angular part of the solution of the Laplace equation is written (in the normalized form) as

$$Y_{\ell,m}(\theta, \phi) = \epsilon_m \sqrt{\frac{2\ell+1}{4\pi} \frac{(\ell-|m|)!}{(\ell+|m|)!}}\, P_{\ell,m}(\cos\theta)\, e^{im\phi}, \tag{3.101}$$

where the phase is conventionally chosen to be $\epsilon_m = (-1)^{\frac{m+|m|}{2}}$. The $Y_{\ell,m}$'s are known as the spherical harmonics and are the eigenfunctions of the angular momentum operator in quantum mechanics. They satisfy the orthonormality relation

$$\int \sin\theta\, d\theta\, d\phi\, Y_{\ell,m}^{*}(\theta, \phi) Y_{\ell',m'}(\theta, \phi) = \delta_{\ell\ell'} \delta_{mm'}. \tag{3.102}$$

Note that when $m = 0$, the spherical harmonics reduce to the Legendre polynomials up to the normalization constant.

Once the angular part of the solution of the Laplace equation is determined, the radial equation (3.89) becomes

$$\frac{1}{R_{\ell}(r)} \frac{d}{dr}\left(r^2 \frac{dR_{\ell}(r)}{dr}\right) = \ell(\ell+1),$$

or, $$\frac{d^2 R_{\ell}(r)}{dr^2} + \frac{2}{r} \frac{dR_{\ell}(r)}{dr} - \frac{\ell(\ell+1)}{r^2} R_{\ell}(r) = 0. \tag{3.103}$$

The two independent solutions of this equation are clearly

$$R_{\ell}(r) = r^{\ell}, \quad \text{or}, \quad R_{\ell}(r) = r^{-(\ell+1)}, \tag{3.104}$$

and the particular choice of the solution depends on the region in which the problem is being investigated. Thus, if we are looking for the solution outside an enclosed region (which includes the origin) such that the potential vanishes asymptotically, it is the second solution that is relevant. On the other hand, if we are looking at the solution in a region which includes the origin and where we assume the potential to be nonsingular, then the first form of the radial solution is the natural choice. In general, of course, the radial solution can be written as a linear superposition of the two independent solutions. In any case, independent of which form of the radial solution we choose, a general solution of the Laplace equation in spherical coordinates takes the form

$$\Phi(r,\theta,\phi) = \sum_{\ell,m} A_{\ell,m}\, R_\ell(r)\, Y_{\ell,m}(\theta,\phi), \tag{3.105}$$

where $A_{\ell,m}$'s are constants, which can be determined from the given boundary conditions of a physical system. We note here that if we have a physical problem where the potential does not depend on the azimuthal angle ϕ (namely, when $m=0$), the corresponding solutions are known as zonal harmonics.

▶ **Example (Sphere in uniform electric field).** As an example of a physical system where spherical solutions of the Laplace equation may be used, let us consider space without any free charge consisting of a uniform background electric field along the z-axis. Thus, the field lines can be drawn as parallel lines of the same magnitude and we can write

$$\mathbf{E}(\mathbf{r}) = E\,\hat{\mathbf{z}}, \tag{3.106}$$

which suggests an associated electrostatic potential of the form

$$\Phi(\mathbf{r}) = -Ez + C = -Er\cos\theta + C = -ErP_1(\cos\theta) + CP_0(\cos\theta), \tag{3.107}$$

where C is a constant. Clearly, this has azimuthal symmetry and as a result there is no dependence on the azimuthal angle ϕ. (Uniform electric fields can be produced by large capacitors with a small separation. Uniform electric fields over the entire space is, therefore, not a physical concept. However, we can think of all space to mean only a relatively large region.) Since the electric field is uniform, it follows that $\nabla \cdot \mathbf{E} = -\nabla^2\Phi = 0$, which can also be seen explicittly from (3.106) and (3.107).

If we now introduce a conducting sphere of radius R, then, the field lines will be distorted around the surface of the conducting sphere (see Fig. 3.9). Far away from the surface of the sphere, the field lines will continue to be uniform with the potential as given in (3.107). However, there will be surface charges induced on the sphere so that the field lines will end on the surface of the sphere and there will be no electric field inside the sphere. The surface of the sphere will, of course, have a constant potential which we will take to be Φ_0. We are interested in determining the potential as well as the electric field outside the sphere.

Because of the azimuthal symmetry present in the problem, it is clear that if we use the spherical solutions of the Laplace equation, we can write the potential

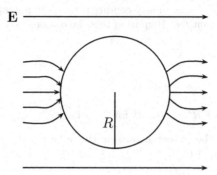

Figure 3.9: A conducting sphere of radius R in a uniform electric field.

outside the sphere as (see (3.105))

$$\Phi(\mathbf{r}) = \sum_{\ell=0}^{\infty} \left(A_\ell\, r^\ell + B_\ell\, r^{-(\ell+1)} \right) P_\ell(\cos\theta). \tag{3.108}$$

Here even though we are interested in the solution outside the sphere (the origin of the coordinate system is chosen to be at the center of the sphere), we have allowed for both the independent forms of the radial solution since the potential for a constant electric field is linear in r in this region. Furthermore, since there is no free charge anywhere (so that Φ satisfies the Laplace equation) and

$$\nabla^2 \left(\frac{1}{r} \right) = -4\pi\delta^3(\mathbf{r}), \tag{3.109}$$

we conclude that $B_0 = 0$. The constants A_ℓ can be determined from the boundary conditions for the problem. First, let us note that since

$$\lim_{r \to \infty} \Phi(\mathbf{r}) \to -Er P_1(\cos\theta) + C P_0(\cos\theta)$$

$$= \lim_{r \to \infty} \sum_{\ell=0}^{\infty} \left(A_\ell r^\ell + B_\ell r^{-(\ell+1)} \right) P_\ell(\cos\theta), \tag{3.110}$$

using the orthonormality of the Legendre polynomials in (3.100) we determine $A_\ell = 0$ for $\ell \geq 2$. Furthermore, from the asymptotic structure of the potential (3.107), we conclude that

$$A_0 = C, \quad A_1 = -E. \tag{3.111}$$

All the terms with coefficients B_ℓ vanish asymptotically and hence there is no constraint on these coefficients from the asymptotic condition. Thus, we can write for $r \geq R$,

$$\Phi(\mathbf{r}) = C - Er\cos\theta + \sum_{\ell=1}^{\infty} B_\ell r^{-(\ell+1)} P_\ell(\cos\theta). \tag{3.112}$$

We still have to satisfy the boundary condition that the potential on the surface of the sphere is a constant Φ_0. Requiring this, we obtain

$$\Phi_0 = C - ER\cos\theta + \sum_{\ell=1}^{\infty} B_\ell R^{-(\ell+1)} P_\ell(\cos\theta), \qquad (3.113)$$

which determines

$$C = \Phi_0, \quad B_1 = ER^3, \quad B_\ell = 0 \text{ for } l > 1, \qquad (3.114)$$

so that we can write the potential outside the sphere, satisfying the boundary conditions, to correspond to

$$\Phi(\mathbf{r}) = \Phi_0 - E\left(1 - \frac{R^3}{r^3}\right) r\cos\theta. \qquad (3.115)$$

The form of the electric field outside the sphere can now be determined to be

$$\mathbf{E}(\mathbf{r}) = -\boldsymbol{\nabla}\Phi(\mathbf{r})$$

$$\rightarrow -\left(\hat{\mathbf{r}}\frac{\partial}{\partial r} + \frac{\hat{\boldsymbol{\theta}}}{r}\frac{\partial}{\partial\theta}\right)\left(\Phi_0 - E\left(1 - \frac{R^3}{r^3}\right) r\cos\theta\right)$$

$$= \hat{\mathbf{r}}\, E\left(1 + \frac{2R^3}{r^3}\right)\cos\theta - \hat{\boldsymbol{\theta}}\, E\left(1 - \frac{R^3}{r^3}\right)\sin\theta. \qquad (3.116)$$

Thus, we see that, in general, the electric field has a radial as well as a θ component. However, on the surface of the sphere ($r = R$), the theta component vanishes, so that the electric field is normal to the surface. Furthermore, it is not a uniform electric field on the surface, rather its value depends on the angle θ. From this, we can determine the surface charge density induced on the sphere to be (we are using the fact that $\mathbf{E} = 0$ inside the condocting sphere)

$$\sigma(R,\theta) = \frac{1}{4\pi}\,\hat{\mathbf{r}}\cdot\mathbf{E}(\mathbf{R}) = \frac{3E}{4\pi}\cos\theta. \qquad (3.117)$$

Integrating this over the surface of the sphere, we obtain

$$Q_{\text{induced}} = \int R^2 \, d\Omega \, \sigma(R,\theta)$$

$$= \frac{3ER^2}{4\pi}\int_0^\pi d\theta \, \sin\theta\cos\theta \int_0^{2\pi} d\phi = 0. \qquad (3.118)$$

This is consistent with our earlier discussion in the last chapter. Namely, the conducting sphere remains neutral, the positive and the negative charges simply rearrange themselves so as to cancel the electric field inside the sphere. Note from (3.116) that asymptotically for large r, the electric field becomes

$$\lim_{r\to\infty} \mathbf{E} \to E(\hat{\mathbf{r}}\cos\theta - \hat{\boldsymbol{\theta}}\sin\theta) = E\hat{\mathbf{z}}, \qquad (3.119)$$

consistent with (3.106). In deriving the surface charge density we have assumed that there is no electric field inside the conducting sphere. However, this analysis can also be carried out in the interior of the sphere as well to show that the potential is a constant Φ_0 and there is no electric field inside which also follow from the general properties of harmonic functions that we discussed earlier. ◀

3.3.4 Circular harmonics. The Laplace equation can also be solved by the method of separation of variables in cylindrical coordinates. In this case, the solutions involve Bessel functions. Instead of going through this complete solution, let us consider the simpler case where the potential is independent of the z coordinate which shows up in many physical problems and, therefore, is more useful. When there is no dependence on the z coordinate, the Laplace equation in cylindrical coordinates (3.62) takes the form (in fact, this is just the Laplace equation in the polar coordinates)

$$\left(\frac{1}{r} \frac{\partial}{\partial r} r \frac{\partial}{\partial r} + \frac{1}{r^2} \frac{\partial^2}{\partial \phi^2} \right) \Phi(r, \phi) = 0. \tag{3.120}$$

Writing a separable solution of the form

$$\Phi(r, \phi) = R(r)Q(\phi), \tag{3.121}$$

and substituting it into the differential equation (3.120), we obtain

$$\frac{r}{R(r)} \frac{d}{dr} \left(r \frac{dR(r)}{dr} \right) = -\frac{1}{Q(\phi)} \frac{d^2 Q(\phi)}{d\phi^2} = k. \tag{3.122}$$

Here, we have used the fact that both sides of (3.122) are functions of independent variables and, therefore, the relation can be satisfied for arbitrary values of r, ϕ only if both sides equal a constant which we have identified with k. The solution for the ϕ equation is straightforward. We note that we will have a single valued function only if $k = n^2, n = 0, 1, \ldots$, with the ϕ solution written as

$$Q_n(\phi) = A_n \cos n\phi + B_n \sin n\phi, \tag{3.123}$$

where A_n, B_n are constants. Furthermore, for this value of the separation constant the radial equation in (3.122) becomes

$$r^2 \frac{d^2 R_n(r)}{dr^2} + r \frac{dR_n(r)}{dr} - n^2 R_n(r) = 0. \tag{3.124}$$

There are two independent solutions of this equation of the forms

$$R_n(r) = r^n, \qquad R_n(r) = r^{-n}, \quad \text{for } n \neq 0, \tag{3.125}$$

so that a general solution for the radial equation (3.124) can be written (for $n \neq 0$) as

$$R_n(r) = C_n r^n + D_n r^{-n}, \tag{3.126}$$

with C_n, D_n constants. Let us note here that when $n = 0$, the solution of the radial equation in (3.124) has the general form

$$R_0(r) = C_0 + D_0 \ln r, \tag{3.127}$$

where C_0, D_0 are constants. Thus, for a given n we can write the solution of the Laplace equation (3.120) as

$$\Phi_n(r, \phi) = R_n(r)Q_n(\phi), \tag{3.128}$$

which are known as circular harmonics with n denoting the degree of the harmonics. A general solution of (3.120), of course, will have the form

$$\Phi(r, \phi) = \sum_n R_n(r)Q_n(\phi). \tag{3.129}$$

▶ **Example (Cylinder in uniform electric field).** As an example of the use of circular harmonics, let us consider the problem of an infinitely long conducting cylinder of radius R in a uniform electric field perpendicular to the axis of the cylinder. Let us assume that the axis of the cylinder is along the z-axis and that the electric field is along the x-axis (see Fig. 3.10). There are no free charges anywhere in space.

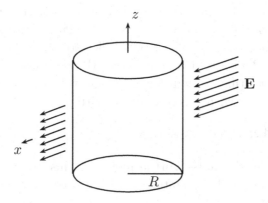

Figure 3.10: A conducting cylinder in a uniform electric field.

Clearly, the potential will be independent of the z coordinate by symmetry and, therefore, we can use circular harmonics. Let us note that since the electric field is uniform along the x-axis, we can write (at least when the cylinder is not present or very far away from the cylinder)

$$\mathbf{E}(\mathbf{r}) = E\hat{x}, \tag{3.130}$$

so that the potential will have the form (in polar coordinates)

$$\Phi(\mathbf{r}) = -Ex + C = -Er \cos \phi + C, \tag{3.131}$$

where C is a constant and r is the radial coordinate on the plane. In the presence of the conducting cylinder, the field lines will be distorted near the surface of the cylinder, but asymptotically they will have the form (3.131). The surface of the cylinder, of course, would be at a constant potential which we take to be Φ_0.

The general solution for the potential in the presence of the cylinder (outside) would have the form (since there are no free charges, the potential will satisfy Laplace equation and in this case will have a solution of the form (3.129))

$$\Phi(\mathbf{r}) = C_0 + D_0 \ln r$$

$$+ \sum_{n=1}^{\infty} (C_n r^n + D_n r^{-n})(A_n \cos n\phi + B_n \sin n\phi). \tag{3.132}$$

However, comparing with the asymptotic form of the potential (3.131) and using the orthonormality relations for the trigonometric functions, we determine that the only coefficients that are nontrivial are A_1, C_0, C_1, D_1 and satisfy

$$C_0 = C, \quad A_1 C_1 = -E, \tag{3.133}$$

so that we can write the form of the potential outside the cylinder to be ($r \geq R$)

$$\Phi(\mathbf{r}) = C - Er \cos \phi + A_1 D_1 r^{-1} \cos \phi. \tag{3.134}$$

We see that the term with the coefficient D_1 vanishes asymptotically and, as a consequence, there is no constraint on this coefficient from the asymptotic condition. (Remember that r is the radial coordinate in cylindrical coordinates and, therefore, there is no inconsistency with the absence of free charges. Furthermore, this term also involves an angular function.)

Let us next impose the boundary condition on the surface of the cylinder. Namely, on the surface of the cylinder ($r = R$) we have,

$$\Phi_0 = C - ER \cos \phi + A_1 D_1 R^{-1} \cos \phi, \tag{3.135}$$

which determines

$$C = \Phi_0, \quad A_1 D_1 = ER^2, \tag{3.136}$$

so that we can write the potential outside the cylinder (3.134) to be

$$\Phi(r, \phi) = \Phi_0 - E \left(r - \frac{R^2}{r} \right) \cos \phi. \tag{3.137}$$

We can now determine the electric field which has the form

$$\mathbf{E}(r, \phi) = -\nabla \Phi(r, \phi)$$

$$= - \left(\hat{\mathbf{r}} \frac{\partial}{\partial r} + \frac{\hat{\phi}}{r} \frac{\partial}{\partial \phi} \right) \left(\Phi_0 - E \left(r - \frac{R^2}{r} \right) \cos \phi \right)$$

$$= \hat{\mathbf{r}} E \left(1 + \frac{R^2}{r^2} \right) \cos \phi - \hat{\phi} E \left(1 - \frac{R^2}{r^2} \right) \sin \phi. \tag{3.138}$$

Once again we see that the electric field outside the cylinder has both a radial and an angular component. However, on the surface of the cylinder ($r = R$), only the normal (radial) component is nontrivial. This allows us to determine the induced surface charge density on the cylinder to be

$$\sigma(R, \phi) = \frac{1}{4\pi} \hat{\mathbf{r}} \cdot \mathbf{E}(R) = \frac{E}{2\pi} \cos \phi, \tag{3.139}$$

where we are assuming that there is no electric field inside the cylinder since the potential is a constant. Consequently, the total induced charge on the surface of the cylinder is obtained to be (L is the length of the cylinder which is to be taken to infinity at the end)

$$Q_{\text{induced}} = \int LR\,d\phi\,\sigma(R,\phi) = \frac{ELR}{2\pi} \int\limits_0^{2\pi} d\phi\,\cos\phi = 0. \tag{3.140}$$

We see that the cylinder remains neutral. The charges simply rearrange themselves on the surface to yield zero electric field inside the cylinder, which can be explicitly checked by carrying out a similar analysis inside the cylinder. Let us also note that asymptotically,

$$\lim_{r\to\infty} \mathbf{E} \to E\left(\hat{\mathbf{r}}\cos\phi - \hat{\boldsymbol{\phi}}\sin\phi\right) = E\,\hat{\mathbf{x}}, \tag{3.141}$$

as we would expect.   ◄

3.4 Solution of the Poisson equation

In the absence of free charges, solutions of the (homogeneous) Laplace equation determine the electrostatic potential and, therefore, the electric field. However, when free electric charges are present we have to solve the (inhomogeneous) Poisson equation

$$\nabla^2\Phi(\mathbf{r}) = -4\pi\rho(\mathbf{r}), \tag{3.142}$$

with $\rho(\mathbf{r})$ representing the charge density in order to determine the potential and the electric field. Let us note that the Laplace equation represents the homogeneous part of the Poisson equation and it is important to know the solutions of the Laplace equation for the determination of the solution of the Poisson equation since we know that a general solution of any inhomogeneous differential equation consists of a sum of the solution of the homogeneous equation as well as the particular solution of the inhomogeneous equation. The freedom of adding a homogeneous part to the solution allows us to satisfy boundary conditions in a simple manner.

Since the solutions of the Laplace equation define a complete basis, one method of solving the Poisson equation is to expand the solution in one such (appropriate) complete basis (namely, Cartesian, spherical, cylindrical etc.) and impose the relevant boundary conditions. There is, however, an alternative and powerful method for solving inhomogeneous differential equations which goes under the name of the method of Green's functions. This method is quite useful in many branches of physics and, therefore, is an important concept which we discuss next.

3.4.1 Green's function. To see how the method of Green's function works, let us replace the inhomogeneous source term on the right hand side of the Poisson equation (3.142) by a delta function (source) and consider the equation

$$\nabla^2 G(\mathbf{r}, \mathbf{r}') = -4\pi\, \delta^3(\mathbf{r} - \mathbf{r}'). \tag{3.143}$$

Namely, $G(\mathbf{r}, \mathbf{r}')$ represents the potential (solution of the Poisson equation) at $\mathbf{r}$ due to a unit point source charge at $\mathbf{r}'$. This is known as the Green's function for the Poisson equation (or the Laplacian operator). It is clear that if we know the Green's function for a given equation, then the particular solution for the inhomogeneous equation can be trivially determined. For example, for the Poisson equation (3.142), we can write the particular solution to be

$$\Phi_{\text{particular}}(\mathbf{r}) = \int d^3 r'\, G(\mathbf{r}, \mathbf{r}')\rho(\mathbf{r}'), \tag{3.144}$$

which can be easily checked to satisfy the Poisson equation, namely,

$$\nabla^2 \Phi_{\text{particular}}(\mathbf{r}) = \int d^3 r' \left(\nabla^2 G(\mathbf{r}, \mathbf{r}')\right) \rho(\mathbf{r}')$$

$$= -4\pi \int d^3 r'\, \delta^3(\mathbf{r} - \mathbf{r}')\rho(\mathbf{r}')$$

$$= -4\pi \rho(\mathbf{r}). \tag{3.145}$$

Thus, it is clear that the knowledge of the Green's function for a given equation is quite essential in solving an inhomogeneous differential equation. The determination of the Green's function requires the knowledge of boundary conditions. For example, we already know of a Green's function for the Poisson equation, namely, in three dimensions we have seen that

$$\nabla^2 \left(\frac{1}{|\mathbf{r} - \mathbf{r}'|}\right) = -4\pi\, \delta^3(\mathbf{r} - \mathbf{r}'). \tag{3.146}$$

Therefore, let us identify

$$G^{(0)}(\mathbf{r}, \mathbf{r}') = \frac{1}{|\mathbf{r} - \mathbf{r}'|}. \tag{3.147}$$

The important question to ask is what is the boundary condition for which this is the appropriate Green's function. In this case, it is obvious from (3.147) that for a fixed $\mathbf{r}'$,

$$\lim_{|\mathbf{r}| \to \infty} G^{(0)}(\mathbf{r}, \mathbf{r}') \to 0. \tag{3.148}$$

Therefore, this is the Green's function corresponding to the boundary condition that the solution vanishes asymptotically for large distances away from the source charge (remember that the Green's function really describes a solution of the Poisson equation for a unit source charge). In this case, we see that the general solution of the Poisson equation, subject to the boundary condition that it vanishes asymptotically, has the form

$$\Phi(\mathbf{r}) = \Phi_{\text{homo}}(\mathbf{r}) + \int_V d^3r' \, G^{(0)}(\mathbf{r}, \mathbf{r}')\rho(\mathbf{r}'), \tag{3.149}$$

where Φ_{homo} denotes a solution of the homogeneous equation (vanshing asymptotically). On the other hand, we have already seen earlier that the solution of the homogeneous equation (the Laplace equation), subject to the boundary condition that it vanishes asymptotically, is identically zero. Therefore, the unique solution of the Poisson equation subject to this boundary condition is

$$\Phi(\mathbf{r}) = \int d^3r' \, G^{(0)}(\mathbf{r}, \mathbf{r}')\rho(\mathbf{r}') = \int d^3r' \, \frac{\rho(\mathbf{r}')}{|\mathbf{r} - \mathbf{r}'|}, \tag{3.150}$$

which is the familiar relation for the potential due to a charge distribution that we have discussed earlier (see (1.60)).

The relation between the solution of an inhomogeneous problem and the Green's function can be easily seen from the Green's identity which we have derived earlier (see (3.67)), namely,

$$\int_V d^3r' \left(A\boldsymbol{\nabla}'^2 B - B\boldsymbol{\nabla}'^2 A\right) = \int_S d\mathbf{s}' \cdot \left(A\boldsymbol{\nabla}'B - B\boldsymbol{\nabla}'A\right). \tag{3.151}$$

In (3.151), if we choose $A = G(\mathbf{r}, \mathbf{r}')$, the generic Green's function for the Laplacian and $B = \Phi(\mathbf{r}')$, the solution of the Poisson equation, we obtain

$$\int_V d^3r'(G\boldsymbol{\nabla}'^2\Phi - \Phi\boldsymbol{\nabla}'^2 G) = \int_S d\mathbf{s}' \cdot (G\boldsymbol{\nabla}'\Phi - \Phi\boldsymbol{\nabla}'G), \tag{3.152}$$

which can be simplified to give

$$4\pi \left(-\int_V d^3r' \, G(\mathbf{r}, \mathbf{r}')\rho(\mathbf{r}') + \Phi(\mathbf{r})\right) = \int_S d\mathbf{s}' \cdot (G\boldsymbol{\nabla}'\Phi - \Phi\boldsymbol{\nabla}'G). \tag{3.153}$$

Here, we have used the Poisson equation (3.142) as well as the equation satisfied by the Green's function (3.143). Furthermore, if we use the notation $\hat{\mathbf{e}}_s \cdot \boldsymbol{\nabla} = \frac{\partial}{\partial n}$ to represent the normal derivative, we can rewrite (3.153) also as

$$\Phi(\mathbf{r}) = \frac{1}{4\pi} \int_S ds' \left(G(\mathbf{r}, \mathbf{r}') \frac{\partial \Phi(\mathbf{r}')}{\partial n'} - \Phi(\mathbf{r}') \frac{\partial G(\mathbf{r}, \mathbf{r}')}{\partial n'} \right)$$

$$+ \int_V d^3 r' \, G(\mathbf{r}, \mathbf{r}') \rho(\mathbf{r}'). \tag{3.154}$$

This can be compared with the form of the result obtained earlier in (3.149). Incidentally, the simplest way to check that the first term in (3.154) must be a solution of the homogeneous equation is to note that

$$\boldsymbol{\nabla}^2 \left(\Phi(\mathbf{r}) - \int_V d^3 r' \, G(\mathbf{r}, \mathbf{r}') \rho(\mathbf{r}') \right) = 0. \tag{3.155}$$

A direct verification of this relation for the surface integral is, however, tricky.

3.4.2 Dirichlet boundary condition. It appears from the general result (3.154) that to determine the unique solution of the Poisson equation in the presence of some conducting surfaces we need the value of both Φ and $\frac{\partial \Phi}{\partial n}$ on the surfaces. But, as we have seen, a unique solution of an elliptic equation requires either Φ (Dirichlet boundary condition) or $\frac{\partial \Phi}{\partial n}$ (Neumann boundary condition) to be specified on the boundary and not both. The way out of this puzzle is that we have not yet specified a boundary condition for the Green's function. For example, suppose we are investigating a Dirichlet boundary value problem, then, the value of $\Phi(\mathbf{r})$ on the surface is known. Consequently, if we require the Green's function to satisfy the homogeneous Dirichlet boundary condition, namely,

$$G_D(\mathbf{r}, \mathbf{r}') = 0, \quad \text{for } \mathbf{r} \text{ or } \mathbf{r}' \in S, \tag{3.156}$$

we see that the solution (3.154) can be written as

$$\Phi(\mathbf{r}) = -\frac{1}{4\pi} \int_S ds' \, \Phi(\mathbf{r}') \frac{\partial G_D(\mathbf{r}, \mathbf{r}')}{\partial n'}$$

$$+ \int_V d^3 r' \, G_D(\mathbf{r}, \mathbf{r}') \rho(\mathbf{r}'), \tag{3.157}$$

which is indeed a well defined solution for the given Dirichlet boundary value problem. It is worth noting that this solution is valid even in the region which does not contain any charges (compare this with the method of images where one does not calculate the potential in the region containing the image charge). In such a case, the second term vanishes and the solution is given completely by the surface integral. (Let us also emphasize here that even though the boundary condition for the solution may be inhomogeneous, the Green's function is required to satisfy only the homogeneous condition.)

The main question to answer now is what is the form of the Green's function satisfying the homogeneous Dirichlet boundary condition (3.156). The exact structure of the Green's function would, of course, depend on the particular problem under consideration. However, let us note that we can write a general Green's function to be of the form

$$G(\mathbf{r}, \mathbf{r}') = G^{(0)}(\mathbf{r}, \mathbf{r}') + H(\mathbf{r}, \mathbf{r}'), \tag{3.158}$$

where

$$\nabla^2 H(\mathbf{r}, \mathbf{r}') = 0. \tag{3.159}$$

Namely, the Green's function is unique only up to addition of a term which satisfies the homogeneous equation. We can, therefore, take advantage of this arbitrariness to impose a given boundary condition on a particular Green's function.

▶ **Example (Point charge above a conducting plane).** As an application of this method, let us study the problem of a point charge above a conducting plane of infinite extent which is maintained at some constant potential Φ_0. We can take the plane to be at $z = 0$, assume that the point charge q is at $z = d$ on the z-axis, and we are interested in determining the potential for $z > 0$ (see Fig. 3.1).

This is a Dirichlet boundary value problem and, as we have seen, to determine the potential we need the Dirichlet Green's function in the region $z, z' \geq 0$, satisfying the homogeneous boundary condition $G_D(\mathbf{r}, \mathbf{r}') = 0$ whenever $z = 0$ or $z' = 0$. To determine this Green's function, let us note from our general discussion that

$$G_D(\mathbf{r}, \mathbf{r}') = G^{(0)}(\mathbf{r}, \mathbf{r}') + H(\mathbf{r}, \mathbf{r}')$$

$$= \frac{1}{\sqrt{(x - x')^2 + (y - y')^2 + (z - z')^2}} + H(\mathbf{r}, \mathbf{r}'). \tag{3.160}$$

We require this Green's function to vanish whenever $z = 0$ or $z' = 0$. It is clear from this requirement that

$$H(\mathbf{r}, \mathbf{r}') = -\frac{1}{\sqrt{(x - x')^2 + (y - y')^2 + (z \pm z')^2}}. \tag{3.161}$$

Although there are two possible choices for $H(\mathbf{r}, \mathbf{r}')$ we note that for the second choice of the sign in the denominator in (3.161) we will have $G_D(\mathbf{r}, \mathbf{r}') = 0$ which is a trivial solution. Therefore, the first choice of the sign is the natural one and we obtain

$$
G_D(\mathbf{r}, \mathbf{r}') = \frac{1}{\sqrt{(x-x')^2 + (y-y')^2 + (z-z')^2}}
$$
$$
- \frac{1}{\sqrt{(x-x')^2 + (y-y')^2 + (z+z')^2}}. \tag{3.162}
$$

There are two things to note from (3.162). First, with our choice of $H(\mathbf{r}, \mathbf{r}')$, we note that

$$
\boldsymbol{\nabla}^2 H(\mathbf{r}, \mathbf{r}') = -\boldsymbol{\nabla}^2 \left(\frac{1}{\sqrt{(x-x')^2 + (y-y')^2 + (z+z')^2}} \right)
$$
$$
= 4\pi \delta(x-x')\delta(y-y')\delta(z+z'). \tag{3.163}
$$

Therefore, it appears that $H(\mathbf{r}, \mathbf{r}')$ does not satisfy the homogeneous equation. However, note that in the region that we are interested in, namely, $z, z' > 0$, the right hand side of (3.163) indeed vanishes and $H(\mathbf{r}, \mathbf{r}')$ satisfies the homogeneous equation in this region. The second thing to note is that the Green's function in (3.162) is reminiscent of the structure of the potential for a grounded plane obtained by the method of images in (3.5). This should not be surprising since the Green's function is the potential for a unit source charge with a homogeneous Dirichlet boundary condition.

Substituting (3.162) into the right hand side of the solution (3.157), we obtain (S in this case is the plane $z' = 0$. Actually, S is the closed surface bounding the upper half plane. However, as is clear from the form of the Green's function, the Green's function as well as its derivative vanish when any of the coordinates is at infinity. Consequently, the surface S is effectively the plane at $z' = 0$.)

$$
\Phi(\mathbf{r}) = -\frac{1}{4\pi} \int_S \mathrm{d}s' \, \Phi(\mathbf{r}') \frac{\partial G_D(\mathbf{r}, \mathbf{r}')}{\partial n'} + \int \mathrm{d}^3 r' \, G_D(\mathbf{r}, \mathbf{r}')\rho(\mathbf{r}')
$$
$$
= \frac{\Phi_0}{4\pi} \int \frac{2z\mathrm{d}x' \, \mathrm{d}y'}{((x-x')^2 + (y-y')^2 + z^2)^{\frac{3}{2}}}
$$
$$
+ \int \mathrm{d}^3 r' \, G_D(\mathbf{r}, \mathbf{r}')q\delta(x')\delta(y')\delta(z'-d)
$$
$$
= \frac{\Phi_0 z}{4\pi} \int \frac{2\mathrm{d}x' \, \mathrm{d}y'}{(x'^2 + y'^2 + z^2)^{\frac{3}{2}}}
$$
$$
+ q \left(\frac{1}{(x^2 + y^2 + (z-d)^2)^{\frac{1}{2}}} - \frac{1}{(x^2 + y^2 + (z+d)^2)^{\frac{1}{2}}} \right). \tag{3.164}
$$

(Note that there is an extra negative sign in the surface term because the surface area points along the negative z-axis so that $\frac{\partial}{\partial n'} = -\frac{\partial}{\partial z'}$.) Here, we have translated the coordinates of integration in the first integral for simplicity and recognize that it is best evaluated in polar coordinates

$$\int_{-\infty}^{\infty} \frac{2dx'\, dy'}{(x'^2 + y'^2 + z^2)^{\frac{3}{2}}} = \int \frac{2r'\, dr'\, d\phi'}{(r'^2 + z^2)^{\frac{3}{2}}}$$

$$= 2\pi \int_{0}^{\infty} \frac{dr'^2}{(r'^2 + z^2)^{\frac{3}{2}}}$$

$$= 2\pi\, (-2) \left. \frac{1}{\sqrt{r'^2 + z^2}} \right|_{0}^{\infty} = \frac{4\pi}{z}. \tag{3.165}$$

Substituting this back into the solution (3.164), we have

$$\Phi(\mathbf{r}) = \Phi_0 + q \left(\frac{1}{(x^2 + y^2 + (z - d)^2)^{\frac{1}{2}}} - \frac{1}{(x^2 + y^2 + (z + d)^2)^{\frac{1}{2}}} \right). \tag{3.166}$$

For $\Phi_0 = 0$, this reduces to the solution (3.5). However, we now have the solution for the case when the conducting plane is not grounded, but is held at a nonzero constant potential. Incidentally, from (3.166) (or (3.164)), we see that the potential in the lower half plane is given by (the second term vanishes in the lower half plane because there is no free charge in that region)

$$\Phi(\mathbf{r}) = \Phi_0, \qquad \text{for} \quad z \leq 0, \tag{3.167}$$

which is consistent with what we expect, namely, that there is no electric field in the lower half plane. All the field lines from the point charge terminate on the conducting plane. (An alternative way to see that the surface term in (3.164) for a constant potential Φ_0 on the boundary is simply equal to Φ_0 is as follows. By definition

$$\boldsymbol{\nabla}'^2 G_D(\mathbf{r}, \mathbf{r}') = -4\pi \delta^3(\mathbf{r} - \mathbf{r}'). \tag{3.168}$$

Consequently, using Gauss' theorem we can write

$$\int_{S} d\mathbf{s}' \cdot \boldsymbol{\nabla}' G_D(\mathbf{r}, \mathbf{r}') = \int d^3 r'\, \boldsymbol{\nabla}'^2 G_D(\mathbf{r}, \mathbf{r}')$$

$$= \int d^3 r'\, (-4\pi) \delta^3(\mathbf{r} - \mathbf{r}') = -4\pi, \tag{3.169}$$

which gives the desired result.)   ◀

▶ **Example (Point charge inside a conducting sphere).** Let us next analyze the problem of a conducting sphere of radius R which contains a point charge q inside, at a distance d from the center of the sphere. The surface of the sphere is maintained at a constant potential Φ_0 and we are interested in determining the potential within the sphere.

For simplicity, let us choose the origin to coincide with the center of the sphere and the point charge q to lie on the z-axis so that we are interested in the solution of the equation

$$\boldsymbol{\nabla}^2 \Phi = -4\pi\, q \delta(x) \delta(y) \delta(z - d), \qquad 0 \leq x, y, z \leq R, \quad d < R, \tag{3.170}$$

subject to the boundary condition that

$$\Phi(R, \theta, \phi) = \Phi_0. \tag{3.171}$$

Therefore, this defines a Dirichlet problem. Once again, the Green's function for the problem, satisfying the homogeneous Dirichlet boundary condition, can be written as (see (3.158))

$$G_D(\mathbf{r}, \mathbf{r}') = \frac{1}{|\mathbf{r} - \mathbf{r}'|} + H(\mathbf{r}, \mathbf{r}'). \tag{3.172}$$

Since $H(\mathbf{r}, \mathbf{r}')$ has to satisfy the homogeneous equation (for $r, r' < R$), we note that we can write it as

$$H(\mathbf{r}, \mathbf{r}') = -\frac{\alpha}{|\mathbf{r} - \mathbf{r}''|}, \tag{3.173}$$

where we assume that α is a constant (namely, independent of $\mathbf{r}$) and $\mathbf{r}'' = \mathbf{r}''(\mathbf{r}')$ and lies outside the sphere (namely, $r'' > R$). Now, requiring that $G_D(\mathbf{r}, \mathbf{r}')|_{r=R} = 0$, we obtain

$$\frac{1}{(R^2 + r'^2 - 2Rr' \cos\gamma')^{\frac{1}{2}}} = \frac{\alpha}{(R^2 + r''^2 - 2Rr'' \cos\gamma'')^{\frac{1}{2}}},$$

$$\text{or,} \quad \alpha^2 (R^2 + r'^2 - 2Rr' \cos\gamma') = (R^2 + r''^2 - 2Rr'' \cos\gamma''). \tag{3.174}$$

Here, we are using the generic notation that

$$\cos\gamma' = \cos\theta \cos\theta' + \sin\theta \sin\theta' \cos(\phi - \phi'). \tag{3.175}$$

It is clear that (3.174) can be satisfied if $\mathbf{r}'$ and $\mathbf{r}''$ lie along the same axis (this also follows from the fact that since $\mathbf{r}''$ is a vector function of $\mathbf{r}'$, it must have the form $\mathbf{r}'$ multiplied by a scalar function of r') and a nontrivial solution is obtained if

$$\alpha = \frac{R}{r'}, \qquad r'' = \alpha^2 r' = \frac{R^2}{r'}, \qquad \gamma'' = \gamma'. \tag{3.176}$$

Using (3.176), then, we can write

$$G_D(\mathbf{r}, \mathbf{r}') = \frac{1}{(r^2 + r'^2 - 2rr' \cos\gamma')^{\frac{1}{2}}} - \frac{R}{(r^2 r'^2 + R^4 - 2R^2 rr' \cos\gamma')^{\frac{1}{2}}}. \tag{3.177}$$

It is manifestly symmetric in $\mathbf{r} \leftrightarrow \mathbf{r}'$ and it is straightforward to verify explicitly that it satisfies the homogeneous Dirichlet boundary condition whenever $\mathbf{r}$ or $\mathbf{r}'$ lies on the surface of the sphere. Furthermore, its structure can be compared with the potential obtained from the method of images.

The solution of the Poisson equation inside the sphere is now straightforward ($r < R$, $\frac{\partial}{\partial n'} = \frac{\partial}{\partial r'}$). (For simplicity of evaluation, we will measure the angle θ' with respect to the vector $\mathbf{r}$ in the first integral so that $\gamma' = \theta'$.)

$$\Phi(\mathbf{r}) = -\frac{1}{4\pi} \int_S ds' \, \Phi(\mathbf{r}') \frac{\partial G_D(\mathbf{r}, \mathbf{r}')}{\partial n'} + \int_V d^3 r' \, G_D(\mathbf{r}, \mathbf{r}') \rho(\mathbf{r}')$$

$$= \frac{\Phi_0 (R^2 - r^2)}{4\pi R} \int_0^\pi \int_0^{2\pi} \frac{R^2 \sin\theta' d\theta' \, d\phi'}{(r^2 + R^2 - 2rR \cos\theta')^{\frac{3}{2}}}$$

$$+ \int d^3 r' \, G_D(\mathbf{r}, \mathbf{r}') q \delta(x') \delta(y') \delta(z' - d)$$

$$= \frac{\Phi_0(R^2 - r^2)R}{2} \int_{-1}^{1} \frac{dx}{(r^2 + R^2 - 2rRx)^{\frac{3}{2}}}$$

$$+ q \left(\frac{1}{(r^2 + d^2 - 2rd\cos\theta)^{\frac{1}{2}}} - \frac{R}{(d^2r^2 + R^4 - 2R^2dr\cos\theta)^{\frac{1}{2}}} \right)$$

$$= \frac{\Phi_0(R^2 - r^2)R}{2} \left(\frac{1}{Rr} \right) \frac{1}{(r^2 + R^2 - 2rRx)^{\frac{1}{2}}} \Big|_{-1}^{1}$$

$$+ q \left(\frac{1}{(r^2 + d^2 - 2rd\cos\theta)^{\frac{1}{2}}} - \frac{R}{(d^2r^2 + R^4 - 2R^2dr\cos\theta)^{\frac{1}{2}}} \right)$$

$$= \Phi_0 + q \left(\frac{1}{(r^2 + d^2 - 2rd\cos\theta)^{\frac{1}{2}}} - \frac{R}{(d^2r^2 + R^4 - 2R^2dr\cos\theta)^{\frac{1}{2}}} \right).$$

$$(3.178)$$

This can be compared with the solution obtained from the method of images.   ◀

3.4.3 Neumann boundary condition. The solution of the Poisson equation subject to Neumann boundary conditions is slightly tricky for a variety of reasons. First, let us note that if

$$\nabla^2 \Phi(\mathbf{r}) = -4\pi\rho(\mathbf{r}), \tag{3.179}$$

then, using Gauss' theorem we can write

$$\int_S d\mathbf{s} \cdot \nabla\Phi = \int_V d^3r \, \nabla^2 \Phi(\mathbf{r}),$$

$$\text{or,} \quad \int_S ds \, \frac{\partial\Phi}{\partial n} = -4\pi \int_V d^3r \, \rho(\mathbf{r}) = -4\pi Q. \tag{3.180}$$

This shows that the normal derivative of Φ cannot be specified arbitrarily on the boundary surface. Rather, it should satisfy the constraint condition (3.180) which does not make this a well posed boundary value problem. Furthermore, as in the earlier case of Dirichlet boundary condition, for a solution of the boundary value problem to exist we must define the Neumann Green's function with the homogeneous condition

$$\frac{\partial G_N(\mathbf{r}, \mathbf{r}')}{\partial n'} = 0, \tag{3.181}$$

on the boundary surface. However, let us note from the definition of the Green's function that

$$\nabla'^2 G_N(\mathbf{r}, \mathbf{r}') = -4\pi\delta^3(\mathbf{r} - \mathbf{r}'). \tag{3.182}$$

Consequently, using Gauss' theorem we obtain

$$\int_S \mathrm{d}s' \cdot \boldsymbol{\nabla}'G_N(\mathbf{r},\mathbf{r}') = \int_V \mathrm{d}^3r' \, \boldsymbol{\nabla}'^2 G_N(\mathbf{r},\mathbf{r}'),$$

or, $\quad \int_S \mathrm{d}s' \, \dfrac{\partial G_N(\mathbf{r},\mathbf{r}')}{\partial n'} = -4\pi \int_V \mathrm{d}^3r' \, \delta^3(\mathbf{r}-\mathbf{r}') = -4\pi. \quad (3.183)$

This is not compatible with the homogeneous Neumann condition (3.181).

This second difficulty can be circumvented by defining the Neumann Green's function as satisfying the differential equation

$$\boldsymbol{\nabla}'^2 G_N(\mathbf{r},\mathbf{r}') = -4\pi\left(\delta^3(\mathbf{r}-\mathbf{r}') - \frac{1}{V}\right), \tag{3.184}$$

where V denotes the volume of the region under consideration. In such a case, use of Gauss' theorem leads to

$$\int_S \mathrm{d}s' \cdot \boldsymbol{\nabla}'G_N(\mathbf{r},\mathbf{r}') = \int_S \mathrm{d}s' \, \frac{\partial G_N(\mathbf{r},\mathbf{r}')}{\partial n'} = 0, \tag{3.185}$$

so that a Green's function satisfying the homogeneous Neumann boundary condition can be defined consistently.

With such a choice of the Green's function we note that Gauss' theorem (Green's identity (3.67)) leads to

$$\int_V \mathrm{d}^3r' \, (G_N \boldsymbol{\nabla}'^2\Phi - \Phi\boldsymbol{\nabla}'^2 G_N) = \int_S \mathrm{d}s' \cdot (G_N\boldsymbol{\nabla}'\Phi - \Phi\boldsymbol{\nabla}'G_N),$$

or, $-4\pi\left(\left(\int_V \mathrm{d}^3r'\left(G_N\rho + \dfrac{\Phi}{V}\right)\right) - \Phi(\mathbf{r})\right) = \int_S \mathrm{d}s' \, G_N\dfrac{\partial\Phi}{\partial n'}.$

$$\tag{3.186}$$

Here, we have used the fact that Φ satisfies the Poisson equation while G_N satisfies (3.184). Noting that for any function $A(\mathbf{r})$ the average over a given volume is defined to be

$$A_{\mathrm{avg}} = \overline{A} = \frac{1}{V}\int_V \mathrm{d}^3r\, A(\mathbf{r}), \tag{3.187}$$

we now obtain the solution from (3.186) to be

$$\Phi(\mathbf{r}) = \overline{\Phi} + \frac{1}{4\pi} \int_S ds' \, G_N(\mathbf{r}, \mathbf{r}') \frac{\partial \Phi(\mathbf{r}')}{\partial n'}$$

$$+ \int_V d^3r' \, G_N(\mathbf{r}, \mathbf{r}') \rho(\mathbf{r}'). \tag{3.188}$$

However, the Neumann boundary condition on $\frac{\partial \Phi}{\partial n}$ still has to be specified consistent with the constraint (3.180) which makes such problems extremely difficult and impractical. If we do not constrain the boundary conditions (data), the Neumann problem may not have a solution. (Incidentally, if V is infinite, the average term will not be present in the solution if the potential falls off asymptotically.)

▶ **Example (Point charge above a conducting plane).** Let us analyze the following example to show how this method works. We have already seen in (3.8) that a point charge q above a (grounded) conducting plane, say, on the z-axis at a height d, induces a surface charge density on the plane given by (here r represents the radial coordinate on the $x - y$ plane)

$$\sigma = \frac{1}{4\pi} \, \hat{\mathbf{z}} \cdot \mathbf{E} \Big| = -\frac{1}{4\pi} \frac{\partial \Phi}{\partial z} \Big| = -\frac{qd}{2\pi(r^2 + d^2)^{\frac{3}{2}}},$$

or, $\quad \dfrac{\partial \Phi}{\partial z} \Big| = \dfrac{2qd}{(r^2 + d^2)^{\frac{3}{2}}}.$ $\hspace{3cm}$ (3.189)

Here the restriction stands for $z = 0$, namely, the location of the plane. Consequently, we can use this as a Neumann boundary condition which automatically satisfies the appropriate constraint and try to solve for the potential of the problem. We are interested in the solution in the upper half plane so that the volume is infinite implying that the average term in (3.188) can be ignored.

The Green's function satisfying the homogeneous Neumann boundary condition can be determined as before and has the form

$$G_N(\mathbf{r}, \mathbf{r}') = \frac{1}{\sqrt{(x - x')^2 + (y - y')^2 + (z - z')^2}}$$

$$+ \frac{1}{\sqrt{(x - x')^2 + (y - y')^2 + (z + z')^2}}. \tag{3.190}$$

The additional term as we have seen earlier in (3.163), is a solution of the homogeneous equation and the sign is chosen such that the z (or z') derivative vanishes if either z or z' vanishes (on the surface of the plane). The solution can now be obtained from (3.188). However, unlike the Dirichlet boundary condition it is clear that the surface term now gives a coordinate dependent term and to compare with what we have already done by the method of images, let us calculate the potential along the z-axis ($x = 0, y = 0, z \geq 0$). For such points we obtain (the negative sign in the surface term arises from the direction of the outward normal to the surface as discussed earlier, namely, $\frac{\partial}{\partial n'} = -\frac{\partial}{\partial z'}$)

$$\Phi(x = 0 = y, z) = \frac{1}{4\pi} \int_S ds' \, G_N(\mathbf{r}, \mathbf{r}') \frac{\partial \Phi}{\partial n'} + \int d^3r' \, G_N(\mathbf{r}, \mathbf{r}') \rho(\mathbf{r}')$$

$$= -\frac{2qd}{4\pi} \int \frac{2r'\,dr'\,d\phi'}{(r'^2 + d^2)^{\frac{3}{2}}(r'^2 + z^2)^{\frac{1}{2}}}$$

$$+ \int d^3r'\, G_N(\mathbf{r}, \mathbf{r}')q\delta(x')\delta(y')\delta(z' - d)$$

$$= \frac{q}{2\pi}(2\pi)\frac{\partial}{\partial d}\int_0^\infty \frac{dr'^2}{(r'^2 + d^2)^{\frac{1}{2}}(r'^2 + z^2)^{\frac{1}{2}}}$$

$$+ q\left(\frac{1}{|z - d|} + \frac{1}{z + d}\right)$$

$$= -\frac{2q}{z + d} + \frac{q}{|z - d|} + \frac{q}{z + d}$$

$$= \frac{q}{|z - d|} - \frac{q}{z + d}, \tag{3.191}$$

which is what we had calculated earlier in (3.5) using the method of images. (See, for example, Gradshteyn and Ryzhik 2.261 for the value of the first integral.) ◀

3.5 Selected problems

1. Consider a point charge q at a distance d from the center of a conducting sphere of radius R, where $d > R$. The surface of the sphere is maintained at a constant potential Φ_0.

   a) Determine the "image" charges needed to study this problem.

   b) Determine the potential at any point outside the sphere.

   c) Determine the induced surface charge density as well as the total induced charge.

2. Consider a hollow metallic sphere of finite thickness, with the inner radius a and the outer radius b. (The coordinate origin is chosen to be at the center of the sphere.) A point charge q is placed inside the sphere at a distance $\frac{a}{2}$ from the center of the sphere (the sphere is insulated so that the charge cannot move).

   a) What is the potential at a point $\mathbf{r}$ outside the sphere ($r \gg b$)?

   b) What are the potentials at the inner ($r = a$) as well as the outer ($r = b$) surfaces of the sphere?

   c) What is the potential at the center of the sphere? (Use the method of images to calculate this result.)

3. Two infinite grounded parallel conducting planes are separated by a distance d. A point charge q is placed between the planes.

Determine the induced surface charge densities as well as the total charges on the two planes.

4. Consider a conducting sphere of radius R whose surface is maintained at a potential $\Phi(\mathbf{R}) = \Phi_0 \cos\theta$. Assuming that there are no free charges present (inside or outside), what is the potential inside and outside the sphere?

5. Consider a cylindrical conducting can of radius R and height h. The side and the bottom walls of the can are grounded while the top face is maintained at $\Phi_0(r, \phi)$. Find the electrostatic potential inside the can. (You will have to solve the Laplace equation in cylindrical coordinates, which lead to Bessel functions.)

6. Let us assume that the potentials Φ_1 and Φ_2 are produced by the charge distributions (ρ_1, σ_1) and (ρ_2, σ_2) respectively. Namely, both volume and surface charge distributions are responsible for the potentials that correspond to solutions of the Poisson equations

$$\nabla^2 \Phi_1 = -4\pi\rho_1,$$

$$\nabla^2 \Phi_2 = -4\pi\rho_2.$$

Using Green's identity discussed in this chapter, prove the reciprocity theorem

$$\int_V d^3r\, \rho_1 \Phi_2 + \int_S ds\, \sigma_1 \Phi_2 = \int_V d^3r\, \rho_2 \Phi_1 + \int_S ds\, \sigma_2 \Phi_1.$$

Using this reciprocity theorem, determine the total induced charges on each of two infinite grounded conducting plates separated by a distance d, with a point charge q in the space between them.

7. Consider the one dimensional wave equation

$$\frac{\partial^2 \Phi}{\partial x^2} - \frac{1}{v^2}\frac{\partial^2 \Phi}{\partial t^2} = 0,$$

where v represents the speed of propagation of the wave. This is a hyperbolic equation, and at every point in space-time, there are two characteristics $\xi(x, t)$ and $\eta(x, t)$.

a) Determine the characteristics as functions of (x, t).

b) Invert the relations for the characteristics to express x and t in terms of the characteristics ξ and η. What is the form of the wave equation written in terms of the characteristics as the independent coordinates.

c) With the Cauchy conditions $\Phi(x, t = 0) = f(x)$ and $\left.\frac{\partial \Phi}{\partial t}\right|_{t=0} = g(x)$, determine the solution to the wave equation.

8. Using Green's identity, as well as other identities discussed in this chapter, show that the Green's function satisfying Dirichlet boundary conditions is symmetric in the interchange of its arguments (namely, $G_D(\mathbf{r}, \mathbf{r}') = G_D(\mathbf{r}', \mathbf{r})$). (In general, Neumann Green's functions are not symmetric.)

CHAPTER 4

Dielectrics

4.1 Electric displacement field

As we have discussed earlier dielectrics are materials that can be polarized. Namely, in the presence of an external electric field the bound electrons and the positively charged nuclei of the atoms of such materials get displaced slightly so that each individual atom behaves like a point dipole and all these dipoles inside the material become aligned to give the material a macroscopic dipole moment. We say that the material becomes polarized in the presence of an electric field.

Experimentally it is observed that the net polarization of a material depends on the applied electric field and the general relation between the two is of the form

$$P_i = \chi_{ij} E_j. \tag{4.1}$$

The χ_{ij} are known as the components of the electric susceptibility tensor of the material (for individual atoms the constant of proportionality is known as the polarizability of the atom) and it can, in principle, depend on the electric field itself. However, for electric fields which are not too strong the components of the tensor are independent of the electric field and can be thought of as constants for a given medium. Furthermore, the electric susceptibility is a second rank tensor and, in fact, it is a symmetric tensor. In general, therefore, it can have at the most six independent components. However, by a suitable coordinate transformation (namely, with a suitable choice of the coordinate axes), it can be brought to a diagonal form with the three diagonal elements, in general, distinct. On the other hand, there is a wide class of dielectric materials which are isotropic (or sometimes also called linear) for which the three diagonal elements are the same. In this case, the relation between the polarization and

121

the electric field is a linear one of the form

$$\mathbf{P} = \chi \mathbf{E}, \tag{4.2}$$

with χ, which is a scalar, representing the electric susceptibility of the linear dielectric material. We will restrict ourselves to such materials for simplicity. (The electric susceptibility is a positive quantity, namely, $\chi \geq 0$ which simply represents the fact that the polarization is along the direction of the applied electric field.)

We have already seen (see (2.60)) that when a dielectric material is polarized it produces a potential of the form

$$\Phi_{\text{dipole}}(\mathbf{r}) = \int_S \frac{\mathrm{d}\mathbf{s}' \cdot \mathbf{P}(\mathbf{r}')}{|\mathbf{r} - \mathbf{r}'|} - \int_V \mathrm{d}^3 r' \frac{\mathbf{\nabla}' \cdot \mathbf{P}(\mathbf{r}')}{|\mathbf{r} - \mathbf{r}'|}, \tag{4.3}$$

which allowed us to conclude that a polarized dielectric develops a volume charge density as well as a surface charge density given respectively by ($\hat{\mathbf{n}}$ is the unit vector normal to the surface)

$$\rho_{\text{b}} = -(\mathbf{\nabla} \cdot \mathbf{P}), \quad \sigma_{\text{b}} = \hat{\mathbf{n}} \cdot \mathbf{P}. \tag{4.4}$$

Here, the subscript "b" simply stands for the fact that these charges are bound and are not free to move around unlike the charges in a conductor. It follows from the identification in (4.4) that the total charge in the dielectric is

$$
\begin{aligned}
Q_{\text{b}} &= \int_V \mathrm{d}^3 r \, \rho_{\text{b}}(\mathbf{r}) + \int_S \mathrm{d}s \, \sigma_{\text{b}} \\
&= -\int_V \mathrm{d}^3 r \, \mathbf{\nabla} \cdot \mathbf{P} + \int_S \mathrm{d}s \, \hat{\mathbf{n}} \cdot \mathbf{P}(\mathbf{r}) \\
&= -\int_V \mathrm{d}^3 r \, \mathbf{\nabla} \cdot \mathbf{P}(\mathbf{r}) + \int_S \mathrm{d}\mathbf{s} \cdot \mathbf{P}(\mathbf{r}) = 0, \tag{4.5}
\end{aligned}
$$

where the last identity follows from Gauss' theorem. This is, of course, what we would expect. Namely, the dielectric is charge neutral, all that happens in the presence of an external electric field is that the charges are displaced slightly to give it a macroscopic dipole moment.

It is intuitively clear that because a polarized dielectric develops a volume as well as a surface density of bound charges, the differential form of Gauss' law satisfied by the electric field in a dielectric would modify. To determine this systematically, let us consider a dielectric of infinite extent with some point charges embedded inside. In this case, it is clear that if we consider a Gaussian surface as shown in Fig. 4.1, Gauss' law (1.28) would lead to

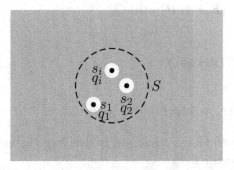

Figure 4.1: A dielectric with a number of embedded free charges. The dashed curve represents the Gaussian surface enclosing the free charges.

$$\int_S d\mathbf{s} \cdot \mathbf{E} = 4\pi(Q + Q_b), \tag{4.6}$$

where Q represents the sum of free charges embedded inside the dielectric and Q_b is the total bound charge within the Gaussian volume (which need not be zero unlike the case of the whole dielectric). By definition (ρ_b is contained only in the volume excluding the free charges)

$$
\begin{aligned}
Q_b &= \int_V d^3r\, \rho_b(\mathbf{r}) + \int_{\sum_i S_i} ds\, \sigma_b(\mathbf{r}) \\
&= -\int_V d^3r\, (\boldsymbol{\nabla} \cdot \mathbf{P}) + \int_{\sum_i S_i} d\mathbf{s} \cdot \mathbf{P} \\
&= -\int_{S+\sum_i S_i} d\mathbf{s} \cdot \mathbf{P} + \int_{\sum_i S_i} d\mathbf{s} \cdot \mathbf{P} \\
&= -\int_S d\mathbf{s} \cdot \mathbf{P}. \tag{4.7}
\end{aligned}
$$

Here S_i represents the surface area of the interface between the charge q_i and the dielectric. Putting this back into the integral form of Gauss' law (4.6), we obtain

$$\int_S d\mathbf{s} \cdot (\mathbf{E} + 4\pi\mathbf{P}) = 4\pi Q,$$

$$\text{or,} \quad \int_S d\mathbf{s} \cdot \mathbf{D} = 4\pi Q,$$

$$\text{or,} \quad \boldsymbol{\nabla} \cdot \mathbf{D}(\mathbf{r}) = 4\pi\rho(\mathbf{r}). \tag{4.8}$$

Here, we have used Gauss' theorem in deriving the last line and have defined a new vector field

$$\mathbf{D}(\mathbf{r}) \equiv \mathbf{E}(\mathbf{r}) + 4\pi\mathbf{P}(\mathbf{r}), \tag{4.9}$$

which is known as the electric displacement vector (electric displacement field). Therefore, in a dielectric it is the flux of the electric displacement field out of a Gaussian surface which equals (4π) times the free charge contained in the Gaussian volume. Correspondingly, it is the divergence of this field that is proportional to the density of free charges which is quite useful because, in reality, we can have information only on the free charges in a system. (The number of bound charges is large and their distribution is clearly an impossible thing to determine.)

Therefore, in the presence of a dielectric the differential form of Gauss' law takes the form

$$\boldsymbol{\nabla} \cdot \mathbf{D} = 4\pi\rho. \tag{4.10}$$

In spite of its similarity with the differential form of Gauss' law (1.30) satisfied by the electric field in the absence of a dielectric, the two vector fields may have quite different characters in general. For example, as we have seen in chapter 1 the electric field is conservative, but $\mathbf{D}$ may not be, namely,

$$\boldsymbol{\nabla} \times \mathbf{E} = 0, \quad \boldsymbol{\nabla} \times \mathbf{D} \neq 0. \tag{4.11}$$

This follows because the polarization vector may not have vanishing curl in general. Furthermore, note that since the polarization vector is parallel to the electric field, (for isotropic dielectrics) we can write

$$\mathbf{D} = \mathbf{E} + 4\pi\mathbf{P} = (1 + 4\pi\chi)\mathbf{E} = \epsilon\,\mathbf{E}, \quad \epsilon = 1 + 4\pi\chi. \tag{4.12}$$

Thus, we see that in an isotropic (linear) dielectric the electric displacement vector is parallel to the electric field and the constant of proportionality ϵ is known as the permittivity or the dielectric constant of the material. (Even when $\mathbf{P}$ is parallel to $\mathbf{E}$, it may still have non-vanishing curl if χ is space dependent.) From the definition of χ it follows that the dielectric constant for a material is greater than unity, $\epsilon \geq 1$. The dielectric constant is unity only for vacuum as we have noted earlier. Note also that even though $\mathbf{D}$ and $\mathbf{E}$ are linearly related, in general

$$\boldsymbol{\nabla} \cdot \mathbf{E} \neq \frac{4\pi}{\epsilon}\, \rho, \tag{4.13}$$

which follows because the dielectric constant may be different for different regions of space (namely, it may be space dependent). However, in a homogeneous region of space (namely, where there is no change in the dielectric constant), we can write the differential form of Gauss' law also in terms of the electric field as

$$\boldsymbol{\nabla} \cdot \mathbf{E} = \frac{4\pi}{\epsilon}\, \rho. \tag{4.14}$$

▶ **Example (Point charge in an isotropic dielectric).** As an example of problems involving dielectric materials, let us consider an isotropic dielectric material of infinite extent and permittivity ϵ with a point charge q embedded inside at a point which we choose to be the origin of the coordinate system (see Fig. 4.2) and we want to calculate the electric as well as the displacement fields. This problem is very simple and yet clarifies some of the essential properties of dielectric materials.

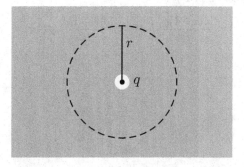

Figure 4.2: A point charge embedded in an isotropic dielectric. The dashed sphere represents the Gaussian surface.

From the (spherical) symmetry of the problem we see that the electric field will be radial (with the charge q at the center) at every point. Furthermore, drawing a Gaussian sphere of radius r we note that the magnitude of the electric field will be the same at every point on the surface of this sphere. Therefore, we determine trivially from (4.8) that (remember that $\mathbf{D}$ is parallel to the electric field and, consequently, is radial as well)

$$4\pi r^2 |\mathbf{D}| = 4\pi q,$$

or, $|\mathbf{D}| = \dfrac{q}{r^2}.$ (4.15)

It follows from this that

$$\mathbf{D} = \frac{q}{r^2}\,\hat{\mathbf{r}},$$

$$\mathbf{E} = \frac{1}{\epsilon}\mathbf{D} = \frac{1}{\epsilon}\frac{q}{r^2}\,\hat{\mathbf{r}},\quad \Phi = \frac{1}{\epsilon}\frac{q}{r}.$$ (4.16)

Namely, the electric field and the potential have the same structures as in vacuum, but are reduced in magnitude (remember that $\epsilon \geq 1$). Furthermore, we can also determine the polarization of the medium from the fact that (see (4.12))

$$\mathbf{P} = \chi\mathbf{E} = \frac{(\epsilon-1)}{4\pi}\,\mathbf{E} = \frac{(\epsilon-1)}{4\pi\epsilon}\,\frac{q}{r^2}\,\hat{\mathbf{r}}.$$ (4.17)

The reason why the strength of the electric field (as well as the potential) is reduced in the presence of the dielectric is easily understood from the fact that the point charge polarizes the medium. In fact, from the structure of the polarization vector in (4.17) as well as using (4.4), we see that (remember $\boldsymbol{\nabla}\cdot\mathbf{r} = 3$ in three dimensions and we are interested in points $\mathbf{r} \neq 0$)

$$\rho_{\mathrm{b}} = -\boldsymbol{\nabla}\cdot\mathbf{P} = -\boldsymbol{\nabla}\cdot\left(\frac{(\epsilon-1)}{4\pi\epsilon}\frac{q\mathbf{r}}{r^3}\right) = 0,$$ (4.18)

so that there is no volume density of bound charges in this case and only a surface charge density is present.

The total bound charge in the Gaussian volume (see Fig. 4.2) can, therefore, be calculated easily to give (the negative sign is because $\hat{\mathbf{n}}$ is inward at the interface opposite to the direction of $\mathbf{P}$)

$$Q_{\mathrm{b}} = \int_S d\mathbf{s}\cdot\mathbf{P} = -\frac{(\epsilon-1)q}{4\pi\epsilon}\int d\Omega = -\frac{(\epsilon-1)q}{4\pi\epsilon}\times(4\pi) = -\frac{(\epsilon-1)}{\epsilon}q.$$ (4.19)

Figure 4.3: The induced negative charges in the interface of the free charge in the dielectric medium.

All of this charge lies on the inner surface as shown in Fig. 4.3 and, as a result, at a distance far away from the origin the total effective (free) charge seen is

$$q_{\text{eff}} = q + Q_{\text{b}} = q \left(1 - \frac{(\epsilon - 1)}{\epsilon} \right) = \frac{q}{\epsilon}. \tag{4.20}$$

In other words, induced bound charges have a tendency to reduce the magnitude of the (free) point charge and, consequently, lead to a weaker electric field. Conventionally, this is known as the screening of a point charge by a dielectric medium.   ◀

▶ **Example (Capacitor filled with dielectric).** Once we have Gauss' law (4.10) in the presence of a dielectric (or the integral form of it in (4.8)) solving electrostatic problems involving dielectrics is no more difficult than what we have already done in chapters **1** and **2**. Let us recall that Gauss' law involves the field **D** and the free charge distribution. Consequently, just as we determined **E** earlier (in chapter **1**) from Gauss' law we can now determine **D**. Furthermore, from the relation between **D** and **E** in (4.12) we can then obtain the electric field as well.

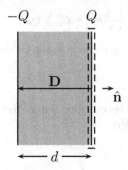

Figure 4.4: A capacitor filled with a dielectric of permittivity ϵ. The dashed rectangle represents the Gaussian surface.

As an example of how this is done, let us consider a large parallel plate capacitor of area A separated by a small distance d. Let us assume that the plates carry charges Q and $-Q$ respectively and that the space between the two capacitor plates is filled with a dielectric material of permittivity ϵ (see Fig. 4.4). Clearly from the symmetry of the problem we conclude, as before, that the charge would distribute itself uniformly over the two plates so that we have two plates with a uniform surface distribution of charge densities σ and $-\sigma$ respectively. We expect the electric field (and, consequently, the electric displacement field where applicable) to be perpendicular to the plates and that it has a nonzero constant value only between the two plates in a direction from the positively charged plate to the negatively charged plate. Calculating the flux out of a rectangular Gaussian surface we conclude that

$$\int \mathbf{ds} \cdot \mathbf{D} = 4\pi Q,$$

or, $\quad |\mathbf{D}| A = 4\pi Q,$

or, $\quad |\mathbf{D}| = \frac{4\pi Q}{A} = 4\pi \sigma. \tag{4.21}$

Thus, we can write the displacement field between the two plates to be

$$\mathbf{D}(\mathbf{r}) = -4\pi\sigma\,\hat{\mathbf{n}}, \tag{4.22}$$

where we have defined $\hat{\mathbf{n}}$ to be the unit vector normal to the plates as in Fig. 4.4 (say, along the z-axis). Here the negative sign arises because the direction of the $\mathbf{D}$ field is opposite to the vector $\hat{\mathbf{n}}$.

The electric field between the plates can now be determined from relation (4.12)

$$\mathbf{E}(\mathbf{r}) = \frac{1}{\epsilon}\,\mathbf{D}(\mathbf{r}) = -\frac{4\pi\sigma}{\epsilon}\,\hat{\mathbf{n}}. \tag{4.23}$$

Thus, as before, we see that the electric field continues to be a constant between the plates. However, its strength is reduced from the case where there was no dielectric between the plates. From (4.23) we determine the potential difference (voltage) between the two plates to be

$$V = |\mathbf{E}|d = \frac{4\pi\sigma}{\epsilon}\,d = \frac{4\pi d}{\epsilon A}\,\sigma A = C_{\text{diel.}}^{-1}\,Q. \tag{4.24}$$

Namely, we determine the capacitance in the presence of the dielectric to be

$$C_{\text{diel.}} = \frac{\epsilon A}{4\pi d}. \tag{4.25}$$

In other words, the capacitance of the system increases in the presence of a dielectric (compare with (2.89)).   ◄

4.2 Boundary conditions in dielectric

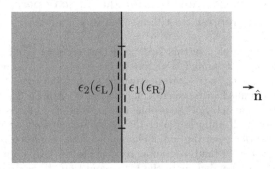

Figure 4.5: The interface of two distinct dielectric media with the dashed curve representing the surface of a rectangular Gaussian volume.

Suppose, we have two distinct dielectric media, characterized by the respective dielectric constants ϵ_1 (or ϵ_R) and ϵ_2 (ϵ_L), separated by a surface (perpendicular to the z-axis). Let us draw a rectangular

Gaussian volume of infinitesimal thickness projecting into both the media as shown in Fig. 4.5.

It is clear now that in the limit of vanishing thickness of the rectangular volume, Gauss' law gives (namely, taking the integral form of Gauss' law in (4.8))

$$\hat{n} \cdot (\mathbf{D}_R - \mathbf{D}_L) = 4\pi\sigma, \tag{4.26}$$

where σ denotes the density of free charges on the surface separating the two media. Thus, the normal component of the $\mathbf{D}$ field is discontinuous across a surface carrying free charges. On the other hand, from the conservative nature of the electric field, it follows (taking the line integral of the electric field or the surface integral of the curl of the electric field) that the tangential component of the electric field is continuous across the boundary, namely,

$$\mathbf{E}_{R,t} - \mathbf{E}_{L,t} = 0. \tag{4.27}$$

Equations (4.26) and (4.27) represent the two boundary conditions that we have to satisfy at an interface of two distinct dielectric materials.

From the fact that the electric field is defined in terms of the potential as

$$\mathbf{E}(\mathbf{r}) = -\boldsymbol{\nabla}\Phi(\mathbf{r}), \tag{4.28}$$

it follows that in the presence of an isotropic dielectric, we have to solve the equation

$$\boldsymbol{\nabla} \cdot \mathbf{D}(\mathbf{r}) = 4\pi\rho(\mathbf{r}),$$

$$\text{or,} \quad \boldsymbol{\nabla} \cdot (\epsilon\,\boldsymbol{\nabla}\Phi(\mathbf{r})) = -4\pi\rho(\mathbf{r}), \tag{4.29}$$

where $\rho(\mathbf{r})$ represents the density of free charges. Furthermore, equation (4.29) needs to be solved subject to the boundary conditions that across a surface separating two dielectric media

$$\epsilon_R \frac{\partial \Phi_R}{\partial n}\bigg|_S = \epsilon_L \frac{\partial \Phi_L}{\partial n}\bigg|_S - 4\pi\sigma, \qquad \frac{\partial}{\partial n} = \hat{n} \cdot \boldsymbol{\nabla},$$

$$\Phi_R(\mathbf{r})|_S = \Phi_L(\mathbf{r})|_S, \tag{4.30}$$

which correspond to the boundary conditions (4.26) and (4.27) respectively in terms of the potential (see the definition in (4.28)).

The boundary conditions in dielectrics lead to some very interesting consequences. For example, let us consider two dielectric media with permittivities ϵ_1 and ϵ_2 separated by a surface which does not contain any free surface charge. Let us further assume that an electric field vector makes an angle θ_1 with the surface in the first dielectric while it makes an angle θ_2 with the surface in the second medium as shown in Fig. 4.6. Denoting the electric fields in the two media by $\mathbf{E}_1$ and $\mathbf{E}_2$ respectively, the boundary conditions (4.26) and (4.27) tell us that

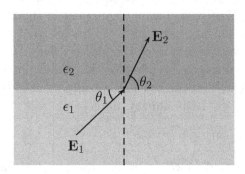

Figure 4.6: An electric field in two distinct dielectric media with permittivities ϵ_1 and ϵ_2.

$$\epsilon_1 |\mathbf{E}_1| \sin \theta_1 = \epsilon_2 |\mathbf{E}_2| \sin \theta_2, \qquad |\mathbf{E}_1| \cos \theta_1 = |\mathbf{E}_2| \cos \theta_2. \quad (4.31)$$

We can take the ratio of the two relations in (4.31) to write

$$\epsilon_1 \tan \theta_1 = \epsilon_2 \tan \theta_2, \qquad\qquad\qquad\qquad\qquad (4.32)$$

which is known as the Snell's law in optics. We will study this in more detail in a later chapter.

▶ **Example (Point charge in semi-infinite dielectric medium).** As an example of applications of the boundary conditions in dielectric media, let us consider a semi-infinite medium of a dielectric material of permittivity ϵ_1. Outside of the dielectric lies vacuum and a point charge of magnitude q is immersed on the boundary between the two media shown in Fig. 4.7. We wish to determine the fields $\mathbf{E}$ and $\mathbf{D}$ everywhere except at the location of the point charge.

Without loss of generality, we can take the origin of the coordinate system to be at the location of the point charge. Furthermore, from the symmetry of the problem we expect both the electric and the displacement fields to be radial and have spherical symmetry in the two regions. Let us denote the fields in the

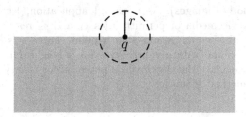

Figure 4.7: A point charge on the interface of a dielectric medium and vacuum with the surface of the dashed sphere representing the Gaussian surface.

dielectric as $\mathbf{E}_1, \mathbf{D}_1$, while we denote them in vacuum as $\mathbf{E}_0, \mathbf{D}_0$. The boundary condition (4.27) for the electric field implies that

$$E_{0,\mathrm{t}}| = E_{1,\mathrm{t}}| . \tag{4.33}$$

On the other hand, since the electric fields are radial, on the boundary surface they are in fact tangential which determines that

$$|\mathbf{E}_0| = |\mathbf{E}_1| . \tag{4.34}$$

Let us next draw a Gaussian sphere (indicated by the dashed sphere in Fig. 4.7) of radius r around the point charge. From Gauss' law we obtain

$$\int d\mathbf{s} \cdot \mathbf{D} = 4\pi q,$$

$$\text{or,} \quad 2\pi r^2 \left(|\mathbf{D}_1| + |\mathbf{D}_0| \right) = 4\pi q,$$

$$\text{or,} \quad (1 + \epsilon_1)|\mathbf{E}_0| = \frac{2q}{r^2},$$

$$\text{or,} \quad \mathbf{E}_0(\mathbf{r}) = \frac{2q\hat{\mathbf{r}}}{(1 + \epsilon_1)r^2}, \tag{4.35}$$

in the upper half plane. Here we have used the fact that the surface area of the hemisphere in both the regions has the value $2\pi r^2$ as well as the relation between the displacement field and the electric field (4.12) in the two regions. It follows now that $((\mathbf{E}_0, \mathbf{D}_0)$ and $(\mathbf{E}_1, \mathbf{D}_1)$ are defined respectively in the upper and lower half planes)

$$\mathbf{E}_1(\mathbf{r}) = \mathbf{E}_0(\mathbf{r}) = \frac{2q\hat{\mathbf{r}}}{(1 + \epsilon_1)r^2},$$

$$\mathbf{D}_0(\mathbf{r}) = \mathbf{E}_0(\mathbf{r}) = \frac{2q\hat{\mathbf{r}}}{(1 + \epsilon_1)r^2},$$

$$\mathbf{D}_1(\mathbf{r}) = \epsilon_1 \mathbf{E}_1(\mathbf{r}) = \frac{2\epsilon_1 q\hat{\mathbf{r}}}{(1 + \epsilon_1)r^2} . \tag{4.36}$$

We note that when $\epsilon_1 = 1$, these fields reduce to the electric field for a point charge in vacuum that we are familiar with. ◀

▶ **Example (Method of images).** As a second application, let us consider two semi-infinite dielectric media of permittivities ϵ_1 and ϵ_2 respectively separated by a plane at $z = 0$. Let us assume that a point charge q is immersed in the second dielectric medium (with permittivity ϵ_2) at a height d from the boundary. We would like to calculate the electrostatic potential for this system in both the regions using the method of images.

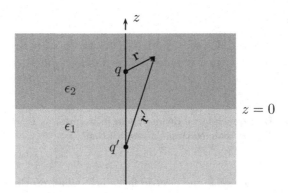

Figure 4.8: A point charge q in the region $z > 0$ with the image charge q' in the region $z < 0$ for calculating the electric field in the upper half plane.

Without loss of generality, we can assume that the point charge lies on the z-axis at a height $z = d$ from the interface. Unlike the case of a grounded conducting plane, here we will have nonvanishing electric fields present in both the dielectric media and, consequently, we need to calculate these in both the regions $z \geq 0$ and $z \leq 0$. Let us recall from our earlier study involving the method of images that we need an image charge in a region where we are not calculating the electric field. As a result, since we have to calculate the potential and the fields in both the regions $z \geq 0$ and $z \leq 0$, we need two sets of image charges - one for each calculation. When we are calculating the field in the region $z \geq 0$, we need an image charge q' located at $z = -d$ as shown in Fig. 4.8. On the other hand, when we calculate the field in the region $z \leq 0$, we need an image charge in the region $z \geq 0$. In fact, if we think for a moment, we realize that the dielectric will be polarized because of the presence of the charge q. As a result, as we have discussed in an earlier example, the effective charge seen in the region $z \leq 0$ will be modified. Consequently, in calculating the field in region $z \leq 0$, we can imagine an image charge present at $z = d$ so as to give rise to an effective charge q'' at that point (namely, the image charge has the value $q'' - q$ at that point), as shown in Fig. 4.9.

With these introductory remarks, the calculation is now straightforward. Let us use cylindrical coordinates for our calculations. From Fig. 4.8, we see that we can write the potential in the region $z \geq 0$ as (see, for example, the discussion

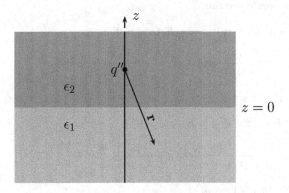

Figure 4.9: For calculating the electric field in the lower half plane the image charge can be chosen to lie on top of q leading to an effective charge q''.

for the electric field in (4.16))

$$\Phi(\mathbf{r}; z \geq 0) = \frac{1}{\epsilon_2} \left(\frac{q}{|\mathbf{r}|} + \frac{q'}{|\mathbf{r}'|} \right)$$

$$= \frac{1}{\epsilon_2} \left(\frac{q}{\sqrt{\rho^2 + (z-d)^2}} + \frac{q'}{\sqrt{\rho^2 + (z+d)^2}} \right). \qquad (4.37)$$

Here ρ represents the radial coordinate on the plane. Similarly, the potential in the region $z \leq 0$ can be determined from Fig. 4.9 to be

$$\Phi(\mathbf{r}; z \leq 0) = \frac{1}{\epsilon_1} \frac{q''}{|\mathbf{r}|} = \frac{1}{\epsilon_1} \frac{q''}{\sqrt{\rho^2 + (d-z)^2}}. \qquad (4.38)$$

The components of the electric fields can now be calculated easily and have the forms

$$E_z(z \geq 0) = -\frac{\partial}{\partial z} \Phi(z \geq 0) = \frac{1}{\epsilon_2} \left[\frac{q(z-d)}{(\rho^2 + (z-d)^2)^{\frac{3}{2}}} + \frac{q'(z+d)}{(\rho^2 + (z+d)^2)^{\frac{3}{2}}} \right],$$

$$E_z(z \leq 0) = -\frac{\partial}{\partial z} \Phi(z \leq 0) = \frac{1}{\epsilon_1} \frac{q''(z-d)}{(\rho^2 + (d-z)^2)^{\frac{3}{2}}},$$

$$E_\rho(z \geq 0) = -\frac{\partial}{\partial \rho} \Phi(z \geq 0) = \frac{1}{\epsilon_2} \left[\frac{q\rho}{(\rho^2 + (z-d)^2)^{\frac{3}{2}}} + \frac{q'\rho}{(\rho^2 + (z+d)^2)^{\frac{3}{2}}} \right],$$

$$E_\rho(z \leq 0) = -\frac{\partial}{\partial \rho} \Phi(z \leq 0) = \frac{1}{\epsilon_1} \frac{q''\rho}{(\rho^2 + (d-z)^2)^{\frac{3}{2}}}. \qquad (4.39)$$

Since there are no free surfaces charges, the boundary conditions (see (4.26) and (4.27)) would require

$$\epsilon_2 E_z(\rho, z = 0^+) = \epsilon_1 E_z(\rho, z = 0^-),$$

or, $$-\frac{d}{(\rho^2 + d^2)^{\frac{3}{2}}}(q - q') = -\frac{d}{(\rho^2 + d^2)^{\frac{3}{2}}} q'',$$

or, $$(q - q') = q'',$$

$$E_\rho(\rho, z = 0^+) = E_\rho(\rho, z = 0^-),$$

or, $$\frac{\rho}{(\rho^2 + d^2)^{\frac{3}{2}}} \frac{(q + q')}{\epsilon_2} = \frac{\rho}{(\rho^2 + d^2)^{\frac{3}{2}}} \frac{q''}{\epsilon_1},$$

or, $$\epsilon_1(q + q') = \epsilon_2 q''. \tag{4.40}$$

The image charges are determined from these relations to be

$$q' = -\frac{\epsilon_1 - \epsilon_2}{\epsilon_1 + \epsilon_2} q, \qquad q'' = \frac{2\epsilon_1}{\epsilon_1 + \epsilon_2} q. \tag{4.41}$$

We note that only when $\epsilon_1 = \epsilon_2$, we have $q'' = q$. Otherwise, the effective charge seen in the region $z \leq 0$ will be different due to polarization of the two media. Once the image charges are determined, the potential as well as the electric fields can be determined in both the regions. Furthermore, the polarizations in the two media can also be calculated using (4.17). In fact, since the permittivity of the two media are different there will be a net polarization charge at the interface, which can be calculated as follows.

We note from our results (as well as (4.17)) that the polarizations on both sides of the interface will have the forms

$$P_z(\rho, z = 0^+) = \frac{(\epsilon_2 - 1)}{4\pi} E_z(\rho, z = 0^+)$$

$$= -\frac{(\epsilon_2 - 1)}{4\pi\epsilon_2} \frac{2\epsilon_1}{\epsilon_1 + \epsilon_2} \frac{qd}{(\rho^2 + d^2)^{\frac{3}{2}}},$$

$$P_z(\rho, z = 0^-) = \frac{(\epsilon_1 - 1)}{4\pi} E_z(\rho, z = 0^-)$$

$$= -\frac{(\epsilon_1 - 1)}{4\pi\epsilon_1} \frac{2\epsilon_1}{\epsilon_1 + \epsilon_2} \frac{qd}{(\rho^2 + d^2)^{\frac{3}{2}}}. \tag{4.42}$$

It follows from this that the net density of surface polarization charge is given by (see (4.4))

$$\sigma_b = \hat{\mathbf{n}} \cdot \mathbf{P} \bigg| = P_z(\rho, z = 0^+) - P_z(\rho, z = 0^-)$$

$$= \frac{(\epsilon_1 - \epsilon_2)}{\epsilon_2(\epsilon_1 + \epsilon_2)} \frac{qd}{2\pi(\rho^2 + d^2)^{\frac{3}{2}}}. \tag{4.43}$$

This shows, in particular, that as long as $\epsilon_1 \neq \epsilon_2$, there will be a net polarization charge at the interface.   ◄

▶ **Example (Dielectric sphere in uniform electric field).** Let us consider a dielectric sphere of radius R and dielectric constant ϵ placed in vacuum in a uniform electric field along the z-axis. Thus, in the absence of the dielectric sphere we have

$$\mathbf{E}(\mathbf{r}) = E\,\hat{\mathbf{z}}, \tag{4.44}$$

leading to the potential

$$\Phi(\mathbf{r}) = -Ez + C = -Er\cos\theta + C, \tag{4.45}$$

where C is a constant and we are assuming the coordinate origin to coincide with the center of the dielectric sphere. When we introduce the dielectric sphere, the sphere will be polarized and would modify the electric field around and within the sphere. However, asymptotically, the form of the potential in (4.45) would continue to hold.

To determine the potential in the presence of the dielectric sphere let us note that outside the dielectric sphere we simply have to solve the Laplace equation in vacuum which has the form

$$\nabla^2\Phi_>(\mathbf{r}) = 0, \qquad r \geq R, \tag{4.46}$$

and whose solutions can be written, in general, as (see (3.105) and note that the present problem has azimuthal symmetry leading to $m = 0$)

$$\Phi_>(\mathbf{r}) = C - Er\,P_1(\cos\theta) + \sum_{\ell=1}^{\infty} A_\ell r^{-(\ell+1)} P_\ell(\cos\theta), \tag{4.47}$$

where we have kept a linear term in r with the asymptotic condition (4.45) in mind. (Here we have also used the fact that for $\Phi_>(\mathbf{r})$ to satisfy the Laplace equation, we must have $A_0 = 0$.) Inside the dielectric sphere we also have to solve the Laplace equation (there are no free charges inside the sphere and note also that within the dielectric sphere ϵ is a constant so that it can be taken out of the gradient operation)

$$\nabla^2\Phi_<(\mathbf{r}) = 0, \qquad r \leq R, \tag{4.48}$$

whose general solution has the form

$$\Phi_<(\mathbf{r}) = \sum_{\ell=0}^{\infty} B_\ell r^\ell P_\ell(\cos\theta). \tag{4.49}$$

We note that in writing these solutions, we have used the fact that the region inside the sphere contains the origin and, consequently, should have regular solutions, while outside the sphere the potential should fall off except for the asymptotic behavior required by a constant electric field.

Now matching the solutions in (4.47) and (4.49) across the surface of the sphere we obtain (see (4.30))

$$\Phi_<(r = R) = \Phi_>(r = R),$$

$$\text{or,} \quad \sum_{\ell=0}^{\infty} B_\ell R^\ell P_\ell(\cos\theta) = C - ERP_1(\cos\theta)$$

$$+ \sum_{\ell=1}^{\infty} A_\ell R^{-(\ell+1)} P_\ell(\cos\theta), \tag{4.50}$$

which determines

$$B_0 = C,$$

$$B_1 R = -ER + A_1 R^{-2},$$

$$B_\ell R^\ell = A_\ell R^{-(\ell+1)}, \qquad \text{for} \quad \ell \geq 2. \tag{4.51}$$

Similarly, the condition on the normal derivatives in (4.30) at the boundary surface gives (remember, there are no free charges on the surface and the space outside the sphere is vacuum)

$$\epsilon \frac{\partial \Phi_<(\mathbf{r})}{\partial r}\bigg|_{r=R} = \frac{\partial \Phi_>(\mathbf{r})}{\partial r}\bigg|_{r=R},$$

or, $\displaystyle\sum_{\ell=1}^{\infty} \epsilon \ell B_\ell R^{(\ell-1)} P_\ell(\cos\theta) = - EP_1(\cos\theta)$

$$- \sum_{\ell=1}^{\infty}(\ell+1)A_\ell R^{-(\ell+2)}P_\ell(\cos\theta), \tag{4.52}$$

which determines

$$\epsilon B_1 = -E - 2A_1 R^{-3},$$

$$\epsilon \ell B_\ell R^{(\ell-1)} = -(\ell+1)A_\ell R^{-(\ell+2)}, \qquad \text{for} \quad \ell \geq 2. \tag{4.53}$$

From the relations (4.51) and (4.53) it is straightforward to determine that

$$A_\ell = 0 = B_\ell, \qquad \text{for} \quad \ell \geq 2,$$

$$B_0 = C,$$

$$A_1 = \frac{\epsilon - 1}{\epsilon + 2} ER^3,$$

$$B_1 = -\frac{3}{\epsilon + 2} E, \tag{4.54}$$

so that we can write

$$\Phi_>(\mathbf{r}) = C - E\left(1 - \frac{\epsilon-1}{\epsilon+2}\left(\frac{R}{r}\right)^3\right) r\cos\theta,$$

$$\Phi_<(\mathbf{r}) = C - \frac{3}{\epsilon+2} Er\cos\theta. \tag{4.55}$$

The electric fields can be determined from these and we note, in particular, that inside the dielectric sphere the electric field is given by ($z = r\cos\theta$)

$$\mathbf{E}_<(\mathbf{r}) = -\nabla\Phi_<(\mathbf{r}) = \frac{3}{\epsilon+2} E\hat{\mathbf{z}}. \tag{4.56}$$

Namely, as a result of the external electric field the electric field present inside the dielectric sphere is uniform. It has a reduced strength, but is along the z-axis like the asymptotic field. This can be contrasted with the case of a conducting sphere where there is no field inside the sphere. (Namely, even though the dielectric is polarized, the polarization is not large enough to completely cancel the field inside.) This shows that the boundary value problems involving dielectrics are solved much the same way by imposing appropriate boundary conditions at the surface separating two dielectric media.

We note here that the problem of a spherical cavity inside an infinite isotropic dielectric medium in the presence of a uniform electric field can also be solved exactly in the same manner. In fact, let us note that the solution is similar except that since the electric field inside a dielectric is reduced compared to that in vacuum, we can obtain the solution simply by letting $\epsilon \to \frac{1}{\epsilon}$. (Namely, the boundary conditions, in this case, lead to an inverted field.)

◀

▶ **Example (Cylindrical electret).** As a final example, let us analyze the electric field associated with a cylindrical electret. We note that, in nature, there are dielectric materials which have a permanent constant polarization such that in the presence of an external field the polarizations orient themselves and retain this polarization even when the external field is switched off. Such materials are known as "electrets" and the associated "retained" polarization is normally referred to as "ferroelectricity".

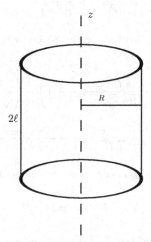

Figure 4.10: A cylindrical electret of radius R and length 2ℓ.

If we have a cylindrical electret of radius R and length 2ℓ along the z-axis with a uniform polarization $\mathbf{P} = P\hat{\mathbf{z}}$ as shown in Fig. 4.10, then by definition (see (4.4)), we have

$$\rho_b = -\boldsymbol{\nabla} \cdot \mathbf{P} = 0, \tag{4.57}$$

so that there is no volume charge density (of bound charges) in the cylinder. At the ends of the cylinder, however, we have a surface density of bound charges given by

$$\sigma_b = \hat{\mathbf{n}} \cdot \mathbf{P} = \pm P. \tag{4.58}$$

Namely, the two ends of the cylinder will have equal, but opposite surface charge densities (because the normal vectors point in opposite directions at the two end surfaces). Thus, we can think of such an electret as the electrostatic equivalent of a bar magnet. For purposes of calculating the electric field, therefore, we can forget about the cylinder and consider the equivalent case of two charged disks separated by a distance of 2ℓ along the z-axis, with surface charge densities given by $\sigma_b = \pm P$.

We have already calculated the electric field for a single charged disk along the z-axis (see (2.29)) and it has the form

$$\mathbf{E}(z) = -2\pi\sigma \left(\frac{z}{\sqrt{R^2 + z^2}} - \text{sgn}(z) \right) \hat{\mathbf{z}}, \tag{4.59}$$

where σ denotes the surface charge density. Choosing $z = 0$ to lie at the midpoint between the two disks, we see that when two disks carrying opposite charge densities are present, for $|z| \geq \ell$, we have

$$\mathbf{E}(z) = -2\pi P \hat{\mathbf{z}} \left[\frac{z - \ell}{\sqrt{R^2 + (z - \ell)^2}} - \frac{z + \ell}{\sqrt{R^2 + (z + \ell)^2}} \right]$$

$$= 2\pi \mathbf{P} \left[\frac{z + \ell}{\sqrt{R^2 + (z + \ell)^2}} - \frac{z - \ell}{\sqrt{R^2 + (z - \ell)^2}} \right]. \tag{4.60}$$

Similarly, the electric field along the z-axis in the region between the two disks is obtained to be ($|z| \leq \ell$)

$$\mathbf{E}(z) = -2\pi P \hat{\mathbf{z}} \left[\left(-\frac{\ell - z}{\sqrt{R^2 + (\ell - z)^2}} + 1 \right) - \left(\frac{\ell + z}{\sqrt{R^2 + (\ell + z)^2}} - 1 \right) \right]$$

$$= 2\pi \mathbf{P} \left[-2 + \frac{z + \ell}{\sqrt{R^2 + (z + \ell)^2}} - \frac{z - \ell}{\sqrt{R^2 + (z - \ell)^2}} \right]. \tag{4.61}$$

We note that as $z \to \ell$, the two expressions give respectively

$$\mathbf{E}(\ell) = 2\pi \mathbf{P} \frac{2\ell}{\sqrt{R^2 + 4\ell^2}}, \qquad \mathbf{E}(\ell) = 2\pi \mathbf{P} \left[-2 + \frac{2\ell}{\sqrt{R^2 + 4\ell^2}} \right]. \tag{4.62}$$

This shows that the electric field is discontinuous across the surface, which is a consequence of the surface polarization charge density on the disk.

◄

4.3 Selected problems

1. Consider a dielectric sphere of radius R and permittivity ϵ placed in vacuum. A point charge q is located outside the sphere at a distance $r = d > R$. Determine the electric field due to this charge both inside and outside the sphere. (A solution using the method of "images" can be found in Am. J. Phys. **61** (1993) 39. However, you can solve the Laplace equation with the appropriate matching condition to determine the solution.)

2. Given a dielectric material and the relation between the polarization vector $\mathbf{P}$ and the bound volume and surface charge densities, namely,

$$\rho_b(\mathbf{r}) = -\boldsymbol{\nabla} \cdot \mathbf{P}(\mathbf{r}), \qquad \sigma_b(\mathbf{r}) = \hat{\mathbf{n}} \cdot \mathbf{P}(\mathbf{r}),$$

where $\hat{\mathbf{n}}$ denotes the unit vector normal to the surface at $\mathbf{r}$, prove the following relationship

$$\int_V \mathrm{d}^3 r\, \mathbf{P}(\mathbf{r}) = \int_V \mathrm{d}^3 r\, \mathbf{r}\, \rho_b(\mathbf{r}) + \int_S \mathrm{d}s\, \mathbf{r}\, \sigma_b(\mathbf{r}).$$

(Physically, this corresponds to the fact that the total polarization is equal to the total dipole moment of the system.)

3. Consider a dielectric sphere of radius R, which carries a bound charge density (in spherical coordinates)

$$\sigma_b = \alpha \cos\theta, \qquad \rho_b = 0,$$

where α is a constant. There are no free charges either inside or outside the sphere. Determine the potential both inside and outside the sphere.

CHAPTER 5

Magnetostatics

5.1 Lorentz force

We have so far discussed only the nature of forces experienced by stationary charges in the presence of electric fields. In addition to such electrostatic forces a moving charged particle also experiences a force in the presence of a magnetic field of the form

$$\mathbf{F}_m = q\frac{\mathbf{v}}{c} \times \mathbf{B} = \frac{q}{c}\mathbf{v} \times \mathbf{B}, \tag{5.1}$$

where $\mathbf{v}$ represents the velocity of the moving charge and $\mathbf{B}$ is known as the magnetic induction vector or simply the magnetic field. The experimentally observed relation (5.1) is known as the Lorentz force law. Note that the magnetic force experienced by a charged particle is perpendicular to its velocity (as well as to the magnetic field) so that

$$\mathbf{v} \cdot \mathbf{F}_m = 0, \tag{5.2}$$

which shows that the magnetic force leads to no work.

There are a couple of things to note from the definition in (5.1). First, it is clear from the definition (5.1) that the unit of the magnetic field in the CGS system, which is Gauss, is the same as the unit of the electric field (compare with (1.8)), namely, $\frac{\text{stat}-\text{Volt}}{\text{cm}}$. This is, of course, the rationale behind the CGS system of units which is to recognize that the electric and the magnetic fields should be treated on an equal footing. (The unit of $\mathbf{B}$ in the MKS system is $\frac{\text{Weber}}{\text{m}^2}$ and $1\frac{\text{Weber}}{\text{m}^2} = 10^4$ Gauss.) The second thing to note is that even though the Lorentz force was originally observed as an empirical law it can also be derived from relativistic invariance which we will discuss later. The total force on a charged particle in the presence of both an electric and a magnetic field is, therefore, given by

$$\mathbf{F} = \mathbf{F}_e + \mathbf{F}_m = q\left(\mathbf{E} + \frac{1}{c}\mathbf{v} \times \mathbf{B}\right). \tag{5.3}$$

This can be taken as the defining relation for the magnetic field in the sense that the magnetic field can be determined from the velocity dependent part of the force experienced by a charged particle.

We have seen that the sources of electric fields are point charges (or monopole charges or charge distributions). Correspondingly, we can ask what are the sources of magnetic fields in nature. Experimentally we know that there are no magnetic monopoles (yet) in nature. The simplest sources of magnetic fields are known as magnetic dipoles. In fact, every magnet in nature including our own earth has two magnetic poles (conventionally called north and south poles) which cannot be separated into magnetic monopoles no matter how hard we try. Thus, the center piece (the basic element) in the study of magnetic phenomena is the magnet or the magnetic dipole. There are some substances in nature which have permanent magnetic moments (or are permanent magnets). One can study the properties of such materials and the effect of the magnetic fields, they produce, on charged particles. However, there is an alternate mechanism for producing magnetic fields which is what we will concern ourselves with. It was observed in a series of experiments by Oersted, Ampere etc. that a current produces a magnetic field. In fact, one can think of closed current loops as magnetic dipoles which is the approach we will take in studying magnetic phenomena.

5.2 Current

Figure 5.1: Charges with a net drift velocity leading to a current.

Charges in motion produce currents. In general, of course, charges inside a conductor may have random motion with a zero net drift velocity. However, if due to some external force charges move with a net drift velocity, then the flow leads to a current as shown in

Fig. 5.1. The current density in a conductor at a given point is simply defined to be the amount of charge crossing a unit cross-sectional area per unit time at that coordinate and is expressed as

$$\mathbf{J} = \rho\mathbf{v}, \tag{5.4}$$

where ρ is the volume charge density and $\mathbf{v}$ represents the net velocity of the charge flow. The current density may or may not be uniform. In either case, the current carried by a conductor (wire) is simply the total charge flowing across a given cross-section of the conductor per unit time

$$I = \int_S d\mathbf{s} \cdot \mathbf{J}. \tag{5.5}$$

For a thin wire, the variation of $\mathbf{J}$ over the cross-sectional area may not be appreciable in which case we can write

$$I = |\mathbf{J}|A, \tag{5.6}$$

where A represents the area parallel to the direction of the current. Most of our discussion would involve thin wires.

When there is a current flow, charge is moving from one section of the conductor to another. Therefore, any cross-sectional area separates a conductor into two parts and the charge has to pass through the cross-sectional area to go from one side to the other. If there is no sudden creation or destruction of charges we expect that charge in one region must decrease as a result of the flow of current out of the region through the cross-sectional area. Thus, drawing an arbitrary volume with the cross-sectional area as the boundary we see that

$$\frac{d}{dt}\int_V d^3r\,\rho = -\int_S d\mathbf{s}\cdot\mathbf{J},$$

$$\text{or,} \quad \int_V d^3r\,\frac{\partial\rho}{\partial t} = -\int_V d^3r\,(\boldsymbol{\nabla}\cdot\mathbf{J}),$$

$$\text{or,} \quad \int_V d^3r\left(\frac{\partial\rho}{\partial t} + \boldsymbol{\nabla}\cdot\mathbf{J}\right) = 0,$$

$$\text{or,} \quad \frac{\partial\rho}{\partial t} + (\boldsymbol{\nabla}\cdot\mathbf{J}) = 0, \tag{5.7}$$

where we have used Gauss' theorem as well as the fact that the integral identity must hold for any volume V and, consequently, the integrand itself must vanish. Equation (5.7) is known as the continuity equation and is one of the fundamental equations in the study of electrodynamics. It expresses the fact that electric charge is conserved which follows from gauge invariance of the theory as we will discuss later. (Incidentally, as we will see later the continuity equation is a relativistic equation, namely, it is invariant under Lorentz transformations.) In the study of magnetostatics we are interested in steady state currents for which

$$\frac{\partial \rho}{\partial t} = 0, \qquad \text{or,} \quad \boldsymbol{\nabla} \cdot \mathbf{J} = 0. \tag{5.8}$$

Namely, in such a case the system has reached equilibrium and as much charge enters a given volume as leaves through a cross-sectional area. In a later chapter, when we study time dependent phenomena we will also analyze currents that are not steady state.

5.3 Force on a current due to a magnetic field

Let us next try to understand some of the properties of current carrying conductors. Once we have understood the magnetic force experienced by a moving point charge (see (5.1)) it is not hard to derive the force experienced by a current in the presence of a magnetic field since a current represents a collection of charges in motion. Let us consider an element $d\boldsymbol{\ell}$ of a current carrying conductor (see Fig. 5.2). We can choose $d\boldsymbol{\ell}$ to lie along the direction of flow of the current so that $d\boldsymbol{\ell} \parallel \mathbf{v}$.

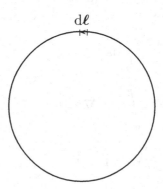

Figure 5.2: An infinitesimal element of a current carrying conductor in a magnetic field.

In this case, the Lorentz force experienced by the charges in the element $d\boldsymbol{\ell}$ is easily seen to be

$$d\mathbf{F} = \frac{\rho A |d\boldsymbol{\ell}|}{c}\,\mathbf{v} \times \mathbf{B} = \frac{1}{c} dV\,\mathbf{J} \times \mathbf{B} = \frac{I}{c} d\boldsymbol{\ell} \times \mathbf{B}, \qquad (5.9)$$

where the last relation is obtained with the assumption that if the cross-sectional area of the conductor (wire) is small, then the current density does not vary appreciably over this area. Integrating (5.9) we obtain the total force on a current carrying conductor due to a magnetic field to be of the form

$$\mathbf{F} = \frac{1}{c}\int d^3r\,\mathbf{J} \times \mathbf{B} = \frac{I}{c}\oint d\boldsymbol{\ell} \times \mathbf{B}. \qquad (5.10)$$

Note that for a constant $\mathbf{B}$ field (uniform) this integral vanishes since the line integral around a closed loop is zero so that a closed loop of wire carrying a current does not feel any force in a uniform magnetic field.

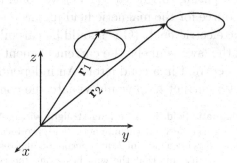

Figure 5.3: Magnetic force between two current carrying loops.

In separate experiments Ampere had discovered that two current carrying loops exert a magnetic force on each other (see Fig. 5.3). If the two loops carry steady currents I_1 and I_2 respectively, then the force on the first loop due to the second was known to be of the form

$$\mathbf{F}_1 = \frac{I_1 I_2}{c^2}\oint\oint \frac{d\boldsymbol{\ell}_1 \times (d\boldsymbol{\ell}_2 \times (\mathbf{r}_1 - \mathbf{r}_2))}{|\mathbf{r}_1 - \mathbf{r}_2|^3}. \qquad (5.11)$$

It is clear that we can write (5.11) in the form of a magnetic force

$$\mathbf{F}_1 = \frac{I_1}{c}\oint d\boldsymbol{\ell}_1 \times \mathbf{B}(\mathbf{r}_1), \qquad (5.12)$$

if we assume that the current in the second loop produces a magnetic field of the form

$$\mathbf{B}(\mathbf{r}) = \frac{I_2}{c} \oint \frac{d\boldsymbol{\ell}_2 \times (\mathbf{r} - \mathbf{r}_2)}{|\mathbf{r} - \mathbf{r}_2|^3}. \tag{5.13}$$

Infinitesimally, we can then write the magnetic field produced at the coordinate $\mathbf{r}$ by an element carrying a current I at a point $\mathbf{r}'$ as given by

$$d\mathbf{B}(\mathbf{r}) = \frac{I}{c} \frac{d\boldsymbol{\ell}' \times (\mathbf{r} - \mathbf{r}')}{|\mathbf{r} - \mathbf{r}'|^3} = \frac{1}{c} dV' \frac{\mathbf{J}(\mathbf{r}') \times (\mathbf{r} - \mathbf{r}')}{|\mathbf{r} - \mathbf{r}'|^3}. \tag{5.14}$$

The infinitesimal as well as the integrated forms of the magnetic field (see (5.13)),

$$\mathbf{B}(\mathbf{r}) = \frac{I}{c} \oint \frac{d\boldsymbol{\ell}' \times (\mathbf{r} - \mathbf{r}')}{|\mathbf{r} - \mathbf{r}'|^3} = \frac{1}{c} \int d^3 r' \frac{\mathbf{J}(\mathbf{r}') \times (\mathbf{r} - \mathbf{r}')}{|\mathbf{r} - \mathbf{r}'|^3}, \tag{5.15}$$

are known as the Biot-Savart law and are observed to hold experimentally. Comparing (5.15) with (1.11) we conclude that an electric current is a source for the magnetic field just as an electric charge is a source for the electric field. (One should be careful with the infinitesimal form of the law. Namely, the current element here is assumed to be a part of a current in a conductor. An independent element which is not part of a current loop would violate the continuity equation.)

▶ **Example (Magnetic field due to a long straight wire).** As an application of the Biot-Savart law let us determine the magnetic field, produced by an infinitely long straight thin wire carrying a current I, at a perpendicular distance r from the wire. For simplicity, let us assume that the wire lies along the z-axis as shown in Fig. 5.4 and that the base of the perpendicular from the point of observation on the z-axis defines the origin.

We know from (5.15) that the magnetic field produced by the current can be written as

$$\mathbf{B}(\mathbf{r}) = \frac{I}{c} \oint \frac{d\boldsymbol{\ell}' \times (\mathbf{r} - \mathbf{r}')}{|\mathbf{r} - \mathbf{r}'|^3}. \tag{5.16}$$

Furthermore, it is obvious from the geometry that since $d\boldsymbol{\ell}'$ is along the z-axis, the magnetic field lines would be along the polar angle $\hat{\phi}$, field lines forming circles surrounding the wire. This can be seen explicitly from the fact that, if we define

$$\mathbf{R} = \mathbf{r} - \mathbf{r}' = \mathbf{r} - z'\hat{z}, \tag{5.17}$$

then (remember that $d\boldsymbol{\ell}'$ is along the z-axis),

$$\hat{z} \times \mathbf{R} = \hat{z} \times (\mathbf{r} - z'\hat{z}) = \hat{z} \times \mathbf{r} = r\,\hat{\phi}. \tag{5.18}$$

The magnitude of the magnetic field can be obtained trivially as

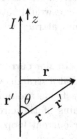

Figure 5.4: The magnetic field due to a current carrying long straight wire carrying current.

$$|\mathbf{B}| = \frac{I}{c} \int\limits_{-\infty}^{\infty} \frac{dz'\, r}{(r^2 + z'^2)^{\frac{3}{2}}} = \frac{2I}{c} \int\limits_{0}^{\infty} \frac{dz'\, r}{(r^2 + z'^2)^{\frac{3}{2}}}$$

$$= \frac{2Ir}{c} \times \frac{1}{r^2} \left. \frac{z'}{\sqrt{r^2 + z'^2}} \right|_{0}^{\infty} = \frac{2I}{cr}. \tag{5.19}$$

Thus, using (5.18) and (5.19), we can write the magnetic field produced by the current as

$$\mathbf{B}(\mathbf{r}) = \frac{2I}{cr}\, \hat{\phi}. \tag{5.20}$$

Incidentally, it is quite easy to see now that if there are two infinite parallel wires separated by a distance r and carrying currents I_1 and I_2 respectively along the same direction (say the z-axis), then there will be a force acting between the two. For, we can think of the current I_2 as producing a magnetic field which gives rise to a magnetic force on the wire with current I_1 and we can write (see (5.12))

$$\mathbf{F} = \frac{I_1}{c} \int d\boldsymbol{\ell} \times \mathbf{B}. \tag{5.21}$$

The magnetic field is along the polar direction $\hat{\phi}$ and the current (or $d\boldsymbol{\ell}$) is along the z-axis. Consequently, the force would be along the radial direction connecting the two wires and would be attractive. Namely, in cylindrical coordinates

$$\hat{\mathbf{r}} \times \hat{\phi} = \hat{\mathbf{z}}, \quad \hat{\phi} \times \hat{\mathbf{z}} = \hat{\mathbf{r}}, \quad \hat{\mathbf{z}} \times \hat{\mathbf{r}} = \hat{\phi}, \tag{5.22}$$

so that we can write the force (5.21) as

$$\mathbf{F} = -\hat{\mathbf{r}} \frac{I_1}{c} \frac{2I_2}{cr} \int\limits_{-L/2}^{L/2} dz = -\frac{2I_1 I_2 L}{c^2 r}\, \hat{\mathbf{r}}. \tag{5.23}$$

Here we have assumed the two wires to be of length L (each) which is to be taken to $L \to \infty$ at the end. Therefore, we obtain the force per unit length between the two currents to be

$$\frac{\mathbf{F}}{L} = -\frac{2I_1 I_2}{c^2 r}\, \hat{\mathbf{r}}. \tag{5.24}$$

The force is clearly attractive. However, if we reverse the direction of one of the currents, then, the direction of the force would reverse as well and leads to the familiar fact that two parallel currents attract while two anti-parallel currents repel each other.   ◄

▶ **Example (Magnetic field due to a circular current loop).** The important result to note from the previous example is that a straight wire carrying a steady current produces concentric circular magnetic fields around the axis of the wire, whose strength falls off inversely as the radius of the circle. Let us next analyze the magnetic field produced by a circular current loop of radius R as shown in Fig. 5.5. We assume that the the current loop is in a plane perpendicular to the z-axis and that the current moves in a clockwise direction when seen from below. The magnetic field due to this current at an arbitrary point is difficult to calculate. Therefore, we will calculate the magnetic field at any point on the z-axis which has a simpler form.

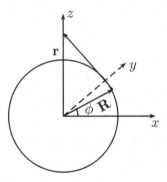

Figure 5.5: The magnetic field due to a circular current loop.

Once again, we use Biot-Savart law (5.15) and choosing the center of the current loop to be the origin of our coordinate system, we have

$$\mathbf{B}(\mathbf{r}) = \frac{I}{c} \oint \frac{d\boldsymbol{\ell} \times (\mathbf{r} - \mathbf{R})}{|\mathbf{r} - \mathbf{R}|^3}. \tag{5.25}$$

The simplest way to evaluate this is to note that ($d\boldsymbol{\ell}$ is orthogonal to $\mathbf{R}$)

$$d\boldsymbol{\ell} = R d\phi\, \hat{\boldsymbol{\phi}}, \quad \hat{\mathbf{R}} = \cos\phi\, \hat{\mathbf{x}} + \sin\phi\, \hat{\mathbf{y}},$$

$$\hat{\boldsymbol{\phi}} \times (\mathbf{r} - \mathbf{R}) = \hat{\boldsymbol{\phi}} \times (z\hat{\mathbf{z}} - R\hat{\mathbf{R}}) = z\hat{\mathbf{R}} + R\hat{\mathbf{z}} = z(\cos\phi\, \hat{\mathbf{x}} + \sin\phi\, \hat{\mathbf{y}}) + R\hat{\mathbf{z}}, \tag{5.26}$$

where we have used the usual rules for cross products, namely, $\hat{\boldsymbol{\phi}} \times \hat{\mathbf{z}} = \hat{\mathbf{R}}, \hat{\mathbf{R}} \times \hat{\boldsymbol{\phi}} = \hat{\mathbf{z}}$. It follows now that

$$\mathbf{B}(z) = \frac{I}{c} \int_0^{2\pi} \frac{R\, d\phi\, [z(\hat{\mathbf{x}}\cos\phi + \hat{\mathbf{y}}\sin\phi) + R\hat{\mathbf{z}}]}{(R^2 + z^2)^{\frac{3}{2}}}$$

$$= \frac{2\pi I R^2}{c(R^2 + z^2)^{\frac{3}{2}}}\, \hat{\mathbf{z}}. \tag{5.27}$$

Thus, along the axis of the loop, the magnetic field is completely parallel to the axis. At the center of the loop ($z = 0$), we note from (5.27) that the magnetic field has the value

$$\mathbf{B}(z = 0) = \frac{2\pi I}{cR}\,\hat{\mathbf{z}}, \qquad (5.28)$$

while very far away, namely, when $z \gg R$, we have

$$\mathbf{B}(z \gg R) \approx \frac{2\pi I R^2}{cz^3}\,\hat{\mathbf{z}}, \qquad (5.29)$$

which is reminiscent of the electric field due to a dipole. Thus, we suspect that a circular current somehow produces a magnetic field which has dipole characteristics. We will see this shortly, but let us note that this suggests that the magnitude of the magnetic dipole moment is proportional to the current times the area enclosed by the current loop.

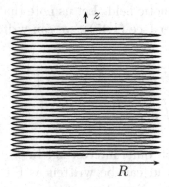

Figure 5.6: A long solenoid of radius R with n turns per unit length.

From (5.27), we can also calculate the magnetic field due to a long solenoid along its axis. Let us consider a solenoid of radius R with n turns per unit length with the z-axis representing the axis of the solenoid (see Fig. 5.6). In a length interval dz there will be $n\,dz$ loops of wire each producing a magnetic field as derived in (5.27). Consequently, the total magnetic field produced by the solenoid along the axis is given by

$$\begin{aligned}
\mathbf{B} &= \frac{2\pi n I R^2}{c}\,\hat{\mathbf{z}} \int_{-\infty}^{\infty} \frac{dz}{(R^2 + z^2)^{\frac{3}{2}}} \\
&= \frac{2\pi n I R^2}{c}\,\hat{\mathbf{z}} \times \frac{1}{R^2}\frac{z}{\sqrt{R^2 + z^2}}\Bigg|_{-\infty}^{\infty} \\
&= \frac{4\pi n I}{c}\,\hat{\mathbf{z}}. \qquad (5.30)
\end{aligned}$$

Namely, the magnetic field is a constant along the axis of an infinitely long solenoid determined completely by the current and the number of turns per unit length. In fact, even though we have not shown this, the magnetic field is really constant at any point inside the solenoid. Long solenoids are often used to produce and maintain constant magnetic fields over short distances.   ◄

5.4 Nature of the magnetic field

We have already seen that the Biot-Savart law (5.15) expresses the magnetic field in terms of the current density as

$$\mathbf{B}(\mathbf{r}) = \frac{1}{c} \int d^3r' \, \frac{\mathbf{J}(\mathbf{r}') \times (\mathbf{r} - \mathbf{r}')}{|\mathbf{r} - \mathbf{r}'|^3}, \tag{5.31}$$

which has a form similar to that of the electric field in electrostatics. In particular, this allows us to think of the current density (current) as the source of the magnetic field. Let us note that using the identities we have derived earlier (see (1.45)), we can write

$$\mathbf{B}(\mathbf{r}) = -\frac{1}{c} \int d^3r' \, \mathbf{J}(\mathbf{r}') \times \boldsymbol{\nabla} \left(\frac{1}{|\mathbf{r} - \mathbf{r}'|} \right)$$

$$= \frac{1}{c} \boldsymbol{\nabla} \times \int d^3r' \, \frac{\mathbf{J}(\mathbf{r}')}{|\mathbf{r} - \mathbf{r}'|}. \tag{5.32}$$

This relation is, in fact, quite interesting. In particular, it shows that since the magnetic field can be written as the curl of a vector its divergence must vanish. Namely, we have

$$\boldsymbol{\nabla} \cdot \mathbf{B}(\mathbf{r}) = 0. \tag{5.33}$$

Comparing this with Gauss' law in electrostatics, we see that this is equivalent to saying that there are no magnetic monopoles (or magnetic monopole charges).

Let us also recall the identity from vector calculus that

$$\boldsymbol{\nabla} \times (\boldsymbol{\nabla} \times \mathbf{A}(\mathbf{r})) = \boldsymbol{\nabla}(\boldsymbol{\nabla} \cdot \mathbf{A}) - \boldsymbol{\nabla}^2 \mathbf{A}. \tag{5.34}$$

Using this as well as (5.32), we now obtain,

$$\boldsymbol{\nabla} \times \mathbf{B}(\mathbf{r}) = \boldsymbol{\nabla} \times \boldsymbol{\nabla} \times \frac{1}{c} \int d^3r' \, \frac{\mathbf{J}(\mathbf{r}')}{|\mathbf{r} - \mathbf{r}'|}$$

$$= \frac{1}{c} \boldsymbol{\nabla} \int d^3r' \, \mathbf{J}(\mathbf{r}') \cdot \boldsymbol{\nabla} \left(\frac{1}{|\mathbf{r} - \mathbf{r}'|} \right)$$

$$- \frac{1}{c} \int d^3r' \, \mathbf{J}(\mathbf{r}') \boldsymbol{\nabla}^2 \left(\frac{1}{|\mathbf{r} - \mathbf{r}'|} \right)$$

$$= -\frac{1}{c} \nabla \int d^3r'\, \mathbf{J}(\mathbf{r}') \cdot \nabla' \left(\frac{1}{|\mathbf{r} - \mathbf{r}'|} \right)$$

$$+ \frac{4\pi}{c} \int d^3r'\, \mathbf{J}(\mathbf{r}') \delta^3(\mathbf{r} - \mathbf{r}')$$

$$= \frac{1}{c} \nabla \int d^3r' \left(\nabla' \cdot \mathbf{J}(\mathbf{r}') \right) \frac{1}{|\mathbf{r} - \mathbf{r}'|} + \frac{4\pi}{c} \mathbf{J}(\mathbf{r})$$

$$= \frac{4\pi}{c} \mathbf{J}(\mathbf{r}), \tag{5.35}$$

where we have used the fact that $\nabla \cdot \mathbf{J} = 0$ for steady currents (see (5.8)). (The surface terms are neglected with the assumption that we only have localized currents.)

Thus, we see that the two fundamental relations of magneto-statics are

$$\nabla \cdot \mathbf{B}(\mathbf{r}) = 0,$$

$$\nabla \times \mathbf{B}(\mathbf{r}) = \frac{4\pi}{c} \mathbf{J}(\mathbf{r}), \tag{5.36}$$

which should be compared with the laws of electrostatics in vacuum,

$$\nabla \cdot \mathbf{E}(\mathbf{r}) = 4\pi \rho(\mathbf{r}),$$

$$\nabla \times \mathbf{E}(\mathbf{r}) = 0. \tag{5.37}$$

Equations (5.36) describe the differential forms of the laws of magnetostatics and with the use of Gauss' and Stokes' theorems, we can obtain the integral representations for them as well. For example, we can write

$$\int_S d\mathbf{s} \cdot \mathbf{B}(\mathbf{r}) = \int_V d^3r\, (\nabla \cdot \mathbf{B}(\mathbf{r})) = 0,$$

$$\oint_C d\boldsymbol{\ell} \cdot \mathbf{B}(\mathbf{r}) = \int_S d\mathbf{s} \cdot (\nabla \times \mathbf{B}(\mathbf{r})) = \frac{4\pi}{c} \int_S d\mathbf{s} \cdot \mathbf{J}(\mathbf{r})$$

$$= \frac{4\pi I}{c}. \tag{5.38}$$

The second of these relations (in either the differential or the integral form) is known as Ampere's law and says that the line integral of the magnetic field around any closed loop is proportional to the current flowing through the cross-sectional area of the loop. It is useful in calculating magnetic fields for problems with symmetry, much like the Gauss' law in calculating electric fields.

► **Example (Magnetic field of a long straight wire).** As an application of Ampere's law, let us calculate the magnetic field due to an infinitely long straight wire carrying a current I along the z-axis as shown in Fig. 5.7. This problem has

Figure 5.7: The magnetic field due to a long straight wire carrying current.

enough symmetry so that to begin with, we know that the magnetic field at any point would be in the tangential direction to the circle drawn around the axis of the wire. Furthermore, the magnitude of the magnetic field would be the same at points (perpendicularly) equidistant from the wire. With this information, let us draw an "Amperian" loop of radius r around the axis of the wire (clockwise as seen from below). Then, according to Ampere's law (5.38), we have

$$\oint d\boldsymbol{\ell} \cdot \mathbf{B} = \frac{4\pi I}{c},$$

$$\text{or,} \quad |\mathbf{B}(\mathbf{r})| \, 2\pi r = \frac{4\pi I}{c},$$

$$\text{or,} \quad |\mathbf{B}(\mathbf{r})| = \frac{2I}{cr}, \tag{5.39}$$

and the magnetic field is along the direction of the polar angle. This is the result we had obtained earlier in (5.20) by explicitly evaluating the integral in the Biot-Savart law. ◄

5.5 Vector potential

In electrostatics we have noted that the electric field is conservative ($\nabla \times \mathbf{E} = 0$) so that we can write it as the (negative) gradient of a scalar potential which also follows from the Helmholtz theorem. In magnetostatics, on the other hand, we find that the magnetic field is not conservative in general. (If there is no current then there is no magnetic field.) Therefore, we do not expect to be able to write the magnetic field in terms of a scalar potential in general. On the other hand, we know that the magnetic field has vanishing divergence (see (5.36)). Consequently, in this case, we expect that we can write it as the curl of a vector and, in fact, we have already seen in (5.32) that this is true, namely,

$$
\begin{aligned}
\mathbf{B}(\mathbf{r}) &= \frac{1}{c} \int \mathrm{d}^3 r' \, \frac{\mathbf{J}(\mathbf{r}') \times (\mathbf{r} - \mathbf{r}')}{|\mathbf{r} - \mathbf{r}'|^3} \\
&= -\frac{1}{c} \int \mathrm{d}^3 r' \, \mathbf{J}(\mathbf{r}') \times \boldsymbol{\nabla} \left(\frac{1}{|\mathbf{r} - \mathbf{r}'|} \right) \\
&= \boldsymbol{\nabla} \times \frac{1}{c} \int \mathrm{d}^3 r' \, \frac{\mathbf{J}(\mathbf{r}')}{|\mathbf{r} - \mathbf{r}'|}.
\end{aligned}
\tag{5.40}
$$

Consequently, let us define

$$
\mathbf{B}(\mathbf{r}) \equiv \boldsymbol{\nabla} \times \mathbf{A}(\mathbf{r}),
\tag{5.41}
$$

where $\mathbf{A}$ is known as the vector potential (it is a vector) and in the particular case of (5.40), we readily identify that

$$
\mathbf{A}(\mathbf{r}) = \frac{1}{c} \int \mathrm{d}^3 r' \, \frac{\mathbf{J}(\mathbf{r}')}{|\mathbf{r} - \mathbf{r}'|}.
\tag{5.42}
$$

We note that writing $\mathbf{B}$ as the curl of a vector automatically satisfies the vanishing divergence equation in (5.36). It is also clear that the vector potential that gives rise to a given magnetic field cannot be unique unless further conditions are specified. Namely, both

$$
\mathbf{A}(\mathbf{r}) \quad \text{and} \quad \mathbf{A}(\mathbf{r}) + \boldsymbol{\nabla} \alpha(\mathbf{r}),
\tag{5.43}
$$

where $\alpha(\mathbf{r})$ is an arbitrary scalar function would give rise to the same magnetic field since the curl of a gradient vanishes. This is the first manifestation of what we would see later as the gauge invariance of Maxwell's equations. For the present, let us simply note that the vector potential $\mathbf{A}$ obtained in (5.42) appears to be unique only because it also satisfies the condition

$$
\begin{aligned}
\boldsymbol{\nabla} \cdot \mathbf{A}(\mathbf{r}) &= \frac{1}{c} \int \mathrm{d}^3 r' \, \mathbf{J}(\mathbf{r}') \cdot \boldsymbol{\nabla} \left(\frac{1}{|\mathbf{r} - \mathbf{r}'|} \right) \\
&= \frac{1}{c} \int \mathrm{d}^3 r' \, \left(\boldsymbol{\nabla}' \cdot \mathbf{J}(\mathbf{r}') \right) \frac{1}{|\mathbf{r} - \mathbf{r}'|} = 0,
\end{aligned}
\tag{5.44}
$$

which follows from the fact that the currents are steady state currents in magnetostatics (the surface terms arising from integration by parts are assumed to vanish for localized currents as in (5.35)). Such conditions, as we will see later, are called gauge conditions and are necessary when dealing with a system of equations which has gauge invariance.

▶ **Example (Vector potential of a long straight wire).** As an example of calculations of the vector potential, let us consider again the example of an infinitely long straight wire carrying a current I along the z-axis as shown in Fig. 5.8. The vector potential is defined in (5.42) to be

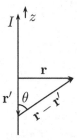

Figure 5.8: Vector potential for a long straight wire carrying current.

$$\mathbf{A}(\mathbf{r}) = \frac{1}{c} \int d^3r' \, \frac{\mathbf{J}(\mathbf{r}')}{|\mathbf{r} - \mathbf{r}'|} = \frac{I}{c} \int \frac{d\mathbf{r}'}{|\mathbf{r} - \mathbf{r}'|}. \tag{5.45}$$

In the present case, the current is along the z-axis. Consequently, only the z-component of the vector potential will be nonzero. If we define the perpendicular distance of a point from the wire as r, then the vector potential takes the form (here we assume that the point at which the field is being calculated lies in the $z = 0$ plane with the origin at the foot of the perpendicular to the wire)

$$\begin{aligned} A_z(\mathbf{r}) &= \frac{I}{c} \int_{-\infty}^{\infty} \frac{dz'}{\sqrt{r^2 + z'^2}} \\ &= \frac{2I}{c} \int_0^{\infty} \frac{dz'}{\sqrt{r^2 + z'^2}} \\ &= \frac{2I}{c} \log \left(\sqrt{r^2 + z'^2} + z' \right) \Big|_0^{\infty} \\ &= -\frac{2I}{c} \log r + \text{constant}, \end{aligned} \tag{5.46}$$

where we have used the standard integration formula (see Gradshteyn and Ryzhik 2.261). Note that the constant of integration is a divergent constant, much like the case of the scalar potential for an infinitely long wire carrying charge. As in the example in electrostatics the constant in (5.46) does not matter in the calculation of physical fields.

From the form of the vector potential (5.46), we obtain the magnetic field to be (in this case, $r = \sqrt{x^2 + y^2}$)

$$
\begin{aligned}
\mathbf{B}(\mathbf{r}) &= \nabla \times \mathbf{A}(\mathbf{r}) \\
&= \frac{\partial A_z}{\partial y}\,\hat{\mathbf{x}} - \frac{\partial A_z}{\partial x}\,\hat{\mathbf{y}} \\
&= \frac{2I}{c}\frac{(-y\,\hat{\mathbf{x}} + x\,\hat{\mathbf{y}})}{r^2} = \frac{2I}{c}\frac{r(-\sin\phi\,\hat{\mathbf{x}} + \cos\phi\,\hat{\mathbf{y}})}{r^2} \\
&= \frac{2I}{cr}\,\hat{\phi}.
\end{aligned} \tag{5.47}
$$

This is, of course, the result we had obtained earlier for the magnetic field by directly evaluating the integral in the Biot-Savart law (see (5.20)).   ◄

5.6 Multipole expansion

Let us consider a small circular current loop with the center at the origin of the coordinate system (see Fig. 5.9). We note from (5.42) that we can write the vector potential as

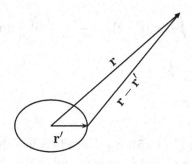

Figure 5.9: The vector potential for a small circular current loop.

$$
\mathbf{A}(\mathbf{r}) = \frac{1}{c}\int \mathrm{d}^3 r'\, \frac{\mathbf{J}(\mathbf{r}')}{|\mathbf{r} - \mathbf{r}'|} = \frac{I}{c}\oint \frac{\mathrm{d}\mathbf{r}'}{|\mathbf{r} - \mathbf{r}'|}, \tag{5.48}
$$

where we are assuming that the wire is thin so that we can assume the current density to be uniform. We note that, as in the case of electrostatics (see, for example, (2.40) and discussions there), we can expand the denominator in (5.48) for $r \gg r'$, so that the vector potential can be written in the form (θ denotes the angle between $\mathbf{r}$ and $\mathbf{r}'$)

$$\mathbf{A}(\mathbf{r}) = \frac{I}{c} \sum_{\ell=0}^{\infty} \frac{1}{r^{\ell+1}} P_\ell(\cos\theta) \oint d\mathbf{r}' \, r'^\ell$$

$$= \frac{I}{cr} \oint d\mathbf{r}' + \frac{I}{cr^2} P_1(\cos\theta) \oint d\mathbf{r}' \, r' + \cdots$$

$$= \frac{I}{cr^3} \oint d\mathbf{r}' \, (\mathbf{r}' \cdot \mathbf{r}) + \cdots, \tag{5.49}$$

where the first term vanishes because the line integral around a closed loop vanishes. Thus, there is no monopole term in the expansion. The dominant term at large distances, therefore, is the dipole term which can be simplified by using some of the vector identities. Note that ($\mathbf{r}$ is a fixed vector)

$$(d\mathbf{r}' \times \mathbf{r}') \times \mathbf{r} = -d\mathbf{r}' (\mathbf{r}' \cdot \mathbf{r}) + \mathbf{r}' (d\mathbf{r}' \cdot \mathbf{r})$$

$$= -2d\mathbf{r}' (\mathbf{r}' \cdot \mathbf{r}) + d(\mathbf{r}'(\mathbf{r}' \cdot \mathbf{r})),$$

$$\text{or,} \quad d\mathbf{r}' (\mathbf{r}' \cdot \mathbf{r}) = \frac{1}{2} d(\mathbf{r}'(\mathbf{r}' \cdot \mathbf{r})) - \frac{1}{2} (d\mathbf{r}' \times \mathbf{r}') \times \mathbf{r}, \tag{5.50}$$

so that keeping only the dipole term in (5.49) we have ($r = |\mathbf{r}|$)

$$\mathbf{A}(\mathbf{r}) \approx \frac{I}{2cr^3} \oint d(\mathbf{r}'(\mathbf{r}' \cdot \mathbf{r})) - \frac{I}{2cr^3} \oint (d\mathbf{r}' \times \mathbf{r}') \times \mathbf{r}$$

$$= -\frac{I}{2cr^3} \oint (d\mathbf{r}' \times \mathbf{r}') \times \mathbf{r}$$

$$= \frac{\mathbf{m} \times \mathbf{r}}{r^3}. \tag{5.51}$$

Here, the first term in (5.51) vanishes because the integral of a total derivative around a closed loop is trivial and we have defined

$$\mathbf{m} = -\frac{I}{2c} \oint d\mathbf{r}' \times \mathbf{r}' = \frac{I}{c} \mathbf{S}, \tag{5.52}$$

where $\mathbf{S}$ represents the area enclosed by the current loop (we do not use the symbol $\mathbf{A}$ since it is used for the vector potential). From the form of $\mathbf{A}$ in (5.51) it is suggestive that we can think of $\mathbf{m}$ as the magnetic dipole moment of the current loop which we will show next.

Once we know the vector potential, we can determine the magnetic field (for r large) to be

$$\mathbf{B}(\mathbf{r}) = \nabla \times \mathbf{A}(\mathbf{r})$$

$$\approx \boldsymbol{\nabla} \times \left(\frac{\mathbf{m} \times \mathbf{r}}{r^3} \right) = -\boldsymbol{\nabla} \times \left(\mathbf{m} \times \boldsymbol{\nabla} \left(\frac{1}{|\mathbf{r}|} \right) \right)$$

$$= -\mathbf{m} \boldsymbol{\nabla}^2 \left(\frac{1}{|\mathbf{r}|} \right) + \boldsymbol{\nabla} \left(\mathbf{m} \cdot \boldsymbol{\nabla} \left(\frac{1}{|\mathbf{r}|} \right) \right)$$

$$= -\boldsymbol{\nabla} \left(\frac{\mathbf{m} \cdot \mathbf{r}}{|\mathbf{r}|^3} \right) = -\frac{\mathbf{m}}{|\mathbf{r}|^3} + \frac{3(\mathbf{m} \cdot \mathbf{r})\mathbf{r}}{|\mathbf{r}|^5}$$

$$= \frac{3(\mathbf{m} \cdot \hat{\mathbf{r}})\hat{\mathbf{r}} - \mathbf{m}}{|\mathbf{r}|^3}, \tag{5.53}$$

which can be compared with the electric field obtained earlier in (2.52) for an electric dipole. (Note that the term $\boldsymbol{\nabla}^2 \left(\frac{1}{|\mathbf{r}|} \right) = -4\pi \delta^3(\mathbf{r})$ vanishes because we are considering points far away from the origin.) It is clear now that $\mathbf{m}$ can indeed be thought of as the magnetic dipole moment of the current loop. This also demonstrates that at large distances the current loop behaves like a magnetic dipole. Considering that there are no magnetic monopoles we see that the basic elements in the study of magnetic phenomena are, therefore, current loops.

5.7 Magnetization

As in the case of electric fields, we note that different materials in nature respond differently to an applied magnetic field and based on their response all materials can be classified broadly into three groups - diamagnetic, paramagnetic and ferromagnetic materials. The response of each of these three kinds of materials can be properly understood within the context of a quantum theory. However, without getting into technical details, let us simply note the broad features of such materials. As we have seen current loops generate a magnetic dipole moment. Every material, as we know, consists of atoms where electrons are moving in orbits and every such electron can be thought of as describing a current loop and, therefore, as producing a magnetic dipole moment. In fact, every atom may have several electrons moving in orbits in a random fashion so that the magnetic moments within an atom may cancel each other leading to a vanishing net magnetic dipole moment for the atom. In the presence of an external magnetic field, however, these individual magnetic moments can align themselves and the material can get magnetized. In fact, they align themselves in such a way as to oppose the external field. This is the basic behavior of a diamagnetic material. We can, of course, also have materials where the magnetic moments due to each of the electron currents in an atom do not quite cancel. In such a case, every atom in the material may have a net magnetic dipole moment. However, the magnetic moments of different atoms in the material may be randomly distributed, leading to a zero net magnetic moment for the material. In the presence of an external magnetic field, once again, the magnetic moments would orient themselves. However, in such a case, the magnetic moments align parallel to the external field and this is the behavior of paramagnetic materials. There are, of course, also ferromagnetic materials. Here, the atoms do have a net magnetic moment like paramagnetic materials. Furthermore, in such materials the atoms are quite close together and the magnetic moments are aligned so that inside the material there are domains with large magnetic moments which, however, are randomly distributed and can give rise to a net zero magnetic moment. In the presence of a magnetic field, however, they all align to give rise to a large magnetic moment which does not vanish even when the magnetic field is switched off. (This phenomenon is known as hysteresis.)

Just as in the case of dielectric materials, for a magnetic material we can define a magnetic dipole moment $\mathbf{M}$ per unit volume (analogous to polarization, see for example, (2.57) and discussions there), which is also known as the magnetization of the material. Experimentally, it is observed, for both diamagnetic and paramagnetic materials, that in the presence of an external magnetic field the magnetization is linearly related to the applied magnetic field so that we can write

$$\mathbf{M} \propto \mathbf{B}. \tag{5.54}$$

For ferromagnetic materials, on the other hand, there is no such simple linear relation. In our discussions, we will not be concerned with ferromagnetic materials at all.

Let us next calculate the magnetic field produced by a magnetized material. If $\mathbf{M}(\mathbf{r})$ represents the magnetic dipole moment per unit volume, then, we see from (5.51) that we can write the vector potential that would be associated with this as

$$\mathbf{A}(\mathbf{r}) = \int d^3 r' \, \frac{\mathbf{M}(\mathbf{r}') \times (\mathbf{r} - \mathbf{r}')}{|\mathbf{r} - \mathbf{r}'|^3} = -\int d^3 r' \, \mathbf{M}(\mathbf{r}') \times \boldsymbol{\nabla} \left(\frac{1}{|\mathbf{r} - \mathbf{r}'|} \right)$$

$$= \int d^3 r' \, \mathbf{M}(\mathbf{r}') \times \boldsymbol{\nabla}' \left(\frac{1}{|\mathbf{r} - \mathbf{r}'|} \right)$$

$$= -\int d^3r' \, \boldsymbol{\nabla}' \times \left(\frac{\mathbf{M}(\mathbf{r}')}{|\mathbf{r} - \mathbf{r}'|} \right) + \int d^3r' \, \frac{\boldsymbol{\nabla}' \times \mathbf{M}(\mathbf{r}')}{|\mathbf{r} - \mathbf{r}'|}$$

$$= \int_V d^3r' \, \frac{\boldsymbol{\nabla}' \times \mathbf{M}(\mathbf{r}')}{|\mathbf{r} - \mathbf{r}'|} - \int_S ds' \, \frac{\hat{\mathbf{n}} \times \mathbf{M}(\mathbf{r}')}{|\mathbf{r} - \mathbf{r}'|}, \tag{5.55}$$

where $\hat{\mathbf{n}}$ represents the unit vector normal to the surface and we have used the identity from vector calculus that for any arbitrary vector $\mathbf{A}$,

$$\int_V d^3r \, \boldsymbol{\nabla} \times \mathbf{A} = \int_S d\mathbf{s} \times \mathbf{A}. \tag{5.56}$$

The relation (5.55) is quite interesting, for it says that the vector potential produced by a magnetized material can be thought of as due to both a volume current density and a surface current density of the forms

$$\mathbf{J}(\mathbf{r}) = c \, \boldsymbol{\nabla} \times \mathbf{M}(\mathbf{r}),$$

$$\mathbf{J}_\sigma(\mathbf{r}) = -c \, \hat{\mathbf{n}} \times \mathbf{M}(\mathbf{r}), \tag{5.57}$$

and the surface current density is tangential to the surface.

Figure 5.10: A sketch of the infinitesimal current loops inside a material.

The current densities in (5.57) can be compared with the charge densities (2.61) or (4.4) which describe a polarized dielectric material. Furthermore, the existence of the volume and surface currents can be understood in the following way. There are many small current loops inside a material. If the current loops are not of the same strength, then inside the material, they would not cancel each other giving rise to a volume current density, while on the surface, of course, there will be a current density since there is nothing to cancel this (see Fig. 5.10 which shows a sketch of current loops in a magnetic material). Let us note that if the small current loops are all uniform, then the current density would cancel in the interior of the material, giving rise to a net vanishing volume current density. We also note from the structure of the surface current density in (5.57) that it vanishes when integrated over any surface, namely,

$$\int_S \mathrm{d}\mathbf{s}\cdot\mathbf{J}_\sigma = -c\int_S \mathrm{d}\mathbf{s}\cdot(\hat{\mathbf{n}}\times\mathbf{M}) = -c\int_S \mathrm{d}s\,\hat{\mathbf{n}}\cdot(\hat{\mathbf{n}}\times\mathbf{M}) = 0, \quad (5.58)$$

which follows from the fact that $\hat{\mathbf{n}}$ is normal to $\hat{\mathbf{n}}\times\mathbf{M}$.

The magnetic field produced by the magnetized material can now be easily obtained from (5.55) to be

$$\mathbf{B}(\mathbf{r}) = \boldsymbol{\nabla}\times\mathbf{A}(\mathbf{r})$$

$$= -\boldsymbol{\nabla}\times\int \mathrm{d}^3r'\,\mathbf{M}(\mathbf{r}')\times\boldsymbol{\nabla}\left(\frac{1}{|\mathbf{r}-\mathbf{r}'|}\right)$$

$$= -\int \mathrm{d}^3r'\left[\mathbf{M}(\mathbf{r}')\,\boldsymbol{\nabla}^2\left(\frac{1}{|\mathbf{r}-\mathbf{r}'|}\right)\right.$$

$$\left. -(\mathbf{M}(\mathbf{r}')\cdot\boldsymbol{\nabla})\boldsymbol{\nabla}\left(\frac{1}{|\mathbf{r}-\mathbf{r}'|}\right)\right]$$

$$= \int \mathrm{d}^3r'\,\mathbf{M}(\mathbf{r}')\,4\pi\delta^3(\mathbf{r}-\mathbf{r}') - \boldsymbol{\nabla}\int \mathrm{d}^3r'\,\frac{\mathbf{M}(\mathbf{r}')\cdot(\mathbf{r}-\mathbf{r}')}{|\mathbf{r}-\mathbf{r}'|^3}$$

$$= 4\pi\,\mathbf{M}(\mathbf{r}) - \boldsymbol{\nabla}\int \mathrm{d}^3r'\,\frac{\mathbf{M}(\mathbf{r}')\cdot(\mathbf{r}-\mathbf{r}')}{|\mathbf{r}-\mathbf{r}'|^3}. \quad (5.59)$$

Here, we have used various identities from vector calculus such as

$$\boldsymbol{\nabla}\times(\mathbf{A}\times\mathbf{C}) = \mathbf{A}(\boldsymbol{\nabla}\cdot\mathbf{C}) + (\mathbf{C}\cdot\boldsymbol{\nabla})\mathbf{A} - (\mathbf{A}\cdot\boldsymbol{\nabla})\mathbf{C} - \mathbf{C}(\boldsymbol{\nabla}\cdot\mathbf{A}),$$

$$\boldsymbol{\nabla}(\mathbf{A}\cdot\mathbf{C}) = (\mathbf{A}\cdot\boldsymbol{\nabla})\mathbf{C} + \mathbf{A}\times(\boldsymbol{\nabla}\times\mathbf{C}) + (\mathbf{C}\cdot\boldsymbol{\nabla})\mathbf{A}$$

$$+ \mathbf{C}\times(\boldsymbol{\nabla}\times\mathbf{A}). \quad (5.60)$$

Thus, we see that, in general, a magnetized material gives rise to a magnetic field which is a sum of two parts – the first is simply the magnetization of the material up to a multiplicative factor and the second is the gradient of a scalar. Let us also note from (5.57) and (5.59) that

$$\boldsymbol{\nabla}\times\mathbf{B} = 4\pi\,\boldsymbol{\nabla}\times\mathbf{M} = \frac{4\pi}{c}\,\mathbf{J}, \quad (5.61)$$

where $\mathbf{J}$ represents the volume current density due to the current loops in the magnetized material.

5.8 Magnetic field intensity

Let us now ask what will be the modifications in the laws of magneto-statics (5.36) in the presence of a magnetic material. The discussion is completely parallel to the case of the laws of electrostatics in the presence of a dielectric material. First, we note that when we have a magnetic material present we will have two kinds of currents, one that is maintained by batteries etc. and the other due to the internal motion of bound electrons inside the magnetic material. Thus, analogous to the case of the dielectrics, let us refer to them as the free and the bound currents respectively. In this case, Ampere's law would say that the magnetic field integrated around any closed curve will be related to the total current,

$$
\oint_C d\boldsymbol{\ell} \cdot \mathbf{B} = \frac{4\pi}{c}(I + I_b), \tag{5.62}
$$

where the right hand side represents the total current through the surface bounded by the curve, with I denoting the free current and I_b representing the current due to the bound electrons of the magnetic material completely in analogy with the case of electrostatics in the presence of a dielectric. We can, of course, write

$$
I_b = \int_S d\mathbf{s} \cdot \mathbf{J}_b, \tag{5.63}
$$

where S is the surface bounded by the closed contour C and $\mathbf{J}_b$ represents the current density due to the bound electrons. As we have seen, this consists of two parts, a volume density and a surface density. However, we also noted in (5.58) that the surface integral of the surface current density actually vanishes so that only the volume current density contributes to the current in (5.63). (The surface current density lies on the surface and is normal to the direction of the surface area.) Furthermore, we had identified the volume current density with $c\boldsymbol{\nabla} \times \mathbf{M}$ (see (5.57)). Consequently, we can write

$$
I_b = \int_S d\mathbf{s} \cdot \mathbf{J}_b = c \int_S d\mathbf{s} \cdot (\boldsymbol{\nabla} \times \mathbf{M}) = c \oint_C d\boldsymbol{\ell} \cdot \mathbf{M}. \tag{5.64}
$$

Using this in (5.62), we see that we can write

$$
\oint_C d\boldsymbol{\ell} \cdot (\mathbf{B} - 4\pi \mathbf{M}) = \frac{4\pi}{c} I,
$$

or, $\oint_C d\ell \cdot \mathbf{H} = \frac{4\pi}{c} I,$ (5.65)

where we have defined (analogous to the case of electrostatics) a new field

$$\mathbf{H} = \mathbf{B} - 4\pi \mathbf{M}.$$ (5.66)

The new field, $\mathbf{H}$, is known as the magnetic field intensity (also known simply as the magnetic field) and is the analog of the electric displacement field. Ampere's law (5.65), in the presence of a magnetic material, is written in terms of this field and the right hand side, in this case, involves only the free currents in a magnetic material. The differential form of Ampere's law can now be obtained using Stokes' theorem and takes the form

$$\nabla \times \mathbf{H} = \frac{4\pi}{c} \mathbf{J},$$ (5.67)

where $\mathbf{J}$ represents the free current density in the system. Thus, the laws of magnetostatics, in the presence of a magnetic material, take the forms

$$\nabla \cdot \mathbf{B} = 0,$$

$$\nabla \times \mathbf{H} = \frac{4\pi}{c} \mathbf{J},$$ (5.68)

where the fields are related as

$$\mathbf{H} = \mathbf{B} - 4\pi \mathbf{M}, \quad \text{or,} \quad \mathbf{B} = \mathbf{H} + 4\pi \mathbf{M}.$$ (5.69)

Since the magnetization is parallel to $\mathbf{B}$ all the vectors $\mathbf{B}, \mathbf{M}$, and $\mathbf{H}$ are parallel. Let us define

$$\mathbf{M} = \chi_m \mathbf{H}.$$ (5.70)

The constant of proportionality χ_m is known as the magnetic susceptibility of the material. For diamagnetic materials it is negative, while it is positive for paramagnetic materials. Furthermore, for both diamagnetic and paramagnetic materials, the magnitude of the magnetic susceptibility is quite small (of the order of $10^{-5} - 10^{-4}$) which should be compared with the electric susceptibility which is positive and is much larger in magnitude. Let us also note that with this definition, we can now write

$$\mathbf{B} = (1 + 4\pi\chi_m) \mathbf{H} = \mu \mathbf{H},$$ (5.71)

where we have identified

$$\mu = 1 + 4\pi\chi_m.$$ (5.72)

This is known as the permeability of a magnetic material.

5.9 Boundary condition

As we have seen, the laws of magnetostatics are given by

$$\nabla \cdot \mathbf{B} = 0,$$

$$\nabla \times \mathbf{H} = \frac{4\pi}{c} \mathbf{J}, \tag{5.73}$$

where $\mathbf{J}$ represents the free current density in the system. All the reference to any bound currents is now completely contained in the definition of the new field $\mathbf{H}$. If we have two different magnetic materials separated by a boundary surface, the magnetic fields in the two media have to satisfy some boundary conditions on the surface. These can be easily determined using the equations (5.73). (They can be derived much the same way as we did for the dielectrics. We do not repeat the discussion here and simply give the results.)

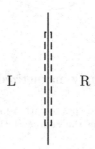

Figure 5.11: The boundary surface of two magnetic materials with distinct permeabilities with the dashed curve representing an Amperian loop.

It is clear from the vanishing of the divergence of $\mathbf{B}$ that the normal component of the $\mathbf{B}$ field must be continuous across the boundary. On the other hand, from Ampere's law in (5.73), we see that if there are free surface currents across a boundary, the tangential component of $\mathbf{H}$ must be discontinuous. We can write the boundary conditions explicitly as

$$B_{\mathrm{R,n}} = B_{\mathrm{L,n}},$$

$$H_{\mathrm{R,t}} = H_{\mathrm{L,t}} + \frac{4\pi}{c} J_{\sigma}, \tag{5.74}$$

where J_σ represents the magnitude of the free surface current density.

▶ **Example (Uniformly magnetized sphere).** With the boundary conditions for the magnetic fields across a surface determined in (5.74), the boundary value problems in magnetostatics can be solved much the same way as in electrostatics. For example, let us consider a sphere of magnetic material of radius R with a uniform magnetization along the z-axis (see Fig. 5.12) and we are interested in calculating the magnetic field produced by the sphere both inside and outside the sphere.

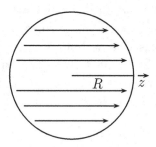

Figure 5.12: A uniformly magnetized sphere of radius R.

Since we have a magnetized sphere with a uniform magnetization along the z-axis, we can write (we assume the center of the sphere to coincide with the origin of the coordinate system)

$$\mathbf{M} = M\hat{\mathbf{z}} = M(\hat{\mathbf{r}}\cos\theta - \hat{\boldsymbol{\theta}}\sin\theta). \tag{5.75}$$

Furthermore, since there are no free currents present in either of the regions (namely, inside and outside the sphere), the equations (5.73) take the forms

$$\boldsymbol{\nabla} \cdot \mathbf{B} = 0 = \boldsymbol{\nabla} \times \mathbf{H}. \tag{5.76}$$

Since the curl of $\mathbf{H}$ vanishes in each of the regions, we can write it as the gradient of a scalar (magnetic) potential (analogous to the case of electrostatics) of the form

$$\mathbf{H}_< = -\boldsymbol{\nabla}\Phi_{m,<}, \quad \mathbf{H}_> = -\boldsymbol{\nabla}\Phi_{m,>}. \tag{5.77}$$

In each of the regions, we see from the divergence equation that we have to solve the Laplace equation in terms of the scalar potential (remember that $\mathbf{B}$ is parallel to $\mathbf{H}$).

We have already discussed the solutions of the Laplace equation in detail. Here we have a system with spherical symmetry. Thus, using spherical coordinates we note that, since there is no dependence on the azimuthal angle, we can write the well behaved solutions in the two regions to be (see (3.105))

$$\Phi_{m,<}(\mathbf{r}) = \sum_{\ell=0}^{\infty} A_\ell r^\ell P_\ell(\cos\theta),$$

$$\Phi_{m,>}(\mathbf{r}) = \sum_{\ell=0}^{\infty} C_\ell r^{-(\ell+1)} P_\ell(\cos\theta). \tag{5.78}$$

We can now match the boundary condition for the tangential components of the $\mathbf{H}$ fields. The tangential component is along the θ direction giving (we can also equivalently write this as $\Phi_{m,>}|_{r=R} - \Phi_{m,<}|_{r=R} = 0$ as in (4.50))

$$H_{>,\theta}|_{r=R} - H_{<,\theta}|_{r=R} = 0,$$

$$\text{or,} \quad \sum_{\ell=1}^{\infty} \left(C_\ell R^{-(\ell+2)} - A_\ell R^{(\ell-1)} \right) \frac{dP_\ell(\cos\theta)}{d\theta} = 0. \tag{5.79}$$

This determines

$$C_\ell = A_\ell R^{2l+1}, \quad \ell \geq 1. \tag{5.80}$$

The matching of the normal component of the $\mathbf{B}$ field can be done as follows. We recall the relation (5.69) between the $\mathbf{B}$ and the $\mathbf{H}$ fields. Since there is no magnetization outside the sphere, we can write (remember that $\mathbf{B} = \mathbf{H} + 4\pi\mathbf{M}$ and that $\mathbf{H} = -\boldsymbol{\nabla}\Phi_m$)

$$B_{>,n}|_{r=R} - B_{<,n}|_{r=R} = 0,$$

$$\text{or,} \quad H_{>,n}|_{r=R} - (H_{<,n} + 4\pi M_n)|_{r=R} = 0,$$

$$\text{or,} \quad \sum_{\ell=0}^{\infty} \left((\ell+1)C_\ell R^{-(\ell+2)} + \ell A_\ell R^{(\ell-1)} \right) P_\ell(\cos\theta) - 4\pi M \cos\theta = 0, \tag{5.81}$$

which (note that $P_0(\cos\theta) = 1, P_1(\cos\theta) = \cos\theta$) determines

$$(\ell+1)C_\ell = -\ell A_\ell R^{2\ell+1}, \quad \ell \neq 1,$$

$$2C_1 = -A_1 R^3 + 4\pi M R^3. \tag{5.82}$$

We can now compare the conditions in (5.80) and (5.82) to determine

$$C_\ell = 0 = A_\ell, \quad \text{for} \quad \ell \geq 2,$$

$$C_0 = 0,$$

$$A_1 = C_1 R^{-3} = \frac{4\pi M}{3}, \tag{5.83}$$

so that, we can write

$$\Phi_{m,>}(\mathbf{r}) = \frac{4\pi M}{3} \left(\frac{R}{r}\right)^3 r\cos\theta,$$

$$\Phi_{m,<}(\mathbf{r}) = \frac{4\pi M}{3} r\cos\theta. \tag{5.84}$$

Here we have used the fact that even though A_0 is an undetermined constant, a constant term in the potential does not influence the fields and accordingly we have chosen to set it to zero. It follows from this that

$$\mathbf{H}_>(\mathbf{r}) = -\boldsymbol{\nabla}\Phi_{m,>}(\mathbf{r}) = \frac{4\pi M}{3} \left(\frac{R}{r}\right)^3 (2\hat{\mathbf{r}}\cos\theta + \hat{\boldsymbol{\theta}}\sin\theta),$$

$$H_<(r) = -\nabla\Phi_<(r) = -\frac{4\pi M}{3}\,\hat{z}. \tag{5.85}$$

It is clear from this that outside the sphere, the magnetic field behaves like that of a magnetic dipole of moment (recall that $3(\hat{z}\cdot\hat{r})\hat{r} - \hat{z} = (2\hat{r}\cos\theta + \hat{\theta}\sin\theta)$)

$$m = \frac{4\pi R^3}{3}\,M = \frac{4\pi R^3}{3}\,M\hat{z}, \quad H_>(r) = \frac{3(m\cdot\hat{r})\hat{r} - m}{r^3}, \tag{5.86}$$

as we would expect (see (5.53)). Inside the magnetic material, however, the magnetic field is anti-parallel to the magnetization M.   ◀

5.10 Faraday's Law

So far, we have discussed problems in electrostatics and magnetostatics. As we have seen, these phenomena are decoupled from each other. Namely, electrostatics is completely described by electric fields while magnetostatics involves only magnetic fields. As a result, one can study such phenomena independently. We have also seen that the work of Ampere, Biot and Savart as well as Oersted showed how a current can produce a magnetic field. Faraday, on the other hand, reasoned that the phenomenon may be reversible in the sense that a magnetic field can possibly produce an electric current as well. Thus, for example, it may be that a magnet placed inside a circular conducting loop would cause an electric current to flow in the loop. Thus, he tried to set up an experiment to study this phenomenon. The experiment was not successful. Namely, he did not find any steady current in the loop when a magnet is placed within the loop. On the other hand, he did observe that as the magnet is brought near the loop, there is a transient current set up in the loop. This led him to believe that it is not the magnetic field which is likely to produce a current, rather it is the change in the magnetic flux through the loop which may be responsible for setting up a current. Several careful experiments, primarily due to Faraday, led to the conclusion that this expectation is indeed true and quantitatively the relation describing this (which is actually due to Maxwell) has the form

$$\oint d\ell\cdot E = -\frac{1}{c}\frac{\partial}{\partial t}\int_S ds\cdot B, \tag{5.87}$$

which is known as the Faraday's law of induction. The presence of c, the speed of light, in the formula is simply understood on dimensional grounds as well as from the structure of the Lorentz force that we have seen earlier.

There are several things to note from this relation. First of all, we have seen in electrostatics that

$$\oint d\boldsymbol{\ell} \cdot \mathbf{E} = 0. \qquad (5.88)$$

Namely, the electric field in electrostatics is conservative. However, if we are dealing with time dependent phenomena, this is no longer true. In fact, in the presence of other forces (mechanical, chemical etc.), the electric field need not be conservative and the integral over a closed contour does not have to vanish. When the contour integral (in (5.88)) does not vanish, its value is known as the electromotive force which is responsible for driving currents in the circuit. Thus, Faraday's law says that when there is a change in the magnetic flux through a circuit the effect is to set up an electromotive force (emf) in the circuit in a direction which will oppose the change in the magnetic flux. Of course, the change in the magnetic flux can happen in several ways. First, the circuit itself may be moving (changing), or the current (in a different circuit) producing the magnetic field may be changing with time or even the magnetic field due to a magnet itself may be changing. If the circuit is fixed and is not changing with time, we can take the time derivative in Faraday's law inside the integral which allows us to write

$$\oint d\boldsymbol{\ell} \cdot \mathbf{E} = -\frac{1}{c} \int_S d\mathbf{s} \cdot \frac{\partial \mathbf{B}}{\partial t},$$

$$\text{or,} \quad \int_S d\mathbf{s} \cdot (\boldsymbol{\nabla} \times \mathbf{E}) = -\frac{1}{c} \int_S d\mathbf{s} \cdot \frac{\partial \mathbf{B}}{\partial t},$$

$$\text{or,} \quad \boldsymbol{\nabla} \times \mathbf{E} = -\frac{1}{c} \frac{\partial \mathbf{B}}{\partial t}. \qquad (5.89)$$

This is the differential form of Faraday's law and shows that when we have time dependent phenomena, electric and magnetic fields essentially become coupled.

5.11 Inductance

Let us consider a single circuit carrying a current I as shown in Fig. 5.13. As we have seen earlier, this will set up a magnetic field and, therefore, a magnetic flux through the area enclosed by the circuit itself. The flux would, of course, depend on the geometry of the circuit as well as on the current in the circuit. In fact, the magnetic field and, therefore, the magnetic flux would be linearly proportional to the current in the circuit (as we have seen from the Biot-Savart law (5.15)). If the circuit is fixed and is not changing with time, the magnetic field and, therefore, the magnetic flux, in such a case, can change with time only if the current changes with time.

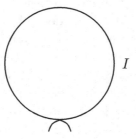

Figure 5.13: A circuit carrying a current I.

If the current changes with time leading to a time dependence in the magnetic flux, according to Faraday's law (5.87), this would produce an emf in the circuit itself leading to a modification of the current in the circuit. Namely, the circuit would act on itself in the following way

$$\text{emf} = \oint d\boldsymbol{\ell} \cdot \mathbf{E} = -\frac{1}{c} \frac{\partial}{\partial t} \int_{S} d\mathbf{s} \cdot \mathbf{B}$$

$$= -\frac{1}{c} \left(\frac{\partial}{\partial I} \int_{S} d\mathbf{s} \cdot \mathbf{B} \right) \frac{dI}{dt}$$

$$= -L \frac{dI}{dt}, \tag{5.90}$$

where we have defined

$$L = \frac{1}{c} \frac{\partial}{\partial I} \int_{S} d\mathbf{s} \cdot \mathbf{B}. \tag{5.91}$$

Here, we see that the parameter L is a property of the circuit (independent of the current if the relation between the magnetic flux and the current is linear) and is known as the self-inductance of the circuit, which determines how a changing current in a circuit acts on itself. Similarly, if we had a number of nearby circuits, each producing a time dependent magnetic flux in one another, that would define a different parameter M, known as the mutual inductance of the circuits, which describes the way different circuits act on one another. Namely, let us assume that there are n current carrying circuits and let us denote generically the magnetic flux as (Φ_m denotes the magnetic flux and not the scalar potential for $\mathbf{H}$ as we had done earlier in the example)

$$\Phi_m = \int_S d\mathbf{s} \cdot \mathbf{B}. \tag{5.92}$$

Then, denoting by $\Phi_{m,i}$ the total magnetic flux through the ith circuit due to the currents in all the circuits, we have

$$\Phi_{m,i} = \sum_{j=1}^{n} \Phi_{m,ij}. \tag{5.93}$$

Here $\Phi_{m,ij}$ represents the magnetic flux through the ith circuit due to the current in the jth circuit,

$$\Phi_{m,ij} = \int_{S_i} d\mathbf{s}_i \cdot \mathbf{B}_j. \tag{5.94}$$

It is clear now from Faraday's law that the emf in the ith circuit due to time dependent currents in the circuits takes the form (recall that $\Phi_{m,ij} \propto I_j$)

$$(\text{emf})_i = -\frac{1}{c} \frac{d\Phi_{m,i}}{dt}$$

$$= -\frac{1}{c} \sum_{j=1}^{n} \frac{\partial \Phi_{m,ij}}{\partial I_j} \frac{dI_j}{dt}$$

$$= -\sum_{j=1}^{n} M_{ij} \frac{dI_j}{dt}, \tag{5.95}$$

where the components M_{ij} represent parameters characteristic of the circuits. In particular, M_{ij} with $i \neq j$, represents the mutual inductance between the circuits i and j. On the other hand, $M_{ii} = L_i$ (no summation), represents the self-inductance of the ith circuit.

5.12 Selected problems

1. A proton of velocity 10^9cm/sec is projected at right angles to a uniform magnetic induction field of 10^3 Gauss.

a) What is the deflection in the path of the particle from a straight line after it has traversed a distance of 1cm?

b) How long would it take the proton to traverse a 90 degree arc?

2. We have seen that, for two wires carrying currents I_1 and I_2 respectively, the force on the first wire due to the current in the second wire is given by

$$\mathbf{F}_1 = \frac{I_1 I_2}{c^2} \oint \oint \frac{d\boldsymbol{\ell}_1 \times (d\boldsymbol{\ell}_2 \times (\mathbf{r}_1 - \mathbf{r}_2))}{|\mathbf{r}_1 - \mathbf{r}_2|^3}.$$

What is the force $\mathbf{F}_2$ on the second wire due to the current in the first wire? Show that these forces satisfy Newton's law, namely,

$$\mathbf{F}_1 + \mathbf{F}_2 = 0.$$

3. Consider two magnetic media with permeabilities μ_1 and μ_2, separated by a boundary surface without any free current. From the boundary conditions satisfied by the magnetic fields derive the "Snell's law" for the present case.

4. Consider a sphere of radius R and permeability μ, placed in a magnetic field in vacuum, which is initially uniform along the z-axis, namely, initially, $\mathbf{B} = B\hat{\mathbf{z}}$. Determine the magnetic field, in the presence of the sphere, both inside and outside, namely, for $r < R$ as well as for $r > R$.

CHAPTER 6

Maxwell's equations

6.1 Generalization of Ampere's law

As we have seen in the last chapter, in the presence of time dependent currents one of the laws of electrostatics changes and the electric field is no longer curl free (conservative). Similarly, Maxwell realized that Ampere's law must also modify in the presence of time dependent sources. In fact, there was no experimental basis for proposing a generalization at the time and yet Maxwell proposed his now famous modification completely from the theoretical consistency of the set of equations.

Let us consider the differential form of Ampere's law (5.65) which says that

$$\boldsymbol{\nabla} \times \mathbf{H} = \frac{4\pi}{c}\, \mathbf{J}. \tag{6.1}$$

From the fact that the divergence of a curl vanishes, (6.1) leads to

$$\boldsymbol{\nabla} \cdot (\boldsymbol{\nabla} \times \mathbf{H}) = \frac{4\pi}{c}\, \boldsymbol{\nabla} \cdot \mathbf{J},$$

$$\text{or,} \quad 0 = \frac{4\pi}{c}\, \boldsymbol{\nabla} \cdot \mathbf{J}. \tag{6.2}$$

Of course, this holds true in magnetostatics because we are interested in steady currents (see (5.8)). However, it is clear immediately from (6.2) that if we are considering a general, time dependent current (whose divergence need not vanish), then, Ampere's law (6.1) cannot hold in the present form since the left hand side is divergence free by definition while the right hand side is not in general.

Maxwell proposed to remove this inconsistency as follows. Let us recall that the continuity equation which describes conservation of electric charge has the form

$$\frac{\partial \rho}{\partial t} + \boldsymbol{\nabla} \cdot \mathbf{J} = 0, \tag{6.3}$$

and must hold for any time dependent sources. Furthermore, Gauss'
law has the general form

$$\nabla \cdot \mathbf{D} = 4\pi\rho. \tag{6.4}$$

Consequently, it is clear that if we generalize Ampere's law as

$$\nabla \times \mathbf{H} = \frac{4\pi}{c}\mathbf{J} + \frac{1}{c}\frac{\partial \mathbf{D}}{\partial t} = \frac{4\pi}{c}\left(\mathbf{J} + \frac{1}{4\pi}\frac{\partial \mathbf{D}}{\partial t}\right), \tag{6.5}$$

then, using (6.3) and (6.4), we obtain

$$\nabla \cdot (\nabla \times \mathbf{H}) = \frac{4\pi}{c}\left(\nabla \cdot \mathbf{J} + \frac{1}{4\pi}\frac{\partial(\nabla \cdot \mathbf{D})}{\partial t}\right)$$

$$= \frac{4\pi}{c}\left(\nabla \cdot \mathbf{J} + \frac{\partial\rho}{\partial t}\right) = 0, \tag{6.6}$$

for general time dependent sources. In particular, when the charge
density has no time dependence so that the current is steady state,
then (6.5) reduces to Ampere's law (6.1) which we know to be valid
for magnetostatics.

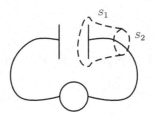

Figure 6.1: Capacitor plates filled with a dielectric and connected to
an alternating power source.

While the modification in (6.5) is quite clear from the theoretical
consistency of the equation, Maxwell also tried to envision an experimental set up where the necessity for such a term would naturally
arise. Consider a time dependent current charging the plates of a capacitor (see Fig. 6.1). (For example, we can think of an alternating
current source connected to the capacitor as a different set up.) Furthermore, let us suppose that the space between the capacitor plates
is filled with a dielectric material. Then, from Ampere's law (6.1) we

know that

$$\oint d\boldsymbol{\ell} \cdot \mathbf{H} = \int_S ds \cdot (\boldsymbol{\nabla} \times \mathbf{H}) = \frac{4\pi}{c} \int_S ds \cdot \mathbf{J}. \tag{6.7}$$

For the same closed contour, if we choose two different surfaces S_1 and S_2 (see Fig. 6.1) – one enclosing one of the capacitor plates and the other without – then, it is clear that we obtain respectively

$$\oint d\boldsymbol{\ell} \cdot \mathbf{H} = \int_{S_1} ds \cdot (\boldsymbol{\nabla} \times \mathbf{H}) = \frac{4\pi}{c} \int_{S_1} ds \cdot \mathbf{J} = 0,$$

$$\oint d\boldsymbol{\ell} \cdot \mathbf{H} = \int_{S_2} ds \cdot (\boldsymbol{\nabla} \times \mathbf{H}) = \frac{4\pi}{c} \int_{S_2} ds \cdot \mathbf{J} = \frac{4\pi I}{c}. \tag{6.8}$$

The first relation follows from the fact that there is no conduction current within the dielectric. The two relations in (6.8) are, therefore, inconsistent. On the other hand, the additional term in (6.5) will remove this inconsistency since inside the dielectric there is a displacement field (even though there is no conduction current) whose time rate of change provides the effect of a current when integrated over a surface, namely, (here we will assume the surface to be closed)

$$\frac{1}{4\pi} \int_S ds \cdot \frac{\partial \mathbf{D}}{\partial t} = \frac{\partial}{\partial t} \left(\frac{1}{4\pi} \int_S ds \cdot \mathbf{D} \right)$$

$$= \frac{d}{dt} \left(\frac{1}{4\pi} \int_V d^3x \, \boldsymbol{\nabla} \cdot \mathbf{D} \right) = \frac{d}{dt} \int_V d^3x \, \rho$$

$$= \frac{dQ}{dt} = I_D. \tag{6.9}$$

We know that a current produces a magnetic field and so, in keeping with this concept Maxwell identified the new term added to the right hand side of Ampere's law with a current (since it contributes to the magnetic field) known as the displacement current,

$$\mathbf{J}_D = \frac{1}{4\pi} \frac{\partial \mathbf{D}}{\partial t}. \tag{6.10}$$

The simplest way to think of this current is to recall that in the presence of an applied field (including an alternating one) the charge centers in a dielectric are displaced leading to the effect of a current.

Such currents are, however, different from the usual conduction currents that we are used to, since the charges never really leave the nucleus (or the atom). They are known as displacement currents since they arise from a displacement of charge centers. (Sometimes, they are also known as polarization currents. Actually, during the time of Maxwell, it was believed that space is filled with ether which acts like a dielectric and Maxwell himself believed that this effect should arise in ether as well.) Maxwell's proposal (6.5), of course, had a purely theoretical origin and the experimental verification of this would not come until two decades later in the experiments of Hertz. The main difficulty in the experimental verification lies in the fact that in a conductor, where we know how to measure a current, the conduction current is overwhelmingly dominant over the displacement current unless the frequency of the time dependent current is extremely high. However, Hertz's experiments clearly demonstrated the existence of a displacement current in dielectrics and the validity of Maxwell's modification of Ampere's law in (6.5).

Together with this modification, we can write all the laws of electricity and magnetism for general time dependent fields (and sources) as

$$\boldsymbol{\nabla} \cdot \mathbf{D} = 4\pi\rho,$$

$$\boldsymbol{\nabla} \cdot \mathbf{B} = 0,$$

$$\boldsymbol{\nabla} \times \mathbf{E} = -\frac{1}{c}\frac{\partial \mathbf{B}}{\partial t},$$

$$\boldsymbol{\nabla} \times \mathbf{H} = \frac{4\pi}{c}\mathbf{J} + \frac{1}{c}\frac{\partial \mathbf{D}}{\partial t} = \frac{4\pi}{c}\left(\mathbf{J} + \mathbf{J}_D\right). \tag{6.11}$$

These are the fundamental laws of electrodynamics and they are known as Maxwell's equations. They hold for both time dependent as well as time independent fields and sources. As we see, these are coupled differential equations (incidentally, one can also write the integral forms for these equations using Gauss' and Stokes' theorems, but we would not go into this), which can be checked to be self-consistent and which become decoupled in the static limit. The equations (6.11), of course, have to be supplemented further by the continuity equation (6.3) as well as various other equations describing the effects of the medium, namely,

$$\frac{\partial \rho}{\partial t} + \boldsymbol{\nabla} \cdot \mathbf{J} = 0,$$

$$\mathbf{J} = \sigma\mathbf{E},$$

$$\mathbf{F} = \begin{cases} q\left(\mathbf{E} + \frac{1}{c}\mathbf{v} \times \mathbf{B}\right) & \text{for point charges,} \\ \rho\mathbf{E} + \frac{1}{c}\mathbf{J} \times \mathbf{B} & \text{for continuous distributions.} \end{cases} \tag{6.12}$$

Here, we note that the second of the supplementary equations in (6.12) is simply Ohm's law with σ representing the conductivity of the conductor, while the third describes the Lorentz force law (for continuous distributions, the third condition corresponds to the force density).

6.2 Plane wave solution

Let us next consider Maxwell's equations in an isotropic and homogeneous dielectric medium of infinite extent. In such a case, we can write

$$\mathbf{D} = \epsilon\mathbf{E}, \qquad \mathbf{B} = \mu\mathbf{H}, \tag{6.13}$$

in the entire space. Let us further assume that there are no free charges or currents present in the medium in which case the set of four Maxwell's equations (6.11) takes the form,

$$\boldsymbol{\nabla} \cdot \mathbf{E} = 0,$$

$$\boldsymbol{\nabla} \cdot \mathbf{B} = 0,$$

$$\boldsymbol{\nabla} \times \mathbf{E} = -\frac{1}{c}\frac{\partial \mathbf{B}}{\partial t},$$

$$\boldsymbol{\nabla} \times \mathbf{B} = \frac{\epsilon\mu}{c}\frac{\partial \mathbf{E}}{\partial t}. \tag{6.14}$$

Let us note the vector identity

$$\boldsymbol{\nabla} \times (\boldsymbol{\nabla} \times \mathbf{A}) = \boldsymbol{\nabla}(\boldsymbol{\nabla} \cdot \mathbf{A}) - \nabla^2\mathbf{A}, \tag{6.15}$$

which holds for any arbitrary vector $\mathbf{A}$. Using this, we obtain

$$\boldsymbol{\nabla} \times (\boldsymbol{\nabla} \times \mathbf{E}) = -\frac{1}{c}\frac{\partial(\boldsymbol{\nabla} \times \mathbf{B})}{\partial t},$$

$$\text{or,} \quad \boldsymbol{\nabla}(\boldsymbol{\nabla} \cdot \mathbf{E}) - \nabla^2\mathbf{E} = -\frac{\epsilon\mu}{c^2}\frac{\partial^2 \mathbf{E}}{\partial t^2},$$

$$\text{or,} \quad \nabla^2\mathbf{E} - \frac{\epsilon\mu}{c^2}\frac{\partial^2 \mathbf{E}}{\partial t^2} = 0. \tag{6.16}$$

Similarly, taking the curl of the equation for the magnetic field (the last equation in (6.14)), we obtain

$$\nabla^2\mathbf{B} - \frac{\epsilon\mu}{c^2}\frac{\partial^2 \mathbf{B}}{\partial t^2} = 0. \tag{6.17}$$

Equations (6.16) and (6.17) show that both the electric field as well as the magnetic field satisfy the three dimensional wave equation, with the velocity (speed) of propagation given by (see (3.48))

$$v = \frac{c}{\sqrt{\epsilon\mu}}. \tag{6.18}$$

In particular, since $\epsilon = 1 = \mu$ in vacuum we see that the speed of propagation of these waves in vacuum coincides with the speed of light. This was the first evidence that light waves arise because of time dependent electric and magnetic fields or that light waves are simply electromagnetic waves.

To further understand the behavior of these waves, let us consider for simplicity a plane wave solution for the electric field. A plane wave by definition is a wave where the wave variable has the same phase at any point on the wavefront which is an infinite plane. Thus, we note that a plane wave solution of the equation involving the electric field (6.16) will have the form

$$\mathbf{E}(\mathbf{x}, t) = \mathbf{E}^{(0)} e^{\mp i\omega t + i\mathbf{k}\cdot\mathbf{x}}, \tag{6.19}$$

where $\mathbf{E}^{(0)}$ is a constant vector, provided the parameters ω and $\mathbf{k}$ satisfy

$$\frac{\omega}{|\mathbf{k}|} = \frac{c}{\sqrt{\epsilon\mu}} = v, \tag{6.20}$$

which is precisely the relation satisfied by traveling waves. Conventionally, we say that $\omega = 2\pi\nu$ represents the angular frequency of the wave, while $|\mathbf{k}| = \frac{2\pi}{\lambda}$ with λ the wavelength denotes the wave number. This is seen by noting that the direction of propagation of the wave is along $\mathbf{k}$ and along that direction, points separated by a distance of λ (the wavelength) are in phase. With this identification, (6.20) leads to the familiar relation of wave phenomena, namely,

$$\nu\lambda = v = \frac{c}{\sqrt{\epsilon\mu}}. \tag{6.21}$$

As we have noted, we are analyzing plane wave solutions. The wavefronts (points where the phases are the same), in this case, are given by

$$\mp\omega t + \mathbf{k}\cdot\mathbf{x} = \text{constant}, \tag{6.22}$$

which, for a fixed time, correspond to planes of infinite extent satisfying

$$\mathbf{k}\cdot\mathbf{x} = \text{constant}. \tag{6.23}$$

The two independent solutions we have found are characteristic of the wave equation which is a second order differential equation. The solution, with the negative sign in the exponent for the first term in (6.19), represents a forward moving wave (namely, a wave moving along **k**) while the one with the positive sign is known as a backward moving wave. This can be seen as follows. The velocity of propagation of a wave is precisely the velocity with which wavefronts (namely, planes with constant phase) move. Thus, defining ξ to represent the component of **x** along **k** (namely, $\xi = \mathbf{x} \cdot \hat{\mathbf{k}}$), we have

$$\mp \omega t + |\mathbf{k}|\xi = \text{constant},$$

$$\text{or,} \quad \frac{d\xi}{dt} = \pm \frac{\omega}{|\mathbf{k}|} = \pm v, \tag{6.24}$$

showing that the velocity of propagation in one case is along **k** while it is in the opposite direction in the other case.

Let us emphasize here that the wave solutions that we have constructed in (6.19) are known as monochromatic plane waves since they involve only a single frequency ω (also known as harmonic waves). A monochromatic plane wave can consist of a linear superposition of both forward and backward moving waves of the same frequency. A general solution, on the other hand, would involve a sum (or an integral) over distinct frequencies as well, in which case it is not a monochromatic wave. The velocity of propagation for a monochromatic wave is known as the phase velocity, while for a wave packet consisting of distinct frequencies the velocity of propagation is known as the group velocity. The two velocities can, in general, be different. Furthermore, let us note that while the four Maxwell's equations (6.14) lead to the wave equations (6.16) and (6.17) (for **E** and **B** fields), the two wave equations are not equivalent to the set of four Maxwell's equations. (Namely, the solutions of (6.16) and (6.17) would not automatically satisfy all the equations in (6.14).) Therefore, a plane wave solution of Maxwell's equations has to satisfy further conditions. Thus, for example, we see from the first equation of (6.14)

$$\nabla \cdot \mathbf{E} = 0, \tag{6.25}$$

that the plane wave solution for the electric field (6.19) must satisfy

$$\mathbf{k} \cdot \mathbf{E} = 0. \tag{6.26}$$

Namely, the electric field must be orthogonal to the direction of propagation of the wave.

A similar analysis holds for the magnetic fields as well, leading to the fact that Maxwell's equations have plane wave solutions for both $\mathbf{E}$ and $\mathbf{B}$ fields of the forms

$$\mathbf{E}(\mathbf{x}, t) = \mathbf{E}^{(0)}\, e^{\mp i\omega t + i\mathbf{k}\cdot\mathbf{x}}, \quad \mathbf{B}(\mathbf{x}, t) = \mathbf{B}^{(0)}\, e^{\mp i\omega t + i\mathbf{k}\cdot\mathbf{x}}, \qquad (6.27)$$

subject to (6.20) and satisfying

$$\mathbf{k} \cdot \mathbf{E} = 0 = \mathbf{k} \cdot \mathbf{B}. \qquad (6.28)$$

Namely, both the electric and the magnetic fields are perpendicular to the direction of propagation of the wave which shows that electromagnetic waves are transverse waves (unlike sound waves which are longitudinal). It is also worth emphasizing here that both the electric and the magnetic fields are real quantities. In writing a solution in the form (6.27), the assumption is that the electric and the magnetic fields correspond to either the real or the imaginary parts of the complex solutions which would respectively give cosine or sine solutions. Furthermore, from the third equation in (6.14)

$$\boldsymbol{\nabla} \times \mathbf{E} = -\frac{1}{c}\frac{\partial \mathbf{B}}{\partial t}, \qquad (6.29)$$

we obtain for the forward moving wave,

$$\mathbf{k} \times \mathbf{E} = \frac{\omega}{c}\, \mathbf{B}. \qquad (6.30)$$

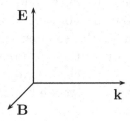

Figure 6.2: The electric and the magnetic fields as well as the direction of propagation as defining an orthogonal system.

In other words, the electric and the magnetic fields are not only perpendicular to the direction of propagation, but they are orthogonal to each other as well (see, for example, Fig. 6.2). It is also clear from

(6.20) and (6.30) that if the wave is propagating along the z-axis, then ($\mathbf{B} = \mu\mathbf{H}$),

$$\frac{|E_x|}{|H_y|} = \frac{|E_y|}{|H_x|} = \frac{\omega\mu}{c|\mathbf{k}|} = \frac{\mu}{\sqrt{\epsilon\mu}} = \sqrt{\frac{\mu}{\epsilon}}, \tag{6.31}$$

which is a property of the material under consideration. Such relations are quite important from the point of machine design.

Electromagnetic waves were originally identified with visible light simply because of the fact that the speed of propagation in vacuum coincides with the speed of light. However, experiments by Hertz showed that the electromagnetic waves are much more than just the visible light. In fact, the visible light forms only a small part of the spectrum of electromagnetic waves. Radio waves, microwaves etc. covering a wider range of frequencies are also governed by Maxwell's equations. The ratio of the speed of propagation of electromagnetic waves in vacuum to that in a dielectric medium is known as the index of refraction of the medium (a term borrowed from optics)

$$n = \frac{c}{v} = \sqrt{\epsilon\mu}, \tag{6.32}$$

which follows from (6.20). As is clear, the index of refraction is completely determined by the dielectric constant and the permeability of the medium. This relation is very well tested in radio waves. However, in the range of optical frequencies, one observes a variation in the value of the refractive index which is understood as follows. The dielectric constant of a medium is really not a constant. Rather, it depends on the frequency of the applied field which leads to significant changes in its value in the range of optical frequencies. Consequently, the index of refraction of a medium also becomes dependent on frequency, a phenomenon known as dispersion.

6.2.1 Polarization. We note that the plane wave solutions of Maxwell's equations (6.27) have two distinct aspects. The exponential characterizes the wave nature of the solution. However, the amplitude is a vector where information about other aspects of Maxwell's equations (such as transversality etc.) are contained. The vector amplitudes $\mathbf{E}^{(0)}$ and $\mathbf{B}^{(0)}$ can, of course, contain constant phases. The directional (vectorial) properties of electromagnetic waves are known as polarization and are completely determined from the structure of the electric field alone. (This is because the directional property of the magnetic field can be determined from a knowledge of the electric field and the direction of propagation, as is clear from (6.30).) For example, we know that the electric field is orthogonal to the direction of propagation of the wave. If we assume that the wave propagates along the z-axis, then it is clear that the electric field must lie in the $x - y$ plane, known as the plane of polarization for the wave. Thus, for a wave propagating along the z-axis, we can write (taking the real part to denote the electric field)

$$\mathbf{E}(z,t) = \text{Re}\left(\left(E_1^{(0)}\hat{\mathbf{x}} + E_2^{(0)}\hat{\mathbf{y}}\right)e^{-i\omega t + ikz}\right). \tag{6.33}$$

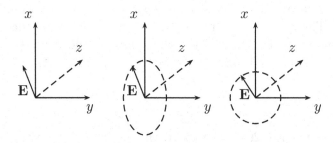

Figure 6.3: Linear, elliptic and circular polarizations of electromagnetic waves.

If we assume that the relative (constant) phase between $E_1^{(0)}$ and $E_2^{(0)}$ is zero, namely, that

$$E_1^{(0)} = |E_1^{(0)}|e^{i\phi}, \qquad E_2^{(0)} = |E_2^{(0)}|e^{i\phi}, \tag{6.34}$$

where ϕ is the constant phase of the two amplitudes, then we can write

$$\mathbf{E}(z,t) = \text{Re}\left(\left(|E_1^{(0)}|\hat{\mathbf{x}} + |E_2^{(0)}|\hat{\mathbf{y}}\right)e^{-i\omega t + ikz + i\phi}\right)$$

$$= \left(|E_1^{(0)}|\hat{\mathbf{x}} + |E_2^{(0)}|\hat{\mathbf{y}}\right)\cos(\omega t - kz - \phi). \tag{6.35}$$

It is clear from (6.35) that the magnitude of the electric field varies between zero and $\sqrt{|E_1^{(0)}|^2 + |E_2^{(0)}|^2}$ with time, but the direction of the vector always lies along $(|E_1^{(0)}|\hat{\mathbf{x}} + |E_2^{(0)}|\hat{\mathbf{y}})$ which is constant. In this case, we say that the wave is linearly polarized along this direction. On the other hand, if the (constant) relative phase between the two components of the electric field is arbitrary, then choosing the phase of the first component to be zero we have

$$\mathbf{E}(z,t) = \mathrm{Re}\left(|E_1^{(0)}|\hat{\mathbf{x}}\,e^{-i\omega t + ikz} + |E_2^{(0)}|\hat{\mathbf{y}}\,e^{-i\omega t + ikz + i\phi}\right)$$

$$= \left(|E_1^{(0)}|\hat{\mathbf{x}}\cos(\omega t - kz) + |E_2^{(0)}|\hat{\mathbf{y}}\cos(\omega t - kz - \phi)\right). \quad (6.36)$$

In this case, it is easy to see that not only the magnitude of the electric field, but also its direction changes with time tracing out an ellipse in the $x - y$ plane and we say that the wave is elliptically polarized. Finally, if $|E_1^{(0)}| = |E_2^{(0)}|$ and the magnitude of the relative phase between the two components is $\frac{\pi}{2}$, then we can write from (6.33)

$$\mathbf{E}(z,t) = \mathrm{Re}\left(|E_1^{(0)}|(\hat{\mathbf{x}} \pm i\hat{\mathbf{y}})e^{-i\omega t + ikz}\right)$$

$$= |E_1^{(0)}|\left(\hat{\mathbf{x}}\cos(\omega t - kz) \pm \hat{\mathbf{y}}\sin(\omega t - kz)\right). \quad (6.37)$$

In this case, the magnitude of the electric field is a constant ($|E_1^{(0)}|$), but the direction changes with time tracing out a circle in the $x - y$ plane and we say that the wave is circularly polarized. If the rotation of the electric field is clockwise to an observer facing the incoming wave, the wave is said to be right circularly polarized. For an opposite rotation, the wave is correspondingly known as left circularly polarized. (Namely, the two terms $\hat{\mathbf{x}} \pm i\hat{\mathbf{y}}$ in (6.37) denote respectively left and right circular polarizations, see Fig. 6.3.) The different polarizations are sketched in Fig. 6.3.

6.3 Boundary conditions

One of the great triumphs of Maxwell's equations is the prediction of electromagnetic waves which can be identified with light waves in the appropriate frequency range. It is, therefore, important to check whether the solutions of Maxwell's equations lead to familiar phenomena observed in optics such as reflection, refraction etc. To be able to study such phenomena, we have to first derive the boundary conditions which electric and magnetic fields have to satisfy in the time dependent case. When we have two distinct dielectric media separated by a boundary surface, then the boundary conditions which time dependent electric and magnetic fields would satisfy can be derived much the same way as we did for the static case. Let us assume that there are no free charges and currents present on the

boundary surface separating the two media. In such a case, the first
two of Maxwell's equations in (6.11)

$$\nabla \cdot \mathbf{D} = \nabla \cdot (\epsilon \mathbf{E}) = 0 = \nabla \cdot \mathbf{B}, \tag{6.38}$$

hold in the two dielectric media and tell us that the normal components of the electric displacement field and the magnetic field must
be continuous across the boundary. Specifically, we have

$$\epsilon_R E_{n,R} = \epsilon_L E_{n,L}, \qquad B_{n,R} = B_{n,L}. \tag{6.39}$$

The other two equations in (6.11), namely,

$$\nabla \times \mathbf{E} = -\frac{1}{c} \frac{\partial \mathbf{B}}{\partial t},$$

$$\nabla \times \mathbf{H} = \nabla \times \left(\frac{1}{\mu} \mathbf{B} \right) = \frac{1}{c} \frac{\partial \mathbf{D}}{\partial t}, \tag{6.40}$$

which are supposed to tell us about the continuity of the tangential
components of the electric and the magnetic fields appear slightly
tricky. However, the boundary conditions for these components can
also be derived in a simple manner.

L   R

Figure 6.4: A boundary surface separating two dielectric media.

Let us consider a rectangular closed loop of infinitesimal width
as shown in Fig. 6.4. Then, integrating the first equation in (6.40)
over the area enclosed by this closed loop, we have

$$\int_S d\mathbf{s} \cdot (\nabla \times \mathbf{E}) = -\frac{1}{c} \int_S d\mathbf{s} \cdot \frac{\partial \mathbf{B}}{\partial t},$$

or, $\qquad \oint d\boldsymbol{\ell} \cdot \mathbf{E} = -\dfrac{1}{c} \dfrac{\partial}{\partial t} \int\limits_{S} d\mathbf{s} \cdot \mathbf{B}.$ $\qquad\qquad\qquad$ (6.41)

In the limit of vanishing width of the rectangular loop, the right hand side of the equation vanishes because the area enclosed by the loop does. In the same limit, the left hand side simply gives the difference in the tangential components of the electric field on the two sides multiplied by the horizontal length of the curve. Therefore, in this limit, (6.41) leads to

$$\mathbf{E}_{t,R} = \mathbf{E}_{t,L}. \qquad\qquad (6.42)$$

In a similar manner, it is straightforward to show that the tangential components of the magnetic fields multiplied by the appropriate inverses of the permeabilities are also continuous across the boundary (namely, it is the tangential components of $\mathbf{H}$ which are continuous across the boundary, $\mathbf{H}_{t,R} = \mathbf{H}_{t,L}$),

$$\frac{1}{\mu_R} \mathbf{B}_{t,R} = \frac{1}{\mu_L} \mathbf{B}_{t,L}. \qquad\qquad (6.43)$$

With these boundary conditions, we are now ready to discuss the problem of reflection and refraction in a dielectric media.

▶ **Example (Normal incidence).** As a simple example, let us consider the question of reflection and refraction for an electromagnetic plane wave incident perpendicularly on the interface of two dielectric media. Thus, let us consider an incident wave moving along the z-axis, incident on the boundary surface between two (homogeneous and isotropic) dielectric media of infinite extent. We assume the boundary surface to be the plane $z = 0$ as in Fig. 6.5. In the region to the left (namely, for $z \leq 0$), we have a forward moving incident wave as well as a backward moving reflected wave, while in the region to the right (namely, for $z \geq 0$), we have only a forward moving transmitted wave. Thus, we can write ($k_L, k_R > 0$)

$$\mathbf{E}_L = \mathbf{E}^{inc} + \mathbf{E}^{Refl},$$

$$\mathbf{E}^{inc} = \operatorname{Re}\left(\mathbf{E}_1^{(0)} e^{-i\omega t + ik_L z}\right) = \mathbf{E}_1^{(0)} \cos(\omega t - k_L z),$$

$$\mathbf{E}^{Refl} = \operatorname{Re}\left(\mathbf{E}_2^{(0)} e^{i\omega t + ik_L z}\right) = \mathbf{E}_2^{(0)} \cos(\omega t + k_L z),$$

$$\mathbf{E}_R = \mathbf{E}^{trans} = \operatorname{Re}\left(\mathbf{E}_3^{(0)} e^{-i\omega t + ik_R z}\right) = \mathbf{E}_3^{(0)} \cos(\omega t - k_R z). \qquad (6.44)$$

Here, we are assuming that the vector amplitudes are real for simplicity. Furthermore, for these to represent solutions of Maxwell's equations in the two media with dielectric constants and permeabilities (ϵ_L, μ_L) and (ϵ_R, μ_R) respectively, we must have (see (6.20))

$$k_L = \frac{\omega \sqrt{\epsilon_L \mu_L}}{c} = n_L k, \qquad k_R = \frac{\omega \sqrt{\epsilon_R \mu_R}}{c} = n_R k, \qquad (6.45)$$

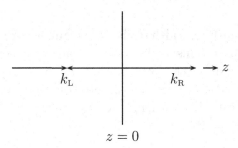

Figure 6.5: An electromagnetic wave at normal incidence on a surface separating two dielectric media.

where k represents the wave number in vacuum. (Incidentally, the fact that the waves are all moving along the z-axis follows from the fact that the tangential components of the fields have to be continuous across the boundary. The coplanarity of the waves is a general property that we will see in more detail in the next example where we consider reflection and refraction for oblique incidence.)

Furthermore, from (6.30)

$$\mathbf{k} \times \mathbf{E} = \frac{\omega}{c} \mathbf{B}, \tag{6.46}$$

as well as (6.45), we obtain

$$\mathbf{B}_{\mathrm{L}} = \mathbf{B}^{\mathrm{inc}} + \mathbf{B}^{\mathrm{Refl}},$$

$$\mathbf{B}^{\mathrm{inc}} = \sqrt{\epsilon_{\mathrm{L}}\mu_{\mathrm{L}}}\ \hat{\mathbf{z}} \times \mathbf{E}^{\mathrm{inc}} = \sqrt{\epsilon_{\mathrm{L}}\mu_{\mathrm{L}}}\ \hat{\mathbf{z}} \times \mathbf{E}_1^{(0)} \cos(\omega t - k_{\mathrm{L}} z),$$

$$\mathbf{B}^{\mathrm{Refl}} = -\sqrt{\epsilon_{\mathrm{L}}\mu_{\mathrm{L}}}\ \hat{\mathbf{z}} \times \mathbf{E}^{\mathrm{refl}} = -\sqrt{\epsilon_{\mathrm{L}}\mu_{\mathrm{L}}}\ \hat{\mathbf{z}} \times \mathbf{E}_2^{(0)} \cos(\omega t + k_{\mathrm{L}} z),$$

$$\mathbf{B}_{\mathrm{R}} = \mathbf{B}^{\mathrm{trans}} = \sqrt{\epsilon_{\mathrm{R}}\mu_{\mathrm{R}}}\ \hat{\mathbf{z}} \times \mathbf{E}^{\mathrm{trans}} = \sqrt{\epsilon_{\mathrm{R}}\mu_{\mathrm{R}}}\ \hat{\mathbf{z}} \times \mathbf{E}_3^{(0)} \cos(\omega t - k_{\mathrm{R}} z). \tag{6.47}$$

We note here that the electric and the magnetic fields are in the plane orthogonal to the direction of propagation which is along the z-axis. Consequently, there are no components of these fields normal to the boundary plane. Matching the tangential components of the electric and the magnetic fields at $z = 0$ (see (6.42) and (6.43)), we obtain from (6.44) and (6.47)

$$\mathbf{E}_1^{(0)} + \mathbf{E}_2^{(0)} = \mathbf{E}_3^{(0)},$$

$$\sqrt{\frac{\epsilon_{\mathrm{L}}}{\mu_{\mathrm{L}}}} \left(\mathbf{E}_1^{(0)} - \mathbf{E}_2^{(0)} \right) = \sqrt{\frac{\epsilon_{\mathrm{R}}}{\mu_{\mathrm{R}}}}\ \mathbf{E}_3^{(0)}. \tag{6.48}$$

We can solve for $\mathbf{E}_2^{(0)}$ and $\mathbf{E}_3^{(0)}$ from the two relations in (6.48) in terms of $\mathbf{E}_1^{(0)}$

and they have the forms

$$\mathbf{E}_2^{(0)} = \frac{\sqrt{\frac{\epsilon_L}{\mu_L}} - \sqrt{\frac{\epsilon_R}{\mu_R}}}{\sqrt{\frac{\epsilon_L}{\mu_L}} + \sqrt{\frac{\epsilon_R}{\mu_R}}} \, \mathbf{E}_1^{(0)},$$

$$\mathbf{E}_3^{(0)} = \frac{2\sqrt{\frac{\epsilon_L}{\mu_L}}}{\sqrt{\frac{\epsilon_L}{\mu_L}} + \sqrt{\frac{\epsilon_R}{\mu_R}}} \, \mathbf{E}_1^{(0)}. \tag{6.49}$$

For optically transparent materials, $\mu \approx 1$, in which case, we can write the refractive index of the material as (see (6.32))

$$n = \sqrt{\epsilon\mu} \approx \sqrt{\epsilon}. \tag{6.50}$$

Thus, for two optically transparent materials separated by an interface, (6.49) leads to

$$\mathbf{E}_2^{(0)} = \frac{n_L - n_R}{n_L + n_R} \, \mathbf{E}_1^{(0)},$$

$$\mathbf{E}_3^{(0)} = \frac{2n_L}{n_L + n_R} \, \mathbf{E}_1^{(0)}. \tag{6.51}$$

◀

▶ **Example (Oblique incidence).** Let us next consider the case where a plane electromagnetic wave is incident on a boundary surface separating two dielectric media at an oblique angle (see Fig. 6.6). We choose the boundary surface to be the plane $z = 0$ and assume that the wave is incident at an angle θ_i (with the z-axis). Without loss of generality, we can assume the plane of incidence to be the $x - z$ plane (plane of incidence is defined to be the plane containing the direction of propagation of the incident wave $\hat{\mathbf{k}}_i$ and the normal to the boundary surface $\hat{\mathbf{n}}$ which corresponds to $\hat{\mathbf{z}}$ in the present example). Thus, as before, we can write the incident wave to have the form (see (6.44))

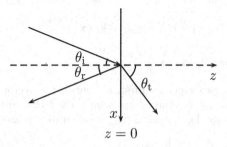

$$z = 0$$

Figure 6.6: An electromagnetic wave incident on the interface of two dielectric media at an oblique angle.

$$\mathbf{E}^{inc} = \mathbf{E}_1^{(0)} \cos(\omega t - \mathbf{k}_i \cdot \mathbf{x}), \tag{6.52}$$

where, as we have noted, $\hat{\mathbf{k}}_i$ denotes the direction of propagation of the incident wave which we have chosen to be in the $x - z$ plane so that it has the form (see Fig. 6.6)

$$\hat{\mathbf{k}}_i = (\sin \theta_i \, \hat{\mathbf{x}} + \cos \theta_i \, \hat{\mathbf{z}}). \tag{6.53}$$

As we have discussed earlier, there will also be a reflected wave in the first medium and a transmitted wave in the second, with the forms (see (6.44))

$$\mathbf{E}^{\text{refl}} = \mathbf{E}_2^{(0)} \cos(\omega t - \mathbf{k}_r \cdot \mathbf{x}),$$

$$\mathbf{E}^{\text{trans}} = \mathbf{E}_3^{(0)} \cos(\omega t - \mathbf{k}_t \cdot \mathbf{x}), \tag{6.54}$$

where $\hat{\mathbf{k}}_r$ and $\hat{\mathbf{k}}_t$ denote respectively the directions of propagation for the reflected as well as the transmitted waves. (The sign of the direction of propagation for the reflected wave has been absorbed into the definition of $\hat{\mathbf{k}}_r$.) The waves need not *a priori* be coplanar. However, we know that for them to satisfy Maxwell's equations, the magnitudes of the wave vectors must satisfy (see (6.20) and (6.32))

$$|\mathbf{k}_i| = |\mathbf{k}_r| = \frac{n_L \omega}{c} = n_L |\mathbf{k}|, \qquad |\mathbf{k}_t| = \frac{n_R \omega}{c} = n_R |\mathbf{k}|, \tag{6.55}$$

where we have defined $|\mathbf{k}|$ to represent the wave number in free space.

Given the electric fields, we can also determine the magnetic fields from (6.30) (or (6.46)) and we have

$$\mathbf{B}^{\text{inc}} = \frac{\mathbf{k}_i}{|\mathbf{k}|} \times \mathbf{E}^{\text{inc}} = n_L \hat{\mathbf{k}}_i \times \mathbf{E}^{\text{inc}},$$

$$\mathbf{B}^{\text{refl}} = \frac{\mathbf{k}_r}{|\mathbf{k}|} \times \mathbf{E}^{\text{refl}} = n_L \hat{\mathbf{k}}_r \times \mathbf{E}^{\text{refl}},$$

$$\mathbf{B}^{\text{trans}} = \frac{\mathbf{k}_t}{|\mathbf{k}|} \times \mathbf{E}^{\text{trans}} = n_R \hat{\mathbf{k}}_t \times \mathbf{E}^{\text{trans}}. \tag{6.56}$$

With these fields, we can now match the boundary conditions (6.39), (6.42) and (6.43). Let us note, for example, that the tangential components of the electric fields have to be continuous across the boundary $z = 0$, namely,

$$\left(\mathbf{E}_1^{(0)} \cos(\omega t - \mathbf{k}_i \cdot \mathbf{x}) + \mathbf{E}_2^{(0)} \cos(\omega t - \mathbf{k}_r \cdot \mathbf{x}) \right)_{z=0}^{\text{tang.}}$$

$$= \left(\mathbf{E}_3^{(0)} \cos(\omega t - \mathbf{k}_t \cdot \mathbf{x}) \right)_{z=0}^{\text{tang.}}. \tag{6.57}$$

Such a condition has two aspects. First, of course, the vector amplitudes have to satisfy some conditions, but more important is the fact that the phases should match as well. For example, matching the phases in (6.57) leads to

$$\mathbf{k}_i \cdot \mathbf{x}|_{z=0} = \mathbf{k}_r \cdot \mathbf{x}|_{z=0} = \mathbf{k}_t \cdot \mathbf{x}|_{z=0}. \tag{6.58}$$

From the fact that $\mathbf{k}_i$ lies in the $x - z$ plane, it now follows from (6.58) that all the three wave vectors must also lie in the $x - z$ plane. This can be seen simply as follows. The matching condition (6.58) explicitly gives

$$k_{ix} x = k_{rx} x + k_{ry} y = k_{tx} x + k_{ty} y, \tag{6.59}$$

which leads to the fact that $k_{ry} = 0 = k_{ty}$. In other words, all the three plane waves have to be coplanar (have to lie on the same plane). Therefore, from the geometry of the problem under study (see Fig. 6.6) we note that the unit vectors have the forms

$$\hat{\mathbf{k}}_i = (\sin\theta_i\,\hat{\mathbf{x}} + \cos\theta_i\,\hat{\mathbf{z}}),$$

$$\hat{\mathbf{k}}_r = (\sin\theta_r\,\hat{\mathbf{x}} - \cos\theta_r\,\hat{\mathbf{z}}),$$

$$\hat{\mathbf{k}}_t = (\sin\theta_t\,\hat{\mathbf{x}} + \cos\theta_t\,\hat{\mathbf{z}}),\tag{6.60}$$

where θ_i, θ_r and θ_t denote respectively the angles of incidence, reflection and transmission.

In addition to determining that the waves have to be coplanar, the matching of the phases at the boundary (6.59) also requires that

$$k_{ix} = k_{rx} = k_{tx}.\tag{6.61}$$

Upon using this in (6.55), it now follows that

$$k_{ix} = k_{rx},$$

$$\text{or,}\quad n_L|\mathbf{k}|\sin\theta_i = n_L|\mathbf{k}|\sin\theta_r,$$

$$\text{or,}\quad \theta_i = \theta_r,\tag{6.62}$$

and similarly,

$$k_{ix} = k_{tx},$$

$$\text{or,}\quad n_L|\mathbf{k}|\sin\theta_i = n_R|\mathbf{k}|\sin\theta_t,$$

$$\text{or,}\quad \frac{\sin\theta_i}{\sin\theta_t} = \frac{n_R}{n_L}.\tag{6.63}$$

Equation (6.62) describes the familiar law (from optics) that the angle of incidence is equal to the angle of reflection, while (6.63) is the Snell's law for refraction.

Let us next come to the vector amplitudes. There are two independent cases to analyze and let us start with the simple case where the electric field for the incident wave is normal to the plane of incidence, namely, let us assume that it lies along the y-axis (so that the electric field has no normal component). Then, it follows from the boundary condition (6.57) that the electric fields of all the three waves will lie along the y-axis and the boundary condition for the normal components of the displacement field will hold automatically. In this case, we have chosen an incident wave polarized along the y-axis. The vector amplitudes for the magnetic fields can now be easily calculated from (6.56) and (6.60)

$$\mathbf{B}_1^{(0)} = n_L E_1^{(0)}\hat{\mathbf{k}}_i \times \hat{\mathbf{y}} = n_L E_1^{(0)}(-\cos\theta_i\,\hat{\mathbf{x}} + \sin\theta_i\,\hat{\mathbf{z}}),$$

$$\mathbf{B}_2^{(0)} = n_L E_2^{(0)}\hat{\mathbf{k}}_r \times \hat{\mathbf{y}} = n_L E_2^{(0)}(\cos\theta_i\,\hat{\mathbf{x}} + \sin\theta_i\,\hat{\mathbf{z}}),$$

$$\mathbf{B}_3^{(0)} = n_R E_3^{(0)}\hat{\mathbf{k}}_t \times \hat{\mathbf{y}} = n_R E_3^{(0)}(-\cos\theta_t\,\hat{\mathbf{x}} + \sin\theta_t\,\hat{\mathbf{z}}).\tag{6.64}$$

Furthermore, since for most optically transparent material $\mu \approx 1$, we will use such a value. In such a case, matching the tangential components of the electric and the magnetic fields at $z = 0$, we obtain (the normal components of the magnetic fields are automatically continuous which can be seen using Snell's law)

$$E_1^{(0)} + E_2^{(0)} = E_3^{(0)},$$

$$n_L \left(E_1^{(0)} - E_2^{(0)} \right) \cos \theta_i = n_R \, E_3^{(0)} \cos \theta_t. \tag{6.65}$$

Using Snell's law (6.63), we can now solve for $E_2^{(0)}$ and $E_3^{(0)}$ in terms of $E_1^{(0)}$ and we obtain

$$\frac{E_2^{(0)}}{E_1^{(0)}} = \frac{n_L \cos \theta_i - n_R \cos \theta_t}{n_L \cos \theta_i + n_R \cos \theta_t} = \frac{\cos \theta_i - \sin \theta_i \cot \theta_t}{\cos \theta_i + \sin \theta_i \cot \theta_t}$$

$$= \frac{\tan \theta_t - \tan \theta_i}{\tan \theta_t + \tan \theta_i},$$

$$\frac{E_3^{(0)}}{E_1^{(0)}} = \frac{2 n_L \cos \theta_i}{n_L \cos \theta_i + n_R \cos \theta_t} = \frac{2 \cos \theta_i}{\cos \theta_i + \sin \theta_i \cot \theta_t}$$

$$= \frac{2 \tan \theta_t}{\tan \theta_t + \tan \theta_i}. \tag{6.66}$$

The other case to analyze is when the electric field is polarized parallel to the plane of incidence. Since the fields have to be perpendicular to the direction of propagation, we can choose (this choice also makes the normal components of the displacement field continuous across the boundary, namely, $n_L^2 \, \hat{\mathbf{z}} \cdot (\mathbf{E}_1^{(0)} + \mathbf{E}_2^{(0)}) = n_R^2 \, \hat{\mathbf{z}} \cdot \mathbf{E}_3^{(0)}$)

$$\mathbf{E}_1^{(0)} = E_1^{(0)}(\cos \theta_i \, \hat{\mathbf{x}} - \sin \theta_i \, \hat{\mathbf{z}}),$$

$$\mathbf{E}_2^{(0)} = -E_2^{(0)}(\cos \theta_i \, \hat{\mathbf{x}} + \sin \theta_i \, \hat{\mathbf{z}}),$$

$$\mathbf{E}_3^{(0)} = E_3^{(0)}(\cos \theta_t \, \hat{\mathbf{x}} - \sin \theta_t \, \hat{\mathbf{z}}), \tag{6.67}$$

which will give rise to the magnetic fields (see (6.56))

$$\mathbf{B}_1^{(0)} = n_L E_1^{(0)} \, \hat{\mathbf{y}},$$

$$\mathbf{B}_2^{(0)} = n_L E_2^{(0)} \, \hat{\mathbf{y}},$$

$$\mathbf{B}_3^{(0)} = n_R E_3^{(0)} \, \hat{\mathbf{y}}. \tag{6.68}$$

In this case, matching the tangential components of the electric and the magnetic fields across the boundary we obtain (we are assuming $\mu \approx 1$)

$$\left(E_1^{(0)} - E_2^{(0)} \right) \cos \theta_i = E_3^{(0)} \cos \theta_t,$$

$$n_L \left(E_1^{(0)} + E_2^{(0)} \right) = n_R E_3^{(0)}. \tag{6.69}$$

Once again, we can solve for $E_2^{(0)}$ and $E_3^{(0)}$ using Snell's law and we obtain

$$\frac{E_2^{(0)}}{E_1^{(0)}} = \frac{n_R \cos \theta_i - n_L \cos \theta_t}{n_R \cos \theta_i + n_L \cos \theta_t} = \frac{\frac{\sin \theta_i \cos \theta_i}{\sin \theta_t} - \cos \theta_t}{\frac{\sin \theta_i \cos \theta_i}{\sin \theta_t} + \cos \theta_t}$$

$$= \frac{\tan(\theta_i - \theta_t)}{\tan(\theta_i + \theta_t)},$$

$$\frac{E_3^{(0)}}{E_1^{(0)}} = \frac{2n_L \cos\theta_i}{n_R \cos\theta_i + n_L \cos\theta_t} = \frac{2\cos\theta_i}{\frac{\sin\theta_i \cos\theta_i}{\sin\theta_t} + \cos\theta_t}$$

$$= \frac{2\cos\theta_i \sin\theta_t}{\sin(\theta_i + \theta_t)\cos(\theta_i - \theta_t)}. \tag{6.70}$$

It is worth noting from (6.70) that when

$$\theta_i + \theta_t = \frac{\pi}{2}, \tag{6.71}$$

there is no reflected wave. In such a case, Snell's law gives

$$\frac{n_R}{n_L} = \frac{\sin\theta_i}{\sin\theta_t} = \tan\theta_i. \tag{6.72}$$

The incident angle for which this holds is known as the Brewster's angle. In general, of course, a wave can be decomposed into a sum of waves polarized parallel and perpendicular to the plane of incidence. What this analysis shows is that at the Brewster's angle of incidence, the component (of the field) polarized parallel to the plane will not be reflected and, consequently, the reflected wave will be polarized perpendicular to the plane of incidence.

The other important observation from all of the above analysis is that if $n_L > n_R$, then, from Snell's law (6.63) we obtain

$$\frac{\sin\theta_i}{\sin\theta_t} = \frac{n_R}{n_L},$$

or, $\theta_t > \theta_i.$ \hfill (6.73)

It follows, therefore, that there is some angle of incidence for which $\theta_t = \frac{\pi}{2}$. Let us call this θ_{int} so that we have from Snell's law

$$\sin\theta_{int} = \frac{n_R}{n_L}. \tag{6.74}$$

Let us note that for $\theta_i = \theta_{int}$, we have $\theta_t = \frac{\pi}{2}$ so that at this angle of incidence

$$\mathbf{E}^{trans} = \mathbf{E}_3^{(0)} \cos(\omega t - \mathbf{k}_t \cdot \mathbf{x})$$

$$= \mathbf{E}_3^{(0)} \cos(\omega t - |\mathbf{k}_t|(\sin\theta_t\, x + \cos\theta_t\, z)) = \mathbf{E}_3^{(0)} \cos(\omega t - |\mathbf{k}_t|x). \tag{6.75}$$

Namely, there is no z dependence in the transmitted wave. In other words, there is no transmitted wave in the $z \geq 0$ region independent of the polarization of the wave. Furthermore, if $\theta_i > \theta_{int}$, then

$$\cos\theta_t = \sqrt{1 - \sin^2\theta_t} = \sqrt{1 - \left(\frac{n_L}{n_R}\right)^2 \sin^2\theta_i} = \sqrt{1 - \frac{\sin^2\theta_i}{\sin^2\theta_{int}}}, \tag{6.76}$$

which becomes imaginary. As a result, the transmitted wave becomes exponentially damped with z and the wave propagates only along the x-axis. Such a wave is conventionally known as a surface wave.   ◄

6.4 Energy and the Poynting vector

We have already seen in the case of electrostatics how we can calculate the energy stored in an electric field configuration (see (1.85)). Let us next ask how we can determine the energy stored in a time dependent electromagnetic field. In order to do that let us start with the discussion of the static case. We have seen in electrostatics that the energy is given by (we are considering a dielectric medium of permittivity ϵ)

$$
\begin{aligned}
W_{\text{elec}} &= \int \mathrm{d}^3x\, w_{\text{elec}} = \frac{1}{2} \int \mathrm{d}^3x\, \rho(\mathbf{x})\Phi(\mathbf{x}) \\
&= \frac{1}{8\pi} \int \mathrm{d}^3x\, (\boldsymbol{\nabla} \cdot \mathbf{D}(\mathbf{x}))\Phi(\mathbf{x}) = -\frac{1}{8\pi} \int \mathrm{d}^3x\, \mathbf{D}(\mathbf{x}) \cdot \boldsymbol{\nabla}\Phi(\mathbf{x}) \\
&= \frac{1}{8\pi} \int \mathrm{d}^3x\, \mathbf{D} \cdot \mathbf{E} = \frac{1}{8\pi} \int \mathrm{d}^3x\, \epsilon\, \mathbf{E}^2.
\end{aligned} \tag{6.77}
$$

Here, we have used Gauss' law as well as integration by parts in the intermediate steps. Therefore, we can talk of an energy density stored in the static electric fields as given by

$$
w_{\text{elec}} = \frac{1}{8\pi} \mathbf{D} \cdot \mathbf{E} = \frac{\epsilon}{8\pi} \mathbf{E}^2. \tag{6.78}
$$

The derivation of the energy density stored in a magnetic field is a bit more involved. However, we can note the following analogy between the electrostatic and the magnetostatic cases to determine the energy density stored in the magnetic fields in a simple manner, namely,

$$
\rho(\mathbf{x}) \leftrightarrow \frac{\mathbf{J}(\mathbf{x})}{c}, \qquad \Phi(\mathbf{x}) \leftrightarrow \mathbf{A}(\mathbf{x}). \tag{6.79}
$$

With this analogy, we can intuitively determine the energy stored in static magnetic fields as

$$
\begin{aligned}
W_{\text{mag}} &= \int \mathrm{d}^3x\, w_{\text{mag}} = \frac{1}{2} \int \mathrm{d}^3x\, \frac{\mathbf{J}(\mathbf{x})}{c} \cdot \mathbf{A}(\mathbf{x}) \\
&= \frac{1}{8\pi} \int \mathrm{d}^3x\, (\boldsymbol{\nabla} \times \mathbf{H}(\mathbf{x})) \cdot \mathbf{A}(\mathbf{x}) \\
&= \frac{1}{8\pi} \int \mathrm{d}^3x\, \mathbf{H}(\mathbf{x}) \cdot (\boldsymbol{\nabla} \times \mathbf{A}(\mathbf{x})) \\
&= \frac{1}{8\pi} \int \mathrm{d}^3x\, \mathbf{H} \cdot \mathbf{B} = \frac{1}{8\pi} \int \mathrm{d}^3x\, \mu\, \mathbf{H}^2,
\end{aligned} \tag{6.80}
$$

so that, we can identify the energy density stored in the static magnetic field as

$$w_{\text{mag}} = \frac{1}{8\pi} \mathbf{H} \cdot \mathbf{B} = \frac{1}{8\pi} \mathbf{B} \cdot \mathbf{H} = \frac{\mu}{8\pi} \mathbf{H}^2. \tag{6.81}$$

Thus, we can write the total energy density stored in the electromagnetic field, in the static case, as given by

$$w = w_{\text{elec}} + w_{\text{mag}} = \frac{\mathbf{D} \cdot \mathbf{E} + \mathbf{B} \cdot \mathbf{H}}{8\pi}. \tag{6.82}$$

This also continues to be the energy density stored in the electromagnetic field in the time dependent case which can be seen as follows and which also brings out the concept of energy conservation within the context of electromagnetic phenomena.

Let us consider a volume V containing electromagnetic fields as well as sources (charges and currents). With time the energy stored in the electromagnetic fields inside the volume would decrease in two possible ways. First, there may be some dissipation of the energy density due to conversion into heat or other forms of mechanical energy. For example, a wire carrying current may heat up due to the resistance and in the process lose energy. Second, electromagnetic waves may leave the volume V carrying with them energy. Let us call the two kinds of energy losses as mechanical and radiation respectively. The mechanical energy loss is easy to calculate directly from the Lorentz force law. The rate at which the electromagnetic field does work on a charged particle moving with velocity $\mathbf{v}$ is given by

$$P_{\text{mech}} = \frac{\mathrm{d}W_{\text{mech}}}{\mathrm{d}t}$$

$$= \int_V \mathrm{d}^3x\, \mathbf{f} \cdot \mathbf{v} = \int_V \mathrm{d}^3x\, \rho \left(\mathbf{E} + \frac{1}{c} \mathbf{v} \times \mathbf{B} \right) \cdot \mathbf{v}$$

$$= \int_V \mathrm{d}^3x\, \mathbf{E} \cdot (\rho \mathbf{v}) = \int_V \mathrm{d}^3x\, \mathbf{E} \cdot \mathbf{J}. \tag{6.83}$$

Therefore, this denotes the magnitude of the rate at which energy stored in the electromagnetic fields is lost to other forms of energy. This can also be obtained from Ohm's law where we know that a wire carrying a current I and maintained at a potential difference $\mathcal{E}$ (the conventional terminology for this is, of course, V which we are

avoiding in order not to have any confusion with the volume V) loses
energy at the rate

$$P_{\text{mech}} = I\mathcal{E} = I \oint d\boldsymbol{\ell} \cdot \mathbf{E} = \int_V d^3x\, \mathbf{E} \cdot \mathbf{J}, \tag{6.84}$$

which is the result obtained in (6.83).

The second form of energy loss is determined as follows. Let us
note the vector identity

$$\boldsymbol{\nabla} \cdot (\mathbf{E} \times \mathbf{H}) = -\mathbf{E} \cdot (\boldsymbol{\nabla} \times \mathbf{H}) + \mathbf{H} \cdot (\boldsymbol{\nabla} \times \mathbf{E})$$

$$= -\mathbf{E} \cdot \left(\frac{4\pi}{c}\mathbf{J} + \frac{1}{c}\frac{\partial \mathbf{D}}{\partial t}\right) + \mathbf{H} \cdot \left(-\frac{1}{c}\frac{\partial \mathbf{B}}{\partial t}\right)$$

$$= -\frac{4\pi}{c}\left(\mathbf{E} \cdot \mathbf{J} + \frac{\partial}{\partial t}\frac{\mathbf{D} \cdot \mathbf{E} + \mathbf{B} \cdot \mathbf{H}}{8\pi}\right)$$

$$= -\frac{4\pi}{c}\left(\mathbf{E} \cdot \mathbf{J} + \frac{\partial w}{\partial t}\right). \tag{6.85}$$

Thus, defining a vector

$$\mathbf{S} = \frac{c}{4\pi}\mathbf{E} \times \mathbf{H}, \tag{6.86}$$

we see that we can write (6.85) as

$$\boldsymbol{\nabla} \cdot \mathbf{S} = -\left(\mathbf{E} \cdot \mathbf{J} + \frac{\partial w}{\partial t}\right),$$

or, $$\frac{\partial w}{\partial t} + \boldsymbol{\nabla} \cdot \mathbf{S} = -\mathbf{E} \cdot \mathbf{J}. \tag{6.87}$$

This indeed has the structure of a continuity equation describing
conservation of energy (see, for example, (6.3)) with the term on the
right hand side representing the dissipation of energy calculated in
(6.83). This shows that w calculated in (6.82) indeed corresponds to
the energy density for time dependent electromagnetic fields. In fact,
integrating this over the volume V, we obtain

$$\frac{dW}{dt} = -(P_{\text{mech}} + P_{\text{rad}}), \tag{6.88}$$

which shows that the rate at which energy is lost has two parts with
P_{rad} denoting the power carried out by the radiation fields and is
given by

$$P_{\text{rad}} = \int_V d^3x\, \boldsymbol{\nabla} \cdot \mathbf{S} = \int_S d\mathbf{s} \cdot \mathbf{S}. \tag{6.89}$$

The vector $\mathbf{S}$ is known as the Poynting vector and represents the density of power carried away by the radiation fields across the surface bounding the volume V. The "continuity" equation (6.87) indeed describes the conservation of energy when electromagnetic fields are involved.

In a similar manner, we can also determine the momentum associated with the electromagnetic fields. Let us consider a system of sources (particles, currents etc) as well as electromagnetic (radiation) fields in a given volume V. Let $\mathbf{p}_{\text{mech}}$ represent the momentum associated with the charge distribution of the system and $\mathbf{p}_{\text{rad}}$ the momentum associated with the radiation fields (electromagnetic fields). Then, from Newton's law as well as the Lorentz force law, we identify

$$
\frac{\mathrm{d}\mathbf{p}_{\text{mech}}}{\mathrm{d}t} = \mathbf{F} = \int_V \mathrm{d}^3x \left(\rho \mathbf{E} + \frac{1}{c} \mathbf{J} \times \mathbf{B} \right)
$$

$$
= \frac{1}{4\pi} \int_V \mathrm{d}^3x \left((\boldsymbol{\nabla} \cdot \mathbf{D})\mathbf{E} + \left(\boldsymbol{\nabla} \times \mathbf{H} - \frac{1}{c} \frac{\partial \mathbf{D}}{\partial t} \right) \times \mathbf{B} \right)
$$

$$
= \frac{1}{4\pi} \int_V \mathrm{d}^3x \left(\mathbf{E}(\boldsymbol{\nabla} \cdot \mathbf{D}) - \frac{1}{c} \frac{\partial (\mathbf{D} \times \mathbf{B})}{\partial t} \right.
$$

$$
\left. + \mathbf{D} \times \frac{1}{c} \frac{\partial \mathbf{B}}{\partial t} - \mathbf{B} \times (\boldsymbol{\nabla} \times \mathbf{H}) \right)
$$

$$
= \frac{1}{4\pi} \int_V \mathrm{d}^3x \left(\mathbf{E}(\boldsymbol{\nabla} \cdot \mathbf{D}) + \mathbf{H}(\boldsymbol{\nabla} \cdot \mathbf{B}) - \mathbf{D} \times (\boldsymbol{\nabla} \times \mathbf{E}) \right.
$$

$$
\left. - \mathbf{B} \times (\boldsymbol{\nabla} \times \mathbf{H}) - \frac{1}{c} \frac{\partial (\mathbf{D} \times \mathbf{B})}{\partial t} \right), \qquad (6.90)
$$

where we have added a term which gives a vanishing contribution since $\boldsymbol{\nabla} \cdot \mathbf{B} = 0$.

Thus, we can write

$$
\frac{\mathrm{d}\mathbf{p}_{\text{mech}}}{\mathrm{d}t} + \frac{\mathrm{d}\mathbf{p}_{\text{rad}}}{\mathrm{d}t}
$$

$$
= \frac{1}{4\pi} \int_V \mathrm{d}^3x \left(\mathbf{E}(\boldsymbol{\nabla} \cdot \mathbf{D}) + \mathbf{H}(\boldsymbol{\nabla} \cdot \mathbf{B}) - \mathbf{D} \times (\boldsymbol{\nabla} \times \mathbf{E}) \right.
$$

$$
\left. - \mathbf{B} \times (\boldsymbol{\nabla} \times \mathbf{H}) \right), \qquad (6.91)
$$

where we have identified

$$\mathbf{P}_{\text{rad}} = \frac{1}{4\pi c} \int_V d^3x \, (\mathbf{D} \times \mathbf{B}) = \frac{\epsilon\mu}{4\pi c} \int_V d^3x \, (\mathbf{E} \times \mathbf{H})$$

$$= \frac{\epsilon\mu}{c^2} \int_V d^3x \, \mathbf{S}. \tag{6.92}$$

Furthermore, we can simplify the integrand on the right hand side of (6.91) as follows. Let us denote the ith component of the integrand by I_i. The integrand is symmetric in the electric and the magnetic fields (namely, under $\mathbf{E} \leftrightarrow \mathbf{H}, \mathbf{D} \leftrightarrow \mathbf{B}$ or under $\mathbf{E} \leftrightarrow \mathbf{H}, \epsilon \leftrightarrow \mu$). Consequently, let us look at only the terms involving the electric fields and we have (recall the identity $\epsilon_{ijk}\epsilon_{k\ell m} = \delta_{i\ell}\delta_{jm} - \delta_{im}\delta_{j\ell}$),

$$I_i^E = \frac{1}{4\pi} \left(E_i \partial_j D_j - \epsilon_{ijk}\epsilon_{klm} D_j \partial_l E_m \right)$$

$$= \frac{1}{4\pi} \left(E_i \partial_j D_j - D_j \partial_i E_j + D_j \partial_j E_i \right)$$

$$= \frac{1}{4\pi} \left(\partial_j (E_i D_j) - \epsilon E_j \partial_i E_j \right)$$

$$= \frac{\epsilon}{4\pi} \left(\partial_j (E_i E_j) - \frac{1}{2} \partial_i \left(E_j^2 \right) \right)$$

$$= \partial_j \left(\frac{\epsilon}{4\pi} \left(E_i E_j - \frac{1}{2} \delta_{ij} E_k^2 \right) \right). \tag{6.93}$$

Adding the magnetic part as well, we see that the integrand (6.91) can be written as a total divergence

$$I_i = \partial_j \left(\frac{1}{4\pi} \left(\epsilon E_i E_j + \mu H_i H_j - \frac{1}{2} \delta_{ij} (\epsilon \mathbf{E}^2 + \mu \mathbf{H}^2) \right) \right)$$

$$= \partial_j T_{ij}, \tag{6.94}$$

where $T_{ij} = T_{ji}$ is known as the stress tensor (spatial components) for the radiation field (Maxwell field).

Using this result, we see that we can write the conservation equation for momentum (6.91) in components as

$$\left(\frac{d\mathbf{p}_{\text{mech}}}{dt} + \frac{d\mathbf{p}_{\text{rad}}}{dt} \right)_i = \int_V d^3x \, \partial_j T_{ij} = \int_S ds_j \, T_{ij}. \tag{6.95}$$

This is like a continuity equation and shows that the components of T_{ij} represent the density of momentum flux through the surface bounding the volume. In an analogous manner, we can derive the angular momentum associated with the radiation field as well.

To close this section, let us simply note that we are using complex notation to represent electric and magnetic fields with the understanding that the fields are real. Similarly, the Poynting vector

$$\mathbf{S} = \frac{c}{4\pi} \mathbf{E} \times \mathbf{H},$$

is understood to be defined in terms of the real electric and magnetic fields. However, if we are dealing with harmonic fields, then it is easy to see that the time averaged value of the Poynting vector can be represented in terms of complex fields as

$$\overline{\mathbf{S}} = \frac{c}{8\pi} \operatorname{Re} \left(\mathbf{E} \times \mathbf{H}^* \right), \tag{6.96}$$

which is quite useful in practical calculations as we will see later.

6.5 Gauge invariance of Maxwell's equations

The set of four Maxwell's equations are given by

$$\boldsymbol{\nabla} \cdot \mathbf{D} = 4\pi \rho,$$

$$\boldsymbol{\nabla} \cdot \mathbf{B} = 0,$$

$$\boldsymbol{\nabla} \times \mathbf{E} = -\frac{1}{c} \frac{\partial \mathbf{B}}{\partial t},$$

$$\boldsymbol{\nabla} \times \mathbf{H} = \frac{4\pi}{c} \mathbf{J} + \frac{1}{c} \frac{\partial \mathbf{D}}{\partial t}. \tag{6.97}$$

In the static case, we saw that we can write the electric and the magnetic fields in terms of scalar and vector potentials and the question is whether we can continue to do so for time dependent fields.

To analyze this, let us note that the second of Maxwell's equations implies that the magnetic field is divergence free. This can, of course, be solved as in the static case to give

$$\mathbf{B}(\mathbf{x}, t) = \boldsymbol{\nabla} \times \mathbf{A}(\mathbf{x}, t). \tag{6.98}$$

The only difference is that the vector potential, in the present case, would be a function of time as well. Putting this back into the third equation of Maxwell, we obtain

$$\boldsymbol{\nabla} \times \mathbf{E} = -\frac{1}{c} \frac{\partial \mathbf{B}}{\partial t} = -\frac{1}{c} \frac{\partial (\boldsymbol{\nabla} \times \mathbf{A})}{\partial t},$$

or, $\quad \mathbf{\nabla} \times \left(\mathbf{E} + \dfrac{1}{c} \dfrac{\partial \mathbf{A}}{\partial t} \right) = 0,$

or, $\quad \mathbf{E} + \dfrac{1}{c} \dfrac{\partial \mathbf{A}}{\partial t} = -\mathbf{\nabla}\Phi,$

or, $\quad \mathbf{E}(\mathbf{x}, t) = -\mathbf{\nabla}\Phi(\mathbf{x}, t) - \dfrac{1}{c} \dfrac{\partial \mathbf{A}(\mathbf{x}, t)}{\partial t}.$ \hfill (6.99)

Once again, here the scalar potential Φ depends on space and time. We note that when $\Phi, \mathbf{A}$ are independent of time (6.98) and (6.99) reduce to our earlier discussion of potentials in the static case. As in the static case, it is clear that the scalar and the vector potentials have an arbitrariness, namely,

$$\Phi' = \Phi + \frac{1}{c} \frac{\partial \Lambda(\mathbf{x}, t)}{\partial t}, \qquad \mathbf{A}' = \mathbf{A} - \mathbf{\nabla}\Lambda(\mathbf{x}, t), \qquad (6.100)$$

where $\Lambda(\mathbf{x}, t)$ is an arbitrary function give the same electric and magnetic fields. Consequently, Maxwell's equations, written in terms of the scalar and the vector potentials, will reflect this arbitrariness. This is known as the gauge invariance of Maxwell's equations which we will study in some detail next.

We see that two of Maxwell's equations (second and third equations in (6.97)) can be solved to express the electric and the magnetic fields in terms of scalar and vector potentials. Let us next substitute the solutions (6.98) and (6.99) into the other two equations. In any medium, Gauss' law takes the form

$$\mathbf{\nabla} \cdot \mathbf{D} = \epsilon \mathbf{\nabla} \cdot \mathbf{E} = 4\pi \rho,$$

or, $\epsilon \mathbf{\nabla} \cdot \left(-\mathbf{\nabla}\Phi - \dfrac{1}{c} \dfrac{\partial \mathbf{A}}{\partial t} \right) = 4\pi \rho,$

or, $\mathbf{\nabla}^2 \Phi + \dfrac{1}{c} \dfrac{\partial(\mathbf{\nabla} \cdot \mathbf{A})}{\partial t} = -\dfrac{4\pi}{\epsilon} \rho,$

or, $\mathbf{\nabla}^2 \Phi - \dfrac{\epsilon\mu}{c^2} \dfrac{\partial^2 \Phi}{\partial t^2} = -\dfrac{4\pi}{\epsilon} \rho - \dfrac{1}{c} \dfrac{\partial}{\partial t} \left(\mathbf{\nabla} \cdot \mathbf{A} + \dfrac{\epsilon\mu}{c} \dfrac{\partial \Phi}{\partial t} \right).$ \hfill (6.101)

Similarly, from the last of Maxwell's equations in (6.97), we obtain

$$\mathbf{\nabla} \times \mathbf{H} = \frac{4\pi}{c} \mathbf{J} + \frac{1}{c} \frac{\partial \mathbf{D}}{\partial t},$$

or, $\dfrac{1}{\mu} \mathbf{\nabla} \times (\mathbf{\nabla} \times \mathbf{A}) = \dfrac{4\pi}{c} \mathbf{J} + \dfrac{\epsilon}{c} \dfrac{\partial}{\partial t} \left(-\mathbf{\nabla}\Phi - \dfrac{1}{c} \dfrac{\partial \mathbf{A}}{\partial t} \right),$

or, $\left(\mathbf{\nabla}(\mathbf{\nabla} \cdot \mathbf{A}) - \mathbf{\nabla}^2 \mathbf{A}\right) = \dfrac{4\pi\mu}{c} \mathbf{J} - \dfrac{\epsilon\mu}{c} \dfrac{\partial(\mathbf{\nabla}\Phi)}{\partial t} - \dfrac{\epsilon\mu}{c^2} \dfrac{\partial^2 \mathbf{A}}{\partial t^2},$

or, $\mathbf{\nabla}^2 \mathbf{A} - \dfrac{\epsilon\mu}{c^2} \dfrac{\partial^2 \mathbf{A}}{\partial t^2} = -\dfrac{4\pi\mu}{c} \mathbf{J} + \mathbf{\nabla}\left(\mathbf{\nabla} \cdot \mathbf{A} + \dfrac{\epsilon\mu}{c} \dfrac{\partial\Phi}{\partial t}\right).$ (6.102)

The two equations, (6.101) and (6.102), appear to be coupled second order equations. However, let us recall that the scalar and the vector potentials are arbitrary up to gauge transformations (6.100). This is also reflected in the fact that the two coupled equations are invariant under a gauge transformation. In such a case, the Cauchy initial value problem cannot be solved uniquely unless we specify some further conditions on the potentials. Let us, therefore, choose the scalar and the vector potentials such that

$$\frac{\epsilon\mu}{c} \frac{\partial\Phi}{\partial t} + \mathbf{\nabla} \cdot \mathbf{A} = 0. \tag{6.103}$$

With such a choice of the potentials, the two equations (6.101) and (6.102) become decoupled and take the forms

$$\mathbf{\nabla}^2\Phi - \frac{\epsilon\mu}{c^2} \frac{\partial^2\Phi}{\partial t^2} = -\frac{4\pi}{\epsilon} \rho,$$

$$\mathbf{\nabla}^2\mathbf{A} - \frac{\epsilon\mu}{c^2} \frac{\partial^2\mathbf{A}}{\partial t^2} = -\frac{4\pi\mu}{c} \mathbf{J}. \tag{6.104}$$

Namely, with such a choice of the potentials, both the scalar and the vector potentials satisfy the wave equation with sources.

The choice of a form of the potentials is known as a choice of gauge. And the particular gauge we have chosen in (6.103) is known as the Lorenz gauge (named after Ludvig Lorenz) which is manifestly relativistic invariant as we will see. The choice of a gauge is subject to the condition that it should be implementable, namely, that we can always find potentials which would satisfy the gauge condition. For the case at hand, for example, suppose our potentials did not satisfy the Lorenz gauge condition, namely, if

$$\mathbf{\nabla} \cdot \mathbf{A} + \frac{\epsilon\mu}{c} \frac{\partial\Phi}{\partial t} \neq 0, \tag{6.105}$$

then, we can make a gauge transformation

$$\Phi' = \Phi + \frac{1}{c} \frac{\partial\Lambda}{\partial t}, \quad \mathbf{A}' = \mathbf{A} - \mathbf{\nabla}\Lambda, \tag{6.106}$$

requiring that the new potentials would satisfy the gauge condition

$$\mathbf{\nabla} \cdot \mathbf{A}' + \frac{\epsilon\mu}{c} \frac{\partial\Phi'}{\partial t} = 0. \tag{6.107}$$

This leads to

$$\boldsymbol{\nabla} \cdot (\mathbf{A} - \boldsymbol{\nabla}\Lambda) + \frac{\epsilon\mu}{c}\frac{\partial}{\partial t}\left(\Phi + \frac{1}{c}\frac{\partial \Lambda}{\partial t}\right) = 0,$$

$$\text{or,} \quad \left(\boldsymbol{\nabla}^2 - \frac{\epsilon\mu}{c^2}\frac{\partial^2}{\partial t^2}\right)\Lambda = \boldsymbol{\nabla} \cdot \mathbf{A} + \frac{\epsilon\mu}{c}\frac{\partial \Phi}{\partial t}. \tag{6.108}$$

This is an inhomogeneous equation and since the inverse (Green's function) of the operator on the left exists, this equation is soluble. Namely, even if our potentials did not satisfy the Lorenz gauge condition, we can always find a Λ and, therefore, a gauge transformation such that the new potentials will satisfy the Lorenz gauge condition. Let us note here that the second order operator on the left is the generalization of the Laplacian to four dimensions including both space and time and is known as the D'Alembertian.

The physical results are, of course, independent of the choice of gauge and one is not forced to choose the Lorenz gauge to study (solve) Maxwell's equations. Other gauge choices may be more suitable to study specific phenomena. One such gauge choice is known as the Coulomb gauge or the transverse gauge where the potentials are required to satisfy

$$\boldsymbol{\nabla} \cdot \mathbf{A} = 0. \tag{6.109}$$

One can see in a straightforward manner as before that this is an implementable gauge. Furthermore, with such a choice of gauge, Gauss' law (6.101) takes the form

$$\boldsymbol{\nabla}^2\Phi = -\frac{4\pi}{\epsilon}\rho, \tag{6.110}$$

which is the Poisson equation. The solution of this, as we have seen before, can be written in the form

$$\Phi(\mathbf{x}, t) = \frac{1}{\epsilon}\int d^3x'\, \frac{\rho(\mathbf{x}', t)}{|\mathbf{x} - \mathbf{x}'|}. \tag{6.111}$$

If we now look at the (modified) Ampere's law (6.102) in this gauge, we have

$$\boldsymbol{\nabla}^2\mathbf{A} - \frac{\epsilon\mu}{c^2}\frac{\partial^2\mathbf{A}}{\partial t^2} = -\frac{4\pi\mu}{c}\mathbf{J} + \frac{\epsilon\mu}{c}\boldsymbol{\nabla}\left(\frac{\partial \Phi}{\partial t}\right)$$

$$= -\frac{4\pi\mu}{c}\mathbf{J} + \frac{\mu}{c}\boldsymbol{\nabla}\int d^3x'\, \frac{\frac{\partial\rho}{\partial t}}{|\mathbf{x} - \mathbf{x}'|}$$

$$= -\frac{4\pi\mu}{c}\mathbf{J} - \frac{\mu}{c}\boldsymbol{\nabla}\int d^3x' \frac{\boldsymbol{\nabla}'\cdot\mathbf{J}}{|\mathbf{x}-\mathbf{x}'|}$$

$$= -\frac{4\pi\mu}{c}\mathbf{J} - \frac{\mu}{c}\boldsymbol{\nabla}\int d^3x' \frac{\boldsymbol{\nabla}'\cdot\mathbf{J}_l}{|\mathbf{x}-\mathbf{x}'|}. \tag{6.112}$$

Here, we have used the continuity equation (6.3) as well as the familiar decomposition (also discussed in connection with the Helmholtz theorem in chapter **1**, see (1.53), (1.54) and the discussion there) that a given vector can be written as a sum of two terms, one longitudinal and the other transverse with respect to the operation of $\boldsymbol{\nabla}$, namely,

$$\mathbf{J} = \mathbf{J}_l + \mathbf{J}_t, \qquad \boldsymbol{\nabla}\cdot\mathbf{J}_t = 0, \qquad \boldsymbol{\nabla}\times\mathbf{J}_l = 0. \tag{6.113}$$

Furthermore, taking $\boldsymbol{\nabla}$ inside the integral in (6.112) and integrating by parts, we obtain

$$\boldsymbol{\nabla}^2\mathbf{A} - \frac{\epsilon\mu}{c^2}\frac{\partial^2\mathbf{A}}{\partial t^2} = -\frac{4\pi\mu}{c}\mathbf{J} - \frac{\mu}{c}\int d^3x' \frac{\boldsymbol{\nabla}'(\boldsymbol{\nabla}'\cdot\mathbf{J}_l)}{|\mathbf{x}-\mathbf{x}'|}$$

$$= -\frac{4\pi\mu}{c}\mathbf{J} - \frac{\mu}{c}\int d^3x' \frac{\boldsymbol{\nabla}'\times(\boldsymbol{\nabla}'\times\mathbf{J}_l) + \boldsymbol{\nabla}'^2\mathbf{J}_l}{|\mathbf{x}-\mathbf{x}'|}$$

$$= -\frac{4\pi\mu}{c}\mathbf{J} - \frac{\mu}{c}\int d^3x' \frac{\boldsymbol{\nabla}'^2\mathbf{J}_l}{|\mathbf{x}-\mathbf{x}'|} = -\frac{4\pi\mu}{c}\mathbf{J} + \frac{4\pi\mu}{c}\mathbf{J}_l$$

$$= -\frac{4\pi\mu}{c}(\mathbf{J} - \mathbf{J}_l) = -\frac{4\pi\mu}{c}\mathbf{J}_t. \tag{6.114}$$

Here, in the intermediate steps, we have used integration by parts as well as the fact that

$$\boldsymbol{\nabla}^2\left(\frac{1}{|\mathbf{x}-\mathbf{x}'|}\right) = -4\pi\delta^3(\mathbf{x}-\mathbf{x}'). \tag{6.115}$$

We have also used the fact that the term $\boldsymbol{\nabla}\times(\boldsymbol{\nabla}\times\mathbf{J}_l)$ vanishes by definition (6.113).

Thus, with the choice of the Coulomb gauge (6.109), we see that the scalar potential satisfies the Poisson equation while the vector potential satisfies the wave equation with a transverse current (source), namely,

$$\boldsymbol{\nabla}^2\Phi = -\frac{4\pi}{\epsilon}\rho,$$

$$\boldsymbol{\nabla}^2\mathbf{A} - \frac{\epsilon\mu}{c^2}\frac{\partial^2\mathbf{A}}{\partial t^2} = -\frac{4\pi\mu}{c}\mathbf{J}_t. \tag{6.116}$$

This is, of course, consistent with the choice of the gauge condition
(6.109). Since the vector potential is transverse in this case, one also
says that this is the transverse gauge (sometimes it is also called the
physical gauge). The Coulomb gauge is quite useful when there are
no sources present, namely, when $\rho = 0 = \mathbf{J}$. In such a case, the
solution of the Poisson equation is trivial and the vector potential
satisfies the free wave equation

$$\boldsymbol{\nabla}^2 \mathbf{A} - \frac{\epsilon\mu}{c^2} \frac{\partial^2 \mathbf{A}}{\partial t^2} = 0. \tag{6.117}$$

It is also worth noting here that even after choosing a gauge,
there is some residual gauge invariance. For example, in the case of
the Lorentz gauge (6.103), even when

$$\frac{\epsilon\mu}{c} \frac{\partial \Phi}{\partial t} + \boldsymbol{\nabla} \cdot \mathbf{A} = 0, \tag{6.118}$$

holds we can still make a gauge transformation preserving this gauge.
In other words, we can define a new set of potentials

$$\Phi' = \Phi + \frac{1}{c} \frac{\partial \tilde{\Lambda}}{\partial t}, \qquad \mathbf{A}' = \mathbf{A} - \boldsymbol{\nabla}\tilde{\Lambda}, \tag{6.119}$$

which would also satisfy the Lorentz condition provided

$$\left(\boldsymbol{\nabla}^2 - \frac{\epsilon\mu}{c^2} \frac{\partial^2}{\partial t^2}\right) \tilde{\Lambda} = 0. \tag{6.120}$$

Unlike the case of the Laplacian, the D'Alembertian operator allows
for oscillatory solutions to this homogeneous equation.

6.6 Lorentz transformation

To appreciate Lorentz transformations and Lorentz invariance properly, it is important to look at the chronological development of
events in physics. Newton's equation was known to be invariant under Galilean transformations or Galilean boosts (in addition to being
covariant under spatial rotations)

$$\mathbf{x}' = \mathbf{x} - \mathbf{v}t,$$

$$t' = t, \tag{6.121}$$

where $\mathbf{v}$ represents the constant velocity with which an observer in
a reference frame is moving. The basic idea following from this invariance led to the understanding that there are an infinite number

of inertial frames which move with constant velocities with respect to one another and physical laws are independent of the choice of the frame. Consequently, while time was assumed to be absolute, velocities were thought of as relative depending on what is the (Galilean) inertial frame being used. Most of the laws of physics known at the time were compatible with invariance under Galilean relativity except for a couple of phenomena.

One such puzzle came from the plane wave solutions of Maxwell's equations which, as we have seen, travel in vacuum with the speed of light which is a constant. An absolute velocity, however, was against the spirit of Galilean relativity. Consequently, there was a major conflict and to avoid this conflict, Maxwell even assumed that the wave solutions of his equations travel in a medium called ether, namely, he tried to promote that the speed with which the waves propagate is really the speed of light in a specific inertial frame represented by ether. On the other hand, Michelson and Morley conclusively showed through their experiments that there is no ether and, therefore, the speed of light is a constant independent of the inertial frame of reference. This was indeed the turning point in thinking, for Maxwell's equations were not invariant under Galilean transformations. (In retrospect even a simple experimental measurement such as the life time of muon, performed decades later, cannot conform to the ideas of Galilean invariance. Experimentally it is measured that the life time of the muon decaying at rest in the laboratory is shorter by an order of magnitude than the life time measured for the muons decaying in the cosmic ray showers ($\tau_{\text{lab}} \approx 10^{-6}$ sec). This cannot be explained with Galilean relativity where time is absolute and does not depend on the choice of the reference frame.)

Until the time of Einstein, it was believed that time is a coordinate very different from space where events take place. Einstein was the first to propose that space and time should really be considered on an equal footing and not distinct from each other. Based on the earlier work of Lorentz, he also proposed that physical laws should be invariant under Lorentz transformations and not under Galilean transformations as was thought to be the case until that time. The Lorentz transformations are very different from the Galilean transformations in that they mix up space and time coordinates. For example, under a Lorentz transformation the space and time coordinates for an observer in an inertial frame, moving with velocity v along the x-axis (with respect to another observer), would be given by

$$t' = \gamma\left(t - \frac{\beta}{c}x\right),$$

$$x' = \gamma(x - \beta ct),$$

$$y' = y,$$

$$z' = z, \tag{6.122}$$

where we have defined

$$\beta = \frac{v}{c}, \quad \gamma = \left(1 - \beta^2\right)^{-\frac{1}{2}}, \quad \gamma^2\left(1 - \beta^2\right) = 1. \tag{6.123}$$

It is interesting to note from (6.122) that in the limit $c \to \infty$, Lorentz transformations reduce to Galilean transformations (6.121). Lorentz transformations, of course, have far reaching consequences. But, one of the interesting consequences that can be immediately derived from the structure of the transformations in (6.122) is that

$$\left(c^2t'^2 - x'^2 - y'^2 - z'^2\right)$$

$$= \gamma^2 c^2\left(t - \frac{\beta}{c}x\right)^2 - \gamma^2(x - \beta ct)^2 - y^2 - z^2$$

$$= \left(c^2t^2 - x^2 - y^2 - z^2\right). \tag{6.124}$$

All of this led Einstein to propose that space and time together should be thought of as defining a four dimensional manifold where events take place and that Lorentz transformations are symmetry transformations which transform the coordinates of this four dimensional manifold, much like rotations transform the coordinates of the three dimensional space. A vector in such a manifold would consist of four components – one time and three space – and is called a four vector, as opposed to the vectors in three dimensional space that we are all familiar with. However, space and time components can be embedded into this four vector in two distinct ways. For example, let us consider the space-time coordinates themselves which define a four vector. We note that we can define a four component vector as

$$x^\mu = (x^0, \mathbf{x}) = (ct, \mathbf{x}) = (ct, x, y, z), \quad \mu = 0, 1, 2, 3. \tag{6.125}$$

Alternatively, we can define a four vector with a relative negative sign between the time and the space components, namely,

$$x_\mu = (x^0, -\mathbf{x}) = (ct, -\mathbf{x}) = (ct, -x, -y, -z). \tag{6.126}$$

It is worth pointing out that on dimensional grounds the time component has to be multiplied by a velocity (to be on the same footing as the space coordinates) and the only velocity that is a constant is the speed of light. Thus, even though in our discussions, we would restrict to free space ($\epsilon = 1 = \mu$), if one is dealing with a dielectric medium, the appropriate velocity should be used.

Under a Lorentz transformation, say, along the x-axis it is clear that the two distinct four vectors would transform as (repeated indices are summed)

$$x'^{\mu} = \Lambda^{\mu}{}_{\nu}\, x^{\nu},$$

$$x'_{\mu} = \widetilde{\Lambda}_{\mu}{}^{\nu}\, x_{\nu}, \tag{6.127}$$

where from the explicit transformations in (6.122), we note that

$$\Lambda^{\mu}{}_{\nu} = \begin{pmatrix} \gamma & -\gamma\beta & 0 & 0 \\ -\gamma\beta & \gamma & 0 & 0 \\ 0 & 0 & 1 & 0 \\ 0 & 0 & 0 & 1 \end{pmatrix},$$

$$\widetilde{\Lambda}_{\mu}{}^{\nu} = \begin{pmatrix} \gamma & \gamma\beta & 0 & 0 \\ \gamma\beta & \gamma & 0 & 0 \\ 0 & 0 & 1 & 0 \\ 0 & 0 & 0 & 1 \end{pmatrix}. \tag{6.128}$$

It is clear from the structure of the matrices in (6.128) that they are inversely related,

$$\Lambda^{\mu}{}_{\nu}\, (\widetilde{\Lambda}_{\lambda}{}^{\nu})^{T} = \delta^{\mu}_{\lambda}. \tag{6.129}$$

The two distinct four vectors, therefore, behave differently under Lorentz transformations. In fact, they transform in an inverse manner under a Lorentz transformation. The four vector x^{μ} is known as a contravariant vector while x_{μ} is known as a covariant vector.

From the structure of these transformation matrices, it is also clear that

$$\det \Lambda^{\mu}{}_{\nu} = 1 = \det \widetilde{\Lambda}_{\mu}{}^{\nu}. \tag{6.130}$$

This is very much like the relation for the rotation matrices in three dimensional space and suggests that Lorentz transformations can be thought of as rotations in the four dimensional space-time manifold.

Let us recall that in three dimensional space, a rotation around the z-axis by an angle θ is defined by

$$\begin{pmatrix} x' \\ y' \\ z' \end{pmatrix} = \begin{pmatrix} \cos\theta & -\sin\theta & 0 \\ \sin\theta & \cos\theta & 0 \\ 0 & 0 & 1 \end{pmatrix} \begin{pmatrix} x \\ y \\ z \end{pmatrix}. \tag{6.131}$$

Keeping this in mind, let us define (β, γ are defined in (6.123))

$$\cosh\theta = \gamma, \qquad \sinh\theta = \gamma\beta, \tag{6.132}$$

so that $\cosh^2\theta - \sinh^2\theta = 1$ (alternatively, $\tanh\theta = \beta$). With this, the Lorentz transformations (namely, $\Lambda^\mu{}_\nu$ and $\widetilde{\Lambda}_\mu{}^\nu$) do actually correspond to rotations, although the angle of rotation appears to be imaginary. This is a consequence of the structure of the four dimensional manifold that we will discuss. Indeed, just as a rotation around the z-axis can be thought of as a rotation in the $1-2$ ($x-y$) plane, similarly a Lorentz transformation (boost) along the x-axis can be thought of as a rotation in the $0-1$ ($t-x$) plane. (In fact, the totality of space rotations and boosts is known as Lorentz transformations.)

Since rotations leave the length of a vector invariant, here, too, we can ask what is the invariant length under a Lorentz transformation. We have already seen in (6.124) that

$$c^2 t^2 - \mathbf{x}^2 = x^\mu x_\mu = x_\mu x^\mu = x^2, \tag{6.133}$$

is invariant under Lorentz transformations and defines the invariant length. However, it is clear now that the reason for this invariance is that a covariant and a contravariant vector transform inversely under a Lorentz transformation. From the structure of the invariant length in (6.133), we can define a metric tensor for the four dimensional manifold, namely,

$$x^2 = \eta_{\mu\nu} x^\mu x^\nu = \eta^{\mu\nu} x_\mu x_\nu, \tag{6.134}$$

where the covariant and the contravariant metric tensors are determined from (6.133) and (6.134) to be

$$\eta_{\mu\nu} = \begin{pmatrix} 1 & 0 & 0 & 0 \\ 0 & -1 & 0 & 0 \\ 0 & 0 & -1 & 0 \\ 0 & 0 & 0 & -1 \end{pmatrix},$$

$$\eta^{\mu\nu} = \begin{pmatrix} 1 & 0 & 0 & 0 \\ 0 & -1 & 0 & 0 \\ 0 & 0 & -1 & 0 \\ 0 & 0 & 0 & -1 \end{pmatrix}. \tag{6.135}$$

It follows that these matrices are inverses of each other, namely,

$$\eta^{\mu\nu}\eta_{\nu\lambda} = \delta^{\mu}_{\lambda}. \tag{6.136}$$

The metric tensor also allows us to raise and lower the Lorentz indices,

$$x^{\mu} = \eta^{\mu\nu}x_{\nu}, \qquad x_{\mu} = \eta_{\mu\nu}x^{\nu}, \tag{6.137}$$

so that the time component of a vector does not change sign under raising or lowering while the space components do. Let us note here that the three dimensional space that we are used to is known as a Euclidean space where the metric tensor is the trivial Kronecker delta function δ_{ij}. Consequently, there is no difference between the covariant and the contravariant vectors there. The four dimensional space-time manifold, on the other hand, has a nontrivial metric tensor giving rise to distinct covariant and contravariant vectors. A manifold with such a metric (as in (6.135)) is known as a Minkowski space. Furthermore, from the definition of the invariant length in (6.134), we see that unlike the three dimensional case, here the length of a nontrivial (four) vector is not necessarily positive. In fact, it can be positive, negative, or zero. If

$$x^2 = \left(x^0\right)^2 - \mathbf{x}^2 > 0, \tag{6.138}$$

we say that the four vector is time-like, while for

$$x^2 = \left(x^0\right)^2 - \mathbf{x}^2 < 0, \tag{6.139}$$

the vector is called space-like. On the other hand, when

$$x^2 = \left(x^0\right)^2 - \mathbf{x}^2 = 0, \tag{6.140}$$

the vector is called light-like. It is along such light-like directions that a light ray travels. (The three types of vectors do not mix under a Lorentz transformation.) Correspondingly, the structure of the Minkowski space (see Fig. 6.7) is quite different from the three dimensional Euclidean space that we are familiar with.

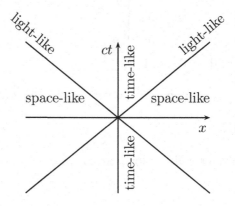

Figure 6.7: Four dimensional Minkowski space-time projected onto two dimensions.

It is now clear that in a Minkowski space we would have two kinds of vectors, the covariant and the contravariant ones. Thus, if A^μ and B^μ denote two arbitrary (contravariant) vectors in this space, we can define an inner product (scalar product) between them as

$$A \cdot B = \eta_{\mu\nu} A^\mu B^\nu = A^0 B^0 - \mathbf{A} \cdot \mathbf{B} = A_\mu B^\mu = A^\mu B_\mu, \quad (6.141)$$

as well as the length of any vector as

$$A^2 = \eta_{\mu\nu} A^\mu A^\nu = \left(A^0\right)^2 - \mathbf{A}^2 = A_\mu A^\mu, \quad (6.142)$$

both of which will be invariant under Lorentz transformations. Any quantity without a free Lorentz index is a Lorentz scalar and is invariant under Lorentz transformations. We have already seen how vectors transform under a Lorentz transformation. Any quantity with more than one free Lorentz index is known as a tensor and its transformation properties follow from the transformation properties of the vectors in a straightforward manner. Thus, for example, a third rank tensor of the form $T^{\mu\nu}{}_\lambda$ (the number of free indices defines the rank of a tensor) would transform under a Lorentz transformation as

$$T^{\mu\nu}{}_\lambda \to T'{}^{\mu\nu}{}_\lambda = \Lambda^\mu{}_{\mu'} \Lambda^\nu{}_{\nu'} \widetilde{\Lambda}_\lambda{}^{\lambda'} T^{\mu'\nu'}{}_{\lambda'}. \quad (6.143)$$

Let us next note that we can define the derivatives (gradients) on this manifold in a standard manner. There will be two kinds of derivatives, contragradient and cogradient, defined respectively as

$$\partial^\mu = \frac{\partial}{\partial x_\mu} = \left(\frac{1}{c}\frac{\partial}{\partial t}, -\nabla\right),$$

$$\partial_\mu = \frac{\partial}{\partial x^\mu} = \left(\frac{1}{c}\frac{\partial}{\partial t}, \nabla\right). \tag{6.144}$$

Note the change in the relative signs between the time and the space components of the gradient vectors compared with the components of the coordinate vectors in (6.125) and (6.126). Since gradients define vectors we can define a scalar (which would be invariant under Lorentz transformations) from these as

$$\Box = \partial^2 = \eta_{\mu\nu}\partial^\mu\partial^\nu = \frac{1}{c^2}\frac{\partial^2}{\partial t^2} - \nabla^2. \tag{6.145}$$

As we have seen, this is the wave operator and is known as the D'Alembertian operator (remember that we have restricted ourselves to free space for which $\epsilon = 1 = \mu$). It is invariant under Lorentz transformations just as the Laplacian is invariant under three dimensional rotations.

There are other familiar quantities from the study of three dimensions which also combine into four vectors. (To be able to combine distinct quantities into a four vector, they must have the right transformation properties under a Lorentz transformation.) Of course, the most familiar is the fact that energy and momentum combine into a four vector such that

$$p^\mu = \left(\frac{E}{c}, \mathbf{p}\right), \qquad p_\mu = \left(\frac{E}{c}, -\mathbf{p}\right). \tag{6.146}$$

The length of this four vector

$$p^2 = \eta_{\mu\nu}p^\mu p^\nu = \frac{E^2}{c^2} - \mathbf{p}^2, \tag{6.147}$$

is Lorentz invariant and we see from this that we can write the Einstein relation as an invariant relation

$$p^2 = \frac{E^2}{c^2} - \mathbf{p}^2 = m^2 c^2,$$

$$\text{or,} \quad E^2 = c^2 \mathbf{p}^2 + m^2 c^4, \tag{6.148}$$

where m is known as the rest mass of the particle.

6.7 Covariance of Maxwell's equations

Just as space-time coordinates as well as energy-momentum combine into four vectors, similarly, let us note that the charge density and the current density also combine into a four vector of the form

$$J^\mu = (c\rho, \mathbf{J}), \qquad J_\mu = (c\rho, -\mathbf{J}), \qquad (6.149)$$

from which it follows that we can write the continuity equation as (see the definition of the gradients in (6.144))

$$\partial_\mu J^\mu = \partial_0 J^0 + \partial_i J^i = \frac{\partial \rho}{\partial t} + \boldsymbol{\nabla} \cdot \mathbf{J} = 0. \qquad (6.150)$$

Namely, the continuity equation is nothing other than the vanishing of the four divergence of the four vector current density. Since the four divergence is a scalar (it has no free Lorentz index), this equation is Lorentz invariant. In other words, the form of the continuity equation is the same in all Lorentz frames. (Incidentally, integrating the continuity equation over a large volume, we obtain the conservation of total charge.)

The scalar and the vector potentials also combine into a four vector known as the (four) vector potential of the form

$$A^\mu = (\Phi, \mathbf{A}), \qquad A_\mu = (\Phi, -\mathbf{A}). \qquad (6.151)$$

Given the vector potential, we can construct a second rank anti-symmetric tensor by taking its four dimensional curl in the following way

$$F_{\mu\nu} = \partial_\mu A_\nu - \partial_\nu A_\mu = -F_{\nu\mu}, \quad \mu, \nu = 0, 1, 2, 3. \qquad (6.152)$$

Being anti-symmetric, this tensor has only six independent components and from the definition of the magnetic and the electric fields in (6.98) and (6.99) in terms of the scalar and the vector potentials, we see that they can be expressed as components of this second rank anti-symmetric tensor in the following way (note that $\mu = 0, 1, 2, 3$ while $i, j = 1, 2, 3$)

$$F_{0i} = \partial_0 A_i - \partial_i A_0$$

$$= -\frac{1}{c}\frac{\partial (\mathbf{A})_i}{\partial t} - \boldsymbol{\nabla}_i \Phi = (\mathbf{E})_i,$$

$$F_{ij} = \partial_i A_j - \partial_j A_i$$

$$= -(\boldsymbol{\nabla}_i(\mathbf{A})_j - \boldsymbol{\nabla}_j(\mathbf{A})_i) = -\epsilon_{ijk}(\boldsymbol{\nabla} \times \mathbf{A})_k$$

$$= -\epsilon_{ijk}(\mathbf{B})_k. \tag{6.153}$$

The anti-symmetric tensor $F_{\mu\nu}$ is known as the field strength tensor since its components are none other than the electric and the magnetic fields.

From the definition of the electric and the magnetic fields in (6.153), it is now straightforward to determine their transformation properties under a Lorentz transformation. Thus, with the explicit form of the Lorentz transformations along the x-axis in (6.128) (or (6.122)), we obtain

$$F_{\mu\nu} \to F'_{\mu\nu} = \widetilde{\Lambda}_\mu{}^{\mu'} \widetilde{\Lambda}_\nu{}^{\nu'} F_{\mu'\nu'}. \tag{6.154}$$

This gives explicitly

$$F'_{0i} = \widetilde{\Lambda}_0{}^0 \widetilde{\Lambda}_i{}^j F_{0j} + \widetilde{\Lambda}_0{}^j \widetilde{\Lambda}_i{}^0 F_{j0} + \widetilde{\Lambda}_0{}^j \widetilde{\Lambda}_i{}^k F_{jk}$$

$$= (\widetilde{\Lambda}_0{}^0 \widetilde{\Lambda}_i{}^j - \widetilde{\Lambda}_0{}^j \widetilde{\Lambda}_i{}^0) F_{0j} + \widetilde{\Lambda}_0{}^j \widetilde{\Lambda}_i{}^k F_{jk},$$

$$F'_{ij} = \widetilde{\Lambda}_i{}^0 \widetilde{\Lambda}_j{}^k F_{0k} + \widetilde{\Lambda}_i{}^k \widetilde{\Lambda}_j{}^0 F_{k0} + \widetilde{\Lambda}_i{}^k \widetilde{\Lambda}_j{}^l F_{kl}$$

$$= (\widetilde{\Lambda}_i{}^0 \widetilde{\Lambda}_j{}^k - \widetilde{\Lambda}_i{}^k \widetilde{\Lambda}_j{}^0) F_{0k} + \widetilde{\Lambda}_i{}^k \widetilde{\Lambda}_j{}^l F_{kl}. \tag{6.155}$$

Using the identifications in (6.153), the transformations of the electric and the magnetic fields then follows

$$\begin{aligned}
E'_x &= E_x, & B'_x &= B_x, \\
E'_y &= \gamma(E_y - \beta B_z), & B'_y &= \gamma(B_y + \beta E_z), \\
E'_z &= \gamma(E_z + \beta B_y), & B'_z &= \gamma(B_z - \beta E_y).
\end{aligned} \tag{6.156}$$

These are indeed the correct transformations for the electric and the magnetic fields under a Lorentz transformation and this shows that the electric and the magnetic fields are really not independent of each other. (That they should have the same dimension also follows.) A more physical way to see this is to note that if a charge is at rest, it produces only an electric field. However, in a different inertial frame the charge would be moving giving rise to a current, which, as we have seen, produces a magnetic field.

In terms of the field strength tensors, two of Maxwell's equations (the first and the last in (6.97) in vacuum) can be written as

$$\partial_\mu F^{\mu\nu} = \frac{4\pi}{c} J^\nu. \tag{6.157}$$

This can be checked as follows. Let $\nu = 0$. In this case, the equation becomes (remember that we are restricting to free space)

$$\partial_i F^{i0} = \frac{4\pi}{c} J^0,$$

or, $\partial_i F_{0i} = \mathbf{\nabla}_i(\mathbf{E})_i = \mathbf{\nabla} \cdot \mathbf{E} = 4\pi\rho,$ (6.158)

which is, of course, the Gauss' law. On the other hand, if we choose $\nu = j$, then (6.157) leads to

$$\partial_0 F^{0j} + \partial_i F^{ij} = -\partial_0 F_{0j} + \partial_i F_{ij} = \frac{4\pi}{c} J^j,$$

or, $-\dfrac{1}{c}\dfrac{\partial(\mathbf{E})_j}{\partial t} + \mathbf{\nabla}_i(-\epsilon_{ijk}(\mathbf{B})_k) = \dfrac{4\pi}{c}(\mathbf{J})_j,$

or, $(\mathbf{\nabla} \times \mathbf{B})_j = \dfrac{4\pi}{c}(\mathbf{J})_j + \dfrac{1}{c}\dfrac{\partial(\mathbf{E})_j}{\partial t},$

or, $\mathbf{\nabla} \times \mathbf{B} = \dfrac{4\pi}{c}\mathbf{J} + \dfrac{1}{c}\dfrac{\partial\mathbf{E}}{\partial t}.$ (6.159)

Clearly, these equations are manifestly Lorentz covariant, since they are expressed in terms of Lorentz covariant quantities. In fact, we note that the left hand side of (6.157) behaves like a vector under a Lorentz transformation (namely, has only one free index) and the same is true for the right hand side as well. Defining the dual of the field strength tensor as

$$\widetilde{F}^{\mu\nu} = \frac{1}{2}\epsilon^{\mu\nu\lambda\rho} F_{\lambda\rho} = -\widetilde{F}^{\nu\mu},$$ (6.160)

where $\epsilon^{\mu\nu\lambda\rho}$ is the completely anti-symmetric (Levi-Civita) tensor in four dimensions with $\epsilon^{0123} = 1$ ($\epsilon^{0ijk} = \epsilon_{ijk}$), we recognize that

$$\widetilde{F}^{0i} = \frac{1}{2}\epsilon^{0ijk} F_{jk} = \frac{1}{2}\epsilon_{ijk}(-\epsilon_{jkl}(\mathbf{B})_l) = -(\mathbf{B})_i,$$

$$\widetilde{F}^{ij} = \epsilon^{ij0k} F_{0k} = \epsilon^{0ijk} F_{0k} = \epsilon_{ijk}(\mathbf{E})_k,$$ (6.161)

namely, the dual interchanges $\mathbf{E}$ and $\mathbf{B}$ fields (up to sign). In terms of the dual, we can write the other two Maxwell's equations as

$$\partial_\mu \widetilde{F}^{\mu\nu} = 0.$$ (6.162)

Note that for $\nu = 0$, (6.162) gives

$$\partial_i \widetilde{F}^{i0} = -\partial_i \widetilde{F}^{0i} = 0,$$

or, $\quad \nabla_i(\mathbf{B})_i = \nabla \cdot \mathbf{B} = 0.$ $\qquad (6.163)$

Similarly, choosing $\nu = j$ in (6.162), we obtain

$$\partial_0 \widetilde{F}^{0j} + \partial_i \widetilde{F}^{ij} = 0,$$

or, $\quad -\dfrac{1}{c}\dfrac{\partial(\mathbf{B})_j}{\partial t} + \nabla_i(\epsilon_{ijk}(\mathbf{E})_k) = 0,$

or, $\quad (\nabla \times \mathbf{E})_j = -\dfrac{1}{c}\dfrac{\partial(\mathbf{B})_j}{\partial t},$

or, $\quad \nabla \times \mathbf{E} = -\dfrac{1}{c}\dfrac{\partial \mathbf{B}}{\partial t}.$ $\qquad (6.164)$

Of course, solving these conditions leads us to the definitions of the electric and the magnetic fields in terms of the scalar and the vector potentials as we have seen before. This demonstrates that the set of four Maxwell's equations is manifestly covariant.

Finally, let us note that the gauge invariance of the system is completely built into the definition of the field strength tensor. In fact, from the definition

$$F_{\mu\nu} = \partial_\mu A_\nu - \partial_\nu A_\mu, \qquad (6.165)$$

we note that the field strength is invariant under (these are precisely the same gauge transformations which we have discussed earlier in (6.100))

$$A_\mu \to A'_\mu = A_\mu + \partial_\mu \Lambda. \qquad (6.166)$$

Namely, under this redefinition,

$$\begin{aligned}
F_{\mu\nu} &= \partial_\mu A_\nu - \partial_\nu A_\mu \\
&\to \partial_\mu(A_\nu + \partial_\nu \Lambda) - \partial_\nu(A_\mu + \partial_\mu \Lambda) \\
&= \partial_\mu A_\nu - \partial_\nu A_\mu = F_{\mu\nu}.
\end{aligned} \qquad (6.167)$$

Furthermore, the Lorenz gauge choice (6.103) can now be seen to be Lorentz invariant from the fact that (remember that we are considering free space)

$$\partial_\mu A^\mu = \dfrac{1}{c}\dfrac{\partial \Phi}{\partial t} + \nabla \cdot \mathbf{A} = 0, \qquad (6.168)$$

is a scalar and, therefore, does not change under a Lorentz transformation.

6.8 Retarded Green's function

As we have seen in (6.104), in the Lorenz gauge both the scalar and the vector potentials satisfy the wave equation with sources, namely, (for simplicity, we are choosing free space)

$$\Box \Phi = \left(\frac{1}{c^2} \frac{\partial^2}{\partial t^2} - \boldsymbol{\nabla}^2 \right) \Phi = 4\pi \rho,$$

$$\Box \mathbf{A} = \left(\frac{1}{c^2} \frac{\partial^2}{\partial t^2} - \boldsymbol{\nabla}^2 \right) \mathbf{A} = \frac{4\pi}{c} \mathbf{J}. \tag{6.169}$$

Of course, we can combine the two equations in (6.169) into a single covariant equation of the form (which is consistent with the Lorenz gauge condition)

$$\Box A_\mu = \left(\frac{1}{c^2} \frac{\partial^2}{\partial t^2} - \boldsymbol{\nabla}^2 \right) A_\mu = \frac{4\pi}{c} J_\mu. \tag{6.170}$$

It is clear, therefore, that we can solve for the potentials if we know the solutions to the equation of the form

$$\Box \Psi = \left(\frac{1}{c^2} \frac{\partial^2}{\partial t^2} - \boldsymbol{\nabla}^2 \right) \Psi(\mathbf{x}, t) = 4\pi f(\mathbf{x}, t). \tag{6.171}$$

As we have discussed earlier, a simple way to solve such inhomogeneous equations is through the method of Green's functions. Namely, let us define the Green's function $G(\mathbf{x}, t; \mathbf{x}', t')$ for the wave operator to satisfy the equation,

$$\Box G = \left(\frac{1}{c^2} \frac{\partial^2}{\partial t^2} - \boldsymbol{\nabla}^2 \right) G(\mathbf{x}, t; \mathbf{x}', t') = -4\pi \, \delta^4(x - x')$$

$$= -4\pi \delta^3(\mathbf{x} - \mathbf{x}') \delta(c(t - t')). \tag{6.172}$$

Then, the particular solution of the inhomogeneous equation (6.171) is easily seen to be (as in (3.145))

$$\Psi(\mathbf{x}, t) = -\int d^4 x' \, G(x, x') f(x')$$

$$= -\int d^3 x' \, c dt' \, G(\mathbf{x}, t; \mathbf{x}', t') f(\mathbf{x}', t'). \tag{6.173}$$

This follows because

$$\Box \Psi(x) = -\Box \int d^4 x' \, G(x, x') f(x')$$

$$= -\int d^4x' \, (\Box G(x,x'))f(x')$$

$$= 4\pi \int d^4x' \, \delta^4(x-x') \, f(x') = 4\pi f(x). \qquad (6.174)$$

Physically, it is clear that the Green's function represents the solution of a given equation for a delta function source.

The Green's function is easily determined by transforming the equation (6.172) into Fourier space. Let us define the Fourier transforms (we are assuming that the Green's function is a function of the difference in the coordinates because of translation invariance)

$$G(x,x') = \int d^4k \, e^{-ik\cdot(x-x')} \, G(k),$$

$$\delta^4(x-x') = \frac{1}{(2\pi)^4} \int d^4k \, e^{-ik\cdot(x-x')}, \qquad (6.175)$$

where we have defined a wave number (four) vector $k^\mu = (\frac{\omega}{c}, \mathbf{k})$. Substituting this into the equation satisfied by the Green's function (6.172), we obtain

$$k^2 G(k) = \frac{4\pi}{(2\pi)^4} = \frac{1}{4\pi^3},$$

$$\text{or,} \quad G(k) = \frac{1}{4\pi^3} \frac{1}{k^2} = \frac{1}{4\pi^3} \frac{1}{(\frac{\omega^2}{c^2} - \mathbf{k}^2)}. \qquad (6.176)$$

This shows the usefulness of the method of Fourier transforms. Basically, the Fourier transform converts a (partial) differential equation into an algebraic equation which is much easier to solve.

We can now determine the Green's function in the coordinate space by substituting (6.176) into the definition of the Fourier transformation in (6.175), namely,

$$G(x,x') = \int d^4k \, e^{-ik\cdot(x-x')} \, G(k)$$

$$= \frac{1}{4\pi^3 c} \int d^3k \, d\omega \, \frac{e^{-i\omega(t-t')+i\mathbf{k}\cdot(\mathbf{x}-\mathbf{x}')}}{\frac{\omega^2}{c^2} - \mathbf{k}^2}. \qquad (6.177)$$

The integrand in (6.177) has poles on the real axis at $\omega = \pm c|\mathbf{k}|$ and in order to evaluate the integral, we have to specify the contour of integration in the complex energy plane. Specifying the contour is equivalent to specifying the boundary condition for the Green's function (solution). There are various possible choices of the contour, just as there are several possible boundary conditions that one can impose on the solutions. However, in classical electrodynamics (or for that matter in classical physics), the boundary condition that is most commonly used is known as the retarded boundary condition. Since the Green's function represents the solution at a point $(\mathbf{x}, t)$ due to a delta function source (disturbance) at $(\mathbf{x}', t')$, on grounds of causality we require that

$$G(\mathbf{x}, t; \mathbf{x}', t') = 0, \qquad \text{for} \quad t < t'. \tag{6.178}$$

In other words, the cause and the effect are related in a retarded manner, namely, the effect cannot precede the cause.

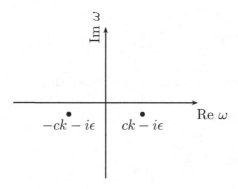

Figure 6.8: Shifted poles in the complex energy plane (with $k = |\mathbf{k}|$) for the retarded Green's function.

The choice of the contour which incorporates the retarded boundary condition is the one which pushes both the poles infinitesimally below the real axis as shown in Fig. 6.8. In other words, let us define

$$G^{(R)}(\mathbf{x}, t; \mathbf{x}', t') = \lim_{\epsilon \to 0} \frac{1}{4\pi^3 c} \int d^3 k d\omega \frac{e^{-i\omega(t-t')+i\mathbf{k}\cdot(\mathbf{x}-\mathbf{x}')}}{\frac{(\omega+i\epsilon)^2}{c^2} - \mathbf{k}^2}. \tag{6.179}$$

In such a case, the poles of the integrand lie at

$$\omega = \pm c|\mathbf{k}| - i\epsilon = \pm ck - i\epsilon, \tag{6.180}$$

both of which are in the lower half of the complex ω plane as in Fig. 6.8. Here we have defined $k = |\mathbf{k}|$ for simplicity.

To see that this indeed satisfies the retarded boundary condition, let us evaluate the ω integral in (6.179) for $t < t'$. In such a case, the exponential in the integrand will be damped on the semi-circle only if the contour is closed in the upper half plane (see Fig. 6.9). However, in such a case, there is no pole inside the contour and, consequently, the integral vanishes by the residue theorem giving

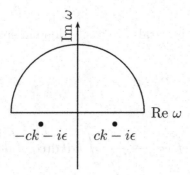

Figure 6.9: Choice of contour in the complex ω-plane for $t - t' < 0$.

$$G^{(R)}(\mathbf{x}, t; \mathbf{x}', t') = 0, \quad \text{for} \quad t < t', \tag{6.181}$$

which indeed defines the retarded Green's function.

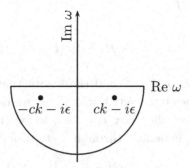

Figure 6.10: Choice of contour in the complex ω-plane for $t - t' > 0$.

On the other hand, when $t > t'$, the exponential will be damped if we close the contour in the lower half plane as in Fig. 6.10. In this case, the contour will enclose both the poles and using Cauchy's method of residues, we obtain (for $t > t'$)

$$G^{(R)}(\mathbf{x}, t; \mathbf{x}', t')$$

$$= -\frac{i}{(2\pi)^2} \int d^3k \left[\frac{e^{-ikc(t-t')+i\mathbf{k}\cdot(\mathbf{x}-\mathbf{x}')}}{k} - \frac{e^{ikc(t-t')+i\mathbf{k}\cdot(\mathbf{x}-\mathbf{x}')}}{k} \right].$$

$$(6.182)$$

The remaining $\mathbf{k}$ integrals are easily done in spherical coordinates. If we define

$$\mathbf{R} = \mathbf{x} - \mathbf{x}', \quad R = |\mathbf{R}|,$$

$$(6.183)$$

we can write

$$G^{(R)}(\mathbf{x}, t; \mathbf{x}', t') = -\frac{i}{(2\pi)^2} \int k \, dk \, d(\cos\theta) d\phi \left[e^{-ikc(t-t')+ikR\cos\theta} \right.$$

$$\left. - e^{ikc(t-t')+ikR\cos\theta} \right]$$

$$= -\frac{1}{2\pi R} \int_0^\infty dk \left[e^{-ik(c(t-t')-R)} - e^{-ik(c(t-t')+R)} \right.$$

$$\left. - e^{ik(c(t-t')+R)} + e^{ik(c(t-t')-R)} \right]$$

$$= -\frac{1}{2\pi R} \int_{-\infty}^\infty dk \left[e^{-ik(c(t-t')-R)} - e^{-ik(c(t-t')+R)} \right]$$

$$= -\frac{1}{R} \left[\delta(c(t-t') - R) - \delta(c(t-t') + R) \right]. \quad (6.184)$$

It is clear that since $t - t' > 0$ (and note that $R > 0$), the second delta function does not contribute. Thus, we determine the retarded Green's function of the wave equation to be (for $t > t'$, a condition which can be implemented through a step function)

$$G^{(R)}(\mathbf{x}, t; \mathbf{x}', t') = -\frac{\delta(c(t-t') - |\mathbf{x} - \mathbf{x}'|)}{|\mathbf{x} - \mathbf{x}'|}. \quad (6.185)$$

With this, we can now obtain a particular solution of the wave equation (6.171) satisfying the retarded boundary condition as (see (6.173))

$$\Psi(\mathbf{x}, t) = -\int d^4x' \, G^{(R)}(x, x') f(x')$$

$$= \int d^3x' \, cdt' \, \frac{\delta(c(t-t') - |\mathbf{x}-\mathbf{x}'|) f(\mathbf{x}',t')}{|\mathbf{x}-\mathbf{x}'|}$$

$$= \int d^3x' \, \frac{f(\mathbf{x}',t')}{|\mathbf{x}-\mathbf{x}'|} \Big|_{t'=t-\frac{|\mathbf{x}-\mathbf{x}'|}{c}}. \tag{6.186}$$

It is clear that this gives a nontrivial solution only for later times, $t = t' + \frac{|\mathbf{x}-\mathbf{x}'|}{c}$ (namely, a retarded solution). Furthermore, from this we can now write down the retarded solutions for the Maxwell's equations (in the Lorenz gauge) (6.170) to be

$$A_\mu(\mathbf{x},t) = \frac{1}{c} \int d^3x' \, \frac{J_\mu(\mathbf{x}',t')}{|\mathbf{x}-\mathbf{x}'|} \Big|_{t'=t-\frac{|\mathbf{x}-\mathbf{x}'|}{c}}, \tag{6.187}$$

which reduces, in the static limit, to the solutions we have determined earlier.

▶ **Example (Lienard-Wiechert potential).** As a simple example of the application of the retarded Green's function, let us calculate the vector potential associated with the fields produced by a slowly moving charged particle. (We will study this problem in more detail in a later chapter.) We have seen in (6.187) that we can write the particular solution for the vector potential, in the Lorenz gauge, as

$$A_\mu(\mathbf{x},t) = \frac{1}{c} \int d^3x' \, \frac{J_\mu(\mathbf{x}',t')}{|\mathbf{x}-\mathbf{x}'|} \Big|_{t'=t-\frac{|\mathbf{x}-\mathbf{x}'|}{c}}$$

$$= \frac{1}{c} \int dt' d^3x' \, \frac{J_\mu(\mathbf{x}',t') \delta\left(t'-t+\frac{|\mathbf{x}-\mathbf{x}'|}{c}\right)}{|\mathbf{x}-\mathbf{x}'|}. \tag{6.188}$$

Let us now consider a point particle with charge q moving along a trajectory $\boldsymbol{\xi}(t)$ so that we can write

$$J_\mu(\mathbf{x},t) = j_\mu(t)\delta^3(\mathbf{x} - \boldsymbol{\xi}(t)), \tag{6.189}$$

where,

$$j^\mu(t) = (cq, q\mathbf{v}) = \left(cq, q\frac{d\boldsymbol{\xi}(t)}{dt}\right). \tag{6.190}$$

Given this, we can compute the potential that such a moving charge would produce using (6.188). Namely,

$$A_\mu(\mathbf{x},t) = \frac{1}{c} \int dt' d^3x' \, \frac{j_\mu(t')\delta^3(\mathbf{x}' - \boldsymbol{\xi}(t'))\delta\left(t'-t+\frac{|\mathbf{x}-\mathbf{x}'|}{c}\right)}{|\mathbf{x}-\mathbf{x}'|}$$

$$= \frac{1}{c} \int dt' \, \frac{j_\mu(t')\delta\left(t'-t+\frac{|\mathbf{x}-\boldsymbol{\xi}(t')|}{c}\right)}{|\mathbf{x}-\boldsymbol{\xi}(t')|}. \tag{6.191}$$

Let us next define

$$\tau = t' - t + \frac{|\mathbf{x} - \boldsymbol{\xi}(t')|}{c}. \tag{6.192}$$

Then, the integration over t' in (6.191) would determine the time coordinate t' to be the one for which the argument of the delta function vanishes, namely,

$$\tau = t' - t + \frac{|\mathbf{x} - \boldsymbol{\xi}(t')|}{c} = 0. \tag{6.193}$$

This can be solved for t' once we know the trajectory of the particle. Furthermore,

$$\frac{d\tau}{dt'} = 1 + \frac{1}{c}\frac{d|\mathbf{x} - \boldsymbol{\xi}(t')|}{dt'}, \tag{6.194}$$

and using the standard formula for integration with a delta function, namely,

$$\int dx\, \delta(f(x))\, g(x) = \frac{1}{\left|\frac{df(x)}{dx}\right|_{x_0}}\, g(x_0), \tag{6.195}$$

where x_0 represents the solution of $f(x) = 0$ (we assume here that there is only one root x_0 of the equation), we obtain

$$A_\mu(\mathbf{x}, t) = \frac{1}{c}\int dt' \frac{j_\mu(t')\delta\left(t' - t + \frac{|\mathbf{x} - \boldsymbol{\xi}(t')|}{c}\right)}{|\mathbf{x} - \boldsymbol{\xi}(t')|}$$

$$= \frac{1}{c} \left. \frac{j_\mu(t')}{|\mathbf{x} - \boldsymbol{\xi}(t')|\left(1 + \frac{1}{c}\frac{d|\mathbf{x} - \boldsymbol{\xi}(t')|}{dt'}\right)}\right|_{\tau=0}$$

$$= \frac{1}{c} \left. \frac{j_\mu(t')}{|\mathbf{x} - \boldsymbol{\xi}(t')| + \frac{1}{2c}\frac{d|\mathbf{x} - \boldsymbol{\xi}(t')|^2}{dt'}}\right|_{\tau=0}. \tag{6.196}$$

These are known as the Lienard-Wiechert potentials. (We note here that for slow moving particles it is not necessary to have the magnitude in the Jacobian coming from the delta function.) From these, we can easily determine the electric and the magnetic fields that a moving charged particle produces which, in turn, are used in the study of radiation due to a moving charged particle. We will discuss this in more detail in a later chapter.   ◀

6.9 Kirchhoff's representation

The Kirchhoff representation is a mathematical statement of Huygen's principle and is the starting point for the discussion of the theory of diffraction. It really follows from a generalization of Green's identity (3.67) to the case where both time and space are involved. Let us note that we have so far determined the particular solution of the wave equation subject to the retarded boundary condition. However, the solution of any differential equation, as we know, consists of a sum of a homogeneous part and the particular integral. In the static case, we saw that the homogeneous solution was the one which allowed us to impose the appropriate Dirichlet boundary condition on the solution. The wave equation, on the other hand, is a hyperbolic equation. In such a case, as we have seen, a unique solution is obtained by imposing Cauchy's initial values, namely, by providing the solution as well as its first derivative with respect to time at $t = 0$. As in the static case, the homogeneous solution allows us to find the solution subject to the given initial value conditions and the generalization of Green's identity to four dimensions is quite crucial in showing this.

Let us consider two arbitrary functions of space and time which we denote by Ψ and Φ. (Here Φ is not necessarily the scalar potential.) It follows now that

$$
\int d^4x' \left(\Phi \Box' \Psi - \Psi \Box' \Phi \right) = \int_{t_i}^{t_f} c dt' \int d^3x' \left(\Phi \partial'^2 \Psi - \Psi \partial'^2 \Phi \right)
$$

$$
= \int_{t_i}^{t_f} c dt' \int d^3x' \, \partial'_\mu \left(\Phi \partial'^\mu \Psi - \Psi \partial'^\mu \Phi \right)
$$

$$
= \int ds'_\mu \left(\Phi \partial'^\mu \Psi - \Psi \partial'^\mu \Phi \right)
$$

$$
= \frac{1}{c} \int d^3x' \left[\Phi \frac{\partial \Psi}{\partial t'} - \Psi \frac{\partial \Phi}{\partial t'} \right]_{t'=t_i}^{t'=t_f}
$$

$$
- c \int_{t_i}^{t_f} dt' \int ds' \cdot \left(\Phi \nabla' \Psi - \Psi \nabla' \Phi \right). \tag{6.197}
$$

Here ds' denotes integration over the two dimensional surface bounding the three dimensional volume. This is the generalization of Green's identity to the four dimensional case for any two arbitrary functions Ψ and Φ.

Let us now specialize to the case where

$$
\Phi = G(\mathbf{x}, t; \mathbf{x}', t'), \quad \Psi = \Psi(\mathbf{x}', t'), \tag{6.198}
$$

such that

$$
\Box \Psi(x) = 4\pi f(x),
$$

$$
\Box G(x, x') = -4\pi \delta^4(x - x'). \tag{6.199}
$$

Substituting this into the identity (6.197) and assuming that $t_i \leq t \leq$

t_f, we obtain,

$$\Psi(\mathbf{x}, t) = -\int d^4 x' \, G(x, x') f(x')$$

$$+ \frac{1}{4\pi c} \int d^3 x' \left[G \frac{\partial \Psi}{\partial t'} - \Psi \frac{\partial G}{\partial t'} \right]_{t'=t_i}^{t'=t_f}$$

$$- \frac{c}{4\pi} \int_{t_i}^{t_f} dt' \int ds' \cdot \left(G \boldsymbol{\nabla}' \Psi(\mathbf{x}', t') - \Psi(\mathbf{x}', t') \boldsymbol{\nabla}' G \right). \qquad (6.200)$$

If we now restrict to retarded solutions, we note that we can identify $G = G^{(R)}$ and that the contribution from the upper limit in the second term on the right in (6.200) vanishes because $t - t' < 0$ at that point. Thus, in such a case, using (6.185) we can write

$$\Psi(\mathbf{x}, t) = -\int d^4 x' \, G^{(R)}(x, x') f(x')$$

$$- \frac{1}{4\pi c} \int d^3 x' \left[G^{(R)} \frac{\partial \Psi}{\partial t'} - \Psi \frac{\partial G^{(R)}}{\partial t'} \right]_{t'=t_i}$$

$$- \frac{c}{4\pi} \int_{t_i}^{t_f} dt' \int ds' \cdot \left(G^{(R)} \boldsymbol{\nabla}' \Psi(\mathbf{x}', t') - \Psi(\mathbf{x}', t') \boldsymbol{\nabla}' G^{(R)} \right)$$

$$= \int d^3 x' \, \frac{f(\mathbf{x}', t')}{|\mathbf{x} - \mathbf{x}'|} \bigg|_{t'=t-\frac{|\mathbf{x}-\mathbf{x}'|}{c}}$$

$$- \frac{1}{4\pi c} \int d^3 x' \left[G^{(R)} \frac{\partial \Psi}{\partial t'} - \Psi \frac{\partial G^{(R)}}{\partial t'} \right]_{t'=t_i}$$

$$- \frac{c}{4\pi} \int_{t_i}^{t_f} dt' \int ds' \cdot \left(G^{(R)} \boldsymbol{\nabla}' \Psi(\mathbf{x}', t') - \Psi(\mathbf{x}', t') \boldsymbol{\nabla}' G^{(R)} \right).$$

$$(6.201)$$

There are two special cases that we will consider now. First, let us consider the case where the volume of space is infinite. In such a case, with the assumptions of asymptotic fall off for the fields (variables), the surface integral in (6.201) vanishes. The remaining terms are determined completely in terms of the initial values of Ψ and $\frac{\partial \Psi}{\partial t}$. Namely, we have a solution of the Cauchy initial value problem subject to the retarded boundary condition. Let us choose $t_i = 0$ and denote

$$\Psi(\mathbf{x}, t = 0) = F(\mathbf{x}), \qquad \left.\frac{\partial \Psi}{\partial t}\right|_{t=0} = H(\mathbf{x}). \tag{6.202}$$

Furthermore, to simplify the evaluation of the integrals, let us choose the point of observation to be $\mathbf{x} = 0$. In such a case, the solution (6.201) becomes

$$\Psi(0, t) = \int d^3 x' \left[\frac{f\left(\mathbf{x}', t' = t - \frac{|\mathbf{x}'|}{c}\right)}{|\mathbf{x}'|} \right.$$

$$\left. + \frac{1}{4\pi c} \left(H(\mathbf{x}') \frac{\delta(ct - |\mathbf{x}'|)}{|\mathbf{x}'|} - F(\mathbf{x}') \frac{\partial}{\partial t'} \frac{\delta(c(t' - t) + |\mathbf{x}'|)}{|\mathbf{x}'|} \right|_{t'=0} \right) \right], \tag{6.203}$$

where we have used the form of the retarded Green's function in (6.185). Going over to the spherical coordinates, this gives

$$\Psi(0, t) = \int d\Omega' \, r' dr' \, f\left(r', \Omega, t' = t - \frac{r'}{c}\right)$$

$$+ \frac{1}{4\pi c} \int d\Omega' \, r' dr' \left(H(r', \Omega') \delta(ct - r') + F(r', \Omega') \frac{\partial \delta(ct - r')}{\partial t} \right)$$

$$= \int d\Omega' \, r' dr' \, f\left(r', \Omega', t' = t - \frac{r'}{c}\right)$$

$$+ \frac{1}{4\pi} \int d\Omega' \left[tH(ct, \Omega') + \frac{\partial(tF(ct, \Omega'))}{\partial t} \right]. \tag{6.204}$$

This gives the solution of the initial value problem once we know the explicit forms of $F(\mathbf{x})$ and $H(\mathbf{x})$.

The second case that we are interested in is when the volume is finite. Furthermore, let us assume that there are no sources present in this volume, namely, $f(x) = 0$ and that the initial values are also trivial. In such a case, (6.201) leads to

$$\Psi(\mathbf{x}, t) = -\frac{c}{4\pi} \int_{t_i}^{t_f} dt' \int ds' \cdot \left(G^{(R)} \nabla' \Psi(\mathbf{x}', t') - \Psi(\mathbf{x}', t') \nabla' G^{(R)} \right). \tag{6.205}$$

Once again, we can use the form of the retarded Green's function in (6.185) and defining, for simplicity $\mathbf{R} = \mathbf{x} - \mathbf{x}'$, we have

$$\boldsymbol{\nabla}'G^{(R)} = -\boldsymbol{\nabla}'\left(\frac{\delta(c(t-t')-R)}{R}\right)$$

$$= -(\boldsymbol{\nabla}'R)\frac{\partial}{\partial R}\left(\frac{\delta(c(t-t')-R)}{R}\right)$$

$$= -\hat{\mathbf{R}}\left(\frac{\delta(c(t-t')-R)}{R^2} - \frac{1}{cR}\frac{\partial\delta(c(t-t')-R)}{\partial t'}\right), \qquad (6.206)$$

where we have used $(\boldsymbol{\nabla}'R) = -\hat{\mathbf{R}}$. Using this, we can do the time integral in (6.205) to obtain

$$\Psi(\mathbf{x},t) = \frac{c}{4\pi}\int\limits_{t_i}^{t_f} dt'\int ds'\cdot\left[\frac{\delta(c(t-t')-R)}{R}\boldsymbol{\nabla}'\Psi\right.$$

$$\left. - \hat{\mathbf{R}}\left(\frac{\delta(c(t-t')-R)}{R^2} - \frac{1}{cR}\frac{\partial\delta(c(t-t')-R)}{\partial t'}\right)\Psi\right]$$

$$= \frac{1}{4\pi}\int ds'\cdot\left[\frac{1}{R}\boldsymbol{\nabla}'\Psi - \frac{\hat{\mathbf{R}}}{R^2}\Psi - \frac{\hat{\mathbf{R}}}{cR}\frac{\partial\Psi}{\partial t'}\right]_{t'=t-\frac{R}{c}}. \qquad (6.207)$$

This is known as Kirchhoff's representation for the solution of the wave equation. It is important to recognize that this does not provide an explicit solution for the equation, since the unknown Ψ appears on the right hand side. Rather, it gives an integral representation for the wave disturbance and this provides a mathematical description of Huygen's principle in the sense that the value of the solution at any point is given completely by its value as well as the derivatives of the function on the surface of a closed volume.

6.10 Selected problems

1. (a) Show that the four Maxwell's equations (in any medium) imply the continuity equation.

    (b) For a system with well localized charges and currents (namely, charges and currents which do not extend to infinity), show that the total charge

$$Q = \int\limits_{\text{all space}} d^3x\,\rho(\mathbf{x},t),$$

of the system does not change with time as a consequence of the continuity equation.

2. Consider the following one dimensional problem. A wire of cross-sectional area A and length L carries a current I and is maintained at a potential difference (voltage) V. Defining the specific resistivity of the material to be ρ which is the inverse of conductivity σ, show that Ohm's law gives

$$J = \sigma E.$$

(Just to remind you, the resistance R is defined as $R = \frac{L}{A}\rho$.)

3. Work out the plane wave solution for a charge neutral (without any free charge) conducting medium, for which you may assume $\mathbf{J} = \sigma \mathbf{E}$. Compare the present solution with the one for the dielectrics obtained in this chapter.

4. Show explicitly that Maxwell's equations in terms of the scalar and the vector potential, namely,

$$\nabla^2 \Phi + \frac{1}{c}\frac{\partial(\nabla \cdot \mathbf{A})}{\partial t} = -\frac{4\pi}{\epsilon}\rho,$$

$$\left(\nabla^2 - \frac{\epsilon\mu}{c^2}\frac{\partial^2}{\partial t^2}\right)\mathbf{A} = -\frac{4\pi\mu}{c}\mathbf{J} + \nabla\left(\nabla \cdot \mathbf{A} + \frac{\epsilon\mu}{c}\frac{\partial\Phi}{\partial t}\right),$$

are invariant under the gauge transformations

$$\Phi \to \Phi + \frac{1}{c}\frac{\partial\Lambda}{\partial t}, \qquad \mathbf{A} \to \mathbf{A} - \nabla\Lambda.$$

5. For a plane wave solution of the Maxwell's equations, calculate the time averaged power radiated through a surface of unit area (namely, calculate the time averaged value of $\hat{\mathbf{k}} \cdot \mathbf{S}$ over one period where $\hat{\mathbf{k}}$ is the direction of propagation).

6. It is possible to define the Poynting vector, $\mathbf{S}$, even in the static case. However, show that, in the static case in the absence of currents, there is no power loss due to radiation through any closed surface.

7. If we write the electric and the magnetic fields in the complex notation, then the time-averaged Poynting vector (in vacuum) can be written in the form

$$\overline{\mathbf{S}} = \frac{c}{8\pi}\operatorname{Re}\left(\mathbf{E} \times \mathbf{H}^*\right) = \frac{c}{8\pi}\operatorname{Re}\left(\mathbf{E} \times \mathbf{B}^*\right).$$

Consider the following three plane wave solutions of Maxwell's equations

$$\mathbf{E}_1 = \hat{\mathbf{x}} E_1 \, e^{-i(\omega t - kz)}, \qquad\qquad \mathbf{B}_1 = \hat{\mathbf{y}} E_1 \, e^{-i(\omega t - kz)},$$

$$\mathbf{E}_2 = \hat{\mathbf{y}} E_2 \, e^{-i(\omega t - kz + \alpha)}, \qquad\quad \mathbf{B}_2 = -\hat{\mathbf{x}} E_2 \, e^{-i(\omega t - kz + \alpha)},$$

$$\mathbf{E}_3 = \hat{\mathbf{x}} E_3 \, e^{-i(\omega t - kz + \alpha)}, \qquad\quad \mathbf{B}_3 = \hat{\mathbf{y}} E_3 \, e^{-i(\omega t - kz + \alpha)},$$

where E_1, E_2, E_3 are real constants.

(a) For a solution which is a superposition of the first two (of equal weight), show that $\overline{\mathbf{S}}_{1+2} = \overline{\mathbf{S}}_1 + \overline{\mathbf{S}}_2$. Explain this result.

(b) For a solution which is a superposition of the first and the third (of equal weight), calculate $\overline{\mathbf{S}}_{1+3}$ and compare with the earlier case.

CHAPTER 7

Wave guides

We have already studied simple solutions of Maxwell's equations and have seen that electromagnetic waves carry power. It is, therefore, an interesting question to ask how this power can be transmitted from one place to another. There are two kinds of transmission problems that may be of interest. First, we may want electric power to be transmitted from a specific point to another without an intermediate party having access to it. The significance of this is clear, for example, from the usual domestic power connection. It is also crucial in the case of telephone conversations, as we may not want a third party to overhear a private conversation. The second kind of transmission corresponds to radio or television transmissions which may not raise such privacy concerns. Here, anyone who can tune to a particular frequency is capable of getting the signal. In this lecture, we will study transmissions of the first kind relegating the second topic to later lectures.

Transmission of electric power from one specific point to another is achieved much the same way as the water supply or the gas supply. The usual transmission of power is through parallel wire lines or through coaxial cables. Such a mode is suitable for low frequency transmissions of less than 200 MHz. For higher frequencies, the physical dimensions of such a carrier system become unrealistic. For transmissions of power of higher frequency one uses wave guides which are basically hollow metal tubes where the metal walls of the tube "guide" the wave along the tube (much like the role of the water pipes). The wave guides can be of any cross-sectional shape. However, the two most commonly used are the rectangular and the cylindrical wave guides. We note that, in discussing wave guides, it is commonly assumed that the metal walls of the wave guide are perfect conductors with $\sigma \to \infty$, where σ represents the conductivity of the metal. In reality, of course, metals have a finite conductivity and electric fields can penetrate inside a metal. However, the skin depth (penetration depth) at high frequencies is quite small. Consequently, the assumption of the walls being perfect conductors works well and the small deviations from this assumption can be calculated systematically if necessary.

7.1 Boundary conditions

In discussing wave guides, it is very important to understand the boundary conditions for the system. Let us recall that Maxwell's equations in an arbitrary region have the form

$$\nabla \cdot \mathbf{D} = 4\pi \rho,$$

$$\nabla \cdot \mathbf{B} = 0,$$

$$\nabla \times \mathbf{E} = -\frac{1}{c} \frac{\partial \mathbf{B}}{\partial t},$$

$$\nabla \times \mathbf{H} = \frac{4\pi}{c} \mathbf{J} + \frac{1}{c} \frac{\partial \mathbf{D}}{\partial t}. \tag{7.1}$$

Various fields satisfy the relations noted earlier in (6.13)

$$\mathbf{D} = \epsilon \, \mathbf{E},$$

$$\mathbf{H} = \frac{1}{\mu} \, \mathbf{B}, \tag{7.2}$$

where ϵ, μ represent respectively the permittivity and the permeability of the medium. In addition, in a metal, we can relate the conduction current to the electric field through Ohm's law as (see the second relation in (6.12))

$$\mathbf{J} = \sigma \, \mathbf{E}, \tag{7.3}$$

with σ representing the conductivity of the metal. (Incidentally, relation (7.3) is true both in the CGS as well as the MKS system of units.)

The boundary conditions in the interface of two different media can be derived from (7.1) in the standard manner as we have done earlier in the case of static problems. Let us note here only the geometrical behavior of the electric and the magnetic fields when time variations are allowed. We note that the first equation of Maxwell (in (7.1)) implies that the electric fields must begin and end on charges while the second implies that magnetic field lines must form closed loops since there is no magnetic charge (magnetic monopole). From

the third equation of Maxwell, we see that electric fields must form
closed loops around time varying magnetic fields. Similarly, the last
equation of Maxwell leads to the fact that magnetic fields must form
closed loops around a conduction current or a "displacement" current
produced by a time varying electric field. This is the general behavior
of electric and magnetic fields when time dependence is present. If we
have a perfect conductor as would be the case for the walls of a wave
guide, then, we realize that the electric and the magnetic fields cannot
penetrate inside the metal. Any change in the external fields would
lead to an instantaneous response whereby charges will move to the
surface of the conductor to prevent any field within. In particular,
we note from (7.3) that, for a perfect conductor with $\sigma \to \infty$, there
cannot be any tangential component of the electric field present on the
surface of the conductor (for any t), which would otherwise imply an
infinite conduction current that is physically untenable. Similarly, the
normal component of the magnetic field must vanish on the surface of
the conductor simply because there are no magnetic monopoles (on
the surface of the conductor). Mathematically, we can write these
boundary conditions as

$$\hat{n} \times \mathbf{E}| = 0,$$

$$\hat{n} \cdot \mathbf{B}| = 0, \tag{7.4}$$

where $\hat{n}$ represents a unit vector normal (outward) to the conducting surface and the restriction implies the validity of these at the
boundary corresponding to the locations of the conducting surfaces.

A wave guide, as we have noted earlier, is simply a hollow metal
tube without any free charge or current inside. Let us assume that
the length of the tube is along the z-axis, which is the direction in
which we would like the electromagnetic wave to be transmitted. The
shape of the transverse cross-section of the wave guide will depend on
whether we have a rectangular or a cylindrical wave guide. Therefore,
to keep things as general as is possible, let us decompose the fields
into components along the z-axis and normal to it as

$$\mathbf{E} = \hat{z} \times (\mathbf{E} \times \hat{z}) + \hat{z} (\hat{z} \cdot \mathbf{E}) = \mathbf{E}_\perp + \mathbf{E}_\| = \mathbf{E}_\perp + \hat{z} E_z,$$

$$\mathbf{B} = \hat{z} \times (\mathbf{B} \times \hat{z}) + \hat{z} (\hat{z} \cdot \mathbf{B}) = \mathbf{B}_\perp + \mathbf{B}_\| = \mathbf{B}_\perp + \hat{z} B_z. \tag{7.5}$$

This follows from the familiar vector identity (if $\mathbf{A}, \mathbf{B}, \mathbf{C}$ involve operators, the order should be maintained)

$$\mathbf{A} \times (\mathbf{B} \times \mathbf{C}) = (\mathbf{A} \cdot \mathbf{C}) \mathbf{B} - (\mathbf{A} \cdot \mathbf{B}) \mathbf{C}. \tag{7.6}$$

Similarly, we can write

$$\mathbf{\nabla} = \mathbf{\nabla}_\perp + \mathbf{\nabla}_\parallel = \mathbf{\nabla}_\perp + \hat{\mathbf{z}} \frac{\partial}{\partial z}. \tag{7.7}$$

Exercise. Although $\hat{\mathbf{z}} \times \mathbf{E}$ gives a component of the $\mathbf{E}$ field normal to the z-axis, show that (7.5) gives the correct decomposition of the fields. Namely, show that any arbitrary vector $\mathbf{V}$ can be written uniquely as

$$\mathbf{V} = \hat{\mathbf{z}} \times (\mathbf{V} \times \hat{\mathbf{z}}) + \hat{\mathbf{z}} (\hat{\mathbf{z}} \cdot \mathbf{V}). \tag{7.8}$$

With the decomposition in (7.5) and (7.7), in a charge free region, the first two equations of (7.1) take the forms

$$\mathbf{\nabla}_\perp \cdot \mathbf{E}_\perp = -\frac{\partial E_z}{\partial z},$$

$$\mathbf{\nabla}_\perp \cdot \mathbf{B}_\perp = -\frac{\partial B_z}{\partial z}. \tag{7.9}$$

The last two equations in (7.1), being vector equations, decompose into two equations each. For example, the third equation in (7.1) can be written as

$$\left(\mathbf{\nabla}_\perp + \hat{\mathbf{z}} \frac{\partial}{\partial z}\right) \times (\mathbf{E}_\perp + \hat{z} E_z) = -\frac{1}{c} \frac{\partial (\mathbf{B}_\perp + \hat{\mathbf{z}} B_z)}{\partial t}, \tag{7.10}$$

$$\text{or,} \quad \mathbf{\nabla}_\perp \times \mathbf{E}_\perp + \hat{\mathbf{z}} \times \left(\frac{\partial \mathbf{E}_\perp}{\partial z} - \mathbf{\nabla}_\perp E_z\right) = -\frac{1}{c} \frac{\partial (\mathbf{B}_\perp + \hat{\mathbf{z}} B_z)}{\partial t}.$$

This, in turn, leads to (taking the dot as well as cross product with $\hat{\mathbf{z}}$)

$$\hat{\mathbf{z}} \cdot (\mathbf{\nabla}_\perp \times \mathbf{E}_\perp) = -\frac{1}{c} \frac{\partial B_z}{\partial t},$$

$$\frac{\partial \mathbf{E}_\perp}{\partial z} - \frac{1}{c} \hat{\mathbf{z}} \times \frac{\partial \mathbf{B}_\perp}{\partial t} = \mathbf{\nabla}_\perp E_z, \tag{7.11}$$

where we have used the fact that $\mathbf{\nabla}_\perp \times \mathbf{E}_\perp$ points along the z-axis as well as (7.6). Similarly, the last equation in (7.1) can be decomposed into

$$\hat{\mathbf{z}} \cdot (\mathbf{\nabla}_\perp \times \mathbf{B}_\perp) = \frac{\epsilon \mu}{c} \frac{\partial E_z}{\partial t},$$

$$\frac{\partial \mathbf{B}_\perp}{\partial z} + \frac{\epsilon \mu}{c} \hat{\mathbf{z}} \times \frac{\partial \mathbf{E}_\perp}{\partial t} = \mathbf{\nabla}_\perp B_z. \tag{7.12}$$

Let us next assume that the fields are harmonically varying with time. Furthermore, since we want the wave to be propagating along the z-axis, we can also extract the z dependence of the fields to write

$$\mathbf{E}(\mathbf{x}_\perp, z, t) = \mathbf{E}(\mathbf{x}_\perp)e^{-i(\omega t - kz)},$$

$$\mathbf{B}(\mathbf{x}_\perp, z, t) = \mathbf{B}(\mathbf{x}_\perp)e^{-i(\omega t - kz)}, \tag{7.13}$$

where ω represents the frequency of the wave and k is a constant parameter, complex, in general. (Recall that the fields are real. When we write it in the form as in (7.13), we are tacitly assuming that we are looking at the real or the imaginary part of it.) Factoring out the (t, z) dependence, the Maxwell's equations, (7.9), (7.11) and (7.12), take the forms

$$\boldsymbol{\nabla}_\perp \cdot \mathbf{E}_\perp = -ikE_z,$$

$$\boldsymbol{\nabla}_\perp \cdot \mathbf{B}_\perp = -ikB_z,$$

$$\hat{\mathbf{z}} \cdot (\boldsymbol{\nabla}_\perp \times \mathbf{E}_\perp) = \frac{i\omega}{c} B_z,$$

$$ik\mathbf{E}_\perp + \frac{i\omega}{c}\hat{\mathbf{z}} \times \mathbf{B}_\perp = \boldsymbol{\nabla}_\perp E_z,$$

$$\hat{\mathbf{z}} \cdot (\boldsymbol{\nabla}_\perp \times \mathbf{B}_\perp) = -\frac{i\epsilon\mu\omega}{c} E_z,$$

$$ik\mathbf{B}_\perp - \frac{i\epsilon\mu\omega}{c}\hat{\mathbf{z}} \times \mathbf{E}_\perp = \boldsymbol{\nabla}_\perp B_z. \tag{7.14}$$

There are several things to note from Eq. (7.14). First, we note that the transverse components of the fields are completely determined from a knowledge of E_z, B_z. For example, from the fourth and the last equations in (7.14), we obtain trivially that (for example, multiply the fourth equation with (ik) and then use the last equation)

$$\mathbf{E}_\perp = \frac{i}{\frac{\epsilon\mu\omega^2}{c^2} - k^2} \left(k\boldsymbol{\nabla}_\perp E_z - \frac{\omega}{c}\hat{\mathbf{z}} \times \boldsymbol{\nabla}_\perp B_z\right),$$

$$\mathbf{B}_\perp = \frac{i}{\frac{\epsilon\mu\omega^2}{c^2} - k^2} \left(k\boldsymbol{\nabla}_\perp B_z + \frac{\epsilon\mu\omega}{c}\hat{\mathbf{z}} \times \boldsymbol{\nabla}_\perp E_z\right). \tag{7.15}$$

(Here, we are considering a wave traveling along the positive z-axis. For a wave traveling in the reverse direction, we simply let $k \to -k$.) Furthermore, from the fourth equation in (7.14) we obtain (using the first as well as the fifth equations)

$$\boldsymbol{\nabla}_\perp \cdot \left(ik\mathbf{E}_\perp + \frac{i\omega}{c} \, \hat{\mathbf{z}} \times \mathbf{B}_\perp \right) = \boldsymbol{\nabla}_\perp^2 E_z,$$

$$\text{or,} \quad \boldsymbol{\nabla}_\perp^2 E_z + \left(\frac{\epsilon\mu\omega^2}{c^2} - k^2 \right) E_z = 0. \tag{7.16}$$

Similarly, it follows from Eq. (7.14) that

$$\boldsymbol{\nabla}_\perp^2 B_z + \left(\frac{\epsilon\mu\omega^2}{c^2} - k^2 \right) B_z = 0. \tag{7.17}$$

The solutions of the wave equation are correspondingly classified into three categories depending on the values of E_z, B_z. When $E_z = 0 = B_z$, namely, if both the electric and the magnetic fields have only transverse components, then the solution is known as the TEM (transverse electromagnetic) solution. This is the case already seen for plane wave traveling solutions following from Maxwell's equations. From Eqs. (7.14) and (7.15), it is clear that such solutions exist only if

$$k = k_0 = \frac{\sqrt{\epsilon\mu}\,\omega}{c}, \tag{7.18}$$

and that in such a case it follows, for example, from the last relation in (7.14) that (remember $B_z = 0$)

$$\mathbf{B}_\perp = \pm \frac{\epsilon\mu\omega}{kc} \, \hat{\mathbf{z}} \times \mathbf{E}_\perp = \pm\sqrt{\epsilon\mu} \, \hat{\mathbf{z}} \times \mathbf{E}_\perp, \tag{7.19}$$

as we have discussed earlier (see, for example, (6.30)). (The two signs correspond to the waves traveling along $\pm z$ directions respectively.) If $B_z = 0$ but $E_z \neq 0$, then the corresponding solution is known as the TM (transverse magnetic) solution since the magnetic field is transverse to the direction of propagation in this case. In this case, from (7.15) (or alternatively from the last relation in (7.14)) we have

$$\mathbf{B}_\perp = \frac{\frac{i\epsilon\mu\omega}{c}}{\frac{\epsilon\mu\omega^2}{c^2} - k^2} \, \hat{\mathbf{z}} \times \boldsymbol{\nabla}_\perp E_z$$

$$= \frac{\frac{i\epsilon\mu\omega}{c}}{\frac{\epsilon\mu\omega^2}{c^2} - k^2} \, \hat{\mathbf{z}} \times \left(\frac{-i(\frac{\epsilon\mu\omega^2}{c^2} - k^2)}{k} \mathbf{E}_\perp \right),$$

$$\text{or,} \quad \mathbf{B}_\perp = \frac{\epsilon\mu\omega}{kc} \, \hat{\mathbf{z}} \times \mathbf{E}_\perp. \tag{7.20}$$

Similarly, if $E_z = 0$ but $B_z \neq 0$, then the corresponding solution is known as the TE (transverse electric) solution because, in this case, the electric field is transverse to the direction of propagation. It follows from (7.15) (or alternatively from the fourth relation in (7.14)) that in this case, we can write

$$\mathbf{B}_\perp = \frac{kc}{\omega} \hat{\mathbf{z}} \times \mathbf{E}_\perp. \tag{7.21}$$

Furthermore, we note that since the direction of propagation is chosen to be along the z-axis, the boundary conditions (7.4) for a wave guide take the simpler form

$$\mathbf{E}_{\text{tan}}| = 0 = \frac{\partial B_z}{\partial n}\bigg|, \tag{7.22}$$

where $\mathbf{E}_{\text{tan}}$ represents the component of the electric field tangential to the boundary surface and $\frac{\partial}{\partial n}$ denotes derivative along the normal direction and the second identity in (7.22) follows from using the second and the last equation in (7.14). Namely, we note that

$$\hat{\mathbf{n}} \cdot (\hat{\mathbf{z}} \times \mathbf{E}_\perp) = -\hat{\mathbf{z}} \cdot (\hat{\mathbf{n}} \times \mathbf{E}_\perp) = -\hat{\mathbf{z}} \cdot (\hat{\mathbf{n}} \times (\mathbf{E} - \hat{\mathbf{z}} E_z))$$

$$= -\hat{\mathbf{z}} \cdot \mathbf{E}_{\text{tan}}, \tag{7.23}$$

which vanishes at the boundary. Furthermore, we recognize that the second boundary condition in (7.4) implies that (recall that $\hat{\mathbf{n}} \cdot \hat{\mathbf{z}} = 0$)

$$\hat{\mathbf{n}} \cdot (\mathbf{B}_\perp + \hat{\mathbf{z}} B_z) = \hat{\mathbf{n}} \cdot \mathbf{B}_\perp, \tag{7.24}$$

which vanishes at the boundary. Using (7.23) and (7.24) in the last equation of (7.14) leads to

$$\hat{\mathbf{n}} \cdot \boldsymbol{\nabla}_\perp B_z| = \hat{\mathbf{n}} \cdot \left(\boldsymbol{\nabla} - \hat{\mathbf{z}} \frac{\partial}{\partial z}\right) B_z\bigg| = \hat{\mathbf{n}} \cdot \boldsymbol{\nabla} B_z|$$

$$= \frac{\partial B_z}{\partial n}\bigg| = 0, \tag{7.25}$$

where we have identified $n = \hat{\mathbf{n}} \cdot \mathbf{x}$. With these basics, we are now ready to analyze the solutions in the case of a rectangular wave guide.

7.2 Rectangular wave guide

Let us consider a rectangular hollow tube along the z-axis and with transverse dimensions satisfying $0 \leq x \leq a, 0 \leq y \leq b$ as shown in

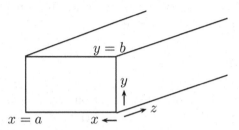

Figure 7.1: A rectangular wave guide along the z-axis.

Fig. 7.1. As we have noted, the walls of the tube are assumed to be perfect conductors. In this case, the solutions of Maxwell's equations have to satisfy the boundary conditions (7.22). It is clear that, in this case, for the wall in the x-z plane, $\hat{n}$ is parallel to $\hat{y}$ and this leads to the boundary conditions (see (7.22))

$$E_x|_{y=0,b} = 0 = E_z|_{y=0,b}, \quad \frac{\partial B_z}{\partial y}\bigg|_{y=0,b} = 0. \tag{7.26}$$

On the other hand, for the wall in the y-z plane, $\hat{n}$ is parallel to $\hat{x}$ leading to the boundary conditions

$$E_y|_{x=0,a} = 0 = E_z|_{x=0,a}, \quad \frac{\partial B_z}{\partial x}\bigg|_{x=0,a} = 0. \tag{7.27}$$

Equations (7.26) and (7.27) define all the boundary conditions in this case. We note that Eq. (7.15) explicitly takes the forms

$$E_x = \frac{i}{\frac{\epsilon\mu\omega^2}{c^2} - k^2} \left(k\frac{\partial E_z}{\partial x} + \frac{\omega}{c}\frac{\partial B_z}{\partial y} \right),$$

$$E_y = \frac{i}{\frac{\epsilon\mu\omega^2}{c^2} - k^2} \left(k\frac{\partial E_z}{\partial y} - \frac{\omega}{c}\frac{\partial B_z}{\partial x} \right),$$

$$B_x = \frac{i}{\frac{\epsilon\mu\omega^2}{c^2} - k^2} \left(k\frac{\partial B_z}{\partial x} - \frac{\epsilon\mu\omega}{c}\frac{\partial E_z}{\partial y} \right),$$

$$B_y = \frac{i}{\frac{\epsilon\mu\omega^2}{c^2} - k^2} \left(k\frac{\partial B_z}{\partial y} + \frac{\epsilon\mu\omega}{c}\frac{\partial E_z}{\partial x} \right), \tag{7.28}$$

while the z-components of the fields satisfy

$$\frac{\partial^2 E_z}{\partial x^2} + \frac{\partial^2 E_z}{\partial y^2} + \left(\frac{\epsilon\mu\omega^2}{c^2} - k^2\right) E_z = 0,$$

$$\frac{\partial^2 B_z}{\partial x^2} + \frac{\partial^2 B_z}{\partial y^2} + \left(\frac{\epsilon\mu\omega^2}{c^2} - k^2\right) B_z = 0. \tag{7.29}$$

Equations (7.28) and (7.29) have to be solved subject to the boundary conditions (7.26) and (7.27).

7.2.1 TM waves. In this case, we assume that $B_z = 0$. The solution for the z-component of the electric field in Eq. (7.29), subject to the boundary conditions in (7.26) and (7.27), yields

$$E_z(\mathbf{x}_\perp) = E_z(x, y) = A \sin\frac{\pi m x}{a} \sin\frac{\pi n y}{b}, \tag{7.30}$$

where A is an arbitrary constant and

$$k^2 = \frac{\epsilon\mu\omega^2}{c^2} - \left(\frac{\pi^2 m^2}{a^2} + \frac{\pi^2 n^2}{b^2}\right), \qquad m, n = 1, 2, \ldots. \tag{7.31}$$

The transverse components of the fields are then determined from (7.28) to be

$$E_x(\mathbf{x}_\perp) = \frac{i}{\frac{\epsilon\mu\omega^2}{c^2} - k^2} \frac{kA\pi m}{a} \cos\frac{\pi m x}{a} \sin\frac{\pi n y}{b},$$

$$E_y(\mathbf{x}_\perp) = \frac{i}{\frac{\epsilon\mu\omega^2}{c^2} - k^2} \frac{kA\pi n}{b} \sin\frac{\pi m x}{a} \cos\frac{\pi n y}{b},$$

$$B_x(\mathbf{x}_\perp) = -\frac{i}{\frac{\epsilon\mu\omega^2}{c^2} - k^2} \frac{\epsilon\mu\omega A\pi n}{cb} \sin\frac{\pi m x}{a} \cos\frac{\pi n y}{b},$$

$$B_y(\mathbf{x}_\perp) = \frac{i}{\frac{\epsilon\mu\omega^2}{c^2} - k^2} \frac{\epsilon\mu\omega A\pi m}{ca} \cos\frac{\pi m x}{a} \sin\frac{\pi n y}{b}. \tag{7.32}$$

It is clear from (7.32) that we can write, in this case,

$$\mathbf{B}_\perp = \frac{\epsilon\mu\omega}{kc} \, \hat{\mathbf{z}} \times \mathbf{E}_\perp, \tag{7.33}$$

consistent with (7.20). Note that all the transverse components of the fields satisfy the boundary conditions in (7.26) and (7.27).

The solutions in (7.30)–(7.32), corresponding to fixed integers, m, n, are known as TM$_{mn}$ modes. From (7.31), it is clear that the constant k depends on ω as well as the values of the integers m, n. For small values of ω, it follows from

$$k = \sqrt{\frac{\epsilon\mu\omega^2}{c^2} - \left(\frac{\pi^2 m^2}{a^2} + \frac{\pi^2 n^2}{b^2}\right)}, \qquad (7.34)$$

that the quantity inside the square root can be negative. In this case, k will become purely imaginary. As a result, there will be no propagation inside the wave guide. Rather, the electromagnetic signal will be attenuated along the tube. Defining

$$\omega_{mn} = \frac{\pi c}{\sqrt{\epsilon\mu}} \sqrt{\frac{m^2}{a^2} + \frac{n^2}{b^2}}, \qquad k = \frac{\sqrt{\epsilon\mu}}{c} \sqrt{\omega^2 - \omega_{mn}^2}, \qquad (7.35)$$

we see that propagation of the TM_{mn} wave can take place inside a rectangular wave guide only if $\omega > \omega_{mn}$ which is known as the cut-off frequency below which propagation of the TM_{mn} wave will not occur. Above this frequency, the TM_{mn} wave will propagate without any attenuation for a wave guide with perfectly conducting walls.

The wavelength of propagation inside the wave guide for the TM_{mn} wave is easily obtained to be

$$\lambda = \frac{2\pi}{k} = \frac{2\pi c}{\sqrt{\epsilon\mu}} \frac{1}{\sqrt{\omega^2 - \omega_{mn}^2}}$$

$$= \frac{2\pi c}{\sqrt{\epsilon\mu}} \frac{1}{\sqrt{\omega^2 - \frac{\pi^2 c^2}{\epsilon\mu}\left(\frac{m^2}{a^2} + \frac{n^2}{b^2}\right)}}. \qquad (7.36)$$

Correspondingly, the velocity of propagation inside the wave guide is given by

$$v = \frac{\omega}{k} = \frac{c}{\sqrt{\epsilon\mu}} \frac{\omega}{\sqrt{\omega^2 - \omega_{mn}^2}}$$

$$= \frac{c}{\sqrt{\epsilon\mu}} \frac{\omega}{\sqrt{\omega^2 - \frac{\pi^2 c^2}{\epsilon\mu}\left(\frac{m^2}{a^2} + \frac{n^2}{b^2}\right)}}. \qquad (7.37)$$

Note that, in deriving all these results, we have assumed the inside of the wave guide to be filled with an arbitrary dielectric. If we assume that it is empty space inside, then in this case we can identify $\epsilon\mu = 1$ and the formulae in this section simplify. With this, we note from Eq. (7.37) that the velocity of propagation of the wave, inside the wave guide, is larger than its value in free space. We note that it is infinite at the cut-off frequency. As the frequency is increased, it decreases and for arbitrarily large frequencies, it approaches c, the speed of light in free space (for $\epsilon\mu = 1$). The fact that the velocity of propagation is larger than the speed of light is not disturbing since this corresponds to the phase velocity of the wave. We note, from (7.31) (or (7.34)), that the group velocity, in this case, is given by

$$v_g = \frac{\partial\omega}{\partial k} = \frac{c^2}{\epsilon\mu}\frac{k}{\omega} = \frac{c^2}{\epsilon\mu}\frac{1}{v},$$

$$\text{or,} \quad v_g v = \frac{c^2}{\epsilon\mu}, \tag{7.38}$$

so that, when the phase velocity is larger than the speed of light, the group velocity is smaller, as it should be.

As is clear, the wave guide can support an infinite number of TM modes. It is for the mode corresponding to $m = 1 = n$ that the cut-off frequency is the smallest and would correspond to the dominant TM mode. In this case, with $\epsilon\mu = 1$, we have

$$\omega_{11} = \frac{\pi c\sqrt{a^2 + b^2}}{ab},$$

$$\lambda = 2\pi \frac{c}{\sqrt{\omega^2 - \frac{\pi^2 c^2(a^2+b^2)}{a^2 b^2}}},$$

$$v = c \frac{\omega}{\sqrt{\omega^2 - \frac{\pi^2 c^2(a^2+b^2)}{a^2 b^2}}}. \tag{7.39}$$

7.2.2 TE waves. In the case of TE modes, we have $E_z = 0$. It follows, then, that the solution of (7.29) subject to the boundary conditions in (7.26) and (7.27) is given by

$$B_z(\mathbf{x}_\perp) = B_z(x, y) = C\cos\frac{\pi m x}{a}\cos\frac{\pi n y}{b}, \tag{7.40}$$

where C is an arbitrary constant and we have, as in Eq. (7.31),

$$k^2 = \frac{\epsilon\mu\omega^2}{c^2} - \left(\frac{\pi^2 m^2}{a^2} + \frac{\pi^2 n^2}{b^2}\right), \quad m, n = 0, 1, 2, \ldots. \tag{7.41}$$

The transverse components of the electric and the magnetic fields, in this case, become

$$E_x(\mathbf{x}_\perp) = -\frac{i}{\frac{\epsilon\mu\omega^2}{c^2} - k^2} \frac{\omega C\pi n}{cb} \cos \frac{\pi m x}{a} \sin \frac{\pi n y}{b},$$

$$E_y(\mathbf{x}_\perp) = \frac{i}{\frac{\epsilon\mu\omega^2}{c^2} - k^2} \frac{\omega C\pi m}{ca} \sin \frac{\pi m x}{a} \cos \frac{\pi n y}{b},$$

$$B_x(\mathbf{x}_\perp) = -\frac{i}{\frac{\epsilon\mu\omega^2}{c^2} - k^2} \frac{kC\pi m}{a} \sin \frac{\pi m x}{a} \cos \frac{\pi n y}{b},$$

$$B_y(\mathbf{x}_\perp) = -\frac{i}{\frac{\epsilon\mu\omega^2}{c^2} - k^2} \frac{kC\pi n}{b} \cos \frac{\pi m x}{a} \sin \frac{\pi n y}{b}. \tag{7.42}$$

It follows from this that, for TE waves, we can write

$$\mathbf{B}_\perp(\mathbf{x}_\perp) = \frac{kc}{\omega} \hat{\mathbf{z}} \times \mathbf{E}_\perp(\mathbf{x}_\perp), \tag{7.43}$$

consistent with (7.21). We see that the transverse components of the fields satisfy the boundary conditions in (7.26).

Like the TM waves, we see that a rectangular wave guide can also support an infinite number of TE modes known as TE_{mn} modes, but unlike the TM modes, here it is possible to have $m = 0$ or $n = 0$. However, from the form of the transverse fields in (7.42), we see that we cannot have $m = 0 = n$ because, in that case, all the transverse fields would vanish and we will have $B_z = $ constant which would correspond to a trivial solution. As in the case of TM modes (see Eq. (7.35)), here, too, there is an analogous cut-off frequency for TE_{mn} waves given by

$$\omega_{mn} = \frac{\pi c}{\sqrt{\epsilon\mu}} \sqrt{\frac{m^2}{a^2} + \frac{n^2}{b^2}}, \tag{7.44}$$

and the discussion of the wave length of propagation as well as the velocity of propagation goes through in a completely parallel manner. However, in this case since we can have $m = 0$ or $n = 0$, the lowest cut-off frequency will be for the TE_{10} mode (where we are assuming that $a > b$) given by

$$\omega_{10} = \frac{\pi c}{\sqrt{\epsilon\mu} a}. \tag{7.45}$$

Correspondingly, the TE_{10} mode is called the dominant mode in a rectangular wave guide. It is also clear from (7.42) that in this case

$E_z = 0 = E_x$. As a result, for this mode the electric field is parallel to the y-axis everywhere. Furthermore, for the dominant TE_{10} mode, we have

$$\lambda = \frac{2\pi}{k} = \frac{2\pi c}{\sqrt{\epsilon\mu}} \frac{1}{\sqrt{\omega^2 - \frac{\pi^2 c^2}{\epsilon\mu a^2}}},$$

$$v = \frac{\omega}{k} = \frac{c}{\sqrt{\epsilon\mu}} \frac{\omega}{\sqrt{\omega^2 - \frac{\pi^2 c^2}{\epsilon\mu a^2}}}. \tag{7.46}$$

7.3 Cylindrical wave guide

The other kind of wave guide that is commonly used is the cylindrical wave guide. Here, we have a hollow cylindrical tube of radius a with a perfectly conducting wall and length of the tube along the z-axis as shown in Fig. 7.2. Because of the symmetry in the problem, it is simpler to study this problem in cylindrical coordinates defined by

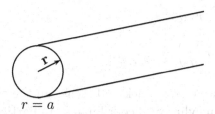

Figure 7.2: A cylindrical wave guide of radius a along the z-axis.

$$x = r\cos\phi, \qquad y = r\sin\phi, \qquad z = z. \tag{7.47}$$

Here, r, ϕ denote the radial and the angular coordinates on a transverse plane. The unit vectors satisfy

$$\hat{\mathbf{r}} \times \hat{\boldsymbol{\phi}} = \hat{\mathbf{z}}, \qquad \hat{\boldsymbol{\phi}} \times \hat{\mathbf{z}} = \hat{\mathbf{r}}, \qquad \hat{\mathbf{z}} \times \hat{\mathbf{r}} = \hat{\boldsymbol{\phi}}.$$

The unit vectors, in non-Cartesian coordinate systems, are not fixed and their variations can be determined from the transformation rules

$$\hat{\mathbf{x}} = \hat{\mathbf{r}} \cos \phi - \hat{\boldsymbol{\phi}} \sin \phi,$$

$$\hat{\mathbf{y}} = \hat{\mathbf{r}} \sin \phi + \hat{\boldsymbol{\phi}} \cos \phi,$$

$$\hat{\mathbf{z}} = \hat{\mathbf{z}}.$$

These can be inverted to give

$$\hat{\mathbf{r}} = \hat{\mathbf{x}} \cos \phi + \hat{\mathbf{y}} \sin \phi,$$

$$\hat{\boldsymbol{\phi}} = -\hat{\mathbf{x}} \sin \phi + \hat{\mathbf{y}} \cos \phi,$$

$$\hat{\mathbf{z}} = \hat{\mathbf{z}}.$$

From these, it follows that the only non-trivial variation of the unit vectors are given by

$$\frac{\partial \hat{\mathbf{r}}}{\partial \phi} = \hat{\boldsymbol{\phi}}, \qquad \frac{\partial \hat{\boldsymbol{\phi}}}{\partial \phi} = -\hat{\mathbf{r}}. \tag{7.48}$$

We also note here that in cylindrical coordinates,

$$\boldsymbol{\nabla}_{\perp} = \hat{\mathbf{r}} \, \frac{\partial}{\partial r} + \frac{\hat{\boldsymbol{\phi}}}{r} \, \frac{\partial}{\partial \phi},$$

$$\boldsymbol{\nabla}_{\perp}^2 = \frac{1}{r} \frac{\partial}{\partial r} \left(r \frac{\partial}{\partial r} \right) + \frac{1}{r^2} \frac{\partial^2}{\partial \phi^2}.$$

With these, we can go back and recast all the equations (7.13)–(7.17) and (7.22) in cylindrical coordinates. In particular, the boundary conditions (7.22) in the case of cylindrical wave guides take the form (here $\hat{\mathbf{n}}$ is parallel to $\hat{\mathbf{r}}$)

$$E_z|_{r=a} = 0 = E_\phi|_{r=a} \,, \qquad \left. \frac{\partial B_z}{\partial r} \right|_{r=a} = 0. \tag{7.49}$$

The relation for the transverse components in (7.15) take the forms

$$E_r(\mathbf{x}_\perp) = E_r(r, \phi) = \frac{i}{\frac{\epsilon \mu \omega^2}{c^2} - k^2} \left(k \frac{\partial E_z}{\partial r} + \frac{\omega}{cr} \frac{\partial B_z}{\partial \phi} \right),$$

$$E_\phi(\mathbf{x}_\perp) = E_\phi(r, \phi) = \frac{i}{\frac{\epsilon \mu \omega^2}{c^2} - k^2} \left(\frac{k}{r} \frac{\partial E_z}{\partial \phi} - \frac{\omega}{c} \frac{\partial B_z}{\partial r} \right),$$

$$B_r(\mathbf{x}_\perp) = B_r(r,\phi) = \frac{i}{\frac{\epsilon\mu\omega^2}{c^2} - k^2}\left(k\frac{\partial B_z}{\partial r} - \frac{\epsilon\mu\omega}{cr}\frac{\partial E_z}{\partial\phi}\right),$$

$$B_\phi(\mathbf{x}_\perp) = B_\phi(r,\phi) = \frac{i}{\frac{\epsilon\mu\omega^2}{c^2} - k^2}\left(\frac{k}{r}\frac{\partial B_z}{\partial\phi} + \frac{\epsilon\mu\omega}{c}\frac{\partial E_z}{\partial r}\right), \quad (7.50)$$

while the equations for the E_z, B_z (see Eqs. (7.16)–(7.17)) take the forms

$$\frac{\partial^2 E_z}{\partial r^2} + \frac{1}{r}\frac{\partial E_z}{\partial r} + \frac{1}{r^2}\frac{\partial^2 E_z}{\partial\phi^2} + \left(\frac{\epsilon\mu\omega^2}{c^2} - k^2\right)E_z = 0,$$

$$\frac{\partial^2 B_z}{\partial r^2} + \frac{1}{r}\frac{\partial B_z}{\partial r} + \frac{1}{r^2}\frac{\partial^2 B_z}{\partial\phi^2} + \left(\frac{\epsilon\mu\omega^2}{c^2} - k^2\right)B_z = 0. \quad (7.51)$$

Physical solutions have to be periodic in the angular variable. Consequently, we can extract the ϕ dependence of the z-components of the fields as

$$E_z(r,\phi) = E_z^{(0)}(r)\left(A_n e^{in(\phi+\phi_0)} + B_n e^{-in(\phi+\phi_0)}\right),$$

$$B_z(r,\phi) = B_z^{(0)}(r)\left(C_n e^{in(\phi+\phi_0)} + D_n e^{-in(\phi+\phi_0)}\right). \quad (7.52)$$

Furthermore, by choosing the arbitrary constants $A_n, B_n, C_n, D_n, \phi_0$ appropriately (which is equivalent to choosing an axis of orientation of the wave), we can write

$$E_z(r,\phi) = E_z^{(0)}(r)\cos n\phi,$$

$$B_z(r,\phi) = B_z^{(0)}(r)\cos n\phi, \quad (7.53)$$

where $n = 0, 1, 2, \ldots$ and the overall constants have been absorbed into the definitions of $E_z^{(0)}, B_z^{(0)}$. Thus, Eq. (7.51) takes the form

$$\frac{d^2 E_z^{(0)}}{dr^2} + \frac{1}{r}\frac{dE_z^{(0)}}{dr} + \left(\frac{\epsilon\mu\omega^2}{c^2} - k^2 - \frac{n^2}{r^2}\right)E_z^{(0)} = 0,$$

$$\frac{d^2 B_z^{(0)}}{dr^2} + \frac{1}{r}\frac{dB_z^{(0)}}{dr} + \left(\frac{\epsilon\mu\omega^2}{c^2} - k^2 - \frac{n^2}{r^2}\right)B_z^{(0)} = 0. \quad (7.54)$$

Each of the equations in (7.54) is a Bessel equation of order n. (Fig. 7.3 shows the behavior of $J_0(x)$ as a function of x.) Consequently, we can write the solutions of (7.51) as

$$E_z(r,\phi) = a_n J_n(hr)\cos n\phi,$$

$$B_z(r,\phi) = b_n J_n(hr)\cos n\phi, \quad (7.55)$$

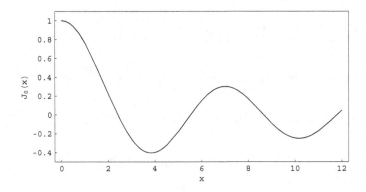

Figure 7.3: The behavior of the Bessel function $J_0(x)$ as a function of x.

where we have defined

$$h = \sqrt{\frac{\epsilon\mu\omega^2}{c^2} - k^2}. \tag{7.56}$$

7.3.1 TM waves. In this case, we assume that $B_z = 0$ (namely, $b_n = 0$), whereas E_z has the form given in (7.55). The boundary condition on E_z, (7.49), implies that

$$J_n(ha) = 0. \tag{7.57}$$

There are an infinite number of zeros of the Bessel function for any n and denoting all such roots as $(ha)_{mn}$ (implying the mth zero of J_n), we note that the first few take the values

$$(ha)_{10} = 2.405, \quad (ha)_{11} = 3.85, \quad (ha)_{12} = 7.02,$$
$$(ha)_{20} = 5.52, \cdots . \tag{7.58}$$

From (7.56), this determines the constant k to be

$$k = \sqrt{\frac{\epsilon\mu\omega^2}{c^2} - h_{mn}^2}. \tag{7.59}$$

The transverse components of the fields can now be determined from (7.50) to be

$$E_r(r,\phi) = \frac{ia_n}{\frac{\epsilon\mu\omega^2}{c^2} - k^2} \, k \, J_n'(hr) \cos n\phi,$$

$$E_\phi(r,\phi) = -\frac{ia_n}{\frac{\epsilon\mu\omega^2}{c^2} - k^2} \, \frac{kn}{r} \, J_n(hr) \sin n\phi,$$

$$B_r(r,\phi) = \frac{ia_n}{\frac{\epsilon\mu\omega^2}{c^2} - k^2} \, \frac{\epsilon\mu\omega n}{cr} \, J_n(hr) \sin n\phi,$$

$$B_\phi(r,\phi) = \frac{ia_n}{\frac{\epsilon\mu\omega^2}{c^2} - k^2} \, \frac{\epsilon\mu\omega}{c} \, J_n'(hr) \cos n\phi. \tag{7.60}$$

Here, a prime denotes a derivative with respect to the radial coordinate. We note from Eq. (7.60) that the transverse field E_ϕ satisfies the boundary conditions in (7.49). It is clear from Eq. (7.60) that we can write, in this case,

$$\mathbf{B}_\perp = \frac{\epsilon\mu\omega}{kc} \, \hat{\mathbf{z}} \times \mathbf{E}_\perp, \tag{7.61}$$

consistent with (7.20).

The TM modes with different values for (ha) (see (7.58)) are known as TM_{mn} modes of the cylindrical wave guide. From Eq. (7.59) we note that, as in the case of the rectangular wave guide, there is also a cut-off frequency here given by

$$\omega_{mn} = \frac{c}{\sqrt{\epsilon\mu}} \, h_{mn}. \tag{7.62}$$

For $\omega < \omega_{mn}$, there is no transmission of the TM_{mn} wave in a cylindrical wave guide. With $\omega > \omega_{mn}$, the wave is propagated without any attenuation for perfectly conducting outer boundaries. The wave length of propagation, in this case, is given by

$$\lambda = \frac{2\pi}{k} = \frac{2\pi c}{\sqrt{\epsilon\mu}} \, \frac{1}{\sqrt{\omega^2 - \omega_{mn}^2}}, \tag{7.63}$$

while the velocity of propagation takes the form

$$v = \frac{\omega}{k} = \frac{c}{\sqrt{\epsilon\mu}} \, \frac{\omega}{\sqrt{\omega^2 - \omega_{mn}^2}}. \tag{7.64}$$

The velocity, as in the case of the rectangular wave guide, is infinitely large near the cut-off frequency and goes to the speed of light in a dielectric for asymptotically large frequencies (in free space, this is just c). As is clear from Eqs. (7.58) and (7.62), the dominant TM mode in a cylindrical wave guide is the TM_{10} mode.

7.3.2 TE waves. In the case of TE waves, we assume that $E_z = 0$ (namely, $a_n = 0$) while B_z has the form given in Eq. (7.55). However, in this case, the boundary condition on the magnetic field, (7.49), requires that

$$J_n'(ha) = 0. \tag{7.65}$$

In contrast to the TM case, here we are not interested in the zeros of the Bessel function, rather the locations where the Bessel function has vanishing slope are important. Once again, for every value of n, there is an infinite number of such points given by

$$(ha)_{10} = 3.83, \quad (ha)_{11} = 1.84, \quad (ha)_{12} = 3.05,$$

$$(ha)_{20} = 7.02, \quad (ha)_{21} = 5.33, \cdots, \tag{7.66}$$

and as in (7.59), the value of k is determined from (7.56) to be

$$k = \sqrt{\frac{\epsilon\mu\omega^2}{c^2} - h_{mn}^2}. \tag{7.67}$$

The transverse components of the fields are now determined from (7.50) and (7.55) to be

$$E_r(r, \phi) = -\frac{ib_n}{\frac{\epsilon\mu\omega^2}{c^2} - k^2} \frac{\omega n}{cr} J_n(hr) \sin n\phi,$$

$$E_\phi(r, \phi) = -\frac{ib_n}{\frac{\epsilon\mu\omega^2}{c^2} - k^2} \frac{\omega}{c} J_n'(hr) \cos n\phi,$$

$$B_r(r, \phi) = \frac{ib_n}{\frac{\epsilon\mu\omega^2}{c^2} - k^2} k\, J_n'(hr) \cos n\phi,$$

$$B_\phi(r, \phi) = -\frac{ib_n}{\frac{\epsilon\mu\omega^2}{c^2} - k^2} \frac{kn}{r} J_n(hr) \sin n\phi. \tag{7.68}$$

It is clear from Eq. (7.68) that E_ϕ satisfies the boundary condition in (7.49) by virtue of (7.65) and that we can write

$$\mathbf{B}_\perp = \frac{kc}{\omega} \hat{\mathbf{z}} \times \mathbf{E}_\perp, \tag{7.69}$$

consistent with (7.21).

The TE modes with different roots in (7.66) are known as TE$_{mn}$ modes and it is clear that, as in the case of the TM modes, the cut-off frequency is given by

$$\omega_{mn} = \frac{c}{\sqrt{\epsilon\mu}} h_{mn},$$

and correspondingly, for the propagating modes, $\omega > \omega_{mn}$, the wave length and the velocity of propagation are given as in Eqs. (7.63) and (7.64). We note from (7.66) that the dominant mode, in a cylindrical wave guide, is the TE_{11} mode. Thus, we see that the rectangular and the cylindrical wave guides have similar qualitative features.

7.4 Impossibility of having TEM waves in a wave guide

Although we have classified electromagnetic waves into three categories, in discussing propagation in a rectangular or a cylindrical wave guide, we have only discussed the TM and TE modes of propagation. The reason for this is that in a hollow wave guide, rectangular or cylindrical, TEM waves cannot be present. This can be easily seen in the following manner. Let us note that by definition, both the electric and the magnetic fields are transverse in a TEM wave. Namely, for such a solution, we must have

$$E_z = 0 = B_z.$$

We know from Maxwell's equation, $\nabla \cdot \mathbf{B} = 0$, that the magnetic field lines must form closed loops. In particular, when $B_z = 0$, they must form closed loops in the transverse plane to the z-axis. Furthermore, from the last equation of Maxwell in (7.1), we see that in the absence of a conduction current

$$\nabla \times \mathbf{B} = \frac{\epsilon \mu}{c} \frac{\partial \mathbf{E}}{\partial t},$$

so that these closed magnetic loops must enclose the "displacement" current. However, since the closed magnetic loops are in the transverse plane, this is possible only if the "displacement" current has a component along the z-axis, which would imply that the electric field itself has a time varying component along the z-axis. This is, however, in contradiction to the requirement of a TEM wave for which $E_z = 0$.

Thus, we see that in a hollow wave guide, there cannot be any TEM wave present. The TM and the TE waves represent all the modes that can be present in such a system. On the other hand, waves guided by two infinite parallel conducting plates, for example, can support TEM modes. This can be seen from our previous analysis of the rectangular wave guides by taking the limit $a \to \infty$. In this limit, since there is no boundary in the x-direction, we can assume that the fields are uniform along this direction (namely, there is no x

dependence because of translation invariance). The TM solution, in this case, is easily seen to lead from (7.29) and (7.14) to

$$B_z = 0,$$

$$E_z(y) = \frac{\pi m C}{b} \sin \frac{\pi m y}{b},$$

$$E_x(y) = 0,$$

$$E_y(y) = ikC \cos \frac{\pi m y}{b},$$

$$B_x(y) = -\frac{i\epsilon\mu\omega C}{c} \cos \frac{\pi m y}{b} = -\frac{\epsilon\mu\omega}{kc} E_y,$$

$$B_y = 0. \tag{7.70}$$

Here, we have

$$k = \sqrt{\frac{\epsilon\mu\omega^2}{c^2} - \frac{\pi^2 m^2}{b^2}}, \tag{7.71}$$

and $m = 0, 1, 2, \ldots$. In fact, when $m = 0$, we see that the field configuration has the form

$$B_z(y) = 0 = E_z(y),$$

$$E_x(y) = 0 = B_y(y),$$

$$E_y(y) = ikC,$$

$$B_x = -\frac{i\epsilon\mu\omega C}{c}, \tag{7.72}$$

with

$$k = \frac{\sqrt{\epsilon\mu}\,\omega}{c}. \tag{7.73}$$

This field configuration is clearly that of a TEM wave and we see that waves guided by two infinite parallel conducting planes can support TEM modes unlike a hollow rectangular wave guide. Furthermore, in this case, we see from (7.72) and (7.73) that we can write

$$\mathbf{B}_\perp(y) = \sqrt{\epsilon\mu}\,\hat{\mathbf{z}} \times \mathbf{E}_\perp, \tag{7.74}$$

as we would expect for a TEM wave (see Eq. (7.19)).

Let us note here that TEM waves can also be present if the wave guide is not completely hollow. For example, in the case of a coaxial cable, we have an inner conducting medium surrounded by an external conducting surface. In such a case, the inner conductor can carry a conduction current along the z-axis. The last equation of Maxwell in (7.1), in this case, has the form

$$\mathbf{\nabla} \times \mathbf{B} = \frac{4\pi\mu}{c}\,\mathbf{J} + \frac{\epsilon\mu}{c}\,\frac{\partial \mathbf{E}}{\partial t},$$

and the closed magnetic loops in the transverse plane can enclose the conduction current and we do not need a z-component of the electric field to be present. In fact, from the radial symmetry of the coaxial cable, we see that the electric field must be radial everywhere. As a result the "displacement" current must also be along the radial direction. Since the conduction current is along the z-axis and the "displacement" current along the radial direction, it follows from the above equation that the $\mathbf{B}$ field cannot have a component along the z-axis since there is no current along the ϕ (angular) direction. This shows that TEM wave is the only wave that can exist in a coaxial cable. A similar conclusion also follows for two wire transmission lines. Furthermore, since TEM waves do not have a cut-off frequency, coaxial cables or two wire transmission lines are used to transmit low frequency electromagnetic waves.

7.5 Wave impedance

Let us recall that the impedance of an electromagnetic wave is related to the proportionality constant between the transverse magnetic and electric fields, namely,

$$\mathbf{B}_\perp = \pm\frac{1}{Z}\,\hat{\mathbf{z}} \times \mathbf{E}_\perp, \tag{7.75}$$

where Z is known as the impedance (and the two signs correspond respectively to forward and backward traveling waves). For example, in free space ($\epsilon\mu = 1$), we see from Eq. (7.19) that the impedance for a TEM wave is unity. In general, in a dielectric medium the impedance for a TEM wave is

$$Z_{\text{TEM}} = \frac{1}{\sqrt{\epsilon\mu}}. \tag{7.76}$$

We can, similarly, determine the impedance for the TM and the TE waves in a rectangular as well as a cylindrical wave guide. From Eqs. (7.33), (7.43), (7.61) and (7.69), we see that the impedance for the rectangular as well as the cylindrical wave guides have similar forms, namely,

$$Z_{\mathrm{TM}} = \frac{kc}{\epsilon\mu\omega} = \frac{c}{\sqrt{\epsilon\mu}} \frac{k}{\omega} \frac{1}{\sqrt{\epsilon\mu}}$$

$$= \sqrt{1 - \left(\frac{\omega_{mn}}{\omega}\right)^2} \, Z_{\mathrm{TEM}},$$

$$Z_{\mathrm{TE}} = \frac{\omega}{kc} = \frac{\sqrt{\epsilon\mu}}{c} \frac{\omega}{k} \frac{1}{\sqrt{\epsilon\mu}}$$

$$= \frac{1}{\sqrt{1 - \left(\frac{\omega_{mn}}{\omega}\right)^2}} \, Z_{\mathrm{TEM}}. \tag{7.77}$$

Of course, the values of ω_{mn} are different for the two wave guides. However, for a given wave guide (for example, rectangular), we see from the forms of the impedances in (7.77) that if the space inside is empty ($\epsilon\mu = 1, Z_{\mathrm{TEM}} = 1$), then we can formally write

$$Z_{\mathrm{TM}} = \frac{1}{Z_{\mathrm{TE}}}. \tag{7.78}$$

In general, though, we have

$$Z_{\mathrm{TM}} = \frac{1}{Z_{\mathrm{TE}}} Z_{\mathrm{TEM}}^2. \tag{7.79}$$

7.6 Attenuation factor in wave guides

Thus far, we have discussed wave guides where the external wall is assumed to be a perfect conductor with $\sigma \to \infty$. In reality, however, the metallic conductor has a finite conductivity, be it very large. Normally, a metal is considered a good conductor if $\sigma \gg \frac{\epsilon\omega}{4\pi}$ (see (8.57)). When the conductivity is finite, as in a realistic conductor, the electromagnetic fields within the wave guide can penetrate inside the conducting walls. The penetration depth, also known as the skin depth, is normally very small and this phenomenon induces a surface current in the metal that plays a very important role. Namely, it leads to heating and Ohmic losses in the conductor. Even though this effect is small, it leads to an attenuation of the electromagnetic fields that are propagated inside the wave guide. This is reflected in the fact that in the regime of propagation (namely, for $\omega > \omega_{mn}$), the wave vector k becomes complex of the form (see discussion in section 8.4)

$$k \rightarrow k + i\alpha, \qquad \alpha > 0, \tag{7.80}$$

so that all the non-vanishing components of the fields have the z dependence of the form

$$\mathbf{E} \sim e^{ikz - \alpha z}, \qquad \mathbf{B} \sim e^{ikz - \alpha z}. \tag{7.81}$$

Here, α is known as the attenuation factor. In this case, the time averaged power transmitted along the z-axis can be obtained from the Poynting vector as

$$P(z) = \int_A da \, \hat{\mathbf{z}} \cdot \mathbf{S} = \frac{c}{8\pi} \int_A da \, \text{Re} \, \hat{\mathbf{z}} \cdot (\mathbf{E} \times \mathbf{H}^*)$$

$$= \frac{c}{8\pi} \int_A da \, \text{Re} \, \hat{\mathbf{z}} \cdot (\mathbf{E}_\perp \times \mathbf{H}_\perp^*) \sim e^{-2\alpha z}, \tag{7.82}$$

where A represents the cross sectional area of the wave guide perpendicular to the z-axis (we do not use the conventional symbol S to avoid confusion with the Poynting vector) and we have used the definition of the time averaged Poynting vector in (6.96)

$$\mathbf{S} = \frac{c}{8\pi} \text{Re} \, (\mathbf{E} \times \mathbf{H}^*). \tag{7.83}$$

It now follows that

$$\frac{dP(z)}{dz} = -2\alpha \, P(z)$$

$$\text{or,} \quad \alpha = \left| \frac{1}{2P(z)} \frac{dP(z)}{dz} \right|$$

$$= \frac{\text{Power lost per unit length}}{2 \times \text{Power transmitted}}. \tag{7.84}$$

Thus, we see that the attenuation factor for the wave guide can be calculated once we know the power lost as well as the power transmitted during the process of propagation.

A rigorous calculation of the attenuation factor can be carried out systematically, starting from Maxwell's equations with modified boundary conditions to take care of the finite conductivity of the metal. However, an approximate calculation that gives very good agreement with the actual results can be described as follows. Once we know the solutions for the case of the perfectly conducting wall,

it is reasonable to assume that the solutions inside the wave guide
are unaffected significantly if the conducting surface has a finite but
large conductivity. This allows us to calculate the power flow along
the wave guide as (see (7.82))

$$P(z) = -\frac{cZ\mu}{8\pi} \int_A da \, \text{Re} \, \hat{z} \cdot ((\hat{z} \times \mathbf{H}_\perp) \times \mathbf{H}_\perp^*)$$

$$= \frac{cZ\mu}{8\pi} \int_A da \, |\mathbf{H}_\perp|^2 , \tag{7.85}$$

where Z represents the impedance and we have used Eqs. (7.75) as
well as (7.2). As we have seen earlier in (7.77),

$$Z_{\text{TM}} = \frac{\sqrt{1 - \left(\frac{\omega_{mn}}{\omega}\right)^2}}{\sqrt{\epsilon\mu}}, \quad Z_{\text{TE}} = \frac{1}{\sqrt{\epsilon\mu\left(1 - \left(\frac{\omega_{mn}}{\omega}\right)^2\right)}}. \tag{7.86}$$

The power loss can be calculated in the following manner. By
assumption, the fields inside the wave guide are affected only slightly
when the conductivity is finite. This is particularly true for the tangential component of the magnetic field (tangential to the the wall,
$\mathbf{H}_{\text{tan}} = \hat{n} \times \mathbf{H}$) when the conductivity is large. Through the boundary conditions, we expect the tangential component of the magnetic
field to be continuous which gives us the tangential component of the
magnetic field on the surface of the conductor. This, in turn, leads
to the induced surface current given by (See discussion in section 8.1,
in particular, Eq. (8.9) as well as section 8.4.)

$$\frac{4\pi}{c} \mathbf{J}_s = \hat{n} \times \mathbf{H} = \mathbf{H}_{\text{tan}}. \tag{7.87}$$

The surface current leads to heating and, consequently, to power loss.
The power loss per unit length (along the z-axis) can be calculated
in the standard manner as (see, for example, (6.83) or (6.84))

$$\text{Power loss per unit length} = R_s \int_{\text{surf}} ds \, |\mathbf{J}_s|^2$$

$$= \frac{c^2 R_s}{16\pi^2} \int_{\text{surf}} ds \, |\mathbf{H}_{\text{tan}}|^2 , \tag{7.88}$$

where we have used (6.12) as well as (7.87) and (see (8.73))

$$R_s = \sqrt{\frac{2\pi\mu\omega}{\sigma c^2}}, \tag{7.89}$$

is the resistive component (real part) of the surface impedance of the conductor which has the form (see (8.72))

$$Z_s = (1 - i)\sqrt{\frac{2\pi\mu\omega}{\sigma c^2}}. \tag{7.90}$$

In (7.88), "surf" represents a surface of the conductor of unit length along the z-axis. Note that, for a perfect conductor, $\sigma \to \infty$ and the power loss vanishes so that waves travel unattenuated. However, when σ is finite, there is a power loss and Eqs. (7.85) and (7.88) determine the attenuation constant through (7.84).

A concept related to the attenuation factor is known as the quality factor or simply the Q factor of the wave guide, defined as

$$Q = \omega \frac{\text{Energy stored per unit length}}{\text{Energy lost per unit length per second}}, \tag{7.91}$$

where ω is the angular frequency of the wave. We note that, by definition,

$$\text{Power transmitted} = v_g \times \text{Energy stored per unit length}, \tag{7.92}$$

where v_g denotes the group velocity of propagation. Using this, we note that the Q factor can be written as

$$Q = \frac{\omega}{v_g} \frac{\text{Power transmitted}}{\text{Power lost per unit length}}$$

$$= \frac{\omega}{v_g} \frac{1}{2\alpha} = \frac{1}{2\alpha} \frac{\epsilon\mu\omega}{c^2} v$$

$$= \frac{\sqrt{\epsilon\mu}}{2\alpha c} \frac{\omega}{\sqrt{1 - \left(\frac{\omega_{mn}}{\omega}\right)^2}}, \tag{7.93}$$

where we have used Eqs. (7.37) and (7.38). Since, in wave guides, the attenuation factor α can be very low, it is possible to construct wave guides with large Q factors. This becomes quite important in the construction of resonating cavities, which we will study next.

7.7 Cavity resonators

Let us next consider a rectangular wave guide of length d along the z-axis. Furthermore, let us close the two ends of the wave guide with perfectly conducting metal walls. Then, it is clear that, in this case, the fields have to satisfy additional boundary conditions on the two new surfaces. From (7.4), we see, for example that the electric field components, E_x, E_y must vanish at $z = 0, d$. As a result, the z dependence of the fields cannot have the simple exponential form, as has been assumed in (7.13). Rather, the fields must be expanded in terms of $\sin kz$ and $\cos kz$, as is suitable for the appropriate boundary conditions. Let us note that, for a TM wave, $B_z = 0$ and the additional boundary condition required is obtained to be

$$\left. \frac{\partial E_z}{\partial z} \right|_{z=0,d} = 0, \tag{7.94}$$

which follows from the first equation of (7.9), since (7.1) requires that $\mathbf{E}_\perp = 0$ at $z = 0, d$. This, then, determines the form of E_z from Eq. (7.16) to be (compare with Eq. (7.30))

$$E_z(x, y, z) = A \sin \frac{\pi m x}{a} \sin \frac{\pi n y}{b} \cos \frac{\pi \ell z}{d}, \tag{7.95}$$

with $\ell = 0, 1, 2, \ldots$ as well as the usual restrictions on m, n and

$$\omega = \frac{\pi c}{\sqrt{\epsilon \mu}} \sqrt{\frac{m^2}{a^2} + \frac{n^2}{b^2} + \frac{\ell^2}{d^2}} = \omega_{mn\ell}, \tag{7.96}$$

which follows from (7.34) with the identification $k = \frac{\pi \ell}{d}$. The transverse field components can now be determined from the Maxwell's equations, (7.9), (7.10), (7.11) and (7.12), to be

$$\mathbf{E}_\perp(\mathbf{x}) = \mathbf{E}_\perp(x, y, z) = \frac{1}{\frac{\epsilon \mu \omega^2}{c^2} - \frac{\pi^2 \ell^2}{d^2}} \nabla_\perp \frac{\partial E_z}{\partial z},$$

$$\mathbf{B}_\perp(\mathbf{x}) = \mathbf{B}_\perp(x, y, z) = \frac{i}{\frac{\epsilon \mu \omega^2}{c^2} - \frac{\pi^2 \ell^2}{d^2}} \frac{\epsilon \mu \omega}{c} \hat{\mathbf{z}} \times \nabla_\perp E_z, \tag{7.97}$$

which lead explicitly to

$$E_x(x, y, z) = -A \frac{\frac{\pi^2 m \ell}{ad}}{\frac{\epsilon \mu \omega^2}{c^2} - \frac{\pi^2 \ell^2}{d^2}} \cos \frac{\pi m x}{a} \sin \frac{\pi n y}{b} \sin \frac{\pi \ell z}{d},$$

$$E_y(x, y, z) = -A \frac{\frac{\pi^2 n \ell}{bd}}{\frac{\epsilon \mu \omega^2}{c^2} - \frac{\pi^2 \ell^2}{d^2}} \sin \frac{\pi m x}{a} \cos \frac{\pi n y}{b} \sin \frac{\pi \ell z}{d},$$

$$B_x(x, y, z) = -iA \frac{\frac{\epsilon \mu \omega \pi n}{cb}}{\frac{\epsilon \mu \omega^2}{c^2} - \frac{\pi^2 \ell^2}{d^2}} \sin \frac{\pi m x}{a} \cos \frac{\pi n y}{b} \cos \frac{\pi \ell z}{d},$$

$$B_y(x, y, z) = iA \frac{\frac{\epsilon \mu \omega \pi m}{ca}}{\frac{\epsilon \mu \omega^2}{c^2} - \frac{\pi^2 \ell^2}{d^2}} \cos \frac{\pi m x}{a} \sin \frac{\pi n y}{b} \cos \frac{\pi \ell z}{d}. \tag{7.98}$$

It is easy to see from the explicit forms of $\mathbf{E}_\perp$ in (7.98) that they satisfy the required boundary condition (7.4) at $z = 0, d$ (namely, they vanish at these boundaries). Parenthetically, let us sketch here how, for example, the first relation in (7.97) will be obtained. From (7.9) we obtain

$$\mathbf{\nabla}_\perp \frac{\partial E_z}{\partial z} = -\mathbf{\nabla}_\perp \left(\mathbf{\nabla}_\perp \cdot \mathbf{E}_\perp \right)$$

$$= -\mathbf{\nabla}_\perp^2 \mathbf{E}_\perp - \mathbf{\nabla}_\perp \times \left(\mathbf{\nabla}_\perp \times \mathbf{E}_\perp \right)$$

$$= -\mathbf{\nabla}_\perp^2 \mathbf{E}_\perp - \frac{i\omega}{c} \mathbf{\nabla}_\perp \times \hat{\mathbf{z}} B_z$$

$$= -\mathbf{\nabla}_\perp^2 \mathbf{E}_\perp = \left(\frac{\epsilon\mu\omega^2}{c^2} - \frac{\pi^2 \ell^2}{d^2} \right) \mathbf{E}_\perp, \tag{7.99}$$

which leads to the first relation in (7.97). Here we have used the harmonic time dependence of the fields as well as $B_z = 0$ for TM waves in the intermediate step. We have also used the wave equation for $\mathbf{E}_\perp$ in the last step.

It is clear from Eq. (7.96) that, in such a set up, electromagnetic fields exist only for a single frequency depending on the given values of m, n, ℓ. This is the behavior of an undamped resonant system such as an oscillator. As a result, such a set up is called a cavity resonator (or a resonant cavity). The cavity can also have resonant TE modes. For the TE modes, we note that $E_z = 0$ and the additional boundary conditions have the form

$$B_z|_{z=0,d} = 0, \tag{7.100}$$

which follows directly from Eq. (7.4). It is now easy to determine B_z from (7.17) (compare with (7.40))

$$B_z(x, y, z) = C \cos \frac{\pi m x}{a} \cos \frac{\pi n y}{b} \sin \frac{\pi \ell z}{d}, \tag{7.101}$$

with $\ell = 1, 2, \ldots$ as well as the usual restrictions on m, n and (see (7.34))

$$\omega = \frac{\pi c}{\sqrt{\epsilon\mu}} \sqrt{\frac{m^2}{a^2} + \frac{n^2}{b^2} + \frac{\ell^2}{d^2}} = \omega_{mn\ell}, \tag{7.102}$$

which has the same form as (7.96).

The transverse field components can now be determined from (7.11) and (7.12) to be

$$\mathbf{E}_\perp(\mathbf{x}) = \mathbf{E}_\perp(x, y, z) = -i\frac{\frac{\omega}{c}}{\frac{\epsilon\mu\omega^2}{c^2} - \frac{\pi^2\ell^2}{d^2}}\, \hat{\mathbf{z}} \times \boldsymbol{\nabla}_\perp B_z,$$

$$\mathbf{B}_\perp(\mathbf{x}) = \mathbf{B}_\perp(x, y, z) = \frac{1}{\frac{\epsilon\mu\omega^2}{c^2} - \frac{\pi^2\ell^2}{d^2}}\, \boldsymbol{\nabla}_\perp\frac{\partial B_z}{\partial z}. \tag{7.103}$$

Explicitly, these give

$$E_x(x, y, z) = -iC\,\frac{\frac{\pi\omega n}{cb}}{\frac{\epsilon\mu\omega^2}{c^2} - \frac{\pi^2\ell^2}{d^2}}\cos\frac{\pi m x}{a}\sin\frac{\pi n y}{b}\sin\frac{\pi \ell z}{d},$$

$$E_y(x, y, z) = iC\,\frac{\frac{\pi\omega m}{ca}}{\frac{\epsilon\mu\omega^2}{c^2} - \frac{\pi^2\ell^2}{d^2}}\sin\frac{\pi m x}{a}\cos\frac{\pi n y}{b}\sin\frac{\pi \ell z}{d},$$

$$B_x(x, y, z) = -C\,\frac{\frac{\pi^2 m \ell}{ad}}{\frac{\epsilon\mu\omega^2}{c^2} - \frac{\pi^2\ell^2}{d^2}}\sin\frac{\pi m x}{a}\cos\frac{\pi n y}{b}\cos\frac{\pi \ell z}{d},$$

$$B_y(x, y, z) = -C\,\frac{\frac{\pi^2 n \ell}{bd}}{\frac{\epsilon\mu\omega^2}{c^2} - \frac{\pi^2\ell^2}{d^2}}\cos\frac{\pi m x}{a}\sin\frac{\pi n y}{b}\cos\frac{\pi \ell z}{d}. \tag{7.104}$$

In a similar manner, we can also find the appropriate solutions for a cylindrical cavity resonator of radius a and length d along the z direction. From the earlier analysis of cylindrical wave guides as well as the discussion of the rectangular cavity resonator, it is easy to see that for the TM waves in such a resonator, we will have

$$B_z(r, \phi, z) = 0,$$

$$E_z(r, \phi, z) = a_n\, J_n(hr)\cos n\phi\cos\frac{\pi \ell z}{d}, \tag{7.105}$$

where

$$h = \sqrt{\frac{\epsilon\mu\omega^2}{c^2} - \frac{\pi^2\ell^2}{d^2}}. \tag{7.106}$$

Furthermore, for the electric fields to vanish at the cylindrical walls, we must have (see (7.57))

$$J_n(ha) = 0, \tag{7.107}$$

which determines that $(ha)_{mn}$ must correspond to the m-th zero of the n-th Bessel function. Some of these are already listed in (7.58). In terms of these, we see from Eq. (7.106) that

$$\omega = \frac{c}{\sqrt{\epsilon\mu}} \sqrt{h_{mn}^2 + \frac{\pi^2 \ell^2}{d^2}} = \omega_{mn\ell}, \qquad (7.108)$$

with $\ell = 0, 1, 2, \dots$ and m, n restricted as in the case of TM waves in the cylindrical wave guide.

Similarly, for the TE waves, we have

$$E_z(r, \phi, z) = 0,$$

$$B_z(r, \phi, z) = b_n J_n(hr) \cos n\phi \sin \frac{\pi \ell z}{d}, \qquad (7.109)$$

with h still defined as in (7.106). However, for TE waves, as we have already seen in (7.65), the radial derivative of the magnetic field at the cylindrical walls must vanish leading to

$$J_n'(ha) = 0. \qquad (7.110)$$

This determines $(ha)_{mn}$ to correspond to the m-th zero of the first derivative of the n-th Bessel function, some of which are listed in Eq. (7.66). In terms of these, we obtain, from Eq. (7.106),

$$\omega = \frac{c}{\sqrt{\epsilon\mu}} \sqrt{h_{mn}^2 + \frac{\pi^2 \ell^2}{d^2}} = \omega_{mn\ell}, \qquad (7.111)$$

with $\ell = 1, 2, \dots$ and m, n restricted as in the case of TE waves in the cylindrical wave guide. Once we have the longitudinal components of the fields, the transverse components can be obtained using the Maxwell's equations as we have done earlier (or see (7.97), (7.103)).

Let us note that a cavity resonator, with a variable length (along the z-axis) can be used as a frequency meter. Namely, by varying the length of the cavity, one can make a signal resonate in the cavity and thereby determine its frequency. While cavity resonators can be of any shape, for such a purpose, the TE_{101} wave in a cylindrical cavity resonator is quite useful. This is because, in this case, the radial component of the electric field vanishes (see, for example, (7.68) with $n = 0$) and the electric fields define concentric circles in a plane with $z = $ constant. As a result, there is no radial current and the movable piston (or the "plunger") does not have to make a rubbing contact with the walls of the cavity.

7.8 Q factor of a cavity

As we have seen, electromagnetic waves can exist inside a cavity only in resonant modes of discrete frequencies. (Any other field can, of course, be expanded in terms of these resonant modes.) However, this result is obtained by assuming that the walls of the cavity are perfectly conducting, which is not the case in realistic cavities. As a result, as in the case of wave guides, there are Ohmic losses and power is lost. A consequence of this is that the resonant frequencies are no longer sharp, rather they spread out a little. The Q factor of the cavity gives a measure of this spread and is defined as

$$Q = \omega_r \, \frac{\text{Energy stored in a cavity}}{\text{Power lost}}, \qquad (7.112)$$

where ω_r represents a resonant frequency (of the original lossless system). Let us note that if U denotes the energy stored in a cavity, then the power lost is the energy lost per unit time. Thus, from the definition (7.112), it follows that

$$\frac{dU}{dt} = -\frac{\omega_r}{Q} \, U$$

or, $\quad U(t) = U(0) \, e^{-\frac{\omega_r t}{Q}}. \qquad (7.113)$

This is clearly defined only for positive times and the energy inside the cavity decreases as time evolves due to the losses at the walls of the cavity (as well as possible losses in the dielectric inside).

It is clear from this that, since there is energy loss in a cavity, if we want to excite a particular resonant mode in the cavity by, say, introducing an external electromagnetic wave into the cavity through a small aperture, the system would behave like a damped oscillator with a time dependent driving force. From the form of the energy U in (7.113), we note that we can write the form of the electric fields inside the cavity to have the form

$$\mathbf{E}(t) = \mathbf{E}(0) \, e^{-i(\bar{\omega}_r - i\frac{\omega_r}{2Q})t}, \qquad (7.114)$$

where we have suppressed the spatial dependence of the fields for simplicity and have defined

$$\bar{\omega}_r = \omega_r + \delta\omega.$$

The quantity, $\delta\omega$, has been introduced to account for the possible smearing of the resonant frequency due to other effects. Taking the

Fourier transform of (7.114), we obtain

$$\mathbf{E}(\omega) = \frac{1}{2\pi} \int_0^\infty dt\, e^{i\omega t}\, \mathbf{E}(t) = \frac{\mathbf{E}(0)}{2\pi} \int_0^\infty dt\, e^{i(\omega - \bar{\omega}_r + i\frac{\omega_r}{2Q})t}$$

$$= \frac{i\mathbf{E}(0)}{2\pi} \frac{1}{(\omega - \bar{\omega}_r) + i\frac{\omega_r}{2Q}}. \tag{7.115}$$

It follows now that

$$|\mathbf{E}(\omega)|^2 = \frac{|\mathbf{E}(0)|^2}{4\pi^2} \frac{1}{(\omega - \bar{\omega}_r)^2 + \frac{\omega_r^2}{4Q^2}}. \tag{7.116}$$

This has the characteristics of a resonant behavior (Breit-Wigner shape) and shows that the electric field no longer has a sharp, discrete frequency. Rather, it is smeared out around the resonant value $\omega = \bar{\omega}_r$ as shown in Fig. 7.4. From (7.116), we see that the intensity has the maximum value at $\omega = \bar{\omega}_r$. It decreases to half of its maximum value at

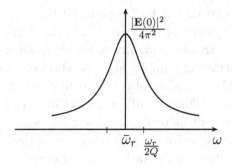

Figure 7.4: The behavior of the absolute square of the electric field as a function of the frequency ω.

$$\omega = \bar{\omega}_r \pm \frac{\omega_r}{2Q}. \tag{7.117}$$

Consequently, the width of the curve at half the peak value is obtained to be

$$\Gamma = \Delta\omega = \frac{\omega_r}{Q},$$

or, $\quad Q = \frac{\omega_r}{\Gamma}.$ \hfill (7.118)

This shows that the Q factor, indeed, measures the spread out in the frequency due to Ohmic losses.

As in the case of the wave guide, the power loss at the walls can be calculated in a simple manner and, for microwave cavities, one can obtain a Q factor as large as ten thousand. This implies that one can construct cavities that can show very sharp resonant behavior.

7.9 Dielectric wave guides (optical fibers)

The perfectly conducting walls of a wave guide basically do not allow the electromagnetic waves to escape from the wave guide. As a result, the waves bounce back and forth at the walls and propagate along the axis of the wave guide. We have also seen earlier that, while electromagnetic waves can be transmitted through a dielectric boundary surface, under appropriate conditions, there can be total internal reflection (without any transmission). Thus, under appropriate conditions, a dielectric slab can also be used to guide waves through successive total internal reflections, much like wave guides with perfectly conducting walls. The first analysis of such a phenomenon was carried out by Debye in 1910.

When the frequency of the wave lies in the optical or in the infrared range, such a transmission line is known as an optical fiber and is extremely important in reliable, high speed telecommunications. A common optical fiber consists of a silicon fiber of small dimensions (of the order $\sim 50 - 100 \mu$m) and an optical fiber cable, typically of the order of a few centimeters, contains many such fibers allowing for multi-mode transmissions. Technology in optical fibers has advanced tremendously over the last couple of decades which signifies the importance of this. Of course, the conditions in optical fibers cannot always be met perfectly so as to have complete total internal reflection. As a result, a small amount of the signal leaks through, leading to losses much like in the metallic wave guides. Nonetheless, this form of transmission is quite important considering that the frequencies involved are large thereby allowing for large bandwidths for transmission.

Although, in practice, one uses cylindrical dielectric wave guides (optical fibers), the mathematical analysis of such systems is rather involved. Therefore, just to get a feeling for the qualitative properties of such a system, let us analyze a rectangular dielectric slab of width $-a \leq x \leq a$ which we choose to be of infinite extension in the y direction (see Fig. 7.5), for simplicity. As before, we will assume the wave to propagate along the z-axis. Because of the infinite extension along the y direction, translations along this direction define a symmetry and, consequently, it follows that the fields cannot depend on the y coordinate. Therefore, we can choose the electric and the magnetic fields to have the forms

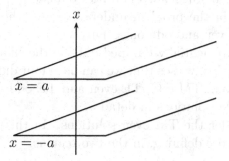

Figure 7.5: An infinite rectangular dielectric slab with a finite width along the x-axis and wave propagation along the z-axis.

$$\mathbf{E}(\mathbf{x}, t) = \mathbf{E}(x) \, e^{-i(\omega t - kz)},$$

$$\mathbf{B}(\mathbf{x}, t) = \mathbf{B}(x) \, e^{-i(\omega t - kz)}. \tag{7.119}$$

Substituting these into (7.16)–(7.17), we obtain

$$\frac{\mathrm{d}^2 E_z(x)}{\mathrm{d}x^2} + \left(\frac{\epsilon\mu\omega^2}{c^2} - k^2 \right) E_z(x) = 0,$$

$$\frac{\mathrm{d}^2 B_z(x)}{\mathrm{d}x^2} + \left(\frac{\epsilon\mu\omega^2}{c^2} - k^2 \right) B_z(x) = 0, \tag{7.120}$$

while Eq. (7.15) leads to

$$E_x(x) = \frac{ik}{\frac{\epsilon\mu\omega^2}{c^2} - k^2} \, \frac{\mathrm{d}E_z(x)}{\mathrm{d}x},$$

$$E_y(x) = -\frac{i\frac{\omega}{c}}{\frac{\epsilon\mu\omega^2}{c^2} - k^2} \, \frac{\mathrm{d}B_z(x)}{\mathrm{d}x},$$

$$B_x(x) = \frac{ik}{\frac{\epsilon\mu\omega^2}{c^2} - k^2} \, \frac{\mathrm{d}B_z(x)}{\mathrm{d}x},$$

$$B_y(x) = \frac{i\frac{\epsilon\mu\omega}{c}}{\frac{\epsilon\mu\omega^2}{c^2} - k^2} \frac{dE_z(x)}{dx}. \tag{7.121}$$

These equations must hold in both the regions – inside and outside the dielectric slab. Let us assume that ϵ_1, μ_1 represent the permittivity and the permeability inside the dielectric $(-a < x < a)$ while ϵ_2, μ_2 represent the same quantities outside $(|x| > a)$. Because of the symmetry in the problem under $x \leftrightarrow -x$, the solutions can be classified into even and odd ones. Furthermore, we would like the solutions to be exponentially damped outside the slab while oscillatory inside. With this, we see that we can have four different kinds of solutions – TM even, TM odd, TE even and TE odd. Let us simply work out one of the solutions in detail.

Let us consider the TM even solutions. In this case, we have $B_z = 0$. Furthermore defining, in the two regions,

$$\alpha = \sqrt{\frac{\epsilon_1\mu_1\omega^2}{c^2} - k^2}, \qquad \gamma = \sqrt{k^2 - \frac{\epsilon_2\mu_2\omega^2}{c^2}}, \tag{7.122}$$

where we assume that both α, γ are real and positive, we see that

$$\alpha^2 + \gamma^2 = (\epsilon_1\mu_1 - \epsilon_2\mu_2)\frac{\omega^2}{c^2} = \Delta \frac{\omega^2}{c^2}. \tag{7.123}$$

Furthermore, we note that because of the symmetry of the solutions, we can restrict ourselves to the positive x-axis only $(x \geq 0)$. Using (7.122) and (7.123), we obtain from Eqs. (7.120) and (7.121) (for $x \geq 0$)

$$
\begin{array}{llll}
E_z^{(1)}(x) & = & A \cos\alpha x, & E_z^{(2)}(x) & = C e^{-\gamma x}, \\
E_x^{(1)}(x) & = & -iA\frac{k}{\alpha}\sin\alpha x, & E_x^{(2)}(x) & = iC\frac{k}{\gamma}e^{-\gamma x}, \\
E_y^{(1)}(x) & = & 0, & E_y^{(2)}(x) & = 0, \\
B_x^{(1)}(x) & = & 0, & B_x^{(2)}(x) & = 0, \\
B_y^{(1)}(x) & = & -iA\frac{\epsilon_1\mu_1\omega}{c\alpha}\sin\alpha x, & B_y^{(2)}(x) & = iC\frac{\epsilon_2\mu_2\omega}{c\gamma}e^{-\gamma x},
\end{array}
\tag{7.124}
$$

where the superscripts $(1), (2)$ denote the two regions $0 \leq x \leq a$ and $x \geq a$ respectively (because of our choice $x \geq 0$).

Since these are even solutions, we can apply the boundary conditions only at $x = a$ (the conditions will then automatically hold at the boundary $x = -a$ by symmetry). We have already worked out the boundary conditions at the interface of two dielectric surfaces (see (6.39), (6.42) and (6.43)) and, in the present case, they take the forms

$$\epsilon_1 E_x^{(1)}(a) = \epsilon_2 E_x^{(2)}(a),$$

$$B_x^{(1)}(a) = B_x^{(2)}(a),$$

$$\mathbf{E}_{\text{tan}}^{(1)}(a) = \mathbf{E}_{\text{tan}}^{(2)}(a),$$

$$\frac{1}{\mu_1} \mathbf{B}_{\text{tan}}^{(1)}(a) = \frac{1}{\mu_2} \mathbf{B}_{\text{tan}}^{(2)}(a). \tag{7.125}$$

In the present case, since E_y, B_x (and B_z) are zero in the two regions, we obtain from the third relation in (7.125) that

$$E_z^{(1)}(a) = E_z^{(2)}(a)$$

or, $\quad A \cos \alpha a = C e^{-\gamma a}. \tag{7.126}$

Similarly, the last relation in (7.125) leads to

$$\frac{1}{\mu_1} B_y^{(1)}(a) = \frac{1}{\mu_2} B_y^{(2)}(a)$$

or, $\quad A \dfrac{\epsilon_1}{\alpha} \sin \alpha a = -C \dfrac{\epsilon_2}{\gamma} e^{-\gamma a}. \tag{7.127}$

It is easy to see that the first relation in (7.125) leads to the same relation as (7.127) and, therefore, (7.126) and (7.127) represent essentially the boundary conditions that need to be satisfied. (The second relation in (7.125) is trivially satisfied.)

Taking the ratio of Eqs. (7.126) and (7.127), we obtain

$$\frac{\alpha}{\epsilon_1} \cot \alpha a = -\frac{\gamma}{\epsilon_2}$$

or, $\quad \alpha \cot \alpha a = -\dfrac{\epsilon_1}{\epsilon_2} \gamma. \tag{7.128}$

Using (7.123) and defining dimensionless variables $\xi = \alpha a, \eta = \gamma a$, we can write (7.128) also as

$$\xi \cot \xi = -\frac{\epsilon_1}{\epsilon_2} \eta = -\frac{\epsilon_1}{\epsilon_2} \sqrt{\frac{\Delta \omega^2 a^2}{c^2} - \xi^2}. \tag{7.129}$$

This is a transcendental equation for ξ (or α) as a function of ω and the solutions can be obtained graphically (see Fig. 7.6), much

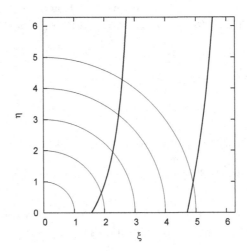

Figure 7.6: The TM even solutions are obtained from the intersections of $\eta = -\frac{\epsilon_2}{\epsilon_1} \xi \cot \xi$ and $\eta^2 + \xi^2 = $ constant.

like in the case of the square well potential in quantum mechanics. In fact, let us note that we can analyze the other modes as well exactly in this manner and the boundary conditions, in each case, would yield

$$\text{TM (odd)}: \qquad \xi \tan \xi = \frac{\epsilon_1}{\epsilon_2} \sqrt{\frac{\Delta \omega^2 a^2}{c^2} - \xi^2},$$

$$\text{TE (even)}: \qquad \xi \cot \xi = -\frac{\mu_1}{\mu_2} \sqrt{\frac{\Delta \omega^2 a^2}{c^2} - \xi^2},$$

$$\text{TE (odd)}: \qquad \xi \tan \xi = \frac{\mu_1}{\mu_2} \sqrt{\frac{\Delta \omega^2 a^2}{c^2} - \xi^2}, \qquad (7.130)$$

with α, γ satisfying (7.122) and (7.123). Thus, in each case, we have a transcendental equation to solve. Much like in quantum mechanical potential problems, here, too, one finds that there is a cut off frequency below which there does not exist any solution. Beyond this, there is a range of values of the frequency for which there exists only one solution and so on. In this respect, it is quite reminiscent of the square well potential in quantum mechanics and this behavior is also qualitatively similar to that of metallic wave guides.

7.10 Selected problems

1. Consider a pair of perfectly conducting parallel plates of infinite dimensions along the y and the z axes, separated by a distance "a" along the x-axis. Determine all the possible solutions for an electromagnetic wave to propagate along the z-axis.

2. Calculate the attenuation factor for the TEM waves in the set up described in the previous question.

3. A rectangular wave guide (with perfectly conducting walls) has cross-sectional dimensions

$$a = 7\text{cm}, \qquad b = 4\text{cm}$$

   Determine all the modes which will propagate at a frequency of (a) 3000MHz, (b) 5000MHz.

4. If λ and $\bar{\lambda}$ represent the wavelengths of the same electromagnetic wave in vacuum and inside a wave guide respectively, then, show that

$$\lambda = \frac{\bar{\lambda}\lambda_c}{\sqrt{\bar{\lambda}^2 + \lambda_c^2}},$$

   where λ_c denotes the wavelength corresponding to the cut-off frequency ω_c of the wave guide.

5. Derive the transcendental equations (discussed in this chapter) determining the TM (odd), TE (even) and the TE (odd) solutions for a wave traveling between two parallel dielectric slabs.

6. A wave guide of right triangular cross section is bounded by perfectly conducting walls at $x = a, y = 0, x = y$. Find the cut-off frequencies and the field modes of such a wave guide.

CHAPTER 8

Propagation through a conducting medium

8.1 Boundary conditions

We have so far discussed the propagation of electromagnetic waves in vacuum or in dielectric media. Let us next analyze the propagation of such waves through a conducting medium. Of course, the essential change, in this case, lies in the boundary conditions. If we look at the Maxwell's equations in an arbitrary medium in the Fourier transformed space (in the time variable), they take the forms

$$\boldsymbol{\nabla} \cdot (\epsilon \mathbf{E}) = 4\pi \rho,$$

$$\boldsymbol{\nabla} \cdot \mathbf{B} = 0,$$

$$\boldsymbol{\nabla} \times \mathbf{E} = ik\,\mathbf{B},$$

$$\boldsymbol{\nabla} \times \left(\frac{1}{\mu} \mathbf{B} \right) = \frac{4\pi}{c} \mathbf{J} - ik\epsilon \, \mathbf{E} = \frac{4\pi\sigma}{c} \mathbf{E} - ik\epsilon \, \mathbf{E}, \tag{8.1}$$

where we have defined, as before, $k = \frac{\omega}{c}$ and have identified

$$\mathbf{J} = \sigma \, \mathbf{E}, \tag{8.2}$$

with σ representing the conductivity of the medium. We note that σ, ϵ and μ are, in general, dependent on the frequency and although for most non-magnetic materials we can set $\mu \approx 1$, we will not do so for completeness. Here all the field variables have the coordinate dependence $\phi(\mathbf{x}, \omega)$, since time has been Fourier transformed. Alternatively one can also think of these equations as those for fields with a harmonic time dependence $(\mathbf{E}(t), \mathbf{B}(t)) = (\mathbf{E}(0), \mathbf{B}(0))e^{-i\omega t}$ with the exponential factored out.

The first two equations of Maxwell lead to the fact that, across a boundary with no free charges, the normal components of both $\epsilon \mathbf{E}$

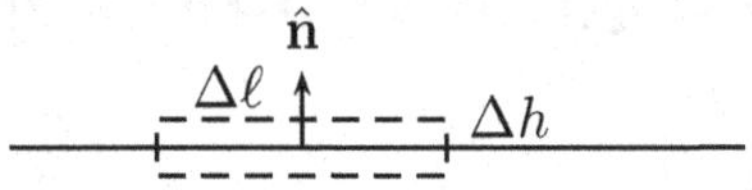

Figure 8.1: Boundary condition for the tangential component of the electric field.

and $\mathbf{B}$ should be continuous. In the presence of free surface charges, the normal component of the $\mathbf{D}$ field will be discontinuous across the surface, the discontinuity being proportional to the surface charge density. This is what we have already seen in the static case. Taking an infinitesimal surface element as shown in Fig. 8.1, we can deduce from the third equation that the tangential component of the electric field must be continuous across the boundary.

$$\int \mathrm{d}\mathbf{s} \cdot (\boldsymbol{\nabla} \times \mathbf{E}) = ik \int \mathrm{d}\mathbf{s} \cdot \mathbf{B}$$

$$\text{or,} \quad \oint \mathrm{d}\boldsymbol{\ell} \cdot \mathbf{E} = ik \int \mathrm{d}\mathbf{s} \cdot \mathbf{B}. \tag{8.3}$$

In the limit of an infinitesimal surface with area going to zero, this leads to

$$\hat{\mathbf{n}} \times \mathbf{E}_1 | = \hat{\mathbf{n}} \times \mathbf{E}_2 |, \tag{8.4}$$

where $\hat{\mathbf{n}}$ represents a unit vector normal to the boundary and the restriction represents the boundary. If the second medium corresponds to a perfect conductor, then there is no electric field in the second medium. Consequently, in such a case, we obtain the boundary condition to be

$$\hat{\mathbf{n}} \times \mathbf{E} | = 0. \tag{8.5}$$

Similarly, taking an infinitesimal surface element across the boundary, we obtain from the last equation in (8.1),

$$\int ds \cdot \left(\boldsymbol{\nabla} \times \frac{1}{\mu} \mathbf{B} \right) = \int ds \cdot \left(\frac{4\pi}{c} \mathbf{J} - ik\epsilon \, \mathbf{E} \right)$$

or, $\quad \oint d\boldsymbol{\ell} \cdot \frac{1}{\mu} \mathbf{B} = \int ds \cdot \left(\frac{4\pi}{c} \mathbf{J} - ik\epsilon \, \mathbf{E} \right).$ \hfill (8.6)

For non-singular field configurations, the right hand side of (8.6) vanishes in the limit of vanishing surface area and the boundary condition takes the form

$$\hat{n} \times \frac{1}{\mu_1} \mathbf{B}_1 \bigg| = \hat{n} \times \frac{1}{\mu_2} \mathbf{B}_2 \bigg|. \hfill (8.7)$$

When the second medium is a perfect conductor, the second term on the right hand side of (8.6) still vanishes in the limit of a vanishing surface area. However, in the presence of time dependent incident electromagnetic fields, a perfect conductor develops singular surface currents so that the first term on the right hand side of (8.6) does not vanish. (What this means is that the surface currents exist only on the surface and, therefore, are described with delta functions which are singular and may not yield a vanishing contribution even in the limit that the surface area vanishes.)

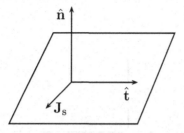

Figure 8.2: The induced surface current on a conductor with $\hat{n}$ and $\hat{t}$ denote respectively the normal to the surface and the direction of $\Delta\ell$.

In fact, let us define the surface current $\mathbf{J}_s$ as the amount of charge crossing the Gaussian surface per second per unit length along the long arm of the surface (see Fig. 8.2). In this case, in the limit of vanishing surface area, (8.6) leads to

$$\hat{n} \times \left(\frac{1}{\mu_1} \mathbf{B}_1 - \frac{1}{\mu_2} \mathbf{B}_2 \right) \Big| \Delta\ell = \frac{4\pi}{c} \mathbf{J}_s \Delta\ell$$

$$\text{or,} \quad \hat{n} \times \frac{1}{\mu_1} \mathbf{B}_1 \Big| = \hat{n} \times \frac{1}{\mu_2} \mathbf{B}_2 \Big| + \frac{4\pi}{c} \mathbf{J}_s. \tag{8.8}$$

Since there is no magnetic field inside a conductor, this condition can also be written as

$$\hat{n} \times \frac{1}{\mu} \mathbf{B} \Big| = \frac{4\pi}{c} \mathbf{J}_s$$

$$\text{or,} \quad \hat{n} \times \mathbf{H} \Big| = \frac{4\pi}{c} \mathbf{J}_s. \tag{8.9}$$

The important thing to note from this relation is that it relates the induced surface current in the conductor to the magnetic field outside. It is, therefore, not a boundary condition. Rather, once all the fields are determined, this relation can be used to calculate the induced surface current.

8.2 Reflection from a perfect conductor at normal incidence

Let us consider a plane wave of frequency ω traveling along the z-axis and let it be incident on a perfectly conducting surface located at $z = 0$ as shown in Fig. 8.3.

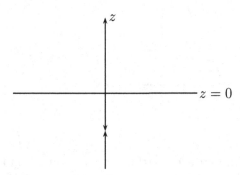

Figure 8.3: Reflection of a wave from a perfectly conducting surface at $z = 0$ at normal incidence.

We assume the first medium to be vacuum for simplicity. Thus, in the region $z < 0$, we expect an incident wave as well as a reflected wave whereas, since the second medium is a perfect conductor, we do not expect any transmitted wave in this case. (This is different from normal incidence in the case of a dielectric medium.) Thus, in region $z < 0$ we have

$$\mathbf{E}(\mathbf{x}, \omega) = \mathbf{E}_{\text{inc}}(\mathbf{x}, \omega) + \mathbf{E}_{\text{refl}}(\mathbf{x}, \omega), \tag{8.10}$$

where the incident and the reflected waves have the coordinate dependence given by

$$\mathbf{E}_{\text{inc}}(\mathbf{x}, \omega) = \mathbf{E}_{\text{inc}}^{(0)} e^{ikz},$$

$$\mathbf{E}_{\text{refl}}(\mathbf{x}, \omega) = \mathbf{E}_{\text{refl}}^{(0)} e^{-ikz}, \tag{8.11}$$

with $k = \frac{\omega}{c}$. (We have factored out the time dependence for simplicity.)

The electric and the magnetic fields, in a plane wave, are orthogonal to the direction of propagation (as well as to each other) and, therefore, are along the surface of the boundary (tangential to the surface). As we see from the boundary condition for a perfect conductor in (8.5),

$$\mathbf{E}(x, y, z = 0, \omega) = 0, \tag{8.12}$$

which determines

$$\mathbf{E}_{\text{refl}}^{(0)} = -\mathbf{E}_{\text{inc}}^{(0)} = -\mathbf{E}^{(0)}, \qquad \hat{\mathbf{z}} \cdot \mathbf{E}^{(0)} = 0. \tag{8.13}$$

Namely, the incident and the reflected electric fields have the same amplitudes, but are out of phase. Thus, we can write

$$\mathbf{E}(\mathbf{x}, \omega) = \mathbf{E}_{\text{inc}}(\mathbf{x}, \omega) + \mathbf{E}_{\text{refl}}(\mathbf{x}, \omega)$$

$$= \mathbf{E}^{(0)}(e^{ikz} - e^{-ikz}) = 2i\mathbf{E}^{(0)} \sin kz. \tag{8.14}$$

Putting in the harmonic time dependence (and recalling that the fields are really defined as the real part of the complex quantities), we obtain the form of the electric field for $z < 0$ to be

$$\mathbf{E}(\mathbf{x}, t, \omega) = \text{Re}\left(2i\mathbf{E}^{(0)} \sin kz\, e^{-i\omega t}\right)$$

$$= 2\mathbf{E}^{(0)} \sin kz \sin \omega t. \tag{8.15}$$

Once we have the electric fields, the magnetic fields can be obtained through the use of the relation (see (7.76) in vacuum)

$$\mathbf{B} = \pm\hat{\mathbf{z}} \times \mathbf{E}, \tag{8.16}$$

where the $\pm$ signs are related to the direction of propagation of the wave. Thus, using (8.16), we obtain

$$\mathbf{B}_{\text{inc}}(\mathbf{x}, \omega) = \hat{\mathbf{z}} \times \mathbf{E}_{\text{inc}}(\mathbf{x}, \omega) = \left(\hat{\mathbf{z}} \times \mathbf{E}^{(0)}\right) e^{ikz},$$

$$\mathbf{B}_{\text{refl}}(\mathbf{x}, \omega) = -\hat{\mathbf{z}} \times \mathbf{E}_{\text{refl}}(\mathbf{x}, \omega) = \left(\hat{\mathbf{z}} \times \mathbf{E}^{(0)}\right) e^{-ikz}. \tag{8.17}$$

As a result, we obtain the total magnetic field in the region $z < 0$ to be

$$\mathbf{B}(\mathbf{x}, \omega) = \mathbf{B}_{\text{inc}}(\mathbf{x}, \omega) + \mathbf{B}_{\text{refl}}(\mathbf{x}, \omega)$$

$$= \left(\hat{\mathbf{z}} \times \mathbf{E}^{(0)}\right) \left(e^{ikz} + e^{-ikz}\right)$$

$$= 2\left(\hat{\mathbf{z}} \times \mathbf{E}^{(0)}\right) \cos kz. \tag{8.18}$$

Putting in the time dependence, we obtain

$$\mathbf{B}(\mathbf{x}, t, \omega) = \text{Re} \left(2(\hat{\mathbf{z}} \times \mathbf{E}^{(0)}) \cos kz \, e^{-i\omega t}\right)$$

$$= 2\left(\hat{\mathbf{z}} \times \mathbf{E}^{(0)}\right) \cos kz \, \cos \omega t. \tag{8.19}$$

There are several things to note from Eqs. (8.14) and (8.18) (or (8.15) and (8.19)). The unique solution to the problem is obtained using only the boundary condition (8.5) (namely, we do not need (8.9) to obtain the solution). Furthermore, the electric and the magnetic fields define standing waves with the electric field vanishing at ($z < 0$)

$$kz = -n\pi, \qquad n = 0, 1, 2, \ldots, \tag{8.20}$$

while the magnetic field vanishes at

$$kz = -\frac{(2n+1)\pi}{2}, \qquad n = 0, 1, 2, \ldots. \tag{8.21}$$

The electric field vanishes at the boundary $z = 0$ simply because the incident and the reflected electric fields are out of phase. However, we note that the magnetic field does not vanish at the boundary. This can, in fact, be physically understood as follows. The (time averaged) radiated power per unit area, in the region $z < 0$, is obtained to be

$$\frac{\mathrm{d}P}{\mathrm{d}a} = \frac{c}{8\pi} \,\mathrm{Re}\, \hat{\mathbf{z}} \cdot (\mathbf{E} \times \mathbf{B}^*)$$

$$= \frac{c}{8\pi} \,\mathrm{Re}\, \left[\hat{\mathbf{z}} \cdot \left(\mathbf{E}^{(0)} \times \left(\hat{\mathbf{z}} \times (\mathbf{E}^{(0)})^*\right)\right)\right.$$

$$\left. \times \left(e^{ikz} - e^{-ikz}\right)\left(e^{-ikz} + e^{ikz}\right)\right]$$

$$= 0. \tag{8.22}$$

The radiated power vanishes simply because the incident and the reflected components of the magnetic field are in phase while those for the electric field are out of phase. Physically, this is clear since there cannot be any transmitted wave inside a perfect conductor and, therefore, there cannot be any power loss.

Let us note that the non-vanishing of the magnetic field, at the boundary, immediately shows that there must be a surface current in the conductor. In fact, from (8.9), we conclude that ($\mu = 1$ and $\hat{\mathbf{z}} \cdot \mathbf{E}^{(0)} = 0$)

$$\frac{4\pi}{c}\, \mathbf{J}_s = \hat{\mathbf{n}} \times \mathbf{B}| = -\,\hat{\mathbf{z}} \times \mathbf{B}|$$

$$= -2\hat{\mathbf{z}} \times \left(\hat{\mathbf{z}} \times \mathbf{E}^{(0)}\right) = 2\mathbf{E}^{(0)},$$

$$\text{or,}\quad \mathbf{J}_s = \frac{c}{2\pi}\, \mathbf{E}^{(0)}. \tag{8.23}$$

This can, therefore, be thought of as the reason for the change in the phase of the reflected component of the electric field.

8.3 Reflection from a perfect conductor at oblique incidence

Let us next consider the reflection of a plane wave of frequency ω from a perfectly conducting surface at oblique incidence as shown in Fig. 8.4. Let us assume that the boundary surface is at $z = 0$ and that the region $z < 0$ corresponds to vacuum. Without loss of generality, we can assume the plane of incidence (the normal to the surface and the direction of propagation define the plane of incidence) to be the $x - z$ plane. Let θ_i be the angle of incidence (namely, the angle between the incident ray and the z-axis).

Therefore, we can write the incident wave in region $z < 0$ to be of the form

$$\mathbf{E}_{\mathrm{inc}}(\mathbf{x}, \omega) = \mathbf{E}_{\mathrm{inc}}^{(0)}\, e^{i\mathbf{k}_i \cdot \mathbf{x}} = \mathbf{E}_{\mathrm{inc}}^{(0)}\, e^{ik_i(\sin\theta_i\, x + \cos\theta_i\, z)}, \tag{8.24}$$

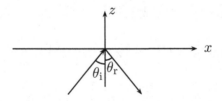

Figure 8.4: Reflection of a wave from a perfectly conducting surface at $z = 0$ at oblique incidence.

where

$$\mathbf{k}_i = k_i (\sin\theta_i\,\hat{\mathbf{x}} + \cos\theta_i\,\hat{\mathbf{z}}), \qquad \hat{\mathbf{k}}_i = \hat{\mathbf{x}}\sin\theta_i + \hat{\mathbf{z}}\cos\theta_i. \qquad (8.25)$$

In the region $z < 0$, we also expect a reflected wave. However, since the region $z > 0$ is perfectly conducting, we do not expect a transmitted wave in this region. We can write the reflected wave to be of the form

$$\mathbf{E}_{\text{refl}}(\mathbf{x},\omega) = \mathbf{E}_{\text{refl}}^{(0)}\,e^{i\mathbf{k}_r\cdot\mathbf{x}} = \mathbf{E}_{\text{refl}}^{(0)}\,e^{ik_r(\sin\theta_r\,x-\cos\theta_r z)}, \qquad (8.26)$$

where

$$\mathbf{k}_r = k_r(\hat{\mathbf{x}}\sin\theta_r - \hat{\mathbf{z}}\cos\theta_r), \qquad \hat{\mathbf{k}}_r = (\hat{\mathbf{x}}\sin\theta_r - \hat{\mathbf{z}}\cos\theta_r). \quad (8.27)$$

Therefore, the total electric field in this region has the form

$$\mathbf{E}(\mathbf{x},\omega) = \mathbf{E}_{\text{inc}}(\mathbf{x},\omega) + \mathbf{E}_{\text{refl}}(\mathbf{x},\omega)$$

$$= \mathbf{E}_{\text{inc}}^{(0)}\,e^{ik_i(\sin\theta_i\,x+\cos\theta_i\,z)} + \mathbf{E}_{\text{refl}}^{(0)}\,e^{ik_r(\sin\theta_r\,x-\cos\theta_r\,z)}. \qquad (8.28)$$

The boundary condition (8.5) now leads to

$$\hat{\mathbf{n}} \times \mathbf{E}(\mathbf{x},\omega)\big|_{z=0} = 0,$$

$$\text{or,} \quad \left(\hat{\mathbf{z}}\times\mathbf{E}_{\text{inc}}^{(0)}\right)e^{ik_i\sin\theta_i\,x} + \left(\hat{\mathbf{z}}\times\mathbf{E}_{\text{refl}}^{(0)}\right)e^{ik_r\sin\theta_r\,x} = 0. \qquad (8.29)$$

Clearly, this is satisfied if

$$k_i = k_r = k,$$

$$\theta_i = \theta_r = \theta,$$

$$\hat{\mathbf{z}} \times \left(\mathbf{E}_{inc}^{(0)} + \mathbf{E}_{refl}^{(0)} \right) = 0. \tag{8.30}$$

This shows that the wave numbers for the incident as well as the reflected waves are the same (they are in the same region) and that the angle of incidence is equal to the angle of reflection, as is also true in the case of reflection from a dielectric surface. However, here we have no transmission.

The last relation in (8.30) allows for two possibilities. First, we can have the electric fields along the y-axis - perpendicular to the plane of incidence. In this case, for the last relation in (8.30) to hold, we must have

$$\mathbf{E}_{refl}^{(0)} = -\mathbf{E}_{inc}^{(0)} = -E^{(0)} \hat{\mathbf{y}}. \tag{8.31}$$

It follows now that

$$\mathbf{E}(\mathbf{x}, \omega) = \mathbf{E}_{inc} + \mathbf{E}_{refl}$$

$$= 2iE^{(0)} \hat{\mathbf{y}} \, e^{ikx \sin \theta} \, \sin(kz \cos \theta). \tag{8.32}$$

Furthermore, using (8.25) and (8.27), we obtain

$$\mathbf{B}_{inc}(\mathbf{x}, \omega) = \hat{\mathbf{k}}_i \times \mathbf{E}_{inc}(\mathbf{x}, \omega)$$

$$= (\hat{\mathbf{x}} \sin \theta + \hat{\mathbf{z}} \cos \theta) \times \hat{\mathbf{y}} \, E^{(0)} \, e^{ik(x \sin \theta + z \cos \theta)}$$

$$= (-\hat{\mathbf{x}} \cos \theta + \hat{\mathbf{z}} \sin \theta) \, E^{(0)} \, e^{ik(x \sin \theta + z \cos \theta)},$$

$$\mathbf{B}_{refl}(\mathbf{x}, \omega) = \hat{\mathbf{k}}_r \times \mathbf{E}_{refl}(\mathbf{x}, \omega)$$

$$= (\hat{\mathbf{x}} \sin \theta - \hat{\mathbf{z}} \cos \theta) \times (-\hat{\mathbf{y}}) \, E^{(0)} \, e^{ik(x \sin \theta - z \cos \theta)}$$

$$= -(\hat{\mathbf{x}} \cos \theta + \hat{\mathbf{z}} \sin \theta) \, E^{(0)} \, e^{ik(x \sin \theta - z \cos \theta)}, \tag{8.33}$$

so that we can write

$$\mathbf{B}(\mathbf{x}, \omega) = \mathbf{B}_{inc}(\mathbf{x}, \omega) + \mathbf{B}_{refl}(\mathbf{x}, \omega)$$

$$= 2E^{(0)} \, e^{ikx \sin \theta} \left[-\hat{\mathbf{x}} \cos \theta \cos(kz \cos \theta) \right.$$

$$\left. + i\hat{\mathbf{z}} \sin \theta \sin(kz \cos \theta) \right]. \tag{8.34}$$

Thus, in this case, we see that while the electric field is perpendicular to the plane of incidence, the magnetic field is parallel to it (lies in the plane of incidence). Furthermore, we have

$$\mathbf{B}(x, y, z = 0, \omega) = -2E^{(0)}\,\hat{\mathbf{x}}\,\cos\theta\,e^{ikx\sin\theta} \neq 0, \qquad (8.35)$$

so that the tangential component of the magnetic field does not vanish at the boundary. This, therefore, leads to a surface current (see (8.9)) of the form

$$\mathbf{J}_s = \frac{c}{4\pi}\,\hat{\mathbf{n}} \times \mathbf{B}| = \frac{c}{4\pi}\,(-\hat{\mathbf{z}} \times \mathbf{B}|)$$

$$= \frac{cE^{(0)}}{2\pi}\,\hat{\mathbf{y}}\,\cos\theta\,e^{ikx\sin\theta}. \qquad (8.36)$$

Furthermore, the time averaged power per unit area is obtained to be

$$\frac{\mathrm{d}P}{\mathrm{d}a} = \frac{c}{8\pi}\,\mathrm{Re}\,\hat{\mathbf{z}} \cdot (\mathbf{E} \times \mathbf{B}^*)$$

$$= \frac{c}{8\pi}\,\mathrm{Re}\,\left(2i|E^{(0)}|^2\,\cos\theta\,\sin(2kz\cos\theta)\right)$$

$$= 0. \qquad (8.37)$$

Namely, since there is no transmission, all the energy that is radiated in the forward direction is reflected back leading to a net zero energy loss.

The other possibility that one can have is that the electric field lies in the plane of incidence. Of course, it has to be orthogonal to the direction of propagation. Therefore, we can choose

$$\mathbf{E}^{(0)}_{\text{inc}} = E^{(0)}\,(\hat{\mathbf{x}}\cos\theta - \hat{\mathbf{z}}\sin\theta),$$

$$\mathbf{E}^{(0)}_{\text{refl}} = -E^{(0)}\,(\hat{\mathbf{x}}\cos\theta + \hat{\mathbf{z}}\sin\theta). \qquad (8.38)$$

This satisfies the last relation in (8.30) and leads to

$$\mathbf{E}(\mathbf{x}, \omega) = \mathbf{E}_{\text{inc}}(\mathbf{x}, \omega) + \mathbf{E}_{\text{refl}}(\mathbf{x}, \omega)$$

$$= 2E^{(0)}\,e^{ikx\sin\theta}\,[i\hat{\mathbf{x}}\,\cos\theta\,\sin(kz\cos\theta)$$

$$-\hat{\mathbf{z}}\,\sin\theta\,\cos(kz\cos\theta)]. \qquad (8.39)$$

In turn, using (8.25) and (8.27), this leads to

$$\mathbf{B}_{\text{inc}}(\mathbf{x}, \omega) = \hat{\mathbf{k}}_i \times \mathbf{E}_{\text{inc}}(\mathbf{x}, \omega)$$

$$= \hat{\mathbf{y}}\, E^{(0)}\, e^{ik(x \sin\theta + z \cos\theta)},$$

$$\mathbf{B}_{\text{refl}}(\mathbf{x}, \omega) = \hat{\mathbf{k}}_r \times \mathbf{E}_{\text{refl}}(\mathbf{x}, \omega)$$

$$= \hat{\mathbf{y}}\, E^{(0)}\, e^{ik(x \sin\theta - z \cos\theta)}. \tag{8.40}$$

As a result, we obtain

$$\mathbf{B}(\mathbf{x}, \omega) = \mathbf{B}_{\text{inc}}(\mathbf{x}, \omega) + \mathbf{B}_{\text{refl}}(\mathbf{x}, \omega)$$

$$= 2E^{(0)}\, \hat{\mathbf{y}}\, e^{ikx \sin\theta}\, \cos(kz \cos\theta). \tag{8.41}$$

In this case, we note that the magnetic field is orthogonal to the plane of incidence while the electric field is parallel to it.

Let us note that

$$\mathbf{B}(x, y, z = 0, \omega) = 2E^{(0)}\, \hat{\mathbf{y}}\, e^{ikx \sin\theta} \neq 0, \tag{8.42}$$

so that the tangential component of the magnetic field does not vanish on the boundary. Consequently, it leads to a surface current of the form

$$\mathbf{J}_s = \frac{c}{4\pi}\, \hat{\mathbf{n}} \times \mathbf{B}| = \frac{c}{4\pi}\, (-\hat{\mathbf{z}} \times \mathbf{B}|)$$

$$= \frac{cE^{(0)}}{2\pi}\, \hat{\mathbf{x}}\, e^{ikx \sin\theta}. \tag{8.43}$$

We can also calculate the power radiated per unit area in region $z < 0$, which takes the form

$$\frac{dP}{da} = \frac{c}{8\pi}\, \text{Re}\, \hat{\mathbf{z}} \cdot (\mathbf{E} \times \mathbf{B}^*)$$

$$= \frac{c}{8\pi}\, \text{Re}\, \left(2i|E^{(0)}|^2 \cos\theta \sin(2kz \cos\theta) \right)$$

$$= 0. \tag{8.44}$$

This shows that all the energy that is incident is reflected back and there is no net loss of energy.

Finally, let us consider, for simplicity, the first solution (where the electric field is orthogonal to the plane of incidence) and note some qualitative features which hold for both the solutions. We see, from Eq. (8.32) that the electric field forms standing waves along the z-axis with nodes at $(z < 0)$

$$kz \cos \theta = -n\pi, \qquad n = 0, 1, 2, \ldots. \tag{8.45}$$

The locations of the nodes depend on the angle of incidence and, in particular, for normal incidence $(\theta = 0)$, we recover the conditions in (8.20). Furthermore, from (8.32), we also note that this entire standing wave travels along the x-axis with a wave number

$$\bar{k} = k \sin \theta. \tag{8.46}$$

Namely, the wave length of propagation along the x-axis is given by

$$\bar{\lambda} = \frac{\lambda}{\sin \theta}, \tag{8.47}$$

corresponding to a propagation velocity

$$\bar{v} = \frac{\omega}{\bar{k}} = \frac{c}{\sin \theta}. \tag{8.48}$$

Thus, we see that the wave length as well as the velocity of propagation are larger than their corresponding values in vacuum. This is very similar to the behavior we have seen in wave guides. In fact, suppose we add a parallel, perfectly conducting surface at $z = -b$, then, from the solution for the electric field in (8.32), we see that this has to satisfy the boundary condition (8.5) at the new boundary leading to

$$kb \cos \theta = n\pi, \tag{8.49}$$

which gives

$$\bar{k} = k \sin \theta = \sqrt{k^2 - k^2 \cos^2 \theta}$$

$$= \sqrt{\frac{\omega^2}{c^2} - \frac{\pi^2 n^2}{b^2}}. \tag{8.50}$$

This shows that for $\omega < \frac{\pi c n}{b}$, there will be no propagation, while for $\omega > \frac{\pi c n}{b}$, there will be propagation of waves. This is, in fact, exactly what we have seen for a wave guide. However, here we only have a pair of parallel conducting surfaces. As we have discussed earlier, a pair of parallel conducting plates can also guide waves and, among other things, has a TEM mode of propagation.

8.4 Reflection from a good conducting surface

We have, thus far, talked about perfect conductors for which $\sigma \to \infty$. In reality, the conductors have a large but finite conductivity. The reflection of an electromagnetic wave from a real conducting surface will, therefore, be different from the earlier results. In this section, we will consider the reflection of a harmonic plane wave from a good (but not perfect) conductor. A conductor is commonly known as a good conductor if $\frac{4\pi\sigma}{\epsilon\omega} \gg 1$, as we will see shortly.

Before analyzing reflection from a good conductor, let us discuss some of the properties of harmonic fields. For harmonic fields (that is when all the field variables have a harmonic time dependence of the form $e^{-i\omega t}$ of a given frequency), the Maxwell's equations take the forms

$$\boldsymbol{\nabla} \cdot \mathbf{D} = 4\pi\rho,$$

$$\boldsymbol{\nabla} \cdot \mathbf{B} = 0,$$

$$\boldsymbol{\nabla} \times \mathbf{E} = \frac{i\omega}{c}\mathbf{B},$$

$$\boldsymbol{\nabla} \times \left(\frac{1}{\mu}\mathbf{B}\right) = \frac{4\pi}{c}\mathbf{J} - \frac{i\omega}{c}\mathbf{D} = \left(\frac{4\pi\sigma}{c} - \frac{i\omega\epsilon}{c}\right)\mathbf{E}. \tag{8.51}$$

We have used here the relation

$$\mathbf{J} = \sigma\,\mathbf{E}, \tag{8.52}$$

as well as

$$\mathbf{D} = \epsilon\mathbf{E}, \tag{8.53}$$

where ϵ represents the permittivity of the medium. Although, for most non-magnetic material, the permeability $\mu \approx 1$, we will keep it in our calculations for completeness.

The important thing to observe from (8.51) is that the first two equations follow as a consequence of the last two and the continuity equation. Furthermore, it follows from the continuity equation that

$$\frac{\partial\rho}{\partial t} + \boldsymbol{\nabla} \cdot \mathbf{J} = 0$$

or, $$\frac{\partial\rho}{\partial t} + \sigma\boldsymbol{\nabla} \cdot \mathbf{E} = 0, \tag{8.54}$$

where we have assumed that the conductivity does not vary with space. Furthermore, using the first of the Maxwell's equations and assuming that the permittivity also is independent of space (this is true when we are considering an isotropic medium), we obtain

$$\frac{\partial \rho}{\partial t} = -\sigma \mathbf{\nabla} \cdot \mathbf{E} = -\frac{4\pi\sigma}{\epsilon}\, \rho,$$

or, $\quad \rho(t) \sim e^{-\frac{4\pi\sigma t}{\epsilon}}.$ $\hspace{4cm}$ (8.55)

This shows that the charge density dissipates with a time scale

$$\tau = \frac{\epsilon}{4\pi\sigma}. \hspace{4cm} (8.56)$$

On the other hand, the only meaningful time scale in a harmonic problem is $\frac{1}{\omega}$ and so, we conclude that if

$$\tau \ll \frac{1}{\omega}$$

or, $\quad \dfrac{4\pi\sigma}{\epsilon\omega} \gg 1,$ $\hspace{4cm}$ (8.57)

the charge will dissipate quite rapidly and this is what defines a good conductor. In such a medium, therefore, we can set the charge density to zero. Note, from (8.55), that for a perfect conductor, $\sigma \to \infty$ so that there is no charge density produced.

Returning to the Maxwell's equations, (8.51), and taking the curl of the last two equations, we note that they lead to

$$\mathbf{\nabla}^2\, \mathbf{E} + \frac{\epsilon\mu\omega^2}{c^2}\left(1 + \frac{4\pi i\sigma}{\epsilon\omega}\right) \mathbf{E} = 0,$$

$$\mathbf{\nabla}^2\, \mathbf{B} + \frac{\epsilon\mu\omega^2}{c^2}\left(1 + \frac{4\pi i\sigma}{\epsilon\omega}\right) \mathbf{B} = 0. \hspace{2cm} (8.58)$$

These are complex equations which we can think of as the wave equations that we have studied earlier, if we allow for a complex permittivity of the form

$$\epsilon_{\text{eff}} = \epsilon\left(1 + \frac{4\pi i\sigma}{\epsilon\omega}\right) = \epsilon + \frac{4\pi i\sigma}{\omega}. \hspace{2cm} (8.59)$$

Since $\sigma = 0$ in vacuum, such a definition is compatible with the vacuum solutions. The equations (8.58) have solutions of the forms

$$\mathbf{E} \sim e^{i\mathbf{K}\cdot\mathbf{x}}, \qquad \mathbf{B} \sim e^{i\mathbf{K}\cdot\mathbf{x}}, \hspace{2cm} (8.60)$$

provided

$$\mathbf{K}^2 = K^2 = \frac{\epsilon\mu\omega^2}{c^2}\left(1 + \frac{4\pi i\sigma}{\epsilon\omega}\right). \tag{8.61}$$

This shows that in a conducting medium, the wave number, in general, becomes complex with a real and an imaginary part. In particular, if we have a good conductor, $\frac{4\pi\sigma}{\epsilon\omega} \gg 1$, we obtain

$$K \approx \frac{\sqrt{\epsilon\mu}\omega}{c}\sqrt{\frac{4\pi i\sigma}{\epsilon\omega}} = (1 + i)\frac{\sqrt{2\pi\sigma\mu\omega}}{c} = k_r + ik_i, \tag{8.62}$$

where we have used

$$\sqrt{i} = \frac{(1 + i)}{\sqrt{2}},$$

and have identified

$$k_r = k_i = \frac{\sqrt{2\pi\sigma\mu\omega}}{c}. \tag{8.63}$$

Thus, in a good conductor, the solutions of the Maxwell's equations traveling along the z-axis, for example, lead to electric and magnetic fields of the forms

$$\mathbf{E}(\mathbf{x}, \omega) = \mathbf{E}^{(0)} e^{ik_r z} e^{-k_i z},$$

$$\mathbf{B}(\mathbf{x}, \omega) = \mathbf{B}^{(0)} e^{ik_r z} e^{-k_i z}. \tag{8.64}$$

This shows that the wave attenuates as it travels because of dissipation in the medium and the length

$$\delta = \frac{1}{k_i} = \frac{c}{\sqrt{2\pi\sigma\mu\omega}}, \tag{8.65}$$

is known as the penetration depth or the skin depth of the medium beyond which the amplitude of the wave becomes negligible. As a result, it follows that the fields and the currents cling close to the surface of a good conductor. For such a wave, it is easy to see from the first two of the Maxwell's equations that

$$E_z = 0 = B_z, \tag{8.66}$$

and the third of Maxwell's equations gives

$$\mathbf{B} = \frac{c}{\omega}(k_r + ik_i)\,\hat{\mathbf{z}} \times \mathbf{E} = \frac{cK}{\omega}\,\hat{\mathbf{z}} \times \mathbf{E}, \tag{8.67}$$

which allows us to identify the impedance for such a TEM wave to be

$$Z = \frac{\omega}{c(k_r + ik_i)} = \frac{\omega}{cK} = (1 - i)\sqrt{\frac{\omega}{8\pi\sigma\mu}}. \tag{8.68}$$

This can be identified with

$$Z = \frac{(1 - i)}{\sqrt{2\epsilon\mu}}\sqrt{\frac{\epsilon\omega}{4\pi\sigma}} = \frac{1}{\sqrt{\epsilon_{\text{eff}}\mu}} \ll 1, \tag{8.69}$$

for a good conductor. We see that the impedance, in this case, is complex and, as a result, the refractive index for such a medium also becomes complex, namely,

$$n = \frac{1}{Z} = \frac{Kc}{\omega} = (1 + i)\sqrt{\frac{2\pi\sigma\mu}{\omega}}, \tag{8.70}$$

which signals absorption by the medium.

Since the electric fields fall off with large values of z, it follows, from (8.52), that the current must also have a similar behavior. This, therefore, allows us to define a surface current density as

$$\begin{aligned} \mathbf{J}_s &= \int_0^\infty dz\, \mathbf{J}(\mathbf{x}, \omega) \\ &= \sigma\, \mathbf{E}^{(0)} \int_0^\infty dz\, e^{iKz} \\ &= \frac{i\sigma\, \mathbf{E}^{(0)}}{K} = \frac{i\sigma}{K}\, \mathbf{E}|, \end{aligned} \tag{8.71}$$

where the restriction refers to the surface assumed to be at $z = 0$. In analogy with electric circuits, the ratio of the tangential component of the electric field at the surface to the surface current density is defined as the surface impedance of the medium, namely,

$$\mathbf{E}| = Z_s\, \mathbf{J}_s,$$

or, $$Z_s = -\frac{iK}{\sigma} = (1 - i)\sqrt{\frac{2\pi\mu\omega}{\sigma c^2}} = (1 - i)\frac{1}{\sigma\delta}, \tag{8.72}$$

where δ represents the skin depth defined in (8.65). The real part of the surface impedance can be thought of as the surface resistance of the medium,

$$R_s = \frac{1}{\sigma\delta} = \sqrt{\frac{2\pi\mu\omega}{\sigma c^2}}. \tag{8.73}$$

Note that when $\sigma \to \infty$, the surface resistance vanishes (as does the surface impedance). It is now straightforward to see from (8.68) and (8.72) that

$$Z = \frac{c}{4\pi\mu} Z_s = (1 - i) \frac{c}{4\pi\mu\sigma\delta}. \tag{8.74}$$

With these, we are now ready to analyze reflection from a good conducting surface. Let us assume that the conducting surface is at $z = 0$ and that an electromagnetic wave is incident from vacuum in $z < 0$ at an angle of incidence θ_i in the $x - z$ plane as shown in Fig. 8.5. In this case, we do expect a reflected wave in the region $z < 0$ with angle of reflection θ_r as well as a transmitted wave in the region $z > 0$ with the angle of transmission θ_t. Of course, the transmitted wave will be highly attenuated and we assume that the thickness of the conducting medium is much larger than the skin depth so that the transmitted wave will practically be a surface wave. We can parameterize the electric fields associated with the three components as

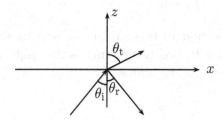

Figure 8.5: Reflection from a good conducting surface at $z = 0$ at oblique incidence.

$$\mathbf{E}_{\text{inc}}(\mathbf{x}, \omega) = \mathbf{E}_i\, e^{ik_i(x \sin \theta_i + z \cos \theta_i)},$$

$$\mathbf{E}_{\text{refl}}(\mathbf{x}, \omega) = \mathbf{E}_r\, e^{ik_r(x \sin \theta_r - z \cos \theta_r)},$$

$$\mathbf{E}_{\text{trans}}(\mathbf{x}, \omega) = \mathbf{E}_t\, e^{ik_t(x \sin \theta_t + z \cos \theta_t)}, \tag{8.75}$$

where we have used the results from our earlier analysis that the propagation has to be in the same plane and have defined here

$$\mathbf{k}_i = k_i(\hat{\mathbf{x}}\sin\theta_i + \hat{\mathbf{z}}\cos\theta_i) = k_i\,\hat{\mathbf{k}}_i,$$

$$\mathbf{k}_r = k_r(\hat{\mathbf{x}}\sin\theta_r - \hat{\mathbf{z}}\cos\theta_r) = k_r\,\hat{\mathbf{k}}_r,$$

$$\mathbf{k}_t = k_t(\hat{\mathbf{x}}\sin\theta_t + \hat{\mathbf{z}}\cos\theta_t) = k_t\,\hat{\mathbf{k}}_t. \tag{8.76}$$

We recall that the wave number k_t inside the conducting medium is complex.

In this case, the boundary condition (8.4) takes the form

$$(\hat{\mathbf{z}}\times\mathbf{E}_i)e^{ik_ix\sin\theta_i} + (\hat{\mathbf{z}}\times\mathbf{E}_r)e^{ik_rx\sin\theta_r} = (\hat{\mathbf{z}}\times\mathbf{E}_t)e^{ik_tx\sin\theta_t}. \tag{8.77}$$

For this to be true, we must have

$$\theta_i = \theta_r = \theta,$$

$$k_i = k_r = k,$$

$$\sin\theta_t = \frac{k}{k_t}\sin\theta$$

$$\hat{\mathbf{z}}\times\mathbf{E}_t = \hat{\mathbf{z}}\times(\mathbf{E}_i + \mathbf{E}_r). \tag{8.78}$$

We note that the first of these relations tells us the familiar result that the angle of reflection is the same as the angle of incidence while the third gives Snell's law, namely

$$\frac{\sin\theta_t}{\sin\theta} = \frac{k}{k_t} = \frac{1}{n_t} = Z_t, \tag{8.79}$$

where n_t represents the index of refraction of the conducting medium and we have used $k = \frac{\omega}{c}$ as well as (8.70). The important thing to note here is that, in this case, the index of refraction is complex signifying dissipation (absorption) of energy in the medium. Furthermore, for a good conductor (see (8.69)),

$$Z_t \ll 1, \tag{8.80}$$

so that we expect

$$\theta_t \approx 0. \tag{8.81}$$

To continue with the analysis, let us assume that the incident electric field is orthogonal to the plane of incidence, namely, it has a component only along the y-axis. It follows, then from the boundary conditions, that all the electric fields will have the same polarization, namely,

$$\mathbf{E}_{\text{inc}}(\mathbf{x}, \omega) = E_{\text{i}} \,\hat{\mathbf{y}}\, e^{ik(x \sin\theta + z \cos\theta)},$$

$$\mathbf{E}_{\text{refl}}(\mathbf{x}, \omega) = E_{\text{r}} \,\hat{\mathbf{y}}\, e^{ik(x \sin\theta - z \cos\theta)},$$

$$\mathbf{E}_{\text{trans}}(\mathbf{x}, \omega) = E_{\text{t}} \,\hat{\mathbf{y}}\, e^{ik_{\text{t}}(x \sin\theta_{\text{t}} + z \cos\theta_{\text{t}})}. \tag{8.82}$$

Furthermore, we can obtain the components of the magnetic fields as

$$\mathbf{B}_{\text{inc}}(\mathbf{x}, \omega) = \frac{1}{Z}\, \hat{\mathbf{k}}_{\text{i}} \times \mathbf{E}_{\text{inc}}$$

$$= (-\hat{\mathbf{x}} \cos\theta + \hat{\mathbf{z}} \sin\theta)\, E_{\text{i}}\, e^{ik(x \sin\theta + z \cos\theta)},$$

$$\mathbf{B}_{\text{refl}}(\mathbf{x}, \omega) = \frac{1}{Z}\, \hat{\mathbf{k}}_{\text{r}} \times \mathbf{E}_{\text{refl}}$$

$$= (\hat{\mathbf{x}} \cos\theta + \hat{\mathbf{z}} \sin\theta)\, E_{\text{r}}\, e^{ik(x \sin\theta - z \cos\theta)},$$

$$\mathbf{B}_{\text{trans}}(\mathbf{x}, \omega) = \frac{1}{Z_{\text{t}}}\, \hat{\mathbf{k}}_{\text{t}} \times \mathbf{E}_{\text{trans}}$$

$$= \frac{1}{Z_{\text{t}}} (-\hat{\mathbf{x}} \cos\theta_{\text{t}} + \hat{\mathbf{z}} \sin\theta_{\text{t}})\, E_{\text{t}}\, e^{ik_{\text{t}}(x \sin\theta + z \cos\theta)}. \tag{8.83}$$

Here we have used $Z = 1$ in vacuum. Matching the tangential components of the electric and the magnetic fields across the boundary (see (8.7) with $\mu \approx 1$ for non-magnetic materials), we obtain (the $\hat{\mathbf{z}}$ component for the magnetic fields does not lead to any new condition because of (8.79))

$$E_{\text{i}} + E_{\text{r}} = E_{\text{t}},$$

$$(E_{\text{i}} - E_{\text{r}}) \cos\theta = \frac{1}{Z_{\text{t}}} E_{\text{t}} \cos\theta_{\text{t}}. \tag{8.84}$$

These can be solved to obtain the coefficients of reflection and refraction in a straightforward manner from

$$\frac{E_{\text{r}}}{E_{\text{i}}} = \frac{Z_{\text{t}} \cos\theta - \cos\theta_{\text{t}}}{Z_{\text{t}} \cos\theta + \cos\theta_{\text{t}}} \approx \frac{Z_{\text{t}} \cos\theta - 1}{Z_{\text{t}} \cos\theta + 1},$$

$$\frac{E_{\text{t}}}{E_{\text{i}}} = \frac{2Z_{\text{t}} \cos\theta}{Z_{\text{t}} \cos\theta + \cos\theta_{\text{t}}} \approx \frac{2Z_{\text{t}} \cos\theta}{Z_{\text{t}} \cos\theta + 1}. \tag{8.85}$$

The other case, where the incident electric field is parallel to the plane of incidence can also be done in a straightforward manner and we only quote the results here. In this case, we have

$$\mathbf{E}_{\text{inc}}(\mathbf{x}, \omega) = (\hat{\mathbf{x}} \cos \theta - \hat{\mathbf{z}} \sin \theta)\, E_i \, e^{ik(x \sin \theta + z \cos \theta)},$$

$$\mathbf{E}_{\text{refl}}(\mathbf{x}, \omega) = -(\hat{\mathbf{x}} \cos \theta + \hat{\mathbf{z}} \sin \theta)\, E_r \, e^{ik(x \sin \theta - z \cos \theta)},$$

$$\mathbf{E}_{\text{trans}}(\mathbf{x}, \omega) = (\hat{\mathbf{x}} \cos \theta_t - \hat{\mathbf{z}} \sin \theta_t)\, E_t \, e^{ik_t(x \sin \theta_t + z \cos \theta_t)}. \qquad (8.86)$$

The magnetic fields are obtained from this to have the forms

$$\mathbf{B}_{\text{inc}}(\mathbf{x}, \omega) = \hat{\mathbf{y}}\, E_i \, e^{ik(x \sin \theta + z \cos \theta)},$$

$$\mathbf{B}_{\text{refl}}(\mathbf{x}, \omega) = \hat{\mathbf{y}}\, E_r \, e^{ik(x \sin \theta - z \cos \theta)},$$

$$\mathbf{B}_{\text{trans}}(\mathbf{x}, \omega) = \hat{\mathbf{y}}\, \frac{E_t}{Z_t} \, e^{ik_t(x \sin \theta_t + z \cos \theta_t)}. \qquad (8.87)$$

Furthermore, the boundary conditions on the tangential components of the electric and the magnetic fields give ($\mu \approx 1$ for non-magnetic materials)

$$(E_i - E_r) \cos \theta = E_t \cos \theta_t,$$

$$E_i + E_r = \frac{1}{Z_t} E_t, \qquad (8.88)$$

and the coefficients of reflection and transmission can be obtained easily from

$$\frac{E_r}{E_i} = \frac{\cos \theta - Z_t \cos \theta_t}{\cos \theta + Z_t \cos \theta_t} \approx \frac{\cos \theta - Z_t}{\cos \theta + Z_t},$$

$$\frac{E_t}{E_i} = \frac{2 Z_t \cos \theta}{\cos \theta + Z_t \cos \theta_t} \approx \frac{2 Z_t \cos \theta}{\cos \theta + Z_t}. \qquad (8.89)$$

We note that, in a good conductor, $\theta_t \approx 0$ independent of the value of the incident angle. Furthermore, the coefficients of reflection reduce to those for reflection from a perfect conductor with $Z_t \to 0$.

8.5 Radiation pressure

As we have studied earlier, electromagnetic fields carry energy and momentum. As a result, when electromagnetic waves are incident on a surface, they can exert a force or pressure on the surface. This is known as radiation pressure (in analogy with statistical mechanics of gases). The examples we have studied in this chapter clearly demonstrate this and bring out the origin behind this (which was also discussed earlier along general lines when we studied the Poynting vector). Let us recall that the momentum density associated with electromagnetic fields is related to the Poynting vector as (see (6.92))

$$\mathbf{p} = \frac{\epsilon\mu}{c^2} \operatorname{Re} \mathbf{S}. \tag{8.90}$$

For harmonic fields, therefore, we can define a time averaged momentum density as

$$\mathbf{p} = \frac{\epsilon\mu}{c^2} \frac{c}{8\pi} \operatorname{Re} (\mathbf{E} \times \mathbf{H}^*) = \frac{\epsilon}{8\pi c} \operatorname{Re} (\mathbf{E} \times \mathbf{B}^*). \tag{8.91}$$

The total momentum exerted by the EM waves on a surface of area "a" in a time interval Δt, therefore, follows to be

$$\Delta\mathbf{p} = \mathbf{p}\, a(c\Delta t) = \frac{\epsilon a(c\Delta t)}{8\pi c} \operatorname{Re} (\mathbf{E} \times \mathbf{B}^*). \tag{8.92}$$

This leads to a pressure exerted by an electromagnetic wave on a surface of the form

$$\mathcal{P} = \frac{\mathbf{F}}{a} = \frac{1}{a} \frac{\Delta\mathbf{p}}{\Delta t} = \mathbf{p}\, c = \frac{\epsilon}{8\pi} \operatorname{Re} (\mathbf{E} \times \mathbf{B}^*). \tag{8.93}$$

Let us now analyze all of this in the case of an electromagnetic wave (in vacuum) at normal incidence on a perfectly conducting surface at $z = 0$ that we have already studied. We have seen that, in this case, the incident electric and the magnetic fields have the forms (see (8.11) and (8.17))

$$\mathbf{E}_{\text{inc}}(\mathbf{x}, \omega) = \mathbf{E}^{(0)}\, e^{ikz},$$

$$\mathbf{B}_{\text{inc}}(\mathbf{x}, \omega) = \left(\hat{\mathbf{z}} \times \mathbf{E}^{(0)}\right) e^{ikz}. \tag{8.94}$$

It follows, therefore, from (8.93) that such a wave will exert a pressure on the surface at $z = 0$ of the form (in vacuum $\epsilon = 1$)

$$\mathcal{P} = \frac{1}{8\pi} \operatorname{Re} \left(\mathbf{E}^{(0)} \times (\hat{\mathbf{z}} \times (\mathbf{E}^{(0)})^*) \right)$$

$$= \hat{\mathbf{z}}\, \frac{1}{8\pi}\, |\mathbf{E}^{(0)}|^2. \tag{8.95}$$

This shows that the incident wave exerts a pressure on the surface along the z-axis. Furthermore, if we have a perfectly conducting surface, the incident wave is totally reflected doubling the pressure so that, for a perfectly conducting surface, we have

$$\mathcal{P}_{\text{total}} = 2\mathcal{P} = \hat{\mathbf{z}}\, \frac{1}{4\pi}\, |\mathbf{E}^{(0)}|^2. \tag{8.96}$$

On the other hand, we note that, for a perfectly absorbing surface such as a blackbody, the pressure will be given by (8.95).

The mechanism leading to the radiation pressure can also be easily understood from this simple example. We have already seen that, in this case, there will be induced surface currents. However, the surface current is uniform so that there is no induced charge density. As a result, when an electromagnetic wave is incident on the surface, the surface current will experience a force (Lorentz force) leading to a time averaged force per unit area as

$$\frac{\mathbf{F}}{a} = \frac{1}{4c} \, \mathrm{Re} \, (\mathbf{J_s} \times \mathbf{B}^*). \tag{8.97}$$

Several comments are in order here. First, the factor of $\frac{1}{4}$ arises as follows. The time averaging leads to $\frac{1}{2}$ whereas the other $\frac{1}{2}$ comes from averaging the current both above and below the surface. (Another way to look at it is to note that the surface current is obtained from the jump in the magnetic field which, in the present case, is twice that associated with the incident magnetic field (see (8.18)).) Second, normally the Lorentz force, given in terms of a current density, gives rise to a force density. However, since we have a surface current (that is already integrated over a unit line), the Lorentz force leads to a force per unit area or pressure. Recalling the definitions of $\mathbf{B}$ and $\mathbf{J}_s$ in Eqs. (8.17) and (8.23) respectively, we obtain

$$\mathcal{P} = \frac{1}{4c} \frac{c}{2\pi} \, \mathrm{Re} \, \left(\mathbf{E}^{(0)} \times \left(\hat{\mathbf{z}} \times \left(\mathbf{E}^{(0)} \right)^* \right) \right)$$

$$= \hat{\mathbf{z}} \, \frac{1}{8\pi} \, |\mathbf{E}^{(0)}|^2. \tag{8.98}$$

It follows now that, for a perfect conductor, the total pressure on the surface due to the incident and the reflected waves is

$$\mathcal{P}_{\text{total}} = 2\mathcal{P} = \hat{\mathbf{z}} \, \frac{1}{4\pi} \, |\mathbf{E}^{(0)}|^2, \tag{8.99}$$

which coincides identically with (8.96).

Thus, we see explicitly the origin of the radiation pressure, namely, the electromagnetic waves exert a force on the charged particles on the surface which generates a pressure.

8.6 Selected problems

1. Determine the reflection coefficient at normal incidence for sea water, fresh water and "good" earth at frequencies 60 Hz, 1

MHz and 1 GHz. Use $\epsilon = 80, \sigma = 3.6 \times 10^{10}/\text{sec}$ for sea water, $\epsilon = 80, \sigma = 4.5 \times 10^7/\text{sec}$ for fresh water and $\epsilon = 15, \sigma = 9 \times 10^7/\text{sec}$ for "good" earth.

2. Show that two dimensional fields that are independent of z can be written as superpositions of the following fields (in vacuum)

$$i) \quad \mathbf{E} = (0, 0, E_z), \qquad \mathbf{H} = (H_x, H_y, 0),$$

with

$$H_x = \frac{i}{k} \frac{\partial E_z}{\partial y}, \quad H_y = -\frac{i}{k} \frac{\partial E_z}{\partial x},$$

$$\frac{\partial^2 E_z}{\partial x^2} + \frac{\partial^2 E_z}{\partial y^2} + k^2 E_z = 0,$$

and

$$ii) \quad \mathbf{E} = (E_x, E_y, 0), \qquad \mathbf{H} = (0, 0, H_z),$$

with

$$E_x = -\frac{i}{k} \frac{\partial H_z}{\partial y}, \quad E_y = \frac{i}{k} \frac{\partial H_z}{\partial x},$$

$$\frac{\partial^2 H_z}{\partial x^2} + \frac{\partial^2 H_z}{\partial y^2} + k^2 H_z = 0.$$

3. Show that, when a uniform plane wave is incident normally on a good conductor, the linear current density, $\mathbf{J_s}$, is essentially independent of the conductivity σ.

CHAPTER 9

Radiation

So far, we have studied some of the properties of the electromagnetic waves in various media without worrying about how they are produced. The sources of electromagnetic waves are, of course, charges and currents. Let us now study, in some detail, the properties of electromagnetic fields produced by a localized distribution of (time dependent) charges and currents. Let us assume that we have a given distribution of charges and currents in a localized volume V. We have already seen (see (6.187) and (6.188)) that the retarded solutions of Maxwell's equations in the Lorentz gauge are given by (in vacuum)

$$
A_\mu(\mathbf{x}, t) = \frac{1}{c} \int d^3 x' \left. \frac{J_\mu(\mathbf{x}', t')}{|\mathbf{x} - \mathbf{x}'|} \right|_{t' = t - \frac{|\mathbf{x} - \mathbf{x}'|}{c}}
$$

$$
= \frac{1}{c} \int d^3 x' dt' \frac{J_\mu(\mathbf{x}', t') \delta \left(t' - t + \frac{|\mathbf{x} - \mathbf{x}'|}{c} \right)}{|\mathbf{x} - \mathbf{x}'|}. \tag{9.1}
$$

In fact, let us note that since in the Lorentz gauge

$$
\partial_\mu A^\mu = \frac{1}{c} \frac{\partial \Phi}{\partial t} + \boldsymbol{\nabla} \cdot \mathbf{A} = 0, \tag{9.2}
$$

the scalar potential is related to the vector potential and it is sufficient for us to study only the (three dimensional) vector potential. From (9.1) we have

$$
\mathbf{A}(\mathbf{x}, t) = \frac{1}{c} \int d^3 x' dt' \frac{\mathbf{J}(\mathbf{x}', t') \delta \left(t' - t + \frac{|\mathbf{x} - \mathbf{x}'|}{c} \right)}{|\mathbf{x} - \mathbf{x}'|}. \tag{9.3}
$$

The space integral here is over the volume V which contains (charges) currents and if we are interested in the fields at points which are far away from the volume containing the (charges) currents, we can make an expansion much like in the static case. This would give rise to the multipole expansion of the vector potential. In general, in the time dependent case this expansion has to be carried out more carefully than in the static case and would contain both electric as well as magnetic multipoles which we will study later. In what follows in this section, we will describe such an expansion for systems with a simple time dependence.

Let us consider the case where the current has a simple harmonic time dependence of the form (remember that the current is real and corresponds to either the real or the imaginary part of this expression)

$$\mathbf{J}(\mathbf{x}, t) = \mathbf{J}(\mathbf{x}) \, e^{-i\omega t}. \tag{9.4}$$

It follows from the continuity equation

$$\frac{\partial \rho}{\partial t} + \boldsymbol{\nabla} \cdot \mathbf{J} = 0,$$

that the charge density must also have exactly the same harmonic time dependence, namely,

$$\rho(\mathbf{x}, t) = \rho(\mathbf{x}) \, e^{-i\omega t}. \tag{9.5}$$

In such a case, the solution for the vector potential becomes

$$\mathbf{A}(\mathbf{x}, t) = \frac{1}{c} \int d^3 x' dt' \, \frac{\mathbf{J}(\mathbf{x}') e^{-i\omega t'} \delta\left(t' - t + \frac{|\mathbf{x} - \mathbf{x}'|}{c}\right)}{|\mathbf{x} - \mathbf{x}'|}$$

$$= \frac{e^{-i\omega t}}{c} \int d^3 x' \, \mathbf{J}(\mathbf{x}') \, \frac{e^{\frac{i\omega |\mathbf{x} - \mathbf{x}'|}{c}}}{|\mathbf{x} - \mathbf{x}'|}. \tag{9.6}$$

Therefore, the (three) vector potential also has a harmonic time dependence and separating out the time dependence of the vector potential,

$$\mathbf{A}(\mathbf{x}, t) = \mathbf{A}(\mathbf{x}) \, e^{-i\omega t}, \tag{9.7}$$

we have

$$\mathbf{A}(\mathbf{x}) = \frac{1}{c} \int d^3 x' \, \mathbf{J}(\mathbf{x}') \, \frac{e^{\frac{i\omega |\mathbf{x} - \mathbf{x}'|}{c}}}{|\mathbf{x} - \mathbf{x}'|}. \tag{9.8}$$

It is clear that all the variables $(\Phi, \mathbf{A})$ and $(\mathbf{E}, \mathbf{B})$ will have the same harmonic time dependence which can be factored out. The space dependent magnetic field can, of course, be determined from (9.8) as

$$\mathbf{B}(\mathbf{x}) = \boldsymbol{\nabla} \times \mathbf{A}(\mathbf{x}), \tag{9.9}$$

while the electric field in the region outside the volume containing charges and currents (in vacuum) can be obtained from the last equation of Maxwell in (7.1) as

$$\mathbf{E}(\mathbf{x}) = \frac{ic}{\omega} \, \boldsymbol{\nabla} \times \mathbf{B}(\mathbf{x}) = \frac{i}{k} \, \boldsymbol{\nabla} \times \mathbf{B}(\mathbf{x}). \tag{9.10}$$

So far, we have made no approximations. But, we recognize that there are now two length scales in the problem (as opposed to the static case, where there was only one length scale), namely, the size d associated with the volume V which contains charges and currents and the wavelength of oscillations $\lambda = \frac{2\pi}{|\mathbf{k}|} = \frac{2\pi c}{\omega}$. (We are assuming that we are in vacuum.) As a result, there are more possibilities in the expansions that we can make. First, let us assume that we are very far away from the sources, namely, $|\mathbf{x}| >> d$ (as also in the static case). In such a case, we can expand $|\mathbf{x} - \mathbf{x}'|$ as

$$|\mathbf{x} - \mathbf{x}'| = (r^2 + r'^2 - 2rr' \cos \theta')^{\frac{1}{2}}$$

$$= r \left(1 - \frac{2r' \cos \theta'}{r} + \frac{r'^2}{r^2} \right)^{\frac{1}{2}}$$

$$= r - r' \cos \theta' + O(\frac{1}{r}). \tag{9.11}$$

Substituting this into the expression for the vector potential in (9.8), we have

$$\mathbf{A}(\mathbf{x}) \approx \frac{1}{c} \int d^3 x' \, \mathbf{J}(\mathbf{x}') \, \frac{e^{i|\mathbf{k}|(r - r' \cos \theta')}}{(r - r' \cos \theta')}$$

$$= \frac{e^{ikr}}{cr} \int d^3 x' \, \mathbf{J}(\mathbf{x}') \left(\frac{e^{-ikr' \cos \theta'}}{1 - \frac{r' \cos \theta'}{r}} \right). \tag{9.12}$$

Here, we have identified $|\mathbf{k}| = k$.

Let us next assume that $d << \lambda$. This is, in fact, an excellent approximation. For, suppose $\nu = \frac{\omega}{2\pi} = 60$ cycles/sec, then, $\lambda = \frac{c}{\nu} = 5 \times 10^8$ cm which is large compared to laboratory sizes. In this case, it follows that for most localized sources we have

$$kd << 1, \tag{9.13}$$

which allows us to make a Taylor expansion of the quantity in the parenthesis in (9.12) as

$$
\left(\frac{e^{-ikr'\cos\theta'}}{1 - \frac{r'\cos\theta'}{r}}\right) = 1 + (-ik)(1 + \frac{i}{kr})r'\cos\theta'
$$

$$
+ \frac{(-ik)^2}{2!}(1 + \frac{2i}{kr} - \frac{2}{(kr)^2})(r'\cos\theta')^2 + \cdots
$$

$$
= \sum_{n=0}^{\infty} X_n(r,k)(r'\cos\theta')^n, \tag{9.14}
$$

where the general form of X_n is easily seen to be

$$
X_n(r,k) = \frac{(-ik)^n}{n!}\left(1 + \frac{a_n^{(1)}}{kr} + \frac{a_n^{(2)}}{(kr)^2} + \cdots + \frac{a_n^{(n)}}{(kr)^n}\right), \tag{9.15}
$$

and $a_n^{(m)}$'s ($m \leq n$) are numerical constants. Since $X_n(r,k)$ is independent of the variables of integration, it can be taken outside the integral in (9.12) leading to an expansion of the vector potential of the form

$$
\mathbf{A}(\mathbf{x}) = \frac{e^{ikr}}{cr}\sum_{n=0}^{\infty} X_n(r,k)\left(\int d^3x'\,\mathbf{J}(\mathbf{x}')(r'\cos\theta')^n\right). \tag{9.16}
$$

There are now two possibilities. Namely, we can either have $r \ll \lambda$ ($kr \ll 1$), or $r \gg \lambda$ ($kr \gg 1$). The two regions are known respectively as the near (static) zone and the far (radiation) zone. In the two cases, the quantities X_n and, therefore, $\mathbf{A}$ behave differently. (In addition, there is also the intermediate zone where $kr \sim 1$ which we will not consider.) Thus, for example, in the near zone, $kr \ll 1$ and we have from (9.15)

$$
X_n(r,k) \approx \frac{(-ik)^n}{n!}\frac{a_n^{(n)}}{(kr)^n} = \frac{\tilde{a}_n^{(n)}}{r^n}, \tag{9.17}
$$

which is independent of k. Substituting this into (9.16), we see that, in such a case, the vector potential takes the form

$$
\mathbf{A}(\mathbf{x}) = \sum_{n=0}^{\infty} \frac{\tilde{a}_n^{(n)}}{cr^{n+1}}\int d^3x'\,\mathbf{J}(\mathbf{x}')\,(r'\cos\theta')^n. \tag{9.18}
$$

In other words, in the near zone, the vector potential oscillates harmonically with time. Otherwise, it has a purely static character (no propagation). (Here we have used the fact that in the near zone, $e^{ikr} \approx 1$.) We note that, since $r' \ll r$, the dominant term in (9.18) is the zeroth order term leading to

$$\mathbf{A}(\mathbf{x}) = \frac{1}{cr} \int d^3x' \, \mathbf{J}(\mathbf{x}').$$
(9.19)

Here, we have used the fact that $\tilde{a}_0^{(0)} = 1$ (see (9.15)).

In contrast, in the far zone ($kr \gg 1$), we see from Eq. (9.15) that we have

$$X_n(r, k) \approx \frac{(-ik)^n}{n!},$$
(9.20)

so that the vector potential in (9.16) takes the form

$$\mathbf{A}(\mathbf{x}) = \frac{e^{ikr}}{cr} \sum_{n=0}^{\infty} \frac{(-ik)^n}{n!} \int d^3x' \, \mathbf{J}(\mathbf{x}')(r' \cos \theta')^n$$

$$= \frac{e^{ikr}}{cr} \int d^3x' \, \mathbf{J}(\mathbf{x}') \sum_{n=0}^{\infty} \frac{(-ikr' \cos \theta')^n}{n!}$$

$$= \frac{e^{ikr}}{cr} \int d^3x' \, \mathbf{J}(\mathbf{x}') \, e^{-ikr' \cos \theta'}.$$
(9.21)

We note that, since $kd \ll 1$ (see Eq. (9.19)), the dominant term in (9.21) leads to

$$\mathbf{A}(\mathbf{x}) = \frac{e^{ikr}}{cr} \int d^3x' \, \mathbf{J}(\mathbf{x}').$$
(9.22)

We see that in the far zone, the vector potential is represented by a spherically outgoing wave. This is because the phase of the complete vector potential is given by $(\omega t - kr)$. As we have learnt earlier, wave fronts are described by surfaces of constant phase. Thus, at any given time, the surfaces of constant phase are given by

$$\omega t - kr = \text{constant}$$

$$\text{or,} \quad kr = \text{constant},$$
(9.23)

which are spherical surfaces of radius r. This is like the plane wave solutions of radiation that we studied earlier, but the waves, in the present case, are spherically outgoing. The other thing to note is that, in the far zone, the higher order terms in the expansion (of the exponential) fall off rapidly, simply because $kd << 1$. Consequently, in the study of radiation, only the first term in the series contributes significantly. We can calculate the electric and the magnetic fields from the potentials (see (9.9), (9.10)). However, we will not go into it now except for the observation that, asymptotically, the electric and the magnetic fields fall off as $\frac{1}{r}$ which is precisely the behavior of radiation fields.

9.1 Electric dipole radiation

In this and in the following sections, we will study the properties of the electric and the magnetic fields produced by some simple charge and current distributions. Let us consider a system consisting of two equal, but opposite charges whose magnitudes oscillate with time. The charges are separated by an infinitesimal distance along the z-axis as shown in Fig. 9.1. This can be thought of as an element of an alternating current circuit (as in the second diagram in Fig. 9.1) and is often called a Hertzian dipole.

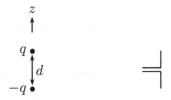

Figure 9.1: A Hertzian dipole element.

Let us further assume that the size of the dipole, d, is very small and that the time dependence of the charge is harmonic as before. Namely,

$$q(t) = q_0 e^{-i\omega t}. \tag{9.24}$$

This shows that

$$I(t) = \frac{dq(t)}{dt} = -i\omega q_0 e^{-i\omega t} = I_0 e^{-i\omega t}. \tag{9.25}$$

Therefore, such a system can be thought of as an alternating current element. Furthermore, both the charge and the current have a simple harmonic dependence on time so that the earlier analysis can be directly applied. In such a case, the expansion of the vector potential, with the assumption that $r \gg d$ can be approximated by the leading term, which is the first term in the expansion. As we have seen in Eqs. (9.19) and (9.22), irrespective of whether we are in the near zone or in the far zone, the vector potential has the leading form (in the near zone $e^{ikr} \approx 1$)

$$\mathbf{A}(\mathbf{x}) = \frac{e^{ikr}}{cr} \int \mathrm{d}^3 x' \, \mathbf{J}(\mathbf{x}'). \tag{9.26}$$

Next, let us use the vector identities $(J_i = -x_i(\partial_j J_j) + \partial_j(x_i J_j))$

$$\int \mathrm{d}^3 x' \, \mathbf{J}(\mathbf{x}') = -\int \mathrm{d}^3 x' \, \mathbf{x}' \, (\boldsymbol{\nabla}' \cdot \mathbf{J}(\mathbf{x}')) + \int (\mathrm{d}\mathbf{s}' \cdot \mathbf{J}(\mathbf{x}')) \mathbf{x}'$$

$$= -i\omega \int \mathrm{d}^3 x' \, \mathbf{x}' \, \rho(\mathbf{x}')$$

$$= -i\omega \mathbf{p}, \tag{9.27}$$

where we have used the fact that the currents are contained within the volume and, consequently, $\hat{\mathbf{n}} \cdot \mathbf{J} = 0$ on the surface. Furthermore, we have used the continuity equation

$$\frac{\partial \rho}{\partial t} + \boldsymbol{\nabla} \cdot \mathbf{J} = 0, \tag{9.28}$$

as well as the definition of the electric dipole moment $\mathbf{p}$ for an arbitrary charge distribution (see (2.49))

$$\mathbf{p} = \int \mathrm{d}^3 x \, \mathbf{x} \, \rho(\mathbf{x}), \tag{9.29}$$

in the above derivation. We note that by construction, the electric dipole moment of our system is along the z-axis.

Substituting (9.27) into (9.26), we see that in such a case, we can write $(k = \frac{\omega}{c})$

$$\mathbf{A}(\mathbf{x}) = \frac{e^{ikr}}{cr}(-i\omega \mathbf{p}) = -ik\mathbf{p}\frac{e^{ikr}}{r}. \tag{9.30}$$

This allows us to calculate the magnetic field directly as

$$\mathbf{B}(\mathbf{x}) = \boldsymbol{\nabla} \times \mathbf{A}(\mathbf{x}) = -ik\,(\hat{\mathbf{r}} \times \mathbf{p})\,\frac{\partial}{\partial r}\left(\frac{e^{ikr}}{r}\right)$$

$$= k^2(\hat{\mathbf{r}} \times \mathbf{p})\left(1 + \frac{i}{kr}\right)\frac{e^{ikr}}{r}. \tag{9.31}$$

On the other hand, the electric field can be calculated from the Ampere's law, namely, (since we are interested in points far away from the sources, $\mathbf{J} = 0$, see also (9.10))

$$\boldsymbol{\nabla} \times \mathbf{B} = \frac{1}{c} \frac{\partial \mathbf{E}}{\partial t}, \tag{9.32}$$

which gives

$$\mathbf{E}(\mathbf{x}) = \frac{i}{k} \left(\boldsymbol{\nabla} \times \mathbf{B}(\mathbf{x}) \right)$$

$$= -k^2 \left[(\hat{\mathbf{r}} \times (\hat{\mathbf{r}} \times \mathbf{p})) \left(1 + \frac{2i}{kr} - \frac{2}{(kr)^2} \right) \right. \tag{9.33}$$

$$\left. - (\hat{\boldsymbol{\theta}} \times (\hat{\boldsymbol{\theta}} \times \mathbf{p}) + \hat{\boldsymbol{\phi}} \times (\hat{\boldsymbol{\phi}} \times \mathbf{p})) \left(\frac{i}{kr} - \frac{1}{(kr)^2} \right) \right] \frac{e^{ikr}}{r},$$

where we have used the fact that the unit vectors in spherical coordinates are not fixed. In fact, while $\frac{\partial \hat{\mathbf{r}}}{\partial r} = 0$, $\frac{\partial \hat{\mathbf{r}}}{\partial \theta} = \hat{\boldsymbol{\theta}}$ and $\frac{\partial \hat{\mathbf{r}}}{\partial \phi} = \hat{\boldsymbol{\phi}} \sin \theta$. Furthermore, recalling that $\mathbf{p}$ is along the z-axis, we can simplify this and write (recall that, in spherical coordinates, $\hat{\mathbf{z}} = \hat{\mathbf{r}} \cos \theta - \hat{\boldsymbol{\theta}} \sin \theta$)

$$\mathbf{E}(\mathbf{x}) = -k^2 \left[(\hat{\mathbf{r}} \times (\hat{\mathbf{r}} \times \mathbf{p})) \right.$$

$$\left. + (3\hat{\mathbf{r}}(\hat{\mathbf{r}} \cdot \mathbf{p}) - \mathbf{p}) \left(\frac{i}{kr} - \frac{1}{(kr)^2} \right) \right] \frac{e^{ikr}}{r}. \tag{9.34}$$

It is clear from (9.31) that the magnetic field is always transverse to the radial vector, namely,

$$\hat{\mathbf{r}} \cdot \mathbf{B} = 0. \tag{9.35}$$

However, it follows from (9.34) that the electric field is not transverse to the radial vector in general. We note that the magnetic field in (9.31) has two terms – one behaving as $\frac{1}{r}$ which dominates for large r, while the second depends on the radial coordinate as $\frac{1}{r^2}$ and, therefore, contributes significantly for small r. The second is known as the static (or induction) term while the first is called the radiation term for reasons that will become clear shortly. Similarly, we see from (9.34) that the electric field has three terms out of which the $\frac{1}{r^3}$ term gives the most contribution for small r (and is called the static field) while the $\frac{1}{r}$ term dominates at large distances and is known as the radiation field.

We note that in the near zone, $kr \ll 1$ and we can write the electric and the magnetic fields as ($e^{ikr} \approx 1$)

$$\mathbf{B}(\mathbf{x}) = \frac{ik(\hat{\mathbf{r}} \times \mathbf{p})}{r^2},$$

$$\mathbf{E}(\mathbf{x}) = \frac{3\hat{\mathbf{r}}(\hat{\mathbf{r}} \cdot \mathbf{p}) - \mathbf{p}}{r^3}. \tag{9.36}$$

This shows that in the near zone, the magnetic field is what would be obtained from the Biot-Savart law for a current element (except for the trivial time dependence that has been factored out). Similarly, in this region, the electric field is that of a static dipole (except for the trivial harmonic time dependence which we have factored out). Furthermore, since in this region $kr \ll 1$, the electric field dominates over the magnetic field. In the far (radiation) zone, on the other hand, we note that $kr \gg 1$ and we can approximate the electric and the magnetic fields in (9.31) and (9.34) as

$$\mathbf{B}(\mathbf{x}) = k^2(\hat{\mathbf{r}} \times \mathbf{p}) \frac{e^{ikr}}{r},$$

$$\mathbf{E}(\mathbf{x}) = -k^2(\hat{\mathbf{r}} \times (\hat{\mathbf{r}} \times \mathbf{p})) \frac{e^{ikr}}{r} = -\hat{\mathbf{r}} \times \mathbf{B}(\mathbf{x}). \tag{9.37}$$

The last relation in (9.37) can also be written as

$$\mathbf{B}(\mathbf{x}) = \hat{\mathbf{r}} \times \mathbf{E}(\mathbf{x}), \tag{9.38}$$

which is the relation for traveling EM waves that we have seen earlier, for example, in (6.30). From (9.37) we see that both the $\mathbf{E}$ and the $\mathbf{B}$ fields fall off as $\frac{1}{r}$ and are transverse to the direction of propagation (as well as to each other) as is expected of radiation fields. We see that the radiation terms are new compared to the behavior of static distributions. We note that, in the static limit ($\omega = 0$ or $k = 0$), the magnetic field identically vanishes everywhere. Furthermore, the radiation component of the electric field also vanishes in this limit. Thus, we see that radiation is an essential feature associated with time varying charges and currents.

Incidentally, although we have discussed a very simple system, it is behind many physical systems such as antennas. In this discussion, we have only retained the lowest order term which, as we see, leads to an electric dipole description (in the static zone). The higher order terms in the expansion, similarly, can be shown to give rise to the description of a magnetic dipole, electric quadrupole etc.

9.1.1 Power radiated by an electric dipole. As we have seen earlier, the energy flow per unit time per unit area or the power flow through a unit area is related to the real part of the Poynting vector, which in the vacuum takes the form (this is the time averaged value defined in (6.96)),

$$\overline{\mathbf{S}} = \frac{c}{8\pi} \, \text{Re} \, (\mathbf{E} \times \mathbf{B}^*). \tag{9.39}$$

If we consider a large sphere of radius R, then, it is clear that the total time averaged power radiated through the surface of this sphere is given by

$$P_{\text{total}} = \int \mathrm{d}a \, \hat{\mathbf{r}} \cdot \overline{\mathbf{S}}, \tag{9.40}$$

where $\mathrm{d}a$ represents a surface element on the sphere of radius R.

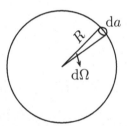

Figure 9.2: The solid angle subtended by an infinitesimal surface area on a sphere at the center.

Since the solid angle subtended by this infinitesimal area element is given by (see Fig. 9.2)

$$\mathrm{d}\Omega = \frac{\mathrm{d}a}{R^2}, \tag{9.41}$$

it follows that the average power radiated through unit solid angle has the form

$$\frac{\mathrm{d}P}{\mathrm{d}\Omega} = R^2 \, \hat{\mathbf{r}} \cdot \overline{\mathbf{S}}. \tag{9.42}$$

If we take the radius of the sphere to infinity, then it is clear from (9.42) that only the field components that decrease as $\frac{1}{r}$ can contribute to the power radiated out to infinity. It is for this reason that these components of the fields are known as the radiation fields. They lead to a radiation of power to infinity (which cannot come back to the system).

If we look at the forms of the radiation components of the electric and the magnetic fields in Eq. (9.37) and recall that the dipole moment is along the z-axis (by assumption) with $\hat{\mathbf{z}} = \hat{\mathbf{r}} \cos\theta - \hat{\boldsymbol{\theta}} \sin\theta$, it follows that in the limit $R \to \infty$,

$$\frac{dP}{d\Omega} = \frac{c}{8\pi} k^4 |\mathbf{p}|^2 \sin^2\theta. \tag{9.43}$$

This shows that the radiated power is highly directional (not uniform, but depends on θ), peaking at $\theta = \frac{\pi}{2}$. This also leads to a total integrated average radiated power of the form

$$P_{\text{total}} = \int d\Omega \, \frac{dP}{d\Omega} = \frac{c}{8\pi} k^4 |\mathbf{p}|^2 \int d\Omega \, \sin^2\theta$$

$$= \frac{c}{8\pi} k^4 |\mathbf{p}|^2 \, \frac{8\pi}{3} = \frac{c k^4 |\mathbf{p}|^2}{3}. \tag{9.44}$$

Using $k = \frac{\omega}{c}$ and the fact that for this simple system we can think of (see also (9.24), (9.25))

$$|\mathbf{p}|^2 = |q_0|^2 d^2 = \frac{|I_0|^2 d^2}{\omega^2}, \tag{9.45}$$

we can also write the (time averaged) total power radiated to infinity, (9.44), as

$$P_{\text{total}} = \frac{\omega^2 d^2}{3c^3} |I_0|^2 = \frac{2\omega^2 d^2}{3c^3} I_{\text{rms}}^2. \tag{9.46}$$

Here, I_0 represents the peak current while I_{rms} denotes the effective current (root mean square). This suggests that we can associate a radiation resistance with the dipole from the standard definitions as

$$R_{\text{rad}} = \frac{2\omega^2 d^2}{3c^3}. \tag{9.47}$$

9.2 Magnetic dipole radiation

Just as we studied the electric and the magnetic fields produced by a short alternating current element (or an electric dipole), we can also do the same for a small alternating current loop. In this case, the

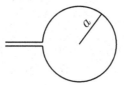

Figure 9.3: An alternating current loop of radius a in the $x-y$ plane.

system would behave like a magnetic dipole. Let us consider a small alternating current loop of radius "a" in the horizontal $(x-y)$ plane as shown in Fig. 9.3.

We will assume, as before, that the current has a harmonic time dependence,

$$I(t) = I_0\, e^{-i\omega t},$$

$$\mathbf{J}(\mathbf{x}, t) = \mathbf{J}(\mathbf{x})e^{-i\omega t} = \hat{\boldsymbol{\phi}}\, I_0\delta(\rho - a)\delta(z)e^{-i\omega t}, \tag{9.48}$$

where ρ denotes the radial coordinate in the $x-y$ plane. Just as in the case of steady currents a closed current loop gives rise to a magnetic dipole moment, here, too, we can show that such a current loop would have a magnetic dipole moment with a harmonic time dependence associated with it,

$$\mathbf{m}(t) = \mathbf{m_0}\, e^{-i\omega t} = \hat{\mathbf{z}}\, I_0\pi a^2\, e^{-i\omega t}. \tag{9.49}$$

Since the current has a harmonic time dependence, we can apply our general analysis and obtain from (9.6) that

$$\mathbf{A}(\mathbf{x}) = \frac{1}{c} \int d^3x'\, \mathbf{J}(\mathbf{x}')\, \frac{e^{ik|\mathbf{x}-\mathbf{x}'|}}{|\mathbf{x}-\mathbf{x}'|}$$

$$= \frac{I_0}{c} \int d^3x'\, \delta(\rho' - a)\delta(z')\, \hat{\boldsymbol{\phi}}'\, \frac{e^{ik|\mathbf{x}-\mathbf{x}'|}}{|\mathbf{x}-\mathbf{x}'|}. \tag{9.50}$$

Let us note that if we write

$$\mathbf{x} = r\left(\hat{\mathbf{x}}\sin\theta\cos\phi + \hat{\mathbf{y}}\sin\theta\sin\phi + \hat{\mathbf{z}}\cos\theta\right),$$

$$\mathbf{x}' = \rho'\left(\hat{\mathbf{x}}\cos\phi' + \hat{\mathbf{y}}\sin\phi'\right), \tag{9.51}$$

then, it follows that (note that $z' = 0$ or equivalently $\theta' = \frac{\pi}{2}$ because of the delta function constraint)

$$|\mathbf{x} - \mathbf{x}'| = (r^2 + \rho'^2 - 2r\rho' \sin\theta \cos(\phi - \phi'))^{\frac{1}{2}}$$

$$\approx r - \rho' \sin\theta \cos(\phi - \phi'). \tag{9.52}$$

Here, we have assumed that the size of the current loop is extremely small compared with the distance of the point of observation ($r \gg \rho'$) which allows us to keep only the lowest order terms in the expansion.

Furthermore, let us assume that we are interested only in the far field approximation (namely, in the radiation zone). In that case, r is very large ($kr \gg 1$) and we can approximate the denominator by the lowest order term. However, we have to be careful with the exponent. Let us also note that we can write $\hat{\phi}' = -\hat{\mathbf{x}} \sin\phi' + \hat{\mathbf{y}} \cos\phi'$. With these, the expression (9.50) for large values of r takes the form (and we assume that a is very small, $ka \ll 1$)

$$\mathbf{A}(\mathbf{x}) \approx \frac{I_0 a e^{ikr}}{cr} \int_0^{2\pi} d\phi' \, (-\hat{\mathbf{x}} \sin\phi' + \hat{\mathbf{y}} \cos\phi') \, e^{-ika \sin\theta \cos(\phi - \phi')}$$

$$\approx \frac{I_0 a e^{ikr}}{cr} \int_0^{2\pi} d\phi' \, (-\hat{\mathbf{x}} \sin\phi' + \hat{\mathbf{y}} \cos\phi')$$

$$\times \left(1 - ika \sin\theta \cos(\phi - \phi') \right)$$

$$= -\frac{ik I_0 a^2 \sin\theta e^{ikr}}{2cr} \int_0^{2\pi} d\phi' \left[-\hat{\mathbf{x}} \left(\sin\phi + \sin(2\phi' - \phi) \right) \right.$$

$$\left. + \hat{\mathbf{y}} \left(\cos\phi + \cos(2\phi' - \phi) \right) \right]$$

$$= -\frac{ik I_0 \pi a^2 \sin\theta e^{ikr}}{cr} (-\hat{\mathbf{x}} \sin\phi + \hat{\mathbf{y}} \cos\phi)$$

$$= -\hat{\phi} \frac{ik I_0 \pi a^2 \sin\theta e^{ikr}}{cr} = -\hat{\phi} \frac{ik |\mathbf{m}_0| \sin\theta e^{ikr}}{cr}. \tag{9.53}$$

Here we have used standard trigonometric identities

$$\sin AB = \frac{1}{2} \left(\sin(A + B) + \sin(A - B) \right),$$

$$\cos AB = \frac{1}{2} \left(\cos(A + B) + \cos(A - B) \right).$$

We note that the form of (9.53) is, in fact, quite similar to the case of the electric dipole radiator studied earlier, with the electric dipole moment replaced by the magnetic dipole moment (along with various constants needed for dimensional reasons).

We note that, unlike the electric dipole case, here the vector potential in the radiation zone is transverse to the direction of propagation, namely,

$$\hat{\mathbf{r}} \cdot \mathbf{A}(\mathbf{x}) = 0. \tag{9.54}$$

The magnetic field can now be calculated easily

$$\mathbf{B}(\mathbf{x}) = \boldsymbol{\nabla} \times \mathbf{A}$$

$$= -\hat{\boldsymbol{\theta}} \, \frac{k^2 I_0 \pi a^2 \sin\theta e^{ikr}}{cr} + O\left(\frac{1}{r^2}\right). \tag{9.55}$$

Similarly, the electric field takes the form

$$\mathbf{E}(\mathbf{x}) = \frac{i}{k} \, \boldsymbol{\nabla} \times \mathbf{B}$$

$$= \hat{\boldsymbol{\phi}} \, \frac{k^2 I_0 \pi a^2 \sin\theta e^{ikr}}{cr} + O\left(\frac{1}{r^2}\right). \tag{9.56}$$

This allows us to calculate the (time averaged) power radiated per unit solid angle in a given direction through the surface of a large sphere to be (see (9.42))

$$\frac{\mathrm{d}P}{\mathrm{d}\Omega} = R^2 \hat{\mathbf{r}} \cdot \overline{\mathbf{S}}$$

$$= \frac{(k^2 I_0 \pi a^2)^2}{8\pi c} \sin^2\theta = \frac{k^4 |\mathbf{m}_0|^2}{8\pi c} \sin^2\theta. \tag{9.57}$$

We see that the angular dependence of the radiated power, in this case, is the same as that in the case of the electric dipole radiation in (9.43). The total power radiated through the surface of a large sphere, averaged over a cycle, is then obtained to be

$$P_{\text{total}} = \frac{\pi^2 k^4 a^4}{3c} \, I_0^2 = \frac{2\pi^2 k^4 a^4}{3c} \, I_{\text{rms}}^2, \tag{9.58}$$

so that we can identify the radiation resistance associated with this system to be

$$R_{\text{rad}} = \frac{2\pi^2 k^4 a^4}{3c}. \tag{9.59}$$

9.3 Center-fed antennas

The previous analysis has direct application in short center-fed antennas. Antennas are open wire systems connected to an alternating current source – a transmitter. The center-fed antenna, for example, can be a parallel transmission line with open wires or wires bent at right angles. Such an antenna of length d (measured from one end to the other even though it is center-fed), is known as a dipole antenna. On the other hand, if one of the transmission lines is grounded (or we have only one wire above a perfectly conducting plane), it also acts like a dipole antenna together with the image inside the earth and such a set up is known as a monopole antenna (see Fig. 9.4).

Figure 9.4: A dipole antenna (on the left) and a monopole antenna (on the right).

In the earlier analysis of the alternating current element, we assumed that the current is constant in the entire length of the wire, which is valid only because we assumed the dipole to have a very small length. In a realistic antenna, however, the current will be different at different points along the length of the wire, as is known from the study of transmission lines. Let us first consider a very short dipole antenna of length d. We assume a harmonic variation of the current with time as before. Since it is center-fed, it is quite reasonable to assume that the current has a maximum at the center and decreases linearly to zero at both ends. (From symmetry, we see that the current pattern will be the same in both the halves of the antenna.) Namely, let us assume that

$$I(z,t) = I_0 \left(1 - \frac{2|z|}{d} \right) e^{-i\omega t}, \qquad -\frac{d}{2} \le z \le \frac{d}{2}. \qquad (9.60)$$

With this, we can repeat the calculation of the earlier sections. There
is a simpler method for obtaining the relevant results by noting that
the average current in the antenna, in this case, follows to be

$$I_{\text{avg}} = \frac{I_0}{2}, \tag{9.61}$$

where we are assuming that I_0 represents the peak current at the
center of the antenna.

With this, we can now extend our previous analysis simply by
letting $I_0 \to \frac{I_0}{2}$ or $I_{\text{rms}} \to \frac{I_{\text{rms}}}{2}$. Thus, for example, for the center-fed
short dipole antenna we obtain (see (9.46)),

$$P_{\text{total}} = \frac{\omega^2 d^2}{6c^3} I_{\text{rms}}^2. \tag{9.62}$$

Correspondingly, the radiation resistance of the center-fed dipole antenna is smaller by a factor of 4,

$$R_{\text{rad}} = \frac{\omega^2 d^2}{6c^3}. \tag{9.63}$$

On the other hand, for a short center-fed monopole antenna,
since only one half of the antenna really radiates, we have

$$P_{\text{total}} = \frac{1}{2} \times \frac{\omega^2 d^2}{6c^3} I_{\text{rms}}^2 = \frac{\omega^2 d^2}{12c^3} I_{\text{rms}}^2, \tag{9.64}$$

and correspondingly, the radiation resistance

$$R_{\text{rad}} = \frac{\omega^2 d^2}{12c^3}, \tag{9.65}$$

is even smaller.

These theoretical predictions work quite well and can be checked
experimentally to hold for short antennas satisfying $d \ll \lambda$ where
λ denotes the wavelength of the signal. In fact, they hold up to
$d \leq \frac{\lambda}{4}$. However, in transmitting radio waves (for radio waves $\nu =$
$300\text{Hz} - 3000\text{GHz}$, $\lambda = 100\text{Km} - 1\text{mm}$), where antennas are primarily
used, it is found that the transmission is better if the dimensions of
the antenna were of the order of the wave length, $d \sim \lambda$ (see Fig. 9.5).
Of course, these are no longer short antennas and the analysis has
to be carried out more carefully. The difficulty really lies in knowing
the distribution of the current along the length of the antenna. If
this distribution is known, the calculation of the fields can be carried
out in principle. Following studies of the transmission lines where

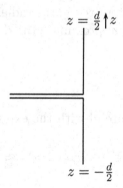

Figure 9.5: A center-fed antenna of length $d \sim \lambda$.

it is known that the current varies sinusoidally with distance, let us assume that the current has the form

$$I(z,t) = I_0 \sin k \left(\frac{d}{2} - |z| \right) e^{-i\omega t}, \qquad -\frac{d}{2} \le z \le \frac{d}{2}. \tag{9.66}$$

Namely, we are assuming the antenna to lie along the z-axis and the current to be sinusoidally varying with z such that the maximum is at the center and the ends have vanishing current. This translates to a current density of the form

$$\mathbf{J}(\mathbf{x}, t) = \mathbf{J}(\mathbf{x})\, e^{-i\omega t} = \hat{\mathbf{z}}\, \delta(x)\delta(y) I(z, t), \tag{9.67}$$

where we identify

$$\mathbf{J}(\mathbf{x}) = \hat{\mathbf{z}}\, \delta(x)\, \delta(y)\, I_0 \sin k \left(\frac{d}{2} - |z| \right), \qquad |z| \le \frac{d}{2}. \tag{9.68}$$

Since this involves a current distribution that is harmonic in time, we can apply our previous analysis and obtain from (9.8) and (9.68)

$$\mathbf{A}(\mathbf{x}) = \frac{1}{c} \int d^3x'\, \mathbf{J}(\mathbf{x}') \frac{e^{ik|\mathbf{x}-\mathbf{x}'|}}{|\mathbf{x}-\mathbf{x}'|}$$

$$= \hat{\mathbf{z}}\, \frac{I_0}{c} \int_{-\frac{d}{2}}^{\frac{d}{2}} dz'\, \sin k \left(\frac{d}{2} - |z'| \right) \frac{e^{ik|\mathbf{x}-z'\hat{\mathbf{z}}|}}{|\mathbf{x}-z'\hat{\mathbf{z}}|}. \tag{9.69}$$

We note that we are interested in the radiation fields for which $r \gg d$. In such a case, we can approximate the denominator in (9.69) by

$$\frac{1}{|\mathbf{x} - z'\hat{\mathbf{z}}|} \approx \frac{1}{r}. \tag{9.70}$$

We have to be more careful with the exponent where we write

$$|\mathbf{x} - z'\hat{\mathbf{z}}| = (r^2 + z'^2 - 2rz' \cos\theta)^{\frac{1}{2}} \approx r - z' \cos\theta. \tag{9.71}$$

Furthermore, let us simplify our calculation by assuming that $d = \frac{\lambda}{2}$ ($kd = \pi$), in which case, we have

$$\sin k \left(\frac{d}{2} - |z'| \right) = \sin \left(\frac{\pi}{2} - k|z'| \right) = \cos k|z'| = \cos kz'. \tag{9.72}$$

Using these in (9.69), we obtain for large r (the other term $\sin(kz' \cos\theta)$ coming from the exponential vanishes by anti-symmetry in z')

$$\mathbf{A}(\mathbf{x}) \approx \hat{\mathbf{z}} \frac{2I_0 e^{ikr}}{cr} \int_0^{\frac{d}{2}} dz' \, \cos kz' \, \cos(kz' \cos\theta)$$

$$= \hat{\mathbf{z}} \frac{I_0 e^{ikr}}{cr} \int_0^{\frac{d}{2}} dz' \, \left(\cos kz'(1 - \cos\theta) + \cos kz'(1 + \cos\theta) \right)$$

$$= \hat{\mathbf{z}} \frac{I_0 e^{ikr}}{cr} \left[\frac{\sin kz'(1 - \cos\theta)}{k(1 - \cos\theta)} + \frac{\sin kz'(1 + \cos\theta)}{k(1 + \cos\theta)} \right]_0^{\frac{d}{2}}$$

$$= \hat{\mathbf{z}} \frac{I_0 e^{ikr}}{ckr} \left[\frac{1}{1 - \cos\theta} + \frac{1}{1 + \cos\theta} \right] \cos \left(\frac{\pi}{2} \cos\theta \right)$$

$$= \hat{\mathbf{z}} \frac{2I_0 e^{ikr}}{ckr} \frac{\cos \left(\frac{\pi}{2} \cos\theta \right)}{\sin^2\theta}. \tag{9.73}$$

Exercise. Show that, for an antenna of arbitrary length d, the vector potential has the large distance behavior given by

$$\mathbf{A}(\mathbf{x}) = -\hat{\mathbf{z}} \frac{2I_0 e^{ikr}}{ckr} \frac{\cos \frac{kd}{2} - \cos \left(\frac{kd}{2} \cos\theta \right)}{\sin^2\theta}.$$

Recalling that $\hat{\mathbf{z}} = \hat{\mathbf{r}} \cos\theta - \hat{\boldsymbol{\theta}} \sin\theta$, we can now calculate the magnetic field for large r from (9.73) to be

$$\mathbf{B}(\mathbf{x}) = \mathbf{\nabla} \times \mathbf{A}(\mathbf{x})$$

$$= \left(\hat{\mathbf{r}} \frac{\partial}{\partial r} + \frac{\hat{\boldsymbol{\theta}}}{r} \frac{\partial}{\partial \theta} + \frac{\hat{\boldsymbol{\phi}}}{r \sin \theta} \frac{\partial}{\partial \phi} \right)$$

$$\times (\hat{\mathbf{r}} \cos \theta - \hat{\boldsymbol{\theta}} \sin \theta) \frac{2 I_0 e^{ikr} \cos \left(\frac{\pi}{2} \cos \theta \right)}{ckr \sin^2 \theta}$$

$$= -\hat{\boldsymbol{\phi}} \frac{2 i I_0 e^{ikr} \cos \left(\frac{\pi}{2} \cos \theta \right)}{cr \sin \theta} + O \left(\frac{1}{r^2} \right). \tag{9.74}$$

This shows that, at large distances, the dominant term in the magnetic field is the radiation term and is along the $\hat{\boldsymbol{\phi}}$ direction. The electric field can also be calculated similarly and for large distances has the form

$$\mathbf{E}(\mathbf{x}) = \frac{i}{k} \mathbf{\nabla} \times \mathbf{B}$$

$$= -\frac{i}{k} \hat{\mathbf{r}} \times \hat{\boldsymbol{\phi}} \frac{\partial}{\partial r} \frac{2 i I_0 e^{ikr} \cos \left(\frac{\pi}{2} \cos \theta \right)}{cr \sin \theta} + O \left(\frac{1}{r^2} \right)$$

$$= -\hat{\boldsymbol{\theta}} \frac{2 i I_0 e^{ikr} \cos \left(\frac{\pi}{2} \cos \theta \right)}{cr \sin \theta} + O \left(\frac{1}{r^2} \right). \tag{9.75}$$

Thus, we see that at large distances, the dominant terms in the electric and the magnetic fields are the radiation terms. They are transverse to the direction of propagation $\hat{\mathbf{r}}$ (since as we recall, at large distances the wave is a spherical wave) as well as to each other which is characteristic of radiation fields. We can now calculate the average power radiated by such a dipole per unit solid angle through the surface of a large sphere using (9.39) and (9.42), which leads to

$$\frac{\mathrm{d}P}{\mathrm{d}\Omega} = R^2 \, \hat{\mathbf{r}} \cdot \overline{\mathbf{S}}$$

$$= \frac{I_0^2}{2\pi c} \frac{\cos^2 \left(\frac{\pi}{2} \cos \theta \right)}{\sin^2 \theta}. \tag{9.76}$$

Similarly, we can evaluate the total power radiated, averaged over a cycle, to be

$$P_{\text{total}} = \frac{2I_0^2}{c} \int\limits_0^{\frac{\pi}{2}} d\theta \, \frac{\cos^2\left(\frac{\pi}{2}\cos\theta\right)}{\sin\theta}$$

$$= \frac{I_0^2}{c} \int\limits_0^{\frac{\pi}{2}} d\theta \, \frac{1 + \cos(\pi\cos\theta)}{\sin\theta}. \tag{9.77}$$

Defining $x = \cos\theta$, this leads to

$$P_{\text{total}} = \frac{I_0^2}{c} \int\limits_0^1 dx \, \frac{1 + \cos\pi x}{1 - x^2}$$

$$= \frac{I_0^2}{2c} \int\limits_0^1 dx \, (1 + \cos\pi x) \left(\frac{1}{1+x} + \frac{1}{1-x}\right)$$

$$= \frac{I_0^2}{2c} \int\limits_{-1}^1 dx \, \frac{1 + \cos\pi x}{1+x}. \tag{9.78}$$

Making a further change of variables, $t = \pi(1+x)$, we obtain

$$P_{\text{total}} = \frac{I_0^2}{2c} \int\limits_0^{2\pi} dt \, \frac{1 - \cos t}{t}$$

$$= \frac{I_0^2}{2c} \left(C + \ln 2\pi - Ci(2\pi)\right). \tag{9.79}$$

Here, $C \approx 0.577$ is the Euler constant and $Ci(x)$ is known as the cosine integral defined as (see, for example, Gradshteyn and Ryzhik)

$$Ci(x) = -\int\limits_x^\infty dt \, \frac{\cos t}{t} = C + \ln x - \int\limits_0^x dt \, \frac{1 - \cos t}{t}. \tag{9.80}$$

The values of $Ci(x)$ are tabulated in various mathematical handbooks and using these one can show that the quantity in the parenthesis in (9.79) has the value 0.609. Therefore, we obtain

$$P_{\text{total}} = \frac{0.609}{2c} I_0^2 = \frac{0.609}{c} I_{\text{rms}}^2, \tag{9.81}$$

so that we can obtain the radiation resistance, associated with such an antenna, to be

$$R_{\text{rad}} = \frac{0.609}{c}. \tag{9.82}$$

For the monopole antenna, the average power radiated would be half of this value and correspondingly the radiation resistance will also decrease by a factor of 2 (since power is radiated only through half the antenna).

9.3.1 Properties of antennas. Radio antennas are primarily used to transmit signal from a given transmitter along some directions suppressing transmission along others. From the results in (9.43) as well as (9.76), we see that, for a short dipole antenna

$$\frac{\mathrm{d}P}{\mathrm{d}\Omega} \sim \sin^2 \theta,$$

whereas for a half wavelength antenna,

$$\frac{\mathrm{d}P}{\mathrm{d}\Omega} \sim \frac{\cos^2 \left(\frac{\pi}{2} \cos \theta\right)}{\sin^2 \theta}.$$

Thus, we see that, in both the cases, the transmitted signal is the strongest along $\theta = \frac{\pi}{2}$ or along the axis perpendicular to the antenna. The detailed patterns of the two antennas are, of course, different. Therefore, unlike a point sound source, which leads to uniform radiation along all directions, antennas have a strong directional property which can be used to focus the transmission along certain directions. This enhancement of the transmitted signal along certain directions is characterized by the antenna gain parameter which is defined as

$$g_d = \frac{4\pi}{P_{\text{total}}} \frac{\mathrm{d}P}{\mathrm{d}\Omega}. \tag{9.83}$$

Namely, it measures the ratio of the power radiated per unit solid angle in a given direction to the average power radiated per unit solid angle. In units of decibels, this can be expressed as

$$G_d = 10 \log_{10} g_d. \tag{9.84}$$

There are also other measures for the antenna gain, but we will not get into these details. Let us simply note here that, by a clever choice of an array of antennas, the antenna gain can be enhanced quite a bit.

9.4 Multipole expansion for electric and magnetic fields

In electrostatics, we obtained a multipole expansion for the electric field by expressing the scalar potential in terms of spherical harmonics. However, in dealing with time dependent Maxwell's equations, we are dealing with vector quantities and correspondingly, the multipole expansion has to be carried out in terms of vector spherical harmonics. Such an expansion is quite useful in solving boundary value problems (including in the study of scattering and diffraction) as is also true in the static case.

To begin with, let us consider a scalar wave equation of the form

$$\mathbf{\nabla}^2 \psi - \frac{1}{c^2} \frac{\partial^2 \psi}{\partial t^2} = 0, \tag{9.85}$$

where, for simplicity, we are assuming wave propagation in vacuum. The conventional way one solves this equation is by Fourier transforming the solution in the time variable

$$\psi(\mathbf{x}, t) = \int d\omega \, e^{-i\omega t} \, \psi(\mathbf{x}, \omega). \tag{9.86}$$

Substituting this into Eq. (9.85), we obtain

$$\left(\mathbf{\nabla}^2 + k^2 \right) \psi(\mathbf{x}, \omega) = 0. \tag{9.87}$$

Here, we have defined, as before, $k = \frac{\omega}{c}$ and this equation is known as the Helmholtz equation. For $k = 0$ ($\omega = 0$) or the static case, this reduces to the Laplace equation which we are quite familiar with. Let us emphasize that we are trying to solve here a general problem and are not assuming a harmonic dependence on time as we had done earlier for simple systems.

The solution of the Helmholtz equation is obvious in Cartesian coordinates. However, our interest is to find solutions in spherical coordinates where the multipole expansion becomes manifest. To this end, drawing from our experience with spherically symmetric equations (say, from quantum mechanics), we write a factorized form for the solution ψ as

$$\psi(\mathbf{x}, \omega) = R_\ell(r, \omega) \, Y_{\ell,m}(\theta, \phi), \tag{9.88}$$

where $Y_{\ell,m}$ represent the spherical harmonics depending only on the angular coordinates while R_ℓ is the radial function which depends on the radial coordinate as well as the energy. Substituting this into (9.87), we find that the Helmholtz equation separates into two equations of the forms

$$\left[\frac{d^2}{dr^2} + \frac{2}{r} \frac{d}{dr} + k^2 - \frac{\ell(\ell+1)}{r^2} \right] R_\ell(r, \omega) = 0,$$

(9.89)

$$- \left[\frac{1}{\sin\theta} \frac{\partial}{\partial\theta} \left(\sin\theta \frac{\partial}{\partial\theta} \right) + \frac{1}{\sin^2\theta} \frac{\partial^2}{\partial\phi^2} \right] Y_{\ell,m} = \ell(\ell+1) Y_{\ell,m}.$$

Here, ℓ takes positive integer values including zero while, for a given ℓ, we have $m = -\ell, -\ell+1, \ldots, \ell$ (as we know from the study of angular momentum in quantum mechanics). The angular functions $Y_{\ell,m}$, the spherical harmonics, are eigenstates of $\mathbf{L}^2$ and L_z.

The radial equation in (9.89) is the equation for the spherical Bessel functions and the two independent solutions can be written as the spherical Bessel functions and the spherical Neumann functions defined as

$$j_\ell(x) = \left(\frac{\pi}{2x} \right)^{\frac{1}{2}} J_{\ell+\frac{1}{2}}(x),$$

$$\eta_\ell(x) = \left(\frac{\pi}{2x} \right)^{\frac{1}{2}} N_{\ell+\frac{1}{2}}(x).$$

(9.90)

The spherical Bessel functions $j_\ell(x)$ are regular at the origin while the spherical Neumann functions $\eta_\ell(x)$ diverge. An alternative way to write the solutions is in terms of the spherical Hankel functions, which are defined as linear combinations of the spherical Bessel functions and the spherical Neumann functions, namely,

$$h_\ell^{(1)}(x) = j_\ell(x) + i\eta_\ell(x),$$

$$h_\ell^{(2)}(x) = j_\ell(x) - i\eta_\ell(x).$$

(9.91)

From the fact that the spherical Bessel functions (as well as the spherical Neumann functions) are real, it follows that the two spherical Hankel functions are complex conjugates of each other. We note that either of the sets in (9.90) or (9.91) can be thought of as an independent set of solutions for the spherical Bessel equation in (9.89). Thus, we can write the most general radial solution of the form

$$R_\ell(r) = a_\ell^{(1)} h_\ell^{(1)}(kr) + a_\ell^{(2)} h_\ell^{(2)}(kr),$$

(9.92)

where $a_\ell^{(1,2)}$ are coordinate independent constants. The full solution for the Helmholtz equation can now be written in the form

$$\psi(\mathbf{x}, \omega) = \sum_{\ell,m} \left(a_\ell^{(1)} h_\ell^{(1)}(kr) + a_\ell^{(2)} h_\ell^{(2)}(kr) \right) Y_{\ell,m}(\theta, \phi).$$

(9.93)

There are several things to note here. First, the spherical harmonics are normalized so that

$$\int \sin\theta d\theta \, d\phi \, Y_{\ell,m}^*(\theta, \phi) Y_{\ell',m'}(\theta, \phi) = \delta_{\ell\ell'}\delta_{mm'}. \tag{9.94}$$

The spherical harmonics reduce to the Legendre polynomials when $m = 0$ and that they are eigenfunctions of the angular momentum operator. Namely, let us define (recall the definition from quantum mechanics)

$$\mathbf{L} = -i(\mathbf{r} \times \boldsymbol{\nabla}) = -i\left(\hat{\boldsymbol{\phi}}\frac{\partial}{\partial\theta} - \frac{\hat{\boldsymbol{\theta}}}{\sin\theta}\frac{\partial}{\partial\phi}\right). \tag{9.95}$$

As is clear, this is the rotation operator (changes only the angles) and it is straightforward to check that

$$\mathbf{L}^2 = -\left[\frac{1}{\sin\theta}\frac{\partial}{\partial\theta}\left(\sin\theta\frac{\partial}{\partial\theta}\right) + \frac{1}{\sin^2\theta}\frac{\partial^2}{\partial\phi^2}\right], \tag{9.96}$$

from which it follows that (see Eq. (9.89))

$$\mathbf{L}^2 Y_{\ell,m}(\theta, \phi) = \ell(\ell + 1)Y_{\ell,m}(\theta, \phi). \tag{9.97}$$

We note that by construction the angular momentum operator is transverse to the radial direction, namely, (as is obvious from (9.95))

$$\hat{\mathbf{r}} \cdot \mathbf{L} = 0. \tag{9.98}$$

Similarly, it satisfies commutation relations analogous to the angular momentum operators in quantum mechanics so that we have

$$\mathbf{L} \times \mathbf{L} = i\mathbf{L}. \tag{9.99}$$

As a result, we also have

$$\hat{\mathbf{r}} \cdot (\mathbf{L} \times \mathbf{L}) = 0. \tag{9.100}$$

With these basics, we are now ready to discuss the multipole expansion for the electric and the magnetic fields. Let us consider an arbitrary distribution of localized charges and currents. Outside the region containing the sources, the Maxwell's equations take the forms (in vacuum, for simplicity)

$$\boldsymbol{\nabla} \cdot \mathbf{E} = 0, \qquad\qquad \boldsymbol{\nabla} \cdot \mathbf{B} = 0,$$
$$\boldsymbol{\nabla} \times \mathbf{E} = -\frac{1}{c}\frac{\partial\mathbf{B}}{\partial t}, \qquad \boldsymbol{\nabla} \times \mathbf{B} = \frac{1}{c}\frac{\partial\mathbf{E}}{\partial t}. \tag{9.101}$$

Fourier transforming the time variable as before, we obtain

$$\nabla \cdot \mathbf{E} = 0, \qquad\qquad \nabla \cdot \mathbf{B} = 0,$$
$$\nabla \times \mathbf{E} = ik\,\mathbf{B}, \qquad\qquad \nabla \times \mathbf{B} = -ik\,\mathbf{E}, \qquad (9.102)$$

where $k = \frac{\omega}{c}$. Here, both the electric and the magnetic fields have a nontrivial dependence on frequency (because we are allowing for an arbitrary time dependence other than the simple harmonic dependence) that is understood.

The Maxwell's equations (9.102) can be written in a simpler form eliminating either the electric or the magnetic field. For example, if we eliminate the electric field from (9.102), we can write

$$\nabla \cdot \mathbf{B} = 0,$$

$$\left(\nabla^2 + k^2\right) \mathbf{B} = 0,$$

$$\mathbf{E} = \frac{i}{k}\, \nabla \times \mathbf{B}. \qquad (9.103)$$

Here, we treat $\mathbf{B}$ as the independent field. Alternatively, if we eliminate the magnetic field from (9.102), then we obtain

$$\nabla \cdot \mathbf{E} = 0,$$

$$\left(\nabla^2 + k^2\right) \mathbf{E} = 0,$$

$$\mathbf{B} = -\frac{i}{k}\, \nabla \times \mathbf{E}, \qquad (9.104)$$

where $\mathbf{E}$ represents the independent field.

Both the sets of equations in (9.103) or (9.104) are equivalent and also equivalent to the Maxwell's equations in (9.102) and we note that the independent field in (9.103) or (9.104) can be solved by solving a Helmholtz equation. However, in the present case, the dynamical variable (the electric or the magnetic field) is a vector, as opposed to the earlier case where ψ was a scalar function. Correspondingly, the solutions can be expressed as before in terms of spherical Hankel functions and spherical harmonics, but with vector coefficients. Thus, for example, the solutions for $\mathbf{B}$ in (9.103) can be written as

$$\mathbf{B}(\mathbf{x}) = \sum_{\ell,m} \left(\mathbf{a}_\ell^{(1)} h_\ell^{(1)}(kr) + \mathbf{a}_\ell^{(2)} h_\ell^{(2)}(kr)\right) Y_{\ell,m}(\theta, \phi), \qquad (9.105)$$

where $\mathbf{a}_\ell^{(1,2)}$ now represent arbitrary vector coefficients. These coefficients are arbitrary except that the magnetic field has to be transverse (see (9.103)) so that we must have

$$\nabla \cdot \mathbf{B} = \nabla \cdot \sum_{\ell,m} \left(\mathbf{a}_\ell^{(1)} h_\ell^{(1)}(kr) + \mathbf{a}_\ell^{(2)} h_\ell^{(2)}(kr) \right) Y_{\ell,m}(\theta, \phi)$$

$$= 0. \tag{9.106}$$

Since $h_\ell^{(1,2)}$ represent independent solutions of the spherical Bessel equation, it follows that, for (9.106) to hold, we must have independently,

$$\nabla \cdot \sum_{\ell,m} \mathbf{a}_\ell^{(i)} h_\ell^{(i)}(kr) Y_{\ell,m}(\theta, \phi) = 0, \qquad i = 1, 2. \tag{9.107}$$

Decomposing the gradient into its radial and angular parts (see (9.95)),

$$\nabla = \hat{\mathbf{r}} \frac{\partial}{\partial r} - \frac{i}{r} \hat{\mathbf{r}} \times \mathbf{L}, \tag{9.108}$$

we obtain from (9.107)

$$\sum_{\ell,m} \left(\hat{\mathbf{r}} \cdot \mathbf{a}_\ell^{(i)} \frac{\mathrm{d} h_\ell^{(i)}(kr)}{\mathrm{d}r} Y_{\ell,m} - \frac{i}{r} h_\ell^{(i)}(kr) \hat{\mathbf{r}} \cdot (\mathbf{L} \times \mathbf{a}_\ell^{(i)}) Y_{\ell,m} \right) = 0, \tag{9.109}$$

where we have used some familiar properties of products of vectors (namely, $(\mathbf{A} \times \mathbf{B}) \cdot \mathbf{C} = \mathbf{A} \cdot (\mathbf{B} \times \mathbf{C})$). From (9.98) and (9.100), we note that a particular solution of (9.109) is given by

$$\mathbf{a}_\ell^{(i)} \sim a_\ell^{(i)} \mathbf{L}. \tag{9.110}$$

The vanishing of the first term in (9.109) is, of course, obvious from (9.98). The vanishing of the second term follows from (9.100), namely,

$$\hat{\mathbf{r}} \cdot (\mathbf{L} \times \mathbf{L}) = 0. \tag{9.111}$$

Thus, we can write a particular solution of the Helmholtz equation in (9.103) satisfying the transversality of the magnetic field as

$$\mathbf{B}(\mathbf{x}) = \sum_{\ell,m} \left(a_\ell^{(1)} h_\ell^{(1)}(kr) + a_\ell^{(2)} h_\ell^{(2)}(kr) \right) (\mathbf{L} Y_{\ell,m}(\theta, \phi)),$$

$$\mathbf{E}(\mathbf{x}) = \frac{i}{k} \nabla \times \mathbf{B}. \tag{9.112}$$

(The fact that this is a solution of the Helmholtz equation follows from the observation that if ψ is a solution of the Helmholtz equation, then so is $(\mathbf{L}\psi)$ since $\mathbf{L}$ commutes with ∇^2 as well as with a scalar.) Such an expansion of the fields leads to the electric multipole fields. Alternatively, if we had started with (9.104), the solution would turn out to be

$$\mathbf{E}(\mathbf{x}) = \sum_{\ell,m} \left(b_\ell^{(1)} h_\ell^{(1)}(kr) + b_\ell^{(2)} h_\ell^{(2)}(kr) \right) (\mathbf{L} Y_{\ell,m}(\theta, \phi)),$$

$$\mathbf{B}(\mathbf{x}) = -\frac{i}{k} \nabla \times \mathbf{E}, \tag{9.113}$$

which leads to the magnetic multipole fields. The reason for this nomenclature will become clear shortly.

Since the combination $(\mathbf{L} Y_{\ell,m})$ arises frequently in the study of electrodynamics, it is given a special name, vector spherical harmonics, and is defined such that it is normalized, namely,

$$\mathbf{Y}_{\ell,m}(\theta, \phi) = \begin{cases} \frac{1}{\sqrt{\ell(\ell+1)}} (\mathbf{L} Y_{\ell,m}(\theta, \phi)), & \ell \neq 0, \\ 0, & \ell = 0. \end{cases} \tag{9.114}$$

The fact that it is normalized follows from

$$\int \sin\theta d\theta \, d\phi \, \mathbf{Y}_{\ell,m}^*(\theta, \phi) \cdot \mathbf{Y}_{\ell',m'}(\theta, \phi)$$

$$= \frac{1}{\sqrt{\ell(\ell+1)\ell'(\ell'+1)}} \int \sin\theta d\theta \, d\phi \, (\mathbf{L}^\dagger Y_{\ell,m}^*) \cdot (\mathbf{L} Y_{\ell',m'})$$

$$= \frac{1}{\sqrt{\ell(\ell+1)\ell'(\ell'+1)}} \int \sin\theta d\theta \, d\phi \, Y_{\ell,m}^*(\theta, \phi)(\mathbf{L}^2 Y_{\ell',m'}(\theta, \phi))$$

$$= \sqrt{\frac{\ell'(\ell'+1)}{\ell(\ell+1)}} \int \sin\theta d\theta \, d\phi \, Y_{\ell,m}^*(\theta, \phi) Y_{\ell',m'}(\theta, \phi)$$

$$= \delta_{\ell\ell'} \delta_{mm'}, \qquad \ell, \ell' \neq 0. \tag{9.115}$$

Here, we have used the fact that $\mathbf{L}$ is a Hermitian operator as well as Eqs. (9.94) and (9.97). With this, we can write the electric multipole fields as

$$\mathbf{B}(\mathbf{x}) = \sum_{\ell,m} \left(a_\ell^{(1)} h_\ell^{(1)}(kr) + a_\ell^{(2)} h_\ell^{(2)}(kr) \right) \mathbf{Y}_{\ell,m}(\theta, \phi),$$

$$\mathbf{E}(\mathbf{x}) = \frac{i}{k} \nabla \times \mathbf{B}, \tag{9.116}$$

while the magnetic multipole fields take the forms

$$\mathbf{E}(\mathbf{x}) = \sum_{\ell,m} \left(b_\ell^{(1)} h_\ell^{(1)}(kr) + b_\ell^{(2)} h_\ell^{(2)}(kr) \right) \mathbf{Y}_{\ell,m}(\theta, \phi),$$

$$\mathbf{B}(\mathbf{x}) = -\frac{i}{k} \boldsymbol{\nabla} \times \mathbf{E}. \tag{9.117}$$

An arbitrary electric field can, of course, be written as a linear superposition of the electric and the magnetic multipole fields.

9.5 Behavior of multipole fields

To understand the nomenclature of electric and magnetic multipole fields, let us analyze the behavior of these fields in the near zone, $kr \ll 1$. For example, for the electric multipole fields in (9.116), we note that the spherical Hankel functions are linear combinations of spherical Bessel functions and spherical Neumann functions. For small values of the arguments, these functions behave as

$$j_\ell(x) \to x^\ell, \qquad \eta_\ell(x) \to \frac{1}{x^{\ell+1}}. \tag{9.118}$$

The spherical Neumann functions diverge at the origin. However, since we are excluding the region containing charges and currents (assumed to be in a finite region near the origin), they are allowed in the solutions. Correspondingly, we see that at small distances, it is the spherical Neumann functions that dominate. Choosing a particular normalization for $a_\ell^{(1,2)}$, we see that, in this case, we can write the magnetic field in the near zone as

$$\mathbf{B}(\mathbf{x}) \to -\sum_{\ell,m} \frac{k}{\ell} \frac{1}{r^{\ell+1}} \mathbf{L} Y_{\ell,m}(\theta, \phi)$$

$$= -\sum_{\ell,m} \frac{k}{\ell} \mathbf{L} \left(\frac{Y_{\ell,m}}{r^{\ell+1}} \right). \tag{9.119}$$

Here, we have used the fact that $\mathbf{L}$ is an angular operator which does not act on the radial coordinate as is clear from (9.95). The behavior of the electric field in the near zone now follows from (9.116) to be

$$\mathbf{E}(\mathbf{x}) = \frac{i}{k} \boldsymbol{\nabla} \times \mathbf{B} \to -\sum_{\ell,m} \frac{i}{\ell} \boldsymbol{\nabla} \times \mathbf{L} \left(\frac{Y_{\ell,m}}{r^{\ell+1}} \right). \tag{9.120}$$

The expression (9.120) can be simplified in the following manner. From the definition of $\mathbf{L}$ in (9.95) and the definition of the cross product (one should be careful about the positions of the operators), we obtain

$$
\begin{aligned}
(\boldsymbol{\nabla} \times \mathbf{L})_i &= -i\,(\boldsymbol{\nabla} \times (\mathbf{r} \times \boldsymbol{\nabla}))_i \\
&= -i\epsilon_{ijk}\epsilon_{kst}\boldsymbol{\nabla}_j r_s \boldsymbol{\nabla}_t \\
&= -i\,(\delta_{is}\delta_{jt} - \delta_{it}\delta_{js})\,\boldsymbol{\nabla}_j r_s \boldsymbol{\nabla}_t \\
&= -i\,(\boldsymbol{\nabla}_j r_i \boldsymbol{\nabla}_j - \boldsymbol{\nabla}_j r_j \boldsymbol{\nabla}_i) \\
&= -i\,(\delta_{ij}\boldsymbol{\nabla}_j + r_i \boldsymbol{\nabla}^2 - 3\boldsymbol{\nabla}_i - r_j \boldsymbol{\nabla}_j \boldsymbol{\nabla}_i) \\
&= -i\,(r_i \boldsymbol{\nabla}^2 - 2\boldsymbol{\nabla}_i - \boldsymbol{\nabla}_i r_j \boldsymbol{\nabla}_j + \delta_{ij}\boldsymbol{\nabla}_j) \\
&= -i\,(r_i \boldsymbol{\nabla}^2 - \boldsymbol{\nabla}_i(1 + \mathbf{r}\cdot\boldsymbol{\nabla}))\,.
\end{aligned}
\tag{9.121}
$$

Thus, we can write

$$
\boldsymbol{\nabla} \times \mathbf{L} = -i\left(\mathbf{r}\boldsymbol{\nabla}^2 - \boldsymbol{\nabla}\left(1 + r\frac{\partial}{\partial r}\right)\right),
\tag{9.122}
$$

where we have used $\mathbf{r}\cdot\boldsymbol{\nabla} = r\frac{\partial}{\partial r}$.

With (9.122), the expression for the electric field in (9.120) for small distances takes the form

$$
\mathbf{E}(\mathbf{x}) \to -\sum_{\ell,m} \frac{1}{\ell}\left(\mathbf{r}\boldsymbol{\nabla}^2 - \boldsymbol{\nabla}\left(1 + r\frac{\partial}{\partial r}\right)\right)\left(\frac{Y_{\ell,m}}{r^{\ell+1}}\right).
\tag{9.123}
$$

Let us recall here (see (9.89) with $k = 0$) that

$$
\boldsymbol{\nabla}^2\left(\frac{Y_{\ell,m}}{r^{\ell+1}}\right) = 0, \quad \boldsymbol{\nabla}^2 = \frac{1}{r^2}\frac{\partial}{\partial r}r^2\frac{\partial}{\partial r} - \frac{1}{r^2}\mathbf{L}^2,
\tag{9.124}
$$

so that the first term on the right hand side of (9.123) vanishes. Furthermore, we observe that ($r\frac{\mathrm{d}}{\mathrm{d}r}$ simply counts the power of r)

$$
\left(1 + r\frac{\mathrm{d}}{\mathrm{d}r}\right)\left(\frac{1}{r^{\ell+1}}\right) = -\frac{\ell}{r^{\ell+1}},
\tag{9.125}
$$

so that the expression for the electric field in (9.123) in the near zone becomes (has the static form of a gradient)

$$
\begin{aligned}
\mathbf{E}(\mathbf{x}) &\to -\sum_{\ell,m} \boldsymbol{\nabla}\left(\frac{Y_{\ell,m}}{r^{\ell+1}}\right) \\
&= -\sum_{\ell,m} \boldsymbol{\nabla}\Phi_{\ell,m} = \sum_{\ell,m} \mathbf{E}_{\ell,m}(\mathbf{x}),
\end{aligned}
\tag{9.126}
$$

where we have identified $\Phi_{\ell,m} = \frac{Y_{\ell,m}}{r^{\ell+1}}$, the factor arising in a multipole expansion. We recognize each term in the sum, namely $\mathbf{E}_{\ell,m}$, to coincide precisely with the static electric multipole moments. For small distances, these behave as $\sim \frac{1}{r^{\ell+2}}$ as opposed to the corresponding terms in the magnetic fields ($\mathbf{B}_{\ell,m} \sim \frac{1}{r^{\ell+1}}$). Therefore, the electric fields dominate over the corresponding magnetic fields. In fact, in the static limit, $k = 0$, the magnetic field vanishes and we are simply left with the electric field which has the correct multipole expansion. It is for these reasons that these solutions are known as the electric multipole fields. (We have already seen this behavior in the case of electric dipole radiation.)

Let us next analyze the behavior of the electric multipole fields in the far away region (radiation zone) where $kr \gg 1$. Asymptotically, for large values of the argument, we know that the spherical Hankel functions behave as

$$h_\ell^{(1)}(x) \to (-i)^{\ell+1} \frac{e^{ix}}{x},$$

$$h_\ell^{(2)}(x) = (h_\ell^{(1)}(x))^* \to (i)^{\ell+1} \frac{e^{-ix}}{x}. \tag{9.127}$$

On the other hand, from our earlier discussion on fields produced by an arbitrary distribution of charges, we know that in the radiation zone the fields have the forms of outgoing spherical waves (see (9.22) and (9.23)). Correspondingly, we conclude that the coefficient $a_\ell^{(2)} = 0$ which leads to (for large values of kr)

$$\mathbf{B}(\mathbf{x}) \to \sum_{\ell,m} \frac{(-i)^{\ell+1}}{k} a_\ell \, \mathbf{L} \left(\frac{e^{ikr} Y_{\ell,m}}{r} \right),$$

$$\mathbf{E}(\mathbf{x}) \to \sum_{\ell,m} \frac{(-i)^\ell}{k^2} a_\ell \, \boldsymbol{\nabla} \times \mathbf{L} \left(\frac{e^{ikr} Y_{\ell,m}}{r} \right) \tag{9.128}$$

$$= \sum_{\ell,m} \frac{(-i)^{\ell+1}}{k^2} a_\ell \left(\mathbf{r} \nabla^2 - \boldsymbol{\nabla} \left(1 + r \frac{\partial}{\partial r} \right) \right) \left(\frac{e^{ikr} Y_{\ell,m}}{r} \right),$$

where we have defined the coefficient a_ℓ by absorbing the normalization factor $\frac{1}{\sqrt{\ell(\ell+1)}}$ coming from the definition of the vector spherical harmonics. The expression for the electric field can be further simplified by noting that

$$\mathbf{r}\nabla^2\left(\frac{e^{ikr}Y_{\ell,m}}{r}\right)$$

$$= \mathbf{r}\left[\frac{1}{r^2}\frac{\partial}{\partial r}\left((ikr-1)e^{ikr}\right)Y_{\ell,m} - \frac{e^{ikr}}{r^3}\mathbf{L}^2Y_{\ell,m}\right]$$

$$= -\hat{\mathbf{r}}\left(k^2 + \frac{\ell(\ell+1)}{r^2}\right)e^{ikr}Y_{\ell,m}. \qquad (9.129)$$

On the other hand, using (9.108) we have

$$\nabla\left(1+r\frac{\partial}{\partial r}\right)\left(\frac{e^{ikr}Y_{\ell,m}}{r}\right)$$

$$= ik\nabla\left(e^{ikr}Y_{\ell,m}\right)$$

$$= -\hat{\mathbf{r}}k^2e^{ikr}Y_{\ell,m} + \frac{ke^{ikr}}{r}\,\hat{\mathbf{r}}\times\mathbf{L}Y_{\ell,m}. \qquad (9.130)$$

Substituting (9.129) and (9.130) into the expression for the electric field in (9.128), we obtain, for large distances, (namely, we are neglecting terms of order $\frac{1}{r^2}$ compared to those of order $\frac{1}{r}$)

$$\mathbf{E}(\mathbf{x}) \rightarrow -\sum_{\ell,m}\frac{(-i)^{\ell+1}}{k}\,a_\ell\,\hat{\mathbf{r}}\times\mathbf{L}\left(\frac{e^{ikr}Y_{\ell,m}}{r}\right) = -\hat{\mathbf{r}}\times\mathbf{B}(\mathbf{x}). \quad (9.131)$$

There are several things to note from the structure of the radiation fields in Eqs. (9.128) and (9.131). We note that the fields in the far off zone do fall off as $\frac{1}{r}$. Furthermore, using (9.98) it is clear that the magnetic field is perpendicular to the direction of propagation $\hat{\mathbf{r}}$. Similarly, the electric field is also perpendicular to the direction of propagation and the electric and the magnetic fields are orthogonal to each other. This is the general characteristic of radiation fields. We can also obtain the small distance as well as the large distance behaviors for the magnetic multipole fields in a similar manner. In fact, looking at Eqs. (9.103) and (9.104), we see that the electric multipole fields go over to the magnetic multipole fields under the transformations (known as duality transformations)

$$\mathbf{E}\rightarrow\mathbf{B}, \qquad \mathbf{B}\rightarrow-\mathbf{E}. \qquad (9.132)$$

As a result, we do not have to carry out a separate analysis for the magnetic multipole fields. Rather, the same analysis can be taken over with the appropriate replacements.

Given the electric and the magnetic fields for either of the multipole expansion, we note that we can write

$$\mathbf{B}(\mathbf{x}) = \sum_{\ell,m} \mathbf{B}_{\ell,m}(\mathbf{x}),$$

$$\mathbf{E}(\mathbf{x}) = \sum_{\ell,m} \mathbf{E}_{\ell,m}(\mathbf{x}). \tag{9.133}$$

Here, $\mathbf{B}_{\ell,m}(\mathbf{x})$ and $\mathbf{E}_{\ell,m}(\mathbf{x})$ are the multipole fields of order (ℓ, m). For the electric multipole fields, in the radiation zone, for example, as we have seen

$$\mathbf{B}_{\ell,m}(\mathbf{x}) = \frac{(-i)^{\ell+1} a_\ell}{k} \mathbf{L} \left(\frac{e^{ikr} Y_{\ell,m}}{r} \right),$$

$$\mathbf{E}_{\ell,m}(\mathbf{x}) = -\hat{\mathbf{r}} \times \mathbf{B}_{\ell,m}(\mathbf{x})$$

$$= -\frac{(-i)^{\ell+1} a_\ell}{k} \hat{\mathbf{r}} \times \mathbf{L} \left(\frac{e^{ikr} Y_{\ell,m}}{r} \right). \tag{9.134}$$

Using (9.134) we can now calculate the time averaged radiated power for the multipole component fields through the surface of a large sphere per unit solid angle along a given direction as

$$\frac{\mathrm{d}P_{\ell,m}}{\mathrm{d}\Omega} = R^2 \hat{\mathbf{r}} \cdot \overline{\mathbf{S}}_{\ell,m} = \frac{cR^2}{8\pi} \, \mathrm{Re} \, \hat{\mathbf{r}} \cdot (\mathbf{E}_{\ell,m} \times \mathbf{B}_{\ell,m}^*)$$

$$= \frac{cR^2}{8\pi} \, (\mathbf{B}_{\ell,m}^* \cdot \mathbf{B}_{\ell,m})$$

$$= \frac{c|a_\ell|^2}{8\pi k^2} \, (\mathbf{L}Y_{\ell,m})^* \cdot (\mathbf{L}Y_{\ell,m}). \tag{9.135}$$

The right hand side can be simplified by noting that

$$(\mathbf{L}Y_{\ell,m})^* \cdot (\mathbf{L}Y_{\ell,m})$$

$$= \frac{1}{2} \left(|L_+ Y_{\ell,m}|^2 + |L_- Y_{\ell,m}|^2 \right) + |L_z Y_{\ell,m}|^2$$

$$= \left[\frac{1}{2} \left((\ell - m)(\ell + m + 1) |Y_{\ell,m+1}|^2 \right. \right.$$

$$\left. \left. + (\ell + m)(\ell - m + 1)|Y_{\ell,m-1}|^2 \right) + m^2 |Y_{\ell,m}|^2 \right], \tag{9.136}$$

where we have defined, as in quantum mechanics, $L_\pm = L_x \pm iL_y$ and used the properties of the action of the angular momentum operators on the eigenstates of angular momenta (namely, $L_\pm Y_{\ell,m} =$

$\sqrt{(\ell \mp m)(\ell \pm m + 1)}Y_{\ell,m\pm 1})$. This, therefore, gives

$$\frac{dP_{\ell,m}}{d\Omega} = \frac{c|a_\ell|^2}{8\pi k^2}\left[\frac{1}{2}\left((\ell - m)(\ell + m + 1)|Y_{\ell,m+1}|^2\right.\right.$$

$$\left.\left.+(\ell + m)(\ell - m + 1)|Y_{\ell,m-1}|^2\right) + m^2|Y_{\ell,m}|^2\right]. \quad (9.137)$$

We see, therefore, that by measuring the angular distribution of the radiated power, it is possible to determine the order of the multipole component of the wave. However, without going into details let us note that the magnetic multipole fields would also lead to the same exact angular pattern for radiated power, since one is obtained from the other by replacing the electric field by the magnetic field (up to a sign). Therefore, it is not possible to determine the nature of the multipole fields (electric or magnetic) from a measurement of the angular distribution of the radiated power alone. However, the two fields have very different behavior under parity (something that we do not get into). Therefore, one needs to measure the polarization of the waves to determine the character of the multipole field components.

Let us note from the definition of the spherical harmonics

$$Y_{\ell,m}(\theta,\phi) = (-1)^{\frac{m+|m|}{2}}\left[\frac{2\ell + 1}{4\pi}\frac{(\ell - |m|)!}{(\ell + |m|)!}\right]^{\frac{1}{2}}\mathcal{P}_{\ell,m}\,e^{im\phi}, \quad (9.138)$$

and the form of (9.137) that the radiated power spectrum is independent of the azimuthal angle ϕ. Therefore, the radiated power per solid angle along a given direction depends only on the angle θ through the associated Legendre polynomials. We note that we have denoted the associated Legendre polynomials by $\mathcal{P}_{\ell,m}$ to avoid confusion with the radiated power for a given multipole field component. The first few associated Legendre polynomials have the forms

$\mathcal{P}_{0,0}(\cos\theta) = \mathcal{P}_0(\cos\theta) = 1,$

$\mathcal{P}_{1,0}(\cos\theta) = \mathcal{P}_1(\cos\theta) = \cos\theta,$

$\mathcal{P}_{1,\pm 1}(\cos\theta) = (1 - \cos^2\theta)^{\frac{1}{2}} = \sin\theta,$

$\mathcal{P}_{2,0}(\cos\theta) = \mathcal{P}_2(\cos\theta) = \frac{1}{2}(3\cos^2\theta - 1),$

$\mathcal{P}_{2,\pm 1}(\cos\theta) = 3\cos\theta(1 - \cos^2\theta)^{\frac{1}{2}} = 3\sin\theta\cos\theta,$

$\mathcal{P}_{2,\pm 2}(\cos\theta) = 3(1 - \cos^2\theta) = 3\sin^2\theta,$

$$\vdots \quad\quad\quad\quad\quad\quad\quad\quad\quad\quad\quad\quad\quad\quad\quad\quad (9.139)$$

Furthermore, by integrating (9.137) over all directions, we obtain the time averaged total power radiated of the form

$$P_{\ell,m}^{\text{total}} = \int d\Omega \, \frac{dP_{\ell,m}}{d\Omega}$$

$$= \frac{c|a_\ell|^2}{8\pi k^2}\left[\frac{1}{2}\left((\ell-m)(\ell+m+1) + (\ell+m)(\ell-m+1)\right) + m^2\right]$$

$$= \frac{c|a_\ell|^2 \ell(\ell+1)}{8\pi k^2}. \tag{9.140}$$

Let us next work out explicitly the radiation pattern for a few low order multipoles. For dipoles, $\ell = 1$ and we obtain, from Eqs. (9.137)–(9.139), that

$$\frac{dP_{1,0}}{d\Omega} = \frac{c|a_1|^2}{8\pi k^2}\frac{3}{4\pi}\sin^2\theta,$$

$$\frac{dP_{1,\pm 1}}{d\Omega} = \frac{c|a_1|^2}{8\pi k^2}\frac{3}{8\pi}(1 + \cos^2\theta). \tag{9.141}$$

We note that for the dipole case, the two distinct possible radiation patterns can be represented as shown in Figs. 9.6 and 9.7. The first describes the polar plot for the case $m = 0$ while the second denotes the case for $m = \pm 1$. For $m = 0$, we see that the maximum power is radiated along $\theta = \frac{\pi}{2}$ while for $m = \pm 1$, it is along $\theta = 0$, as is also clear from (9.141).

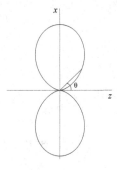

Figure 9.6: Dipole radiation pattern for $m = 0$.

For the quadrupole radiation, $\ell = 2$ and the possible forms for

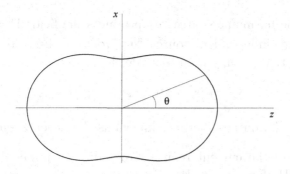

Figure 9.7: Dipole radiation pattern for $m = \pm 1$.

the differential radiated power are given by

$$\frac{dP_{2,0}}{d\Omega} = \frac{c|a_2|^2}{8\pi k^2} \frac{45}{4\pi} \sin^2\theta \cos^2\theta,$$

$$\frac{dP_{2,\pm 1}}{d\Omega} = \frac{c|a_2|^2}{8\pi k^2} \frac{15}{8\pi} \left(1 - 3\cos^2\theta + 4\cos^4\theta\right),$$

$$\frac{dP_{2,\pm 2}}{d\Omega} = \frac{c|a_2|^2}{8\pi k^2} \frac{15}{8\pi} \left(1 - \cos^4\theta\right). \tag{9.142}$$

Similarly, the angular distributions of radiated power for higher multipole fields can also be calculated.

9.6 Selected problems

1. A spherical balloon carries a charge Q uniformly distributed on its surface. The balloon pulsates with a frequency "ν" and amplitude "a" so that its radius is given by

$$r(t) = r_0 + a \sin 2\pi\nu t.$$

Calculate the rate of radiation of electromagnetic energy.

2. Determine the angular distribution of radiated power from a dipole antenna of arbitrary length "d".

3. Show that if ψ is a solution of the Helmholtz equation

$$\left(\nabla^2 + k^2\right)\psi = 0,$$

then, so is $(\mathbf{r} \times \nabla)\psi$.

4. A time harmonic source "a" produces the fields $\mathbf{E}_\omega^{(a)}, \mathbf{B}_\omega^{(a)}$ and another independent source "b" produces the fields $\mathbf{E}_\omega^{(b)}, \mathbf{B}_\omega^{(b)}$. Show that, at any point outside the sources,

$$\boldsymbol{\nabla} \cdot (\mathbf{E}_\omega^{(a)} \times \mathbf{B}_\omega^{(b)}) = \boldsymbol{\nabla} \cdot (\mathbf{E}_\omega^{(b)} \times \mathbf{B}_\omega^{(a)}).$$

This relation is sometimes known as the Lorentz lemma.

5. a) Find the current required to radiate a power of 100W at 100MHz from a 1cm Hertzian dipole.

b) Find the magnitudes of $\mathbf{E}, \mathbf{B}$ at a distance 10^4cm at $\theta = 0°, 90°$.

6. For a harmonically oscillating dipole of moment $\mathbf{p}$, the Hertz vector is defined to be

$$\mathcal{H} = \frac{\mathbf{p}}{r} e^{-i\omega(t-\frac{r}{c})},$$

at a distance r from the dipole.

a) Show that, in terms of the Hertz vector, the potentials can be written as (up to possible multiplicative constants)

$$\mathbf{A} = \frac{1}{c} \frac{\partial \mathcal{H}}{\partial t}, \qquad \Phi = -\boldsymbol{\nabla} \cdot \mathcal{H}.$$

b) Show from this that, at large distances from the dipole, the fields can be written as (up to possible multiplicative constants)

$$\mathbf{E} = \frac{1}{4\pi c r^3} \left\{ \mathbf{r} \times \left[\mathbf{r} \times \frac{d^2}{dt^2} \left(\mathbf{p} e^{-i\omega(t-\frac{r}{c})} \right) \right] \right\},$$

$$\mathbf{B} = -\frac{1}{4\pi c r^2} \left[\mathbf{r} \times \frac{d^2}{dt^2} \left(\mathbf{p} e^{-i\omega(t-\frac{r}{c})} \right) \right].$$

CHAPTER 10

Electromagnetic fields of currents

10.1 Lienard-Wiechert potential

Let us consider in this section the electromagnetic fields associated with the simplest of physical systems. Namely, let us consider a point particle carrying charge q and moving along a trajectory $\boldsymbol{\xi}(t)$ under the action of some forces that we do not specify. A moving charge, of course, produces a current with the current density of the form

$$J_\mu(\mathbf{x}, t) = j_\mu(t)\delta(\mathbf{x} - \boldsymbol{\xi}(t)), \tag{10.1}$$

where

$$j^\mu(t) = (cq, q\mathbf{v}) = \left(cq, q\,\frac{\mathrm{d}\boldsymbol{\xi}(t)}{\mathrm{d}t}\right). \tag{10.2}$$

We have already seen that the retarded solution for the vector potential for an arbitrary distribution of charges and currents in the Lorentz gauge has the form (in vacuum)

$$A_\mu(\mathbf{x}, t) = \frac{1}{c} \int \mathrm{d}^3x' \, \frac{J_\mu(\mathbf{x}', t')}{|\mathbf{x} - \mathbf{x}'|}\bigg|_{t' = t - \frac{|\mathbf{x} - \mathbf{x}'|}{c}}$$

$$= \frac{1}{c} \int \mathrm{d}^3x' \, \mathrm{d}t' \, \frac{J_\mu(\mathbf{x}', t')\,\delta\left(t' - t + \frac{|\mathbf{x} - \mathbf{x}'|}{c}\right)}{|\mathbf{x} - \mathbf{x}'|}, \tag{10.3}$$

where the space integral is over the volume containing the charges and the currents. We can now apply this to the case of a point charge moving along a trajectory, in which case using the form of the current density in (10.1), we obtain

$$A_\mu(\mathbf{x}, t) = \frac{1}{c} \int \mathrm{d}^3x' \, \mathrm{d}t' \, \frac{j_\mu(t')\,\delta\left(\mathbf{x}' - \boldsymbol{\xi}(t')\right)\,\delta(t' - t + \frac{|\mathbf{x} - \mathbf{x}'|}{c})}{|\mathbf{x} - \mathbf{x}'|}$$

$$= \frac{1}{c} \int \mathrm{d}t' \, \frac{j_\mu(t')\,\delta\left(t' - t + \frac{|\mathbf{x} - \boldsymbol{\xi}(t')|}{c}\right)}{|\mathbf{x} - \boldsymbol{\xi}(t')|}. \tag{10.4}$$

To simplify this a little further, let us define

$$\mathbf{R}(t') = \mathbf{x} - \boldsymbol{\xi}(t'),$$

$$\tau = t' - t + \frac{|\mathbf{x} - \boldsymbol{\xi}(t')|}{c} = t' - t + \frac{R(t')}{c}, \tag{10.5}$$

where we have identified $R(t') = |\mathbf{R}(t')|$. Then, the integration over t' in (10.4) would have contribution only from that value of the time coordinate for which the argument of the delta function vanishes, namely,

$$\tau = t' - t + \frac{R(t')}{c} = 0. \tag{10.6}$$

This can, of course, be solved for t' in principle, once we know the trajectory of the particle. Furthermore, we note that

$$s(t') = \frac{d\tau}{dt'} = 1 + \frac{1}{c}\frac{dR(t')}{dt'}$$

$$= 1 + \frac{1}{cR(t')}\,\mathbf{R}(t') \cdot \frac{d\mathbf{R}(t')}{dt'} = 1 - \widehat{\mathbf{R}}(t') \cdot \frac{\mathbf{v}(t')}{c}, \tag{10.7}$$

where we have used the definition of $\mathbf{R}$ in (10.5) and $\widehat{\mathbf{R}} = \frac{\mathbf{R}}{R}$. We note that $s(t')$ is a positive quantity whenever the speed of the particle is less than the speed of light. (This does not hold for Čerenkov radiation which we will discuss later.) We note the standard formula for the delta function integral,

$$\int dx\, \delta(f(x))\, g(x) = \frac{g(x)}{\left|\frac{df}{dx}\right|}\Bigg|_{x=x_0}, \tag{10.8}$$

where x_0 represents the solution of $f(x) = 0$. If there are more than one solution to this equation, of course, one has to sum over all the solutions. Using (10.7) and (10.8), the vector potential in (10.4) takes the form

$$A_\mu(\mathbf{x}, t) = \frac{1}{c}\int dt' \frac{j_\mu(t')\,\delta\left(t' - t + \frac{R(t')}{c}\right)}{R(t')}$$

$$= \frac{1}{c}\frac{j_\mu(t')}{s(t')R(t')}\Bigg|_{\tau=0}. \tag{10.9}$$

These are known as the Lienard-Wiechert potentials which we had studied briefly earlier. In deriving (10.9) we are assuming that $\tau =$

0 has only one solution. (We are also assuming that the speed of the particle is less than the speed of light which is always true in vacuum.) If there are more than one solution, we must sum over all the contributions. Furthermore, the electric and the magnetic fields associated with a moving charge can be calculated from the vector potentials in the following manner.

We note that we can rewrite the scalar and the vector potentials explicitly in a form convenient for our purpose as

$$\Phi(\mathbf{x}, t) = q \int dt' \frac{\delta\left(t' - t + \frac{R(t')}{c}\right)}{R(t')},$$

$$\mathbf{A}(\mathbf{x}, t) = q \int dt' \frac{\mathbf{v}(t')}{c} \frac{\delta\left(t' - t + \frac{R(t')}{c}\right)}{R(t')}. \tag{10.10}$$

The electric field is defined in terms of these potentials to be

$$\mathbf{E}(\mathbf{x}, t) = -\boldsymbol{\nabla}\Phi - \frac{1}{c}\frac{\partial \mathbf{A}}{\partial t}. \tag{10.11}$$

There are several things to note here. First of all, since Φ and $\mathbf{A}$ depend on $\mathbf{x}$ only through their dependence on R, it can be easily checked that acting on these functions, the effect of the gradient inside the integral can be represented as $((\boldsymbol{\nabla}R) = \widehat{\mathbf{R}})$

$$\boldsymbol{\nabla}f(R) = (\boldsymbol{\nabla}R)\frac{\partial f(R)}{\partial R} = \widehat{\mathbf{R}}\,\frac{\partial f(R)}{\partial R}. \tag{10.12}$$

Using this, we obtain,

$$\mathbf{E}(\mathbf{x}, t) = -q \int dt' \,\widehat{\mathbf{R}}\,\frac{\partial}{\partial R}\left(\frac{\delta\left(t' - t + \frac{R}{c}\right)}{R}\right)$$

$$- \frac{q}{c}\frac{\partial}{\partial t}\int dt' \,\frac{\mathbf{v}}{c}\frac{\delta\left(t' - t + \frac{R}{c}\right)}{R}$$

$$= -q \int dt' \,\widehat{\mathbf{R}}\left[-\frac{\delta\left(t' - t + \frac{R}{c}\right)}{R^2} - \frac{1}{cR}\frac{\partial}{\partial t}\delta\left(t' - t + \frac{R}{c}\right)\right]$$

$$- \frac{q}{c}\frac{\partial}{\partial t}\int dt' \,\frac{\mathbf{v}}{c}\frac{\delta\left(t' - t + \frac{R}{c}\right)}{R}$$

$$= q \int dt' \, \hat{\mathbf{R}} \, \frac{\delta \left(t' - t + \frac{R}{c} \right)}{R^2}$$

$$+ \frac{q}{c} \frac{\partial}{\partial t} \int dt' \, \frac{\left(\hat{\mathbf{R}} - \frac{\mathbf{v}}{c} \right) \delta \left(t' - t + \frac{R}{c} \right)}{R}$$

$$= \frac{q\hat{\mathbf{R}}}{sR^2} \bigg|_{\tau=0} + \frac{q}{c} \frac{\partial}{\partial t} \left[\frac{\left(\hat{\mathbf{R}} - \frac{\mathbf{v}}{c} \right)}{sR} \right]_{\tau=0} . \qquad (10.13)$$

Here, we have used the relation (10.8).

The time derivative of a quantity with a restriction such as in (10.13) is tricky and can be carried out in the following manner. Let us note that the restriction $\tau = 0$ can be formally written as

$$t = t' + \frac{R(t')}{c} = A(t')$$

$$\text{or,} \quad \frac{dt}{dt'} = \frac{dA(t')}{dt'} = 1 + \frac{1}{c} \frac{dR(t')}{dt'} = s(t'), \qquad (10.14)$$

where we have used (10.7). If the relation between t' and t is locally invertible (which we assume for a solution to exist), we can write

$$t' = B(t),$$

$$\text{or,} \quad \frac{dt'}{dt} = \frac{dB(t)}{dt} = \frac{1}{s(t')} \bigg|_{t'=B(t)} . \qquad (10.15)$$

With this, we can now show that

$$\frac{\partial}{\partial t} \left[f(t') \right]_{t'=B(t)} = \frac{\partial}{\partial t} f(B(t))$$

$$= \frac{dB(t)}{dt} \, f'(B(t))$$

$$= \frac{\frac{df(t')}{dt'}}{s(t')} \bigg|_{t'=B(t)} , \qquad (10.16)$$

where we have used (10.15).

With the help of these, the expression in (10.13) can be simplified as follows.

$$
\mathbf{E}(\mathbf{x}, t) = \frac{q\widehat{\mathbf{R}}}{sR^2}\bigg|_{t'=B(t)} + \frac{q}{c}\frac{\partial}{\partial t}\left[\frac{\left(\widehat{\mathbf{R}} - \frac{\mathbf{v}}{c}\right)}{sR}\right]_{t'=B(t)}
$$

$$
= q\left[\frac{\widehat{\mathbf{R}}}{sR^2} + \frac{1}{c}\left(\frac{\left(\dot{\widehat{\mathbf{R}}} - \frac{\dot{\mathbf{v}}}{c}\right)}{s^2R} - \frac{\left(\widehat{\mathbf{R}} - \frac{\mathbf{v}}{c}\right)\dot{s}}{s^3R} - \frac{\left(\widehat{\mathbf{R}} - \frac{\mathbf{v}}{c}\right)\dot{R}}{s^2R^2}\right)\right]_{t'=B(t)}.
$$

$$(10.17)$$

From the definitions of $\mathbf{R}$ in (10.5) and s in (10.7), we note that

$$
\dot{\mathbf{R}} = -\mathbf{v},
$$

$$
\dot{R} = -\widehat{\mathbf{R}} \cdot \mathbf{v},
$$

$$
\dot{\widehat{\mathbf{R}}} = \frac{d}{dt'}\left(\frac{\mathbf{R}(t')}{R(t')}\right) = \frac{1}{R}\left(\widehat{\mathbf{R}}\,(\widehat{\mathbf{R}} \cdot \mathbf{v}) - \mathbf{v}\right),
$$

$$
\dot{s} = -\dot{\widehat{\mathbf{R}}} \cdot \frac{\mathbf{v}}{c} - \widehat{\mathbf{R}} \cdot \frac{\dot{\mathbf{v}}}{c}
$$

$$
= -\frac{1}{c}\frac{d\widehat{\mathbf{R}}(t')}{dt'} \cdot \mathbf{v} = -\left(\frac{1}{R}\left(\widehat{\mathbf{R}}(\widehat{\mathbf{R}} \cdot \mathbf{v}) - \mathbf{v}\right) \cdot \frac{\mathbf{v}}{c} + \widehat{\mathbf{R}} \cdot \frac{\dot{\mathbf{v}}}{c}\right).
$$

$$(10.18)$$

Using the definition in (10.7) and the relations in (10.18), as well as some vector identities, the expression for the electric field, (10.17), becomes

$$
\mathbf{E}(\mathbf{x}, t) = q\left[\frac{\left(\widehat{\mathbf{R}} - \frac{\mathbf{v}}{c}\right)\left(1 - (\frac{\mathbf{v}}{c})^2\right)}{s^3R^2}\right.
$$

$$
\left. + \frac{\widehat{\mathbf{R}} \times \left(\left(\widehat{\mathbf{R}} - \frac{\mathbf{v}}{c}\right) \times \frac{\dot{\mathbf{v}}}{c}\right)}{cs^3R}\right]_{t'=B(t)}.
$$

$$(10.19)$$

Similarly, we can also calculate the magnetic field as

$$
\mathbf{B}(\mathbf{x}, t) = \boldsymbol{\nabla} \times \mathbf{A}(\mathbf{x}, t)
$$

$$
= \frac{q}{c}\int dt'\,\boldsymbol{\nabla} \times \frac{\mathbf{v}}{R}\delta\left(t' - t + \frac{R}{c}\right)
$$

$$= \frac{q}{c} \int dt' \, (\widehat{\mathbf{R}} \times \mathbf{v}) \frac{\partial}{\partial R} \frac{\delta \left(t' - t + \frac{R}{c} \right)}{R}$$

$$= -\frac{q}{c} \left[\int dt' \, \frac{(\widehat{\mathbf{R}} \times \mathbf{v}) \, \delta \left(t' - t + \frac{R}{c} \right)}{R^2} \right.$$

$$\left. + \frac{\partial}{\partial t} \int dt' \, \frac{\left(\widehat{\mathbf{R}} \times \mathbf{v} \right) \delta \left(t' - t + \frac{R}{c} \right)}{cR} \right]$$

$$= -\frac{q}{c} \left[\frac{\left(\widehat{\mathbf{R}} \times \mathbf{v} \right)}{sR^2} \bigg|_{t'=B(t)} + \frac{\partial}{\partial t} \left[\frac{\left(\widehat{\mathbf{R}} \times \mathbf{v} \right)}{csR} \right]_{t'=B(t)} \right]. \quad (10.20)$$

This can be further simplified by using (10.16) and (10.18). Furthermore, using some vector identities as well as the form for the electric field in (10.19), it is straightforward to show that

$$\mathbf{B}(\mathbf{x}, t) = \widehat{\mathbf{R}} \bigg|_{t'=B(t)} \times \mathbf{E}(\mathbf{x}, t). \quad (10.21)$$

There are several things to note from the forms of the electric and the magnetic fields in Eqs. (10.19) and (10.21) respectively. We see that we need to solve for $t' = B(t)$ to determine the electromagnetic fields. (t' is known as the retarded time.) This can, in principle, be done once we know the trajectory of the particle. We note from (10.19) that, for a charged particle at rest, $\mathbf{v} = 0 = \dot{\mathbf{v}}$, in which case ($s = 1$ in such a case)

$$\mathbf{E}^{(\mathrm{rest})}(\mathbf{x}, t) = \frac{q\widehat{\mathbf{R}}}{R^2},$$

$$\mathbf{B}^{(\mathrm{rest})}(\mathbf{x}, t) = \widehat{\mathbf{R}} \times \mathbf{E}^{\mathrm{rest}}(\mathbf{x}, t) = 0, \quad (10.22)$$

which is what we will expect from our studies in electrostatics. In this case,

$$\mathbf{R} = \mathbf{x} - \boldsymbol{\xi}, \quad (10.23)$$

where $\boldsymbol{\xi}$ represents the fixed location (independent of time) of the charged particle. In general, though, both the electric and the magnetic fields depend on a term of the form $\frac{1}{R^2}$ as well as $\frac{1}{R}$. For small values of R it is the term corresponding to $\frac{1}{R^2}$ that dominates while, for large R it is the one with $\frac{1}{R}$ that dominates. The fields in the far zone have the right characteristics of radiation fields and will lead to power loss through radiation. However, we note that the term corresponding to the radiation field ($\frac{1}{R}$) is proportional to $\dot{\mathbf{v}}$. Correspondingly, we see that a particle moving in vacuum will radiate power only if it is being accelerated. Conversely, an accelerated charged particle will lose energy through radiation. On the other hand, there is no radiation of power if the particle is not being accelerated (in vacuum). Let us note, from (10.19), that the amplitude of the radiation term dominates if

$$\left. \frac{R|\dot{\mathbf{v}}|}{c^2} \right|_{t'=B(t)} \gg 1. \tag{10.24}$$

Thus, as long as $\dot{\mathbf{v}}$ is nonzero, we can always find a distance large enough for (10.24) to be true. This is the analogue of the far zone condition $kr \gg 1$ that we have studied earlier for systems with a harmonic time dependence.

10.2 Uniform linear motion

As we have seen in Eqs. (10.19) and (10.20) (or (10.21)), in order to derive the fields, we have to determine the retarded time $t' = B(t)$. This is not always easy and, therefore, the fields can be determined in closed form only for a few special classes of motions. The simplest motion that we can think of is, of course, a charged particle moving in vacuum along a trajectory with a uniform velocity. In this case, therefore, the acceleration vanishes and the radiation term is not present in the expression for the fields. For example, the expression for the electric field in (10.19), in this case, becomes

$$\mathbf{E}^{(\text{uniform})}(\mathbf{x}, t) = \left. \frac{q\left(\widehat{\mathbf{R}} - \frac{\mathbf{v}}{c}\right)\left(1 - \left(\frac{\mathbf{v}}{c}\right)^2\right)}{s^3 R^2} \right|_{t'=B(t)}. \tag{10.25}$$

The restriction implies that the retarded time should be expressed in terms of variables at the present time. The velocity is uniform so that

$$\mathbf{v}|_{t'=B(t)} = \mathbf{v} = \text{constant}.$$

We can solve for the trajectory of the particle easily in this case,

$$\boldsymbol{\xi}(t) = \mathbf{v}t, \tag{10.26}$$

where, for simplicity, we have chosen the particle to be at the coordinate origin at time $t = 0$.

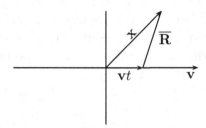

Figure 10.1: A charged particle in uniform linear motion.

From the definition in (10.5), we note that we can define (see Fig. 10.1)

$$\overline{\mathbf{R}} = \mathbf{R}(t) = \mathbf{x} - \mathbf{v}t. \tag{10.27}$$

It follows, therefore, that we can write

$$\mathbf{x} = \overline{\mathbf{R}} + \mathbf{v}t, \quad \mathbf{R}(t') = \mathbf{x} - \mathbf{v}t' = \overline{\mathbf{R}} - \mathbf{v}(t' - t). \tag{10.28}$$

Using this, we can solve for the retarded time

$$t' = t - \frac{R(t')}{c},$$

or, $\quad c(t' - t) = -R(t') = -|\overline{\mathbf{R}} - \mathbf{v}(t' - t)|,$

or, $\quad (c^2 - v^2)(t' - t)^2 + 2\overline{R}v\cos\theta(t' - t) - \overline{R}^2 = 0. \tag{10.29}$

Here, v and $\overline{R}$ stand respectively for the magnitudes of $\mathbf{v}$ and $\overline{\mathbf{R}}$ and θ denotes the angle between $\overline{\mathbf{R}}$ and $\mathbf{v}$ (see also Fig. 10.2). This is a quadratic equation which can be easily solved by the standard method yielding

$$(t' - t) = -\frac{\overline{R}}{c\left(1 - \frac{v^2}{c^2}\right)} \left(\frac{v\cos\theta}{c} \mp \sqrt{1 - \left(\frac{v\sin\theta}{c}\right)^2}\right). \tag{10.30}$$

We note that, since t' represents a retarded time and we assume the particle to be moving with a speed less than the speed of light, it is the second root that is physical, namely,

$$(t' - t) = -\frac{\overline{R}}{c\left(1 - \frac{v^2}{c^2}\right)}\left(\frac{v\cos\theta}{c} + \sqrt{1 - \left(\frac{v\sin\theta}{c}\right)^2}\right). \quad (10.31)$$

Since $\overline{R}$ and θ are actually functions of time t, Eq. (10.31) gives the necessary relation $t' = B(t)$.

Next, let us note from (10.28) and (10.31) that

$$\mathbf{R}(t') = \overline{\mathbf{R}} - \mathbf{v}(t' - t)$$

$$= \overline{\mathbf{R}} + \frac{\overline{R}\frac{\mathbf{v}}{c}}{1 - \frac{v^2}{c^2}}\left(\frac{v\cos\theta}{c} + \sqrt{1 - \left(\frac{v\sin\theta}{c}\right)^2}\right), \quad (10.32)$$

where we have used the physical solution in (10.31). We can calculate and show from (10.32) that

$$R(t') = \frac{\overline{R}}{1 - \frac{v^2}{c^2}}\left(\frac{v\cos\theta}{c} + \sqrt{1 - \left(\frac{v\sin\theta}{c}\right)^2}\right). \quad (10.33)$$

It follows from (10.32) and (10.33) that

$$\widehat{\mathbf{R}}(t') = \frac{\left(1 - \frac{v^2}{c^2}\right)\widehat{\overline{\mathbf{R}}}}{\frac{v\cos\theta}{c} + \sqrt{1 - \left(\frac{v\sin\theta}{c}\right)^2}} + \frac{\mathbf{v}}{c}$$

or, $\quad \widehat{\mathbf{R}}(t') - \frac{\mathbf{v}}{c} = \frac{\widehat{\overline{\mathbf{R}}}\,\overline{R}}{R(t')}. \quad (10.34)$

Similarly, using (10.34), it is easily calculated that

$$s(t') = 1 - \widehat{\mathbf{R}}(t') \cdot \frac{\mathbf{v}}{c} = \left(1 - \frac{v^2}{c^2}\right)\left(1 - \frac{\frac{v\cos\theta}{c}}{\frac{v\cos\theta}{c} + \sqrt{1 - \left(\frac{v\sin\theta}{c}\right)^2}}\right)$$

$$= \frac{\left(1 - \frac{v^2}{c^2}\right)\sqrt{1 - \left(\frac{v\sin\theta}{c}\right)^2}}{\frac{v\cos\theta}{c} + \sqrt{1 - \left(\frac{v\sin\theta}{c}\right)^2}}. \quad (10.35)$$

It follows, therefore, from (10.33) and (10.35) that

$$s(t')R(t') = \overline{R}\left(1 - \left(\frac{v\sin\theta}{c}\right)^2\right)^{\frac{1}{2}}. \quad (10.36)$$

Using (10.34) and (10.36) in the expression for the electric field (10.25), we obtain

$$\mathbf{E}^{(\text{uniform})}(\mathbf{x}, t) = \left. \frac{q\left(\widehat{\mathbf{R}} - \frac{\mathbf{v}}{c}\right)\left(1 - \frac{v^2}{c^2}\right)}{s^3 R^2} \right|_{t'=B(t)}$$

$$= \left. \frac{q\left(\widehat{\mathbf{R}} - \frac{\mathbf{v}}{c}\right) R\left(1 - \frac{v^2}{c^2}\right)}{(sR)^3} \right|_{t'=B(t)}$$

$$= \frac{q\widehat{\overline{\mathbf{R}}}\,\overline{R}\left(1 - \frac{v^2}{c^2}\right)}{\overline{R}^3 \left(1 - \left(\frac{v\sin\theta}{c}\right)^2\right)^{\frac{3}{2}}}$$

$$= \frac{q\widehat{\overline{\mathbf{R}}}\left(1 - \frac{v^2}{c^2}\right)}{\overline{R}^2 \left(1 - \left(\frac{v\sin\theta}{c}\right)^2\right)^{\frac{3}{2}}}. \tag{10.37}$$

The magnetic field can also be calculated for this case using (10.21) and (10.34) and it turns out to be

$$\mathbf{B}^{(\text{uniform})}(\mathbf{x}, t) = \left. \widehat{\mathbf{R}}(t') \right|_{t'=B(t)} \times \mathbf{E}^{(\text{uniform})}(\mathbf{x}, t)$$

$$= \frac{\frac{\mathbf{v}}{c} \times q\widehat{\overline{\mathbf{R}}}\left(1 - \frac{v^2}{c^2}\right)}{\overline{R}^2 \left(1 - \left(\frac{v\sin\theta}{c}\right)^2\right)^{\frac{3}{2}}}$$

$$= \frac{\mathbf{v}}{c} \times \mathbf{E}^{(\text{uniform})}(\mathbf{x}, t). \tag{10.38}$$

This is, in fact, what we would expect from our earlier studies in electrostatics. To see that, let us specialize to the case of particle motion along the x-axis and the observation point $\mathbf{x}$ lying in the $x-y$ plane for simplicity, as shown in Fig. 10.2.

Introducing the conventional notations,

$$\beta = \frac{v}{c},$$

$$\gamma = \frac{1}{\sqrt{1 - \beta^2}} = \frac{1}{\sqrt{1 - \frac{v^2}{c^2}}}, \tag{10.39}$$

we note that we can write the components of the electric and the

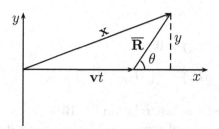

Figure 10.2: A charged particle moving along the x-axis with the observation point in the $x - y$ plane.

magnetic fields in (10.37) and (10.38) to be

$$
\begin{array}{ll}
E_x = \dfrac{q(x-vt)}{\gamma^2 \overline{R}^3 (1-\beta^2 \sin^2 \theta)^{\frac{3}{2}}}, & B_x = 0, \\[3mm]
E_y = \dfrac{qy}{\gamma^2 \overline{R}^3 (1-\beta^2 \sin^2 \theta)^{\frac{3}{2}}}, & B_y = 0, \\[3mm]
E_z = 0, & B_z = \beta E_y \\[3mm]
 & \quad = \dfrac{q\beta y}{\gamma^2 \overline{R}^3 (1-\beta^2 \sin^2 \theta)^{\frac{3}{2}}}.
\end{array}
\tag{10.40}
$$

In this case, we note that

$$
\overline{R} = \left((x - vt)^2 + y^2 \right)^{\frac{1}{2}},
$$

$$
\sin \theta = \frac{y}{\overline{R}} = \frac{y}{\left((x - vt)^2 + y^2 \right)^{\frac{1}{2}}},
$$

$$
\overline{R}^2 (1 - \beta^2 \sin^2 \theta) = \overline{R}^2 - \beta^2 y^2 = (x - vt)^2 + \frac{y^2}{\gamma^2}.
\tag{10.41}
$$

If we now go to a Lorentz frame where the particle is at rest, the electric and the magnetic fields would transform as

$$
E'_x = E_x = \frac{q(x - vt)}{\gamma^2 \overline{R}^3 (1 - \beta^2 \sin^2 \theta)^{\frac{3}{2}}},
$$

$$
E'_y = \gamma (E_y - \beta B_z) = \gamma (1 - \beta^2) E_y = \frac{1}{\gamma} E_y
$$

$$
= \frac{qy}{\gamma^3 \overline{R}^3 (1 - \beta^2 \sin^2 \theta)^{\frac{3}{2}}},
$$

$$E'_z = \gamma(E_z + \beta B_y) = 0,$$

$$B'_x = B_x = 0,$$

$$B'_y = \gamma(B_y + \beta E_z) = 0,$$

$$B'_z = \gamma(B_z - \beta E_y) = \gamma(\beta E_y - \beta E_y) = 0, \tag{10.42}$$

where we have used the last relation in (10.40).

These will be the electric and the magnetic fields in the rest frame of the particle, but still expressed in the old coordinates. We can transform them to the coordinates of the rest frame using

$$x' = \gamma(x - \beta ct),$$

$$y' = y,$$

$$z' = z,$$

$$t' = \gamma\left(t - \frac{\beta x}{c}\right). \tag{10.43}$$

In particular, this leads to

$$(x - vt) = (x - \beta ct) = \frac{x'}{\gamma},$$

$$\overline{R}(1 - \beta^2 \sin^2 \theta)^{\frac{1}{2}} = \frac{r'}{\gamma}, \tag{10.44}$$

where we have defined r' to be the distance of the observation point in the rest frame of the particle, namely,

$$r' = (x'^2 + y'^2)^{\frac{1}{2}}.$$

Using (10.41), (10.43) and (10.44), it follows that

$$E'_x = \frac{qx'}{r'^3},$$

$$E'_y = \frac{qy'}{r'^3},$$

$$E'_z = 0 = B'_x = B'_y = B'_z. \tag{10.45}$$

This shows that, in the rest frame of the charged particle, the magnetic field is zero and the electric field has the form that we will expect for a point charge at rest. (There is no z-component of the

electric field at the observation point since it lies in the x-y plane.)
For a charged particle moving with a uniform velocity in vacuum, we
do not expect any power loss due to radiation, since the radiation
components vanish in this case.

Let us now analyze Eqs. (10.37) and (10.38) in some detail.
First, we see that, for very small velocities, the electric and the magnetic fields have the forms following from Coulomb's law and Biot-
Savart's law respectively. However, as the velocity increases and approaches the speed of light, we see, from Eq. (10.37), that the electric
field becomes negligible along the direction of motion (along $\theta = 0$).
Its magnitude is larger as θ increases from zero and peaks at $\theta = \frac{\pi}{2}$,
namely, at a direction perpendicular to the motion as shown in Fig.
10.3. (Namely, the electric field is dominantly transverse to the direction of motion.)

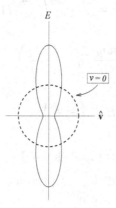

Figure 10.3: The electric field as a function of the velocity of the
particle. The dashed circle represents the case when the particle is
at rest.

The magnetic field is, of course, always orthogonal to the direction of motion as well as to the electric field as is clear from (10.38).
Thus, we see that the fields of an extremely relativistic charged particle in uniform motion along a straight line behave like an electromagnetic plane wave (where both **E** as well as **B** are perpendicular
to the direction of motion as well as to themselves). This can, in fact,
be seen in a more quantitative manner as follows.

Let us assume, as we have done earlier, that the particle is moving along the x-axis. Then, we see from (10.40), that only the z component of the magnetic field is non-zero. Furthermore, using (10.41)
and taking the Fourier transform in the t variable (so that we go to

the frequency space), we obtain

$$E_x(\mathbf{x}, \omega) = \frac{1}{2\pi} \int\limits_{-\infty}^{\infty} dt \, E_x(\mathbf{x}, t) \, e^{i\omega t}$$

$$= \frac{q\gamma}{2\pi} \int\limits_{-\infty}^{\infty} dt \, \frac{(x - vt) \, e^{i\omega t}}{(\gamma^2(x - vt)^2 + y^2)^{\frac{3}{2}}}$$

$$= \frac{q \, e^{\frac{i\omega x}{v}}}{2\pi\gamma vy} \int\limits_{-\infty}^{\infty} d\xi \, \frac{\xi \, e^{-\frac{i\omega y\xi}{\gamma v}}}{(1 + \xi^2)^{\frac{3}{2}}}$$

$$= \frac{q \, e^{\frac{i\omega x}{v}}}{2\pi\gamma vy} \left(\frac{i\gamma v}{\omega} \right) \frac{d}{dy} \int\limits_{-\infty}^{\infty} d\xi \, \frac{e^{-\frac{i\omega y\xi}{\gamma v}}}{(1 + \xi^2)^{\frac{3}{2}}}$$

$$= \frac{q \, e^{\frac{i\omega x}{v}}}{2\pi\gamma vy} \frac{\pi\omega y}{\gamma v} H_0^{(1)} \left(\frac{i\omega y}{\gamma v} \right)$$

$$= \frac{q\omega \, e^{\frac{i\omega x}{v}}}{2\gamma^2 v^2} H_0^{(1)} \left(\frac{i\omega y}{\gamma v} \right), \tag{10.46}$$

where we have used (10.41) and defined $\xi = \frac{\gamma(x-vt)}{y}$ in the intermediate steps. We note that $H_n^{(1)}$ represents the nth Hankel function of the first kind and we have used some standard relations from the table of integrals (see, for example, Gradshteyn and Ryzhik 8.407.1, 8.432.5, 8.472.1). Similarly, the Fourier transform of the y component of the electric field leads to

$$E_y(\mathbf{x}, \omega) = \frac{1}{2\pi} \int\limits_{-\infty}^{\infty} dt \, E_y(\mathbf{x}, t) \, e^{i\omega t}$$

$$= \frac{q \, e^{\frac{i\omega x}{v}}}{2\pi vy} \int\limits_{-\infty}^{\infty} d\xi \, \frac{e^{-\frac{i\omega y\xi}{\gamma v}}}{(1 + \xi^2)^{\frac{3}{2}}}$$

$$= \frac{q \, e^{\frac{i\omega x}{v}}}{2\pi vy} \left(-\frac{\pi\omega y}{\gamma v} \right) H_1^{(1)} \left(\frac{i\omega y}{\gamma v} \right)$$

$$= -\frac{q\omega \, e^{\frac{i\omega x}{v}}}{2\gamma v^2} H_1^{(1)} \left(\frac{i\omega y}{\gamma v} \right). \tag{10.47}$$

It also follows from (10.40) that

$$B_z(\mathbf{x}, \omega) = \beta E_y(\mathbf{x}, \omega) = -\frac{q\beta\omega\, e^{\frac{i\omega x}{v}}}{2\gamma v^2}\, H_1^{(1)}\left(\frac{i\omega y}{\gamma v}\right). \qquad (10.48)$$

Let us recall that asymptotically for large values of the argument (*i.e.* when $\frac{\omega y}{\gamma v}$ large), the Hankel function behaves as

$$H_n^{(1)}\left(\frac{i\omega y}{\gamma v}\right) \to \sqrt{\frac{2\gamma v}{i\pi\omega y}}\, e^{-\frac{\omega y}{\gamma v} - i(2n+1)\frac{\pi}{4}}, \qquad (10.49)$$

while near the origin, the behavior is of the form

$$H_0^{(1)}\left(\frac{i\omega y}{\gamma v}\right) \to -\frac{2i}{\pi}\left(\ln\frac{2\gamma v}{i\omega y} - \ln 1.781\right),$$

$$H_1^{(1)}\left(\frac{i\omega y}{\gamma v}\right) \to -\frac{2\gamma v}{\pi\omega y}. \qquad (10.50)$$

Let us note that for fixed ω, y, as $v \to c$, $\gamma \to \infty$. We see from (10.46) and (10.47) that since E_x is suppressed by a factor of γ (compared to E_y), it will be negligible. For large values of the frequency, all the field components will be exponentially damped. Furthermore, as $v \to c$, we see from the asymptotic forms in (10.50) that (namely, we are looking at small ω)

$$E_x(\mathbf{x}, \omega) \to 0,$$

$$E_y(\mathbf{x}, \omega) \to \frac{q}{\pi v y}\, e^{\frac{i\omega x}{v}} \to \frac{q}{\pi c y}\, e^{\frac{i\omega x}{c}}. \qquad (10.51)$$

The non-vanishing magnetic field along the z-axis also has the same form and magnitude as E_y. Thus, we see that every frequency component of the fields behaves like a plane wave when $v \to c$.

10.3 Method of virtual photons

The analysis of the fields in the Fourier transformed space, as discussed in the previous section is very important and leads to a very useful technique called the "method of virtual photons", originally due to Fermi. Let us note from (10.49) that for fixed y, v, both E_x, E_y are exponentially damped for $\omega > \frac{\gamma v}{y}$. For small values of ω, it follows from (10.46) and (10.50) that E_x is negligible while E_y dominates. Thus, we can assume that E_x is negligible for any frequency and write

$$E_y(\mathbf{x}, \omega) = 0, \qquad\qquad \omega > \frac{\gamma v}{y},$$
$$E_y(\mathbf{x}, \omega) = \frac{q}{\pi v y} e^{\frac{i\omega x}{v}}, \qquad \omega < \frac{\gamma v}{y}. \tag{10.52}$$

This allows us to reconstruct the original fields through an inverse Fourier transformation as

$$\mathbf{E}(\mathbf{x}, t) = \int d\omega \, \mathbf{E}(\mathbf{x}, \omega) \, e^{-i\omega t}. \tag{10.53}$$

This shows that the original field associated with the particle can be thought of as an electromagnetic pulse.

It is, of course, much easier to solve for the fields for a given frequency and once this is known, the original field can be reconstructed by taking a linear superposition of the form in (10.53). An electromagnetic field configuration corresponding to a given frequency can, of course, be associated with a photon of energy $\hbar\omega$. The electromagnetic fields associated with a moving charged particle can, therefore, be thought of as resulting from emission and reabsorption of "virtual" photons by the charged particle. The word "virtual" (also in the method of virtual photons) comes from the following quantum mechanical correspondence. Quantum mechanics allows a charged particle to emit and reabsorb photons while it travels as shown in Fig. 10.4.

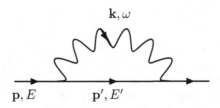

Figure 10.4: An electron emitting and absorbing a virtual photon of wave number $\mathbf{k}$ and frequency ω.

Such a process is, of course, not allowed classically. For example, if $\mathbf{k}$ represents the wave number of the emitted photon and ω its frequency (see Fig. 10.4), then conservation of energy and momentum would require

$$\mathbf{p} - \mathbf{p}' = \hbar\mathbf{k},$$
$$E - E' = \hbar\omega. \tag{10.54}$$

It now follows from (10.54) as well as from Einstein's relation that

$$\mu^2 c^4 = (\hbar\omega)^2 - (\hbar\mathbf{k})^2 c^2 = (E - E')^2 - (\mathbf{p} - \mathbf{p}')^2 c^2$$

$$= (E^2 - \mathbf{p}^2 c^2) + (E'^2 - \mathbf{p}'^2 c^2) - 2EE' + 2\mathbf{p} \cdot \mathbf{p}' c^2$$

$$= 2m^2 c^4 \left(1 - \gamma\gamma'(1 - \beta\beta' \cos\theta)\right)$$

$$\neq 0, \tag{10.55}$$

where θ represents the angle between $\mathbf{p}$ and $\mathbf{p}'$ and we have used the fact that for a real particle, we can write

$$E = mc^2\gamma, \qquad |\mathbf{p}| = \frac{\beta E}{c} = mc\gamma\beta. \tag{10.56}$$

It is clear, therefore, that energy and momentum conservation will be violated in a process where a classical charged particle emits a photon. Quantum mechanically, however, energy can be uncertain due to quantum mechanical fluctuations and satisfies a relation of the form

$$\Delta E \Delta t \geq \frac{\hbar}{2}. \tag{10.57}$$

Therefore, during a time interval of

$$\Delta t \geq \frac{\hbar}{2\Delta E} \sim \frac{\hbar}{2\mu c^2},$$

such a process can take place and would involve a "virtual" photon (it cannot be a real photon since the rest mass cannot be zero by energy-momentum conservation). This is the reason that this method is known as the "method of virtual photons" and the quantity μ is known as the invariant mass of the photon.

Let us note that the electric and the magnetic fields carry energy. In vacuum, the contribution of the electric and the magnetic energies are equal and, therefore, the total energy associated with the transverse components of the fields (volume is considered in cylindrical coordinates with the axis of the cylinder along the x-axis) can be calculated easily to be

$$U = \frac{1}{4\pi} \int d^3x \, |E_y(\mathbf{x}, t)|^2$$

$$= \frac{v}{4\pi} \int (2\pi y \, dy) \, dt \, |E_y(\mathbf{x}, t)|^2$$

$$= \frac{v}{2} \int dy \, y \int_{-\infty}^{\infty} dt \, |E_y(\mathbf{x}, t)|^2. \tag{10.58}$$

Here, we have used the fact that the particle is moving along the x-axis with a uniform velocity v, to convert the x integral to a time integral. Furthermore, from the definitions of the Fourier transform as well as the delta function,

$$\mathbf{E}(\mathbf{x}, t) = \int_{-\infty}^{\infty} d\omega \, \mathbf{E}(\mathbf{x}, \omega) \, e^{-i\omega t},$$

$$\delta(\omega) = \frac{1}{2\pi} \int_{-\infty}^{\infty} dt \, e^{i\omega t} = \frac{1}{2\pi} \int_{-\infty}^{\infty} dt \, e^{-i\omega t}, \tag{10.59}$$

we obtain

$$U = \frac{v}{2} \int dy \, y \int dt \, d\omega \, d\omega' \, E_y(\mathbf{x}, \omega) E_y^*(\mathbf{x}, \omega') \, e^{-i(\omega - \omega')t}$$

$$= \pi v \int dy \, y \int d\omega \, d\omega' \, E_y(\mathbf{x}, \omega) E_y^*(\mathbf{x}, \omega') \, \delta(\omega - \omega')$$

$$= \pi v \int dy \, y \int_{-\infty}^{\infty} d\omega \, |E_y(\mathbf{x}, \omega)|^2$$

$$= 2\pi v \int dy \, y \int_{0}^{\infty} d\omega \, |E_y(\mathbf{x}, \omega)|^2$$

$$= 2\pi v \int_{0}^{\infty} d\omega \int_{b}^{\frac{\gamma v}{\omega}} dy \, \frac{q^2}{\pi^2 v^2 y}$$

$$= \frac{2q^2}{\pi v} \int_{0}^{\infty} d\omega \int_{b}^{\frac{\gamma v}{\omega}} \frac{dy}{y}. \tag{10.60}$$

Here, we have used (10.52) and have introduced a lower cut-off on the y integral since it diverges. The value of this cut-off, which can be thought of as a minimum impact parameter, can be fixed later on physical grounds.

We recognize that we can write

$$U = \int_{0}^{\infty} d\omega \, U(\omega), \tag{10.61}$$

where $U(\omega)$ is the energy carried by a particular frequency mode of the field components. From (10.60), we see that we can write

$$U(\omega) = \frac{2q^2}{\pi v} \int_b^{\frac{\gamma v}{\omega}} \frac{dy}{y} = \frac{2q^2}{\pi v} \ln \frac{\gamma v}{\omega b}. \tag{10.62}$$

Assuming that there are $N(\omega)$ photons with energy $\hbar\omega$ (associated with frequency ω), we can now obtain

$$\hbar\omega\, N(\omega) = U(\omega) = \frac{2q^2}{\pi v} \ln \frac{\gamma v}{\omega b},$$

$$\text{or,} \quad N(\omega) = \frac{2q^2}{\pi v \hbar \omega} \ln \frac{\gamma v}{\omega b}. \tag{10.63}$$

Let us assume that the charged particle is an electron with charge $q = -e$. Then, we note that the fine structure constant has the form

$$\alpha = \frac{e^2}{\hbar c}. \tag{10.64}$$

Furthermore, identifying the minimum impact parameter from the uncertainty relation as

$$b \sim \frac{\hbar}{|\Delta p|} = \frac{\hbar}{|\mathbf{p} - \mathbf{p}'|}, \tag{10.65}$$

as well as using (10.54), we can determine

$$N(\omega) \sim \frac{2\alpha}{\pi \beta \omega} \ln \frac{\gamma \beta c |\mathbf{p} - \mathbf{p}'|}{(E - E')}$$

$$= \frac{2\alpha\hbar}{\pi\beta(E - E')} \ln \left(\frac{E}{mc^2} \frac{\beta c |\mathbf{p} - \mathbf{p}'|}{(E - E')} \right), \tag{10.66}$$

where we have used (10.56). This gives the flux of transverse "virtual" photons. Note that, since the fine structure constant is a small number, the number of photons associated with an electron is also small.

10.4 Asymptotic values of the fields

As is clear from the previous example, even in the simple case of a charged particle moving with a uniform velocity, the calculation of the electric and the magnetic fields is nontrivial. This is primarily because it is not easy, in general, to find a relation between the retarded time and the instantaneous (or observation) time in a form that is convenient for manipulations. It is for this reason that one often uses approximate methods to determine the forms of the fields (when nonuniform motion is involved). The approximations are quite analogous to the ones we have made earlier in connection with systems with a harmonic time dependence and let us discuss these in some detail.

As before, let us assume that the system is characterized by a size d. For a particle in bound motion, the meaning of this size is, of course, quite clear, namely, it corresponds to the size of the bound system. However, for unbounded motion extending to infinity, the meaning of a size is not quite clear. In this case, of course, we are not looking at the motion of the particle along the entire trajectory, rather, we are interested in a finite segment of the trajectory of the particle (that the particle traverses during the time that we need to make the observations). The size d can then be associated with such a segment of the trajectory of the particle. In either case, let us assume that we are interested in observations far away from the source, namely, in the radiation zone. Correspondingly, we can assume that

$$R(t') \approx r \gg d. \tag{10.67}$$

Under this approximation, we see that only the second term in (10.19) would dominate and the electric field would have the form (we are now preparing to consider nonuniform motion)

$$\mathbf{E}(\mathbf{x}, t) = \left. \frac{q\,\widehat{\mathbf{R}} \times ((\widehat{\mathbf{R}} - \boldsymbol{\beta}) \times \dot{\boldsymbol{\beta}})}{cs^3 R} \right|_{t'=B(t)}. \tag{10.68}$$

To evaluate the quantities at the retarded time, let us note that we do not expect the retarded time to be very different from the present time. Therefore, we can make an approximate expansion which leads to (since by assumption $r \gg d$)

$$t' = t - \frac{|\mathbf{x} - \boldsymbol{\xi}(t')|}{c} \approx t - \frac{r}{c},$$

$$\text{or,} \quad (t' - t) \approx -\frac{r}{c}, \tag{10.69}$$

where we have identified $r = |\mathbf{x}|$.

With this approximation, we can now calculate various quantities of interest.

$$\mathbf{R}(t') = \mathbf{x} - \boldsymbol{\xi}(t') \approx \mathbf{x},$$

$$R(t') \approx r = |\mathbf{x}|,$$

$$\widehat{\mathbf{R}}(t') = \frac{\mathbf{R}(t')}{R(t')} \approx \hat{\mathbf{x}},$$

$$\boldsymbol{\beta}(t') \approx \boldsymbol{\beta}(t) + (t' - t)\dot{\boldsymbol{\beta}}(t) \approx \bar{\boldsymbol{\beta}} - \frac{r}{c}\dot{\bar{\boldsymbol{\beta}}} \approx \bar{\boldsymbol{\beta}},$$

$$s(t') = 1 - \widehat{\mathbf{R}}(t') \cdot \boldsymbol{\beta}(t')$$

$$\approx 1 - \hat{\mathbf{x}} \cdot \bar{\boldsymbol{\beta}} + \frac{r}{c}\hat{\mathbf{x}} \cdot \dot{\bar{\boldsymbol{\beta}}}$$

$$\approx 1 - \hat{\mathbf{x}} \cdot \bar{\boldsymbol{\beta}}, \tag{10.70}$$

where we are using our earlier convention of defining quantities with a bar to denote quantities at the instantaneous time. We also note that we are assuming $\frac{r}{c}$ to be small (namely, the retarded time is assumed not to be very different) and, as a result, we have neglected some of the terms. Using (10.70) the electric field in the radiation zone, (10.68), can now be calculated. However, since the form is complicated, we do not give the explicit results. Rather, we would like to discuss several special cases of this approximation.

10.4.1 Dipole approximation. As a first application of this approximation method, let us consider the case of a non-relativistic particle, for which we can assume $|\boldsymbol{\beta}| \ll 1$. The acceleration, even if small, is assumed to be nonzero. In this case, we see from (10.70) that

$$\mathbf{R}(t') \approx \overline{\mathbf{R}} \approx \mathbf{x},$$

$$R(t') \approx \overline{R} \approx r,$$

$$\widehat{\mathbf{R}}(t') \approx \widehat{\overline{\mathbf{R}}} \approx \hat{\mathbf{x}},$$

$$|\boldsymbol{\beta}(t')| \approx |\bar{\boldsymbol{\beta}}| \ll 1,$$

$$s(t') \approx 1. \tag{10.71}$$

In such a case, therefore, the dominant term in the electric field in (10.68) has the form

$$\mathbf{E}^{(\text{dipole})}(\mathbf{x}, t) = \left. \frac{q\,\widehat{\mathbf{R}} \times ((\widehat{\mathbf{R}} - \boldsymbol{\beta}) \times \dot{\boldsymbol{\beta}})}{cs^3 R} \right|_{t'=B(t)}$$

$$\approx \frac{q\,\hat{\mathbf{x}} \times (\hat{\mathbf{x}} \times \bar{\dot{\boldsymbol{\beta}}})}{cr}$$

$$= \frac{q\,(\hat{\mathbf{x}}(\hat{\mathbf{x}} \cdot \bar{\dot{\boldsymbol{\beta}}}) - \bar{\dot{\boldsymbol{\beta}}})}{cr}$$

$$= \frac{qa\sin\theta}{c^2 r}\,\hat{\boldsymbol{\theta}}, \tag{10.72}$$

where we have defined $a = c|\bar{\dot{\boldsymbol{\beta}}}|$ to be the magnitude of the acceleration. Here, θ is the angle between $\hat{\mathbf{x}}$ and $\dot{\mathbf{v}}$ (see Fig. 10.5) and $\hat{\boldsymbol{\theta}}$ represents the unit vector along this direction (which is not fixed) so that $\bar{\dot{\boldsymbol{\beta}}} = \hat{\mathbf{x}}(\hat{\mathbf{x}} \cdot \bar{\dot{\boldsymbol{\beta}}}) - \hat{\boldsymbol{\theta}}|\bar{\dot{\boldsymbol{\beta}}}|\sin\theta$. It follows now that

$$\mathbf{B}^{(\text{dipole})}(\mathbf{x}, t) = \left. \widehat{\mathbf{R}}(t') \right|_{t'=B(t)} \times \mathbf{E}^{(\text{dipole})}(\mathbf{x}, t)$$

$$\approx \frac{qa\sin\theta}{c^2 r}\,(\hat{\mathbf{x}} \times \hat{\boldsymbol{\theta}}). \tag{10.73}$$

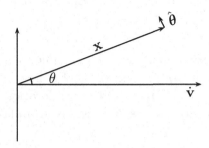

Figure 10.5: The angle θ between $\mathbf{x}$ and $\dot{\mathbf{v}}$.

A particle which is being accelerated radiates. Thus, we can calculate the power loss through radiation as follows. First, we note that the Poynting vector (we are not time averaging in this case and, therefore, the factor differs from the earlier formula by two), in this case, has the form

$$
\mathbf{S} = \frac{c}{4\pi} \,\mathrm{Re}\,\left(\mathbf{E}^{(\mathrm{dipole})} \times \mathbf{H}^{(\mathrm{dipole})\,*}\right)
$$

$$
= \frac{q^2 a^2 \sin^2\theta}{4\pi c^3 r^2}\, \hat{\boldsymbol\theta} \times (\hat{\mathbf{x}} \times \hat{\boldsymbol\theta})
$$

$$
= \frac{q^2 a^2 \sin^2\theta}{4\pi c^3 r^2}\, \hat{\mathbf{x}}. \tag{10.74}
$$

Here, we have used the identities involving vector products as well as the fact that $\hat{\mathbf{x}}$ and $\hat{\boldsymbol\theta}$ are orthogonal. We have also assumed that we are in vacuum where $\mathbf{H} = \mathbf{B}$. Thus, we see that the radiated power along $\hat{\mathbf{x}}$, through a sphere of large radius r, is given by

$$
\frac{\mathrm{d}P}{\mathrm{d}\Omega} = r^2 \hat{\mathbf{x}} \cdot \mathbf{S} = \frac{q^2 a^2}{4\pi c^3} \sin^2\theta. \tag{10.75}
$$

This is exactly the angular distribution of power radiated from a dipole (see (9.43) as well as Fig. 10.6) and for this reason, this approximation is conventionally called the dipole approximation.

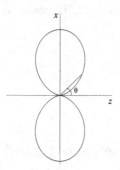

Figure 10.6: Radiation pattern for a dipole with $m = 0$ as in Fig. 9.6.

Let us note that, if we write

$$
\mathbf{p} = q\boldsymbol\xi, \qquad \mathbf{a} = \ddot{\boldsymbol\xi} = \frac{\ddot{\mathbf{p}}}{q}, \tag{10.76}
$$

then, the power radiated can also be written as

$$
\frac{\mathrm{d}P}{\mathrm{d}\Omega} = \frac{|\ddot{\mathbf{p}}|^2}{4\pi c^3} \sin^2\theta. \tag{10.77}
$$

We see that this is exactly the same as in (9.43) if we remember that, for a harmonic time dependence,

$$c = \frac{\omega}{k}, \qquad \ddot{\mathbf{p}} = -\omega^2 \mathbf{p}.$$

The extra factor of two in the denominator in (9.43) comes from the time averaging.

We can integrate (10.75) over the entire solid angle of a large sphere to obtain the total power radiated, which has the form

$$P_{\text{total}} = \frac{q^2 a^2}{4\pi c^3} \, 2\pi \int d\theta \, \sin^3 \theta = \frac{2q^2 a^2}{3c^3}, \tag{10.78}$$

which is also known as the Larmour formula.

10.4.2 Linear acceleration. Let us next consider the case where the particle is extremely relativistic and is subjected to an acceleration that is small during the time scale that observations are made. We also assume that the particle is moving in a linear trajectory so that the acceleration is parallel to the velocity. Therefore, we have

$$\dot{\boldsymbol{\beta}} \parallel \boldsymbol{\beta}, \qquad |\dot{\boldsymbol{\beta}}| \ll 1. \tag{10.79}$$

In this case, therefore, from (10.70) we have

$$R(t') \approx r,$$

$$\widehat{\mathbf{R}}(t') \approx \hat{\mathbf{x}},$$

$$\boldsymbol{\beta}(t') \approx \bar{\boldsymbol{\beta}},$$

$$s(t') \approx 1 - \hat{\mathbf{x}} \cdot \bar{\boldsymbol{\beta}}. \tag{10.80}$$

Furthermore, since the acceleration is parallel to the velocity, we have

$$\boldsymbol{\beta} \times \dot{\boldsymbol{\beta}} = 0. \tag{10.81}$$

With these, we see that at large distances, the electric field takes the form (see (10.68))

$$\mathbf{E}^{(\text{linear})}(\mathbf{x}, t) = \left. \frac{q\,\widehat{\mathbf{R}} \times ((\widehat{\mathbf{R}} - \boldsymbol{\beta}) \times \dot{\boldsymbol{\beta}})}{cs^3 R} \right|_{t'=B(t)}$$

$$\approx \frac{q\,\hat{\mathbf{x}} \times (\hat{\mathbf{x}} \times \bar{\dot{\boldsymbol{\beta}}})}{cr(1 - \hat{\mathbf{x}} \cdot \bar{\boldsymbol{\beta}})^3}$$

$$= \frac{qa\sin\theta}{c^2 r(1 - \bar{\beta}\cos\theta)^3} \, \hat{\boldsymbol{\theta}}, \tag{10.82}$$

where, as in the case of the dipole approximation, we have defined θ to be the angle between $\hat{\mathbf{x}}$ and $\bar{\vec{\beta}}$ (or $\bar{\beta}$ since they are parallel). We have also defined $a = c|\bar{\vec{\beta}}|$ as before. We see that the form of the electric field is very similar to the one in the case of the dipole approximation except for the factors in the denominator. From this, we can determine the magnetic field to be

$$
\mathbf{B}^{(\mathrm{linear})}(\mathbf{x}, t) = \left. \widehat{\mathbf{R}}(t') \right|_{t'=B(t)} \times \mathbf{E}^{(\mathrm{linear})}(\mathbf{x}, t)
$$

$$
\approx \frac{qa \sin \theta}{c^2 r (1 - \bar{\beta} \cos \theta)^3} \, (\hat{\mathbf{x}} \times \hat{\boldsymbol{\theta}}). \tag{10.83}
$$

The power radiated along $\hat{\mathbf{x}}$ through the surface of a sphere of large radius r (in vacuum) can now be calculated easily (not time averaged)

$$
\frac{dP}{d\Omega} = r^2 \hat{\mathbf{x}} \cdot \mathbf{S} = \frac{c r^2}{4\pi} \operatorname{Re} \hat{\mathbf{x}} \cdot \left(\mathbf{E}^{(\mathrm{linear})} \times \mathbf{H}^{(\mathrm{linear})\,*} \right)
$$

$$
\approx \frac{q^2 a^2}{4\pi c^3 (1 - \bar{\beta} \cos \theta)^6} \, \sin^2 \theta. \tag{10.84}
$$

Thus, we see that the radiated power has an angular distribution very much like the dipole approximation except that it is modulated by the relativistic correction $\frac{1}{(1-\bar{\beta}\cos\theta)^6}$ as shown in Fig. 10.7.

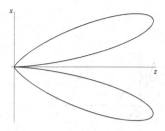

Figure 10.7: Modulated pattern for a dipole radiation.

The total power radiated through the entire surface of a large sphere, then, is obtained by integrating (10.84) over all solid angles, which gives

$$
P_{\text{total}} = \frac{q^2 a^2}{4\pi c^3} \, 2\pi \int\limits_0^\pi d\theta \, \frac{\sin^3 \theta}{(1 - \bar\beta \cos\theta)^6}
$$

$$
= \frac{q^2 a^2}{4\pi c^3} \, 2\pi \int\limits_{-1}^1 dx \, \frac{1 - x^2}{(1 - \bar\beta x)^6}
$$

$$
= \frac{q^2 a^2}{4\pi c^3} \frac{8\pi(5 + \bar\beta^2)}{15(1 - \bar\beta^2)^4} = \frac{2q^2 a^2}{3c^3} \frac{(5 + \bar\beta^2)}{5(1 - \bar\beta^2)^4}. \tag{10.85}
$$

Note that this reduces to (10.78) when $\bar\beta \ll 1$.

10.4.3 Uniform circular motion. As another application, let us analyze the motion of a charged particle moving in a circle. This is a case of harmonic motion and we assume that the particle motion is in the $x - y$ plane with the radius of the circle a small compared to the distance where we are observing the fields. Furthermore, without loss of generality, we can assume that the observation point lies in the $x - z$ plane. Then, we have

$$
\hat{\mathbf{x}} = \hat{\mathbf{e}}_x \sin\theta + \hat{\mathbf{e}}_z \cos\theta, \tag{10.86}
$$

where θ is the angle $\mathbf{x}$ makes with the z-axis (see Fig. 10.8, it is different from the θ in the earlier examples) and $\hat{\mathbf{e}}_{\{x,y,z\}}$ represent the Cartesian unit vectors along the three axes.

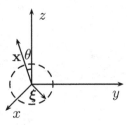

Figure 10.8: Circular motion of a particle in the $x - y$ plane with the observation point lying in the $x - z$ plane.

If the angular frequency associated with the particle motion is ω, then we can write

$$\boldsymbol{\xi}(t) = a\left(\hat{\mathbf{e}}_x \cos \omega t + \hat{\mathbf{e}}_y \sin \omega t\right),$$

$$\boldsymbol{\beta} = \frac{\dot{\boldsymbol{\xi}}}{c} = \frac{\omega a}{c}\left(-\hat{\mathbf{e}}_x \sin \omega t + \hat{\mathbf{e}}_y \cos \omega t\right)$$

$$= \beta\left(-\hat{\mathbf{e}}_x \sin \omega t + \hat{\mathbf{e}}_y \cos \omega t\right),$$

$$\dot{\boldsymbol{\beta}} = -\frac{\omega^2 a}{c}\left(\hat{\mathbf{e}}_x \cos \omega t + \hat{\mathbf{e}}_y \sin \omega t\right)$$

$$= -\frac{c\beta^2}{a}\left(\hat{\mathbf{e}}_x \cos \omega t + \hat{\mathbf{e}}_y \sin \omega t\right)$$

$$= -\dot{\beta}\left(\hat{\mathbf{e}}_x \cos \omega t + \hat{\mathbf{e}}_y \sin \omega t\right). \tag{10.87}$$

Here we have defined

$$\beta = |\boldsymbol{\beta}| = \frac{\omega a}{c}, \qquad \dot{\beta} = |\dot{\boldsymbol{\beta}}| = \frac{c\beta^2}{a}. \tag{10.88}$$

Using Eqs. (10.86) and (10.87), it is easy to obtain

$$\hat{\mathbf{x}} \cdot \boldsymbol{\beta} = -\beta \sin \theta \sin \omega t,$$

$$\hat{\mathbf{x}} \cdot \dot{\boldsymbol{\beta}} = -\dot{\beta} \sin \theta \cos \omega t,$$

$$\boldsymbol{\beta} \cdot \dot{\boldsymbol{\beta}} = 0. \tag{10.89}$$

Here, the magnitude of the velocity is constant and the acceleration is orthogonal to the direction of the velocity, a case complementary to the earlier example where the acceleration was along the direction of the velocity.

With the approximations in (10.70), in this case, we have

$$R(t') \approx r,$$

$$\widehat{\mathbf{R}}(t') \approx \hat{\mathbf{x}},$$

$$\boldsymbol{\beta}(t') = \boldsymbol{\beta},$$

$$s(t') \approx 1 - \hat{\mathbf{x}} \cdot \boldsymbol{\beta} = 1 + \beta \sin \theta \sin \omega t. \tag{10.90}$$

It is now easy to calculate the electric field in (10.19) for large distances, which takes the form

$$\mathbf{E}^{\text{(circular)}}(\mathbf{x}, t) \approx \frac{q}{cs^3 r} \left(\hat{\mathbf{x}} \times ((\hat{\mathbf{x}} - \boldsymbol{\beta}) \times \dot{\boldsymbol{\beta}}) \right)$$

$$= \frac{q}{cs^3 r} \left((\hat{\mathbf{x}} \cdot \dot{\boldsymbol{\beta}})(\hat{\mathbf{x}} - \boldsymbol{\beta}) - (\hat{\mathbf{x}} \cdot (\hat{\mathbf{x}} - \boldsymbol{\beta}))\dot{\boldsymbol{\beta}} \right)$$

$$= \frac{q}{cs^3 r} \left((\hat{\mathbf{x}} \cdot \dot{\boldsymbol{\beta}})(\hat{\mathbf{x}} - \boldsymbol{\beta}) - s\dot{\boldsymbol{\beta}} \right). \tag{10.91}$$

The magnetic field can also be calculated using (10.21), which for large distances takes the form

$$\mathbf{B}^{\text{(circular)}}(\mathbf{x}, t) \approx \hat{\mathbf{x}} \times \mathbf{E}^{\text{(circular)}}(\mathbf{x}, t)$$

$$= -\frac{q}{cs^3 r} \left((\hat{\mathbf{x}} \cdot \dot{\boldsymbol{\beta}})(\hat{\mathbf{x}} \times \boldsymbol{\beta}) + s(\hat{\mathbf{x}} \times \dot{\boldsymbol{\beta}}) \right). \tag{10.92}$$

It follows now that (in vacuum)

$$\mathbf{S} = \frac{c}{4\pi} \operatorname{Re} \mathbf{E}^{\text{(circular)}} \times \mathbf{B}^{*\,\text{(circular)}}$$

$$= -\frac{q^2}{4\pi cs^6 r^2} \left((\hat{\mathbf{x}} \cdot \dot{\boldsymbol{\beta}})(\hat{\mathbf{x}} - \boldsymbol{\beta}) - s\dot{\boldsymbol{\beta}} \right)$$

$$\times \left((\hat{\mathbf{x}} \cdot \dot{\boldsymbol{\beta}})(\hat{\mathbf{x}} \times \boldsymbol{\beta}) + s(\hat{\mathbf{x}} \times \dot{\boldsymbol{\beta}}) \right)$$

$$= \hat{\mathbf{x}} \frac{q^2}{4\pi cr^2} \left(\frac{\dot{\boldsymbol{\beta}}^2}{s^4} - \frac{(1 - \beta^2)(\hat{\mathbf{x}} \cdot \dot{\boldsymbol{\beta}})^2}{s^6} \right), \tag{10.93}$$

where we have used (10.88) and (10.89).

We can now calculate the power radiated along $\hat{\mathbf{x}}$ through the surface of a large sphere of radius r as (not time averaged)

$$\frac{dP}{d\Omega} = r^2 \, \hat{\mathbf{x}} \cdot \mathbf{S}$$

$$= \frac{cq^2 \beta^4}{4\pi a^2} \frac{(1 - \beta^2)\cos^2 \theta + (\beta + \sin\theta \sin\omega t)^2}{(1 + \beta \sin\theta \sin\omega t)^6}. \tag{10.94}$$

This is, of course, time dependent and since the motion is harmonic, we can average over one cycle of the motion. In doing so, however, we have to be careful and note from (10.14) that

$$dt = s\, dt' \approx s dt, \tag{10.95}$$

so that the time averaged radiated power along $\hat{\mathbf{x}}$ is given by

$$
\left\langle \frac{dP}{d\Omega} \right\rangle = \frac{cq^2\beta^4}{4\pi a^2} \frac{\omega}{2\pi} \int_0^{\frac{2\pi}{\omega}} dt \, \frac{(1-\beta^2)\cos^2\theta + (\beta + \sin\theta \sin\omega t)^2}{(1+\beta \sin\theta \sin\omega t)^5}
$$

$$
= \frac{cq^2\beta^4}{8\pi^2 a^2} \int_0^{2\pi} d\phi \, \frac{(1-\beta^2)\cos^2\theta + (\beta + \sin\theta \sin\phi)^2}{(1+\beta \sin\theta \sin\phi)^5}
$$

$$
= \frac{cq^2\beta^2}{8\pi^2 a^2} \int_0^{2\pi} d\phi \left[\frac{(1-\beta^2)(1-\beta^2 \sin^2\theta)}{(1+\beta \sin\theta \sin\phi)^5} \right.
$$

$$
\left. - \frac{2(1-\beta^2)}{(1+\beta \sin\theta \sin\phi)^4} + \frac{1}{(1+\beta \sin\theta \sin\phi)^3} \right].
$$

$$(10.96)$$

Thus, we see that evaluating the integral basically reduces to evaluating an integral of the form

$$
I_n = \int_0^{2\pi} d\phi \, \frac{1}{(1+\beta \sin\theta \sin\phi)^n}, \tag{10.97}
$$

which can be done in a standard manner. Let us define

$$
z = e^{i\phi}. \tag{10.98}
$$

Then, we can write

$$
I_n = (-i) \left(\frac{2i}{\beta \sin\theta} \right)^n \oint dz \, \frac{z^{n-1}}{\left(z^2 + \frac{2i}{\beta \sin\theta} z - 1 \right)^n}, \tag{10.99}
$$

where the integration is over a unit circle. The integrand has n-th order poles at

$$
z = \frac{i}{\beta \sin\theta} \left(-1 \pm \sqrt{1 - \beta^2 \sin^2\theta} \right). \tag{10.100}
$$

We note that only the first root lies within the unit circle and, consequently, the integral can be evaluated in a straightforward (but tedious manner since it is a higher order pole) using the method of residues. Let us simply note the results that

$$
I_3 = \pi \left[-\frac{1}{(1 - \beta^2 \sin^2 \theta)^{\frac{3}{2}}} + \frac{3}{(1 - \beta^2 \sin^2 \theta)^{\frac{5}{2}}} \right],
$$

$$
I_4 = \pi \left[-\frac{3}{(1 - \beta^2 \sin^2 \theta)^{\frac{5}{2}}} + \frac{5}{(1 - \beta^2 \sin^2 \theta)^{\frac{7}{2}}} \right],
$$

$$
I_5 = \frac{\pi}{4} \left[\frac{3}{(1 - \beta^2 \sin^2 \theta)^{\frac{5}{2}}} - \frac{30}{(1 - \beta^2 \sin^2 \theta)^{\frac{7}{2}}} \right.
$$

$$
\left. + \frac{35}{(1 - \beta^2 \sin^2 \theta)^{\frac{9}{2}}} \right]. \tag{10.101}
$$

Using (10.101) in (10.96), we obtain

$$
\left\langle \frac{\mathrm{d}P}{\mathrm{d}\Omega} \right\rangle = \frac{cq^2 \beta^4}{8\pi a^2 (1 - \beta^2 \sin^2 \theta)^{\frac{7}{2}}}
$$

$$
\times \left[(1 + \cos^2 \theta) - \frac{\beta^2}{4} (1 + 3\beta^2) \sin^4 \theta \right]. \tag{10.102}
$$

We see that, for non-relativistic motion,

$$
\left\langle \frac{\mathrm{d}P}{\mathrm{d}\Omega} \right\rangle \approx \frac{cq^2 \beta^4}{8\pi a^2} (1 + \cos^2 \theta) = \frac{q^2 a^2 \omega^4}{8\pi c^3} (1 + \cos^2 \theta), \tag{10.103}
$$

where we have used (10.88). This shows that the power radiated peaks along the z-axis ($\theta = 0$). In contrast, in the relativistic case we see from (10.102) that the radiated power peaks at $\theta = \frac{\pi}{2}$. The study of this system is particularly useful in the analysis of synchrotron radiation in accelerators.

10.5 Čerenkov effect

Earlier, we saw that particles moving with a uniform velocity in vacuum do not give rise to radiation. For radiation in vacuum, the particles have to be accelerated. Let us note that in vacuum, a particle cannot travel faster than the speed of light c for causality to hold. However, in a material medium, the speed of light changes by the index of refraction as $c' = \frac{c}{n}$ which can be smaller than c depending on the index of refraction of the medium. Here, $n = n(\omega) = \sqrt{\epsilon \mu}$ is the index of refraction of the medium and is a function of the frequency of the traveling wave (this is also true of the speed of light in the medium). A charged particle traveling in a medium can, therefore, travel faster than the speed of light in the medium without violating causality provided $c' < v < c$. In such a case, radiation is produced even when the particle is not being accelerated and this effect, known as the Čerenkov effect, provides an important tool in detecting high energy particles.

The Čerenkov effect is an interesting phenomenon where the charged particle does not lead to radiation directly. Rather, it is a collective phenomenon where the radiation is produced by the medium through which the charged particle moves. Quantum mechanically, the phenomenon can be understood as follows. The charged particle moving through the medium excites the electrons in the atoms which, upon return to their original state, emit a coherent radiation. Macroscopically, the phenomenon is analogous to the production of sound waves (shock waves) in the case of supersonic motion in a medium, where the fluctuations in the density of the medium produce a sound wave. In fact, geometrically, it is easy to see that if the particle travels with $v < c'$, then the information spheres (spherical wave fronts traveling with the speed of light in the medium), originating at later times, are contained inside the earlier ones. In this case, it is easy to see geometrically that there is only one unique retarded time for every point as shown in Fig. 10.9. (Basically, what this means is that any observation time can lie only on the surface of a single information sphere and, therefore, would correspond to a unique retarded time.)

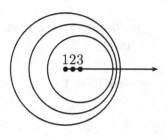

Figure 10.9: Information spheres for a particle travelling with $v < c'$.

On the other hand, if $v > c'$ the particle is moving faster than the speed with which the information spheres travel. Consequently, the information spheres originating at later times overtake the earlier ones (see Fig. 10.10). (In this case, the information spheres intersect leading to the fact that any observation time may lie on the surface of two or more information spheres and, therefore, may correspond to two or more retarded times.) The surfaces of the information spheres define a cone with the charged particle at the vertex and as these waves become more and more dense, a shock wave is produced traveling perpendicular to the conical surface. It is easy to see geometrically, that if θ_c denotes the angle between the axis of the cone and the perpendicular to the surface of the cone, then,

$$\cos \theta_c = \frac{c'}{v}. \tag{10.104}$$

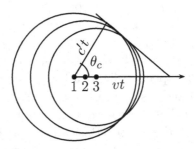

Figure 10.10: Information spheres for a particle travelling with $v > c'$.

Defining $\beta' = \frac{v}{c'} = \beta n(\omega)$, we see that for such a phenomenon to take place, we must have

$$\frac{1}{\beta'} = \frac{1}{\beta n(\omega)} = \cos \theta_c < 1$$

or, $\quad \beta > \dfrac{1}{n(\omega)},$ \hfill (10.105)

which also implies $\beta' > 1$. This is the condition for the emission of Čerenkov radiation. Geometrically, one can see that, in this case, there can be more than one retarded times associated with any given point.

To understand the Čerenkov radiation more quantitatively, let us note that, in an arbitrary medium, the vector potential, in the

Lorentz gauge, is given by (This follows from the Maxwell's equations and so far we had been considering the medium to be vacuum, for which $\epsilon = \mu = 1$.)

$$A_\mu(\mathbf{x}, t) = \frac{\mu}{c'} \int d^3x' \left. \frac{J_\mu(\mathbf{x}', t')}{|\mathbf{x} - \mathbf{x}'|} \right|_{t'=t-\frac{|\mathbf{x}-\mathbf{x}'|}{c'}}$$

$$= \frac{\mu}{c'} \int d^3x' \left. \frac{J_\mu(\mathbf{x}', t')}{|\mathbf{x} - \mathbf{x}'|} \right|_{t'=t-\frac{n(\omega)|\mathbf{x}-\mathbf{x}'|}{c}}, \tag{10.106}$$

where the refractive index of the medium is defined to be

$$n(\omega) = \sqrt{\epsilon\mu} = \frac{c}{c'}. \tag{10.107}$$

Let us assume that we have an isotropic, non-magnetic medium (which is fairly general) for which $\mu = 1$ and, consequently,

$$n(\omega) = \sqrt{\epsilon} = \frac{c}{c'}. \tag{10.108}$$

As we have discussed earlier, in the Lorentz gauge, the scalar and the vector potentials are related to each other so that it is sufficient to study only the vector potential. If we have a charged particle in uniform motion, we know that we can write the current as

$$\mathbf{J}(\mathbf{x}, t) = q\mathbf{v}\,\delta^3(\mathbf{x} - \boldsymbol{\xi}(t)), \tag{10.109}$$

where $\boldsymbol{\xi}(t)$ represents the trajectory of the charged particle. Using this as well as our earlier observations, we can write

$$\mathbf{A}(\mathbf{x}, t) = \frac{1}{c'} \int dt'\, d^3x'\, \frac{q\mathbf{v}\,\delta^3(\mathbf{x}' - \boldsymbol{\xi}(t'))\delta\left(t' - t + \frac{|\mathbf{x}-\mathbf{x}'|}{c'}\right)}{|\mathbf{x} - \mathbf{x}'|}$$

$$= \frac{q\mathbf{v}}{c'} \int dt'\, \frac{\delta\left(t' - t + \frac{|\mathbf{x}-\boldsymbol{\xi}(t')|}{c'}\right)}{|\mathbf{x} - \boldsymbol{\xi}(t')|}$$

$$= \frac{q}{c'} \left. \frac{\mathbf{v}}{R\left|1 - \frac{\hat{\mathbf{R}}\cdot\mathbf{v}}{c'}\right|} \right|_{t'=t-\frac{R}{c'}}, \tag{10.110}$$

where we have defined as before, $\mathbf{R} = \mathbf{x} - \boldsymbol{\xi}$. Equation (10.110) is exactly like the earlier cases (see (10.9)) except that we have the modulus of s in the denominator, which is not necessary when $v < c'$ (for which s is positive definite).

For uniform linear motion, the retarded time can be determined as before. In fact, following the discussion up to (10.30) we see that, in this case, we can write

$$c'(t' - t) = -\frac{\overline{R}}{(1 - \beta'^2)} \left[\beta' \cos\theta \mp \sqrt{(\beta' \cos\theta)^2 + (1 - \beta'^2)}\right].$$

$$(10.111)$$

Here θ is the angle between $\overline{\mathbf{R}} = \mathbf{R}(t)$ and $\mathbf{v}$ (see Fig. 10.2). For t' to represent a retarded time, as we have argued before, only the second root is allowed when $\beta' < 1$. As we will see now, for $\beta' > 1$, both the roots are allowed leading to two retarded times for a given t.

For $\beta' > 1$, we note from (10.111) that real roots will exist only if

$$(\beta' \cos\theta)^2 + (1 - \beta'^2) > 0$$

$$\text{or,} \quad \cos^2\theta > \left(1 - \frac{1}{\beta'^2}\right), \qquad (10.112)$$

which leads to the two solutions

$$\cos\theta > \left(1 - \frac{1}{\beta'^2}\right)^{\frac{1}{2}}, \quad \text{or,} \quad \cos\theta < -\left(1 - \frac{1}{\beta'^2}\right)^{\frac{1}{2}}. \quad (10.113)$$

On the other hand, for t' to represent a retarded time, we see from (10.111) that $\cos\theta$ must be negative (because of the factor in the denominator). Therefore, we choose θ to be an obtuse angle and note from (10.113) that the allowed root corresponds to

$$\arccos\left(-\left(1 - \frac{1}{\beta'^2}\right)^{\frac{1}{2}}\right) < \theta < \pi. \qquad (10.114)$$

In this case, it is easy to see from (10.111) that both the solutions lead to retarded times and provide the two retarded times for this problem.

Even though the retarded times are determined (and now the contributions from both these solutions must be added in (10.110)), evaluating the fields is extremely complicated. In what follows, we will present an alternate derivation of the fields as well as the energy radiated, using the method of Fourier transforms, which also brings out some other interesting features. To keep the discussion parallel to the earlier discussion of uniform linear motion in vacuum, let us choose the charged particle to be traveling along the x-axis. In this case, therefore, the current has a non-vanishing component only along the x-axis given by

$$J_x(\mathbf{x}, t) = qv\, \delta(x - vt)\delta(y)\delta(z). \qquad (10.115)$$

Taking the Fourier transform of this (into the energy-momentum space), we have

$$J_x(\mathbf{k}, \omega) = \frac{1}{(2\pi)^4} \int dt \, d^3x \, J_x(\mathbf{x}, t) \, e^{i(\omega t - \mathbf{k} \cdot \mathbf{x})}$$

$$= \frac{qv}{(2\pi)^4} \int dt \, dx \, \delta(x - vt) \, e^{i(\omega t - k_x x)}$$

$$= \frac{q}{(2\pi)^4} \int dx \, e^{i(\frac{\omega}{v} - k_x)x}$$

$$= \frac{q}{(2\pi)^3} \, \delta\left(k_x - \frac{\omega}{v}\right). \tag{10.116}$$

In the Lorentz gauge, Maxwell's equations (in the presence of currents) in an arbitrary medium (with $\mu = 1$) lead to the inhomogeneous wave equation (recall that in a medium $\partial_\mu = (\frac{1}{c'}\frac{\partial}{\partial t}, \boldsymbol{\nabla})$)

$$\left(\frac{1}{c'^2}\frac{\partial^2}{\partial t^2} - \boldsymbol{\nabla}^2\right) \mathbf{A}(\mathbf{x}, t) = \frac{4\pi}{c'} \mathbf{J}(\mathbf{x}, t), \tag{10.117}$$

which in the Fourier transformed space becomes

$$\left(-\frac{\omega^2}{c'^2} + \mathbf{k}^2\right) \mathbf{A}(\mathbf{k}, \omega) = \frac{4\pi}{c'} \mathbf{J}(\mathbf{k}, \omega),$$

$$\text{or,} \quad \mathbf{A}(\mathbf{k}, \omega) = \frac{4\pi}{c'} \frac{\mathbf{J}(\mathbf{k}, \omega)}{\mathbf{k}^2 - \frac{\omega^2}{c'^2}}. \tag{10.118}$$

It is now straightforward to obtain

$$\mathbf{A}(\mathbf{x}, \omega) = \frac{4\pi}{c'} \int d^3k \, \frac{\mathbf{J}(\mathbf{k}, \omega)}{\mathbf{k}^2 - \frac{\omega^2}{c'^2}} \, e^{i\mathbf{k}\cdot\mathbf{x}}. \tag{10.119}$$

The magnetic field, at any point, can now be obtained as

$$\mathbf{B}(\mathbf{x}, \omega) = \boldsymbol{\nabla} \times \mathbf{A}(\mathbf{x}, \omega)$$

$$= \frac{4\pi i}{c'} \int d^3k \, \frac{\mathbf{k} \times \mathbf{J}(\mathbf{k}, \omega)}{\mathbf{k}^2 - \frac{\omega^2}{c'^2}} \, e^{i\mathbf{k}\cdot\mathbf{x}}. \tag{10.120}$$

Putting in the form of the current in (10.116), we see that the magnetic field is orthogonal to the direction of motion of the charged particle (namely, $B_x = 0$) and has the form

$$\mathbf{B}(\mathbf{x}, \omega) = \frac{iq\, e^{\frac{i\omega x}{v}}}{2\pi^2 c'} \int dk_y\, dk_z\, \frac{(k_z \hat{\mathbf{e}}_y - k_y \hat{\mathbf{e}}_z)}{k_y^2 + k_z^2 - \frac{\omega^2}{c'^2}\left(1 - \frac{1}{\beta'^2}\right)}\, e^{i(k_y y + k_z z)}.$$

$$(10.121)$$

The integral over k_z in (10.121) can be carried out using the method of residues. We note that the integrand has poles at

$$k_* = k_z = \pm \sqrt{\frac{\omega^2}{c'^2}\left(1 - \frac{1}{\beta'^2}\right) - k_y^2} = \pm \sqrt{\omega^2 \left(\frac{1}{c'^2} - \frac{1}{v^2}\right) - k_y^2}.$$

$$(10.122)$$

If $\beta' < 1$ $(v < c')$, we note that the quantity inside the square root is negative and, therefore, the poles are on the imaginary axis. Enclosing the contour in the upper half plane would pick up the contribution of the pole on the positive imaginary axis and it is easy to check that the integrals, in this case, lead to fields of the forms discussed in (10.47) and (10.48).

On the other hand, if $\beta' > 1$ $(v > c')$, then, the quantity inside the square root in (10.122) can be positive for some values of k_y, while it will be negative for other values. When k_y is such that the quantity is negative, the poles will lie on the imaginary axis and the integration can be carried out, much like in the earlier case, by enclosing the contour in the upper half plane. For values of k_y for which the quantity inside the square root is positive, the poles will lie on the real axis. However, since we are interested in retarded solutions, as discussed in (6.179), the proper prescription is obtained by letting (this gives the pole prescription for the retarded Green's function or solution)

$$\omega \to \omega + i\epsilon,$$

whose effect, in the present case, is to push the pole at the positive value of k_z to the upper half plane while pushing the other pole at the negative value to the lower half plane. Thus, enclosing the contour in the upper half plane would pick up two contributions – one from the pole on the positive imaginary axis and the other from the pole slightly above the positive real axis. This is the basic difference from the case $\beta' < 1$ and is a reflection of the existence of two retarded times in this language. Denoting by k_* the generic location of the poles, we obtain from (10.121)

$$B_y(\mathbf{x}, \omega) = \frac{iq\, e^{\frac{i\omega x}{v}}}{2\pi^2 c'} (2\pi i) \int dk_y\, \frac{k_*}{2k_*}\, e^{i(k_y y + k_* z)}$$

$$= -\frac{q\, e^{\frac{i\omega x}{v}}}{2\pi c'} \int dk_y\, e^{i(k_y y + k_* z)},$$

$$B_z(\mathbf{x}, \omega) = -\frac{iq\, e^{\frac{i\omega x}{v}}}{2\pi^2 c'} (2\pi i) \int dk_y\, \frac{k_y}{2k_*}\, e^{i(k_y y + k_* z)}$$

$$= \frac{q\, e^{\frac{i\omega x}{v}}}{2\pi c'} \int dk_y\, \frac{k_y}{k_*}\, e^{i(k_y y + k_* z)}. \tag{10.123}$$

The electric fields can now be calculated from Maxwell's equations, which implies that, away from the trajectory of the charged particle, we have (recall that we are assuming $\mu = 1$)

$$\boldsymbol{\nabla} \times \mathbf{B}(\mathbf{x}, t) = \frac{\epsilon}{c'} \frac{\partial \mathbf{E}(\mathbf{x}, t)}{\partial t}, \tag{10.124}$$

leading to

$$\mathbf{E}(\mathbf{x}, \omega) = \frac{ic'}{\omega \epsilon} \boldsymbol{\nabla} \times \mathbf{B}(\mathbf{x}, \omega). \tag{10.125}$$

It follows now in a straightforward manner from (10.123) that

$$E_x(\mathbf{x}, \omega) = \frac{ic'}{\omega \epsilon} \left(\frac{\partial B_z}{\partial y} - \frac{\partial B_y}{\partial z} \right)$$

$$= -\frac{q\omega\, e^{\frac{i\omega x}{v}}}{2\pi c'^2} \left(1 - \frac{c'^2}{v^2} \right) \int dk_y\, \frac{e^{i(k_y y + k_* z)}}{k_*},$$

$$E_y(\mathbf{x}, \omega) = \frac{ic'}{\omega \epsilon} \left(\frac{\partial B_x}{\partial z} - \frac{\partial B_z}{\partial x} \right)$$

$$= \frac{c'}{v\epsilon} B_z,$$

$$E_z(\mathbf{x}, \omega) = \frac{ic'}{\omega \epsilon} \left(\frac{\partial B_y}{\partial x} - \frac{\partial B_x}{\partial y} \right)$$

$$= -\frac{c'}{v\epsilon} B_y. \tag{10.126}$$

Equations (10.123) and (10.126) give the electric and the magnetic fields produced by an unaccelerated charged particle moving in a medium with a speed larger than the speed of light in the medium. To see that there is radiation, in such a case, let us calculate the energy radiated through the conical surface along the $\hat{\mathbf{z}}$ direction as the particle travels a distance d along the trajectory. Let us consider a surface parallel to the $x - y$ plane defined by $z = z_0$. Furthermore, let us consider a strip of the surface of width d along the x-axis ($x_0 \leq x \leq x_0 + d$). Then, the total energy radiated through this strip (considering contributions from both the surfaces at $\pm z_0$, as we will see shortly, the contribution is independent of the value of z_0) is given by

$$\mathcal{E} = 2\operatorname{Re} \int\limits_{-\infty}^{\infty} dt \int\limits_{-\infty}^{\infty} dy \int\limits_{x_0}^{x_0+d} dx \, \frac{c'}{4\pi} \, \hat{\mathbf{z}} \cdot (\mathbf{E}(\mathbf{x}, t) \times \mathbf{H}^*(\mathbf{x}, t))$$

$$= \operatorname{Re} \int\limits_{-\infty}^{\infty} dt \int\limits_{-\infty}^{\infty} dy \int\limits_{x_0}^{x_0+d} dx \, \frac{c'}{2\pi} \, E_x(\mathbf{x}, t) B_y^*(\mathbf{x}, t)$$

$$= 4\pi \operatorname{Re} \int\limits_{0}^{\infty} d\omega \int\limits_{-\infty}^{\infty} dy \int\limits_{x_0}^{x_0+d} dx \, \frac{c'}{2\pi} \, E_x(\mathbf{x}, \omega) B_y^*(\mathbf{x}, \omega)$$

$$= 2\operatorname{Re} \int\limits_{0}^{\infty} d\omega \int\limits_{-\infty}^{\infty} dy \int\limits_{x_0}^{x_0+d} dx \, c' \, E_x(\mathbf{x}, \omega) B_y^*(\mathbf{x}, \omega), \qquad (10.127)$$

where we have used the fact that $\mathbf{B} = \mathbf{H}$ when $\mu = 1$ as well as the fact that $B_x = 0$.

From the forms of the fields in (10.123) and (10.126), we see that since both E_x, B_y have the same x dependence in the phase, the integrand in (10.127) is independent of x and, consequently, the x integration is trivial leading to a factor of d. Furthermore, defining

$$k_* = \sqrt{\frac{\omega^2}{c'^2}\left(1 - \frac{1}{\beta'^2}\right) - k_y^2},$$

$$k'_* = \sqrt{\frac{\omega^2}{c'^2}\left(1 - \frac{1}{\beta'^2}\right) - k_y'^2}, \qquad (10.128)$$

we note from (10.127) that we can write

$$\mathcal{E} = 2d \operatorname{Re} \int\limits_{0}^{\infty} d\omega \int\limits_{-\infty}^{\infty} dy \, c' \, E_x(\mathbf{x}, \omega) B_y^*(\mathbf{x}, \omega)$$

$$= 2d \left(-\frac{q}{2\pi} \right)^2 \text{Re} \int\limits_0^\infty d\omega \int\limits_{-\infty}^\infty dy \int\limits_{-\infty}^\infty dk_y \int\limits_{-\infty}^\infty dk_y'$$

$$\times \frac{\omega}{c'^2} \left(1 - \frac{c'^2}{v^2} \right) \frac{e^{i(k_y - k_y')y + i(k_* - k_*')z_0}}{k_*}$$

$$= \frac{q^2 d}{\pi} \text{Re} \int\limits_0^\infty d\omega \int\limits_{-\infty}^\infty dk_y \int\limits_{-\infty}^\infty dk_y' \frac{\omega}{c'^2} \left(1 - \frac{c'^2}{v^2} \right)$$

$$\times \frac{\delta \left(k_y - k_y' \right) e^{i(k_* - k_*')z_0}}{k_*}$$

$$= \frac{q^2 d}{\pi} \text{Re} \int\limits_0^\infty d\omega \frac{\omega}{c'^2} \left(1 - \frac{c'^2}{v^2} \right) \int\limits_{-\infty}^\infty dk_y \frac{1}{k_*}. \tag{10.129}$$

As noted earlier, this expression is independent of z_0.

The k_y integration can be done in a trivial manner. We note from the definition in (10.128) that k_* is imaginary for

$$|k_y| > k_0 = \frac{\omega}{c'} \left(1 - \frac{1}{\beta'^2} \right)^{\frac{1}{2}}. \tag{10.130}$$

Therefore, we can cutoff the integral over k_y at these limiting values and obtain

$$\int\limits_{-k_0}^{k_0} \frac{dk_y}{\sqrt{k_0^2 - k_y^2}} = \pi, \tag{10.131}$$

where we have used standard results from the table of integrals (see, for example, Gradshteyn and Ryzhik, 2.274 or this can also be done by elementary methods). With this, we obtain

$$\mathcal{E} = q^2 d \int\limits_0^\infty d\omega \frac{\omega}{c'^2} \left(1 - \frac{c'^2}{v^2} \right)$$

$$= q^2 d \int\limits_0^\infty d\omega \, \omega \left(\frac{1}{c'^2} - \frac{1}{v^2} \right). \tag{10.132}$$

We can also define the energy radiated per unit length as

$$W = \frac{\mathcal{E}}{d} = q^2 \int\limits_0^\infty d\omega \, \omega \left(\frac{1}{c'^2} - \frac{1}{v^2} \right). \tag{10.133}$$

This can be thought of as the energy radiated per unit length of the trajectory of the charged particle. Thus, we see that, even though the charged particle is not being accelerated, by virtue of the fact that it travels faster than the speed of light in the medium, it leads to emission of radiation. This effect is known as the Čerenkov radiation (effect). (Equation (10.133) shows that if $v > c'$, energy is radiated since $\mathcal{E} > 0$. On the other hand, if $v < c'$, no energy is radiated since $\mathcal{E} < 0$.)

10.6 Self-force

So far, we have talked about a point charge which is an idealization. In fact, classically, we know that a point charge leads to singular field configurations (and energy) and, therefore, a better way is to think of a classical charged particle as an extended object of small dimensions with a charge distribution. In electrostatics, for example, one calculates the fields and energy of such a system by assuming a spherical charge distribution of a small radius. Such a description of a classical charged particle leads to interesting effects when the particle is not at rest. As we have seen, for a moving charged particle, the fields at any point are determined by the retarded time (and not by the instantaneous time). If the particle has an extension, then, of course, each element of the object will exert a force on every other element. If the particle is not moving with a uniform velocity, the retarded times associated with the action (force) will be different from that of the reaction (force). As a result, there will be a net force acting on the charged particle, commonly known as the "self-force" which will be proportional to the acceleration of the particle. Intuitively, we can see that something like this should happen from the following simple argument. As we have seen, an accelerated charged particle radiates energy and thereby loses kinetic energy. Therefore, radiation of energy leads to a deceleration implying that there must be a force acting on the particle due to the fields it produces. This is the "self-force" or sometimes also known as the radiation reaction.

The self-force was first studied by Lorentz and was later generalized by Abraham. The derivation of the self-force is quite technical, but let us discuss it within a simple context. Let us assume that the charged particle consists of a charge distribution of small dimension d (in the limiting case of a point particle, we can take this to be zero). We will also assume that we are in a frame where the particle is instantaneously at rest. As before, we will express quantities at the instantaneous time (observation time) with a bar over them. Since the size of the charge distribution is small, in calculating the effect that one element of the charge distribution produces on another, we can Taylor expand the terms in powers of the size (note that both $\mathbf{x}$ and $\boldsymbol{\xi}$ denote points inside the charge distribution whose dimension is small). For example, we note that we can write

$$\mathbf{R}(t') = \overline{\mathbf{R}} + (t' - t)\left.\frac{\mathrm{d}\mathbf{R}(t')}{\mathrm{d}t'}\right|_{t'=t} + \frac{1}{2!}(t'-t)^2\left.\frac{\mathrm{d}^2\mathbf{R}(t')}{\mathrm{d}t'^2}\right|_{t'=t}$$

$$+ \frac{1}{3!}(t'-t)^3\left.\frac{\mathrm{d}^3\mathbf{R}(t')}{\mathrm{d}t'^3}\right|_{t'=t} + \cdots$$

$$= \overline{\mathbf{R}} + (t'-t)(-\overline{\mathbf{v}}) + \frac{(t'-t)^2}{2}(-\dot{\overline{\mathbf{v}}})$$

$$+ \frac{(t'-t)^3}{6}(-\ddot{\overline{\mathbf{v}}}) + \cdots, \tag{10.134}$$

where $\mathbf{R}(t') = \mathbf{x} - \boldsymbol{\xi}(t')$. Let us recall that the retarded time is defined as

$$t' = t - \frac{R(t')}{c}. \tag{10.135}$$

Using this as well as the fact that the particle is instantaneously at rest ($\overline{\mathbf{v}} = 0$), we obtain, from (10.134),

$$\mathbf{R}(t') = \overline{\mathbf{R}} - \frac{R^2(t')}{2c^2}\dot{\overline{\mathbf{v}}} + \frac{R^3(t')}{6c^3}\ddot{\overline{\mathbf{v}}} + \cdots. \tag{10.136}$$

We are assuming here as before that $\frac{R}{c}$ (and, therefore, $t'-t$) is small and, consequently, a Taylor expansion as in (10.136) is meaningful. Furthermore, we keep expansions up to order R^3 because, as we will see shortly, that is sufficient for our purposes.

Equation (10.136) leads to

$$R^2(t') = \overline{R}^2 - \frac{R^2(t')}{c^2}\overline{\mathbf{R}}\cdot\dot{\overline{\mathbf{v}}} + \frac{R^3(t')}{3c^3}\overline{\mathbf{R}}\cdot\ddot{\overline{\mathbf{v}}} + \frac{R^4(t')}{4c^4}\dot{\overline{\mathbf{v}}}\cdot\dot{\overline{\mathbf{v}}} + \cdots, \tag{10.137}$$

which can be rearranged to the form

$$R^2(t')\left(1 + \frac{\overline{\mathbf{R}}\cdot\dot{\overline{\mathbf{v}}}}{c^2}\right) \approx \overline{R}^2 + \frac{R^3(t')\overline{\mathbf{R}}\cdot\ddot{\overline{\mathbf{v}}}}{3c^3} + \frac{R^4(t')\dot{\overline{\mathbf{v}}}\cdot\dot{\overline{\mathbf{v}}}}{4c^4},$$

$$\text{or, } R(t') \approx \left(1 + \frac{\overline{\mathbf{R}}\cdot\dot{\overline{\mathbf{v}}}}{c^2}\right)^{-\frac{1}{2}}\left(\overline{R}^2 + \frac{R^3(t')\overline{\mathbf{R}}\cdot\ddot{\overline{\mathbf{v}}}}{3c^3} + \frac{R^4(t')\dot{\overline{\mathbf{v}}}\cdot\dot{\overline{\mathbf{v}}}}{4c^4}\right)^{\frac{1}{2}}. \tag{10.138}$$

Iterating (10.138) and keeping terms to order $\overline{R}^3$, we obtain

$$R(t') \approx \overline{R}\left(1 - \frac{\mathbf{R}\cdot\bar{\mathbf{v}}}{2c^2} + \frac{3(\mathbf{R}\cdot\bar{\mathbf{v}})^2}{8c^4} + \frac{\overline{R}(\mathbf{R}\cdot\bar{\bar{\mathbf{v}}})}{6c^3} + \frac{\overline{R}^2(\bar{\mathbf{v}}\cdot\bar{\mathbf{v}})}{8c^4}\right).$$

(10.139)

Using this, we can rewrite (10.136) to leading orders as

$$\mathbf{R}(t') = \overline{\mathbf{R}} - \frac{\overline{R}^2\bar{\mathbf{v}}}{2c^2} + \frac{\overline{R}^2(\mathbf{R}\cdot\bar{\mathbf{v}})\bar{\mathbf{v}}}{2c^4} + \frac{\overline{R}^3\bar{\bar{\mathbf{v}}}}{6c^3} + \cdots.$$

(10.140)

In a similar manner, we can also Taylor expand (note that $\bar{\mathbf{v}} = 0$)

$$\mathbf{v}(t') = \bar{\mathbf{v}} + (t'-t)\,\dot{\bar{\mathbf{v}}} + \frac{(t'-t)^2}{2!}\,\ddot{\bar{\mathbf{v}}} + \cdots$$

$$\approx -\frac{\overline{R}\bar{\mathbf{v}}}{c} + \frac{\overline{R}(\mathbf{R}\cdot\bar{\mathbf{v}})\bar{\mathbf{v}}}{2c^3} + \frac{\overline{R}^2\bar{\bar{\mathbf{v}}}}{2c^2}$$

$$= \frac{\overline{R}}{c}\left(-\bar{\mathbf{v}} + \frac{(\mathbf{R}\cdot\bar{\mathbf{v}})\bar{\mathbf{v}}}{2c^2} + \frac{\overline{R}\bar{\bar{\mathbf{v}}}}{2c}\right),$$

$$\dot{\mathbf{v}}(t') \approx \bar{\mathbf{v}} + (t'-t)\bar{\bar{\mathbf{v}}} = \bar{\mathbf{v}} - \frac{\overline{R}}{c}\,\bar{\bar{\mathbf{v}}}.$$

(10.141)

As we have seen in (10.19), the electric field of a point charge can be written as

$$\mathbf{E}(\mathbf{x},t) = q\left[\frac{\left(\mathbf{R} - \frac{R\mathbf{v}}{c}\right)\left(1-(\frac{\mathbf{v}}{c})^2\right)}{(sR)^3} + \frac{\mathbf{R}\times\left(\left(\mathbf{R} - \frac{R\mathbf{v}}{c}\right)\times\frac{\dot{\mathbf{v}}}{c}\right)}{c(sR)^3}\right]_{t'=B(t)}.$$

(10.142)

Thus, using (10.139)-(10.141), we can expand

$$s(t')R(t') = R(t') - \frac{\mathbf{R}(t')\cdot\mathbf{v}(t')}{c}$$

$$= \overline{R}\left[1 + \frac{\mathbf{R}\cdot\bar{\mathbf{v}}}{2c^2} - \frac{(\mathbf{R}\cdot\bar{\mathbf{v}})^2}{8c^4} - \frac{\overline{R}(\mathbf{R}\cdot\bar{\bar{\mathbf{v}}})}{3c^3} - \frac{3\overline{R}^2(\bar{\mathbf{v}}\cdot\bar{\mathbf{v}})}{8c^4} + \cdots\right],$$

(10.143)

and, similarly,

$$\mathbf{R}(t') - \frac{R(t')\mathbf{v}(t')}{c} = \overline{\mathbf{R}} + \frac{\overline{R}^2\bar{\mathbf{v}}}{2c^2} - \frac{\overline{R}^3\bar{\bar{\mathbf{v}}}}{3c^3} - \frac{\overline{R}^2(\mathbf{R}\cdot\bar{\mathbf{v}})\bar{\mathbf{v}}}{2c^4} + \cdots,$$

$$\left(1 - (\frac{\mathbf{v}(t')}{c})^2\right) = 1 - \frac{\overline{R}^2(\bar{\mathbf{v}}\cdot\bar{\mathbf{v}})}{c^4} + \cdots.$$

(10.144)

The reason for keeping terms up to order $\overline{R}^3$ in the expansion is now clear. Since $sR \sim \overline{R}$ and $(sR)^3$ occurs in the denominator (in both the terms), the dominant terms would come from terms up to order $\overline{R}^3$ in the numerator.

Using (10.143) and (10.144), we now obtain

$$
\left. \frac{\left(\mathbf{R} - \frac{R\mathbf{v}}{c}\right)\left(1 - \left(\frac{v}{c}\right)^2\right)}{(sR)^3} \right| = \frac{1}{\overline{R}^3}\left[\mathbf{R}\left(1 - \frac{3(\overline{\mathbf{R}} \cdot \dot{\mathbf{v}})}{2c^2} + \frac{\overline{R}^2(\dot{\mathbf{v}} \cdot \dot{\mathbf{v}})}{8c^4} \right.\right.
$$
$$
\left. + \frac{15(\overline{\mathbf{R}} \cdot \dot{\mathbf{v}})^2}{8c^4} + \frac{\overline{R}(\overline{\mathbf{R}} \cdot \ddot{\mathbf{v}})}{c^3}\right)
$$
$$
\left. + \dot{\mathbf{v}}\left(\frac{\overline{R}^2}{2c^2} - \frac{5\overline{R}^2(\overline{\mathbf{R}} \cdot \dot{\mathbf{v}})}{4c^4}\right) - \frac{\overline{R}^3}{3c^3}\ddot{\mathbf{v}} + \cdots \right],
$$

$$
\left. \frac{\mathbf{R} \times \left(\left(\mathbf{R} - \frac{R\mathbf{v}}{c}\right) \times \frac{\dot{\mathbf{v}}}{c}\right)}{c(sR)^3} \right| = \frac{1}{c\overline{R}^3}\left[\mathbf{R}\left((\overline{\mathbf{R}} \cdot \dot{\mathbf{v}}) - \frac{\overline{R}(\overline{\mathbf{R}} \cdot \ddot{\mathbf{v}})}{c}\right.\right.
$$
$$
\left. - \frac{\overline{R}^2(\dot{\mathbf{v}} \cdot \dot{\mathbf{v}})}{2c^2} - \frac{3(\overline{\mathbf{R}} \cdot \dot{\mathbf{v}})^2}{2c^2}\right)
$$
$$
\left. + \dot{\mathbf{v}}\left(-\overline{R}^2 + \frac{2\overline{R}^2(\overline{\mathbf{R}} \cdot \dot{\mathbf{v}})}{c^2}\right) + \frac{\overline{R}^3}{c}\ddot{\mathbf{v}} + \cdots \right].
$$
$$
\tag{10.145}
$$

If we assume the charge density $\rho(\boldsymbol{\xi})$ of the extended particle to be spherically symmetric, then the electric field produced at $\mathbf{x}$ (inside the particle) due to all the other charge elements of the particle would be given by (see (10.142))

$$
\mathbf{E}(\mathbf{x}, t) = \int d^3\xi\, \rho(\boldsymbol{\xi}) \left[\frac{\left(\mathbf{R} - \frac{R\mathbf{v}}{c}\right)\left(1 - \left(\frac{v}{c}\right)^2\right)}{(sR)^3} \right.
$$
$$
\left. + \frac{\mathbf{R} \times \left(\left(\mathbf{R} - \frac{R\mathbf{v}}{c}\right) \times \frac{\dot{\mathbf{v}}}{c}\right)}{c(sR)^3} \right]_{t'=B(t)}. \tag{10.146}
$$

Thus, the self-force acting on the particle is obtained to be

$$
\mathbf{F}_{\text{self}} = \int d^3x\, \rho(\mathbf{x})\, \mathbf{E}(\mathbf{x}, t)
$$

$$= \int d^3x \, d^3\xi \, \rho(\mathbf{x})\rho(\boldsymbol{\xi}) \left[\frac{\left(\mathbf{R} - \frac{R\mathbf{v}}{c}\right)\left(1 - \left(\frac{\mathbf{v}}{c}\right)^2\right)}{(sR)^3} \right.$$

$$\left. + \frac{\mathbf{R} \times \left(\left(\mathbf{R} - \frac{R\mathbf{v}}{c}\right) \times \frac{\dot{\mathbf{v}}}{c}\right)}{c(sR)^3} \right]\Bigg|. \qquad (10.147)$$

We can now use the expansions in (10.145) and note that, for spherically symmetric charge distributions, the terms in the integrand which are odd in $\overline{\mathbf{R}}$ would vanish (recall that $\mathbf{R} = \mathbf{x} - \boldsymbol{\xi}$). As a result, we obtain

$$\mathbf{F}_{\text{self}} = \int d^3x \, d^3\xi \, \rho(\mathbf{x}) \, \rho(\boldsymbol{\xi}) \left[-\frac{(\overline{\mathbf{R}} \cdot \dot{\overline{\mathbf{v}}})\overline{\mathbf{R}}}{2c^2\overline{R}^3} - \frac{\dot{\overline{\mathbf{v}}}}{2c^2\overline{R}} + \frac{2\ddot{\overline{\mathbf{v}}}}{3c^3} \right]$$

$$= -\frac{2\dot{\overline{\mathbf{v}}}}{3c^2} \int d^3x \, d^3\xi \, \frac{\rho(\mathbf{x}) \, \rho(\boldsymbol{\xi})}{\overline{R}} + \frac{2q^2}{3c^3} \ddot{\overline{\mathbf{v}}}$$

$$= -\frac{4U}{3c^2} \dot{\overline{\mathbf{v}}} + \frac{2q^2}{3c^3} \ddot{\overline{\mathbf{v}}}. \qquad (10.148)$$

Here, we have used symmetric integration in the intermediate steps and have identified the self-energy of the system as

$$U = \frac{1}{2} \int d^3x \, d^3\xi \, \frac{\rho(\mathbf{x}) \, \rho(\boldsymbol{\xi})}{\overline{R}} = \frac{1}{2} \int d^3x \, d^3\xi \, \frac{\rho(\mathbf{x}) \, \rho(\boldsymbol{\xi})}{|\mathbf{x} - \boldsymbol{\xi}|}. \qquad (10.149)$$

In his studies of the forces acting on a charged particle, Lorentz had already argued that, when a particle is accelerated, there must be other forces acting on the particle besides the usual "Lorentz" force. In fact, he had already studied as an additional force precisely the second term on the right hand side of (10.148). This was further generalized by Abraham, following the works of Larmour, Heaviside and others, and correspondingly, the self-force in (10.148) is also known as the Abraham-Lorentz force. If there is an external force acting on an electron, then, together with the self-force, the equation of motion for an electron can be written as

$$m_{\mathrm{I}}\dot{\mathbf{v}} = \mathbf{F}_{\text{ext}} + \mathbf{F}_{\text{self}} = \mathbf{F}_{\text{ext}} - \frac{4U}{3c^2} \dot{\mathbf{v}} + \frac{2e^2}{3c^3} \ddot{\mathbf{v}},$$

$$\text{or,} \quad (m_{\mathrm{I}} + m_{\text{em}}) \, \dot{\mathbf{v}} = \mathbf{F}_{\text{ext}} + \frac{2e^2}{3c^3} \ddot{\mathbf{v}},$$

$$\text{or,} \quad \dot{\mathbf{v}} = \frac{1}{m} \mathbf{F}_{\text{ext}} + \frac{2e^2}{3mc^3} \ddot{\mathbf{v}}, \qquad (10.150)$$

where we have identified, for simplicity, $\bar{\mathbf{v}} = \dot{\mathbf{v}}, \bar{\dot{\mathbf{v}}} = \ddot{\mathbf{v}}$ and have defined

$$m_{\text{em}} = \frac{4U}{3c^2}, \qquad m = m_{\text{I}} + m_{\text{em}}. \tag{10.151}$$

We can think of m_{I} as the inertial mass of the electron while m_{em} can be thought of as the electromagnetic mass of the electron (charged particle). Neither is individually observable. Experimentally, one can measure only the observable mass m.

In deriving Eq. (10.148), we have neglected higher order terms which would vanish in the limit that the particle has no structure. However, in that limit, the self-energy of the electron (particle) in (10.149) and, therefore, the electromagnetic mass diverges. In a quantum theory, such a phenomenon is handled through renormalization. Classically we can think of the electron not as a point particle, rather as one with a structure of the size of about (experiments put an upper bound of 10^{-17} cm on the size of the electron)

$$r_{\text{e}} = \frac{e^2}{mc^2} \approx 3 \times 10^{-13} \text{ cm.} \tag{10.152}$$

This is also known as the Lorentz radius or the Thomson scattering length. Consequently, the self-energy is finite. Furthermore, this distance scale also defines a time scale

$$\tau = \frac{2r_{\text{e}}}{3c} = \frac{2e^2}{3mc^3} \approx 10^{-24} \text{ sec,} \tag{10.153}$$

where we have put in the factor of $\frac{2}{3}$ in the definition of the time scale for later convenience (see also the second term on the right hand side of (10.150)). This time scale is tiny showing that the expansion used is convergent. Moreover, since such a time scale is in the domain of quantum mechanics, we recognize that we can, at best, think of the classical equation in (10.150) as an approximate equation. If not, Eq. (10.150) leads to conceptual problems. For, suppose there is no external force present, namely, $\mathbf{F}_{\text{ext}} = 0$, then we see that

$$\dot{\mathbf{v}} = \tau \ddot{\mathbf{v}}$$

$$\text{or,} \quad \mathbf{v}(t) = \mathbf{v}(0) \, e^{\frac{t}{\tau}} + \text{constant.} \tag{10.154}$$

Namely, we have a velocity that grows exponentially with time implying that the particle self-accelerates to infinite velocities.

The problem with this run away solution can be fixed if we assume that the self-force exists only in the presence of other external

forces and that the radiation loss due to the self-force is small compared with the energy of the particle. We note that these assumptions
are quite reasonable. First of all, the self-force as we see from (10.148)
is proportional to the acceleration (as well as higher derivative terms)
and, consequently, cannot exist in the absence of an external force.
We also note that a charged particle of finite size cannot be in stable
equilibrium unless external forces are applied. Therefore, the first
assumption is quite reasonable. Since τ is very small, the energy loss
due to radiation in any finite amount of time can only be small. To
see this quantitatively, let us consider an one dimensional charged
oscillator in the presence of the self-force.

$$m\ddot{x} + m\omega_0^2 x = m\tau\dddot{v},$$

$$\text{or,} \quad \ddot{x} + \omega_0^2 x = \tau\dddot{v}, \tag{10.155}$$

where ω_0 is the natural frequency of the oscillator. This is a third
order equation in the time derivatives (recall that $v = \dot{x}$). Choosing
a solution of the form

$$x(t) = x(0)\, e^{-i\omega t}, \tag{10.156}$$

we obtain, from (10.155)

$$\omega^3\tau - i\omega^2 + i\omega_0^2 = 0. \tag{10.157}$$

This can also be rewritten as

$$(\omega\tau)^3 - i(\omega\tau)^2 + i(\omega_0\tau)^2 = 0. \tag{10.158}$$

We note that the natural dimensionless variables in this equation are
$\omega\tau$ and $\omega_0\tau$. Assuming that $\omega_0\tau \ll 1$, we can obtain the solution to
the cubic equation in (10.158) perturbatively as

$$\omega\tau = \pm\omega_0\tau - \frac{i}{2}(\omega_0\tau)^2 + O\left((\omega_0\tau)^3\right),$$

$$\omega\tau = i + i(\omega_0\tau)^2 + O\left((\omega_0\tau)^3\right). \tag{10.159}$$

We can discard the last solution as unphysical since it leads to an
exponentially growing solution. The other two solutions, on the other
hand, lead to

$$x(t) = x(0)\, e^{\mp i\omega_0 t}\, e^{-\frac{1}{2}\omega_0^2\tau t}, \qquad t > 0. \tag{10.160}$$

Both these solutions are exponentially damped. We recall that the
radiation component of the electric field is proportional to the acceleration $(\ddot{x})$ which is proportional to ω_0^2 and, consequently we expect

the radiation loss to be small. Furthermore, taking the Fourier transformation, we find

$$x(\omega) = \frac{1}{2\pi} \int_0^\infty dt \, e^{i\omega t} \, x(t)$$

$$= \frac{x(0)}{2\pi} \int_0^\infty dt \, e^{i\left(\omega \mp \omega_0 + \frac{i}{2}\omega_0^2 \tau\right)t}$$

$$= \frac{ix(0)}{2\pi \left((\omega \mp \omega_0) + \frac{i}{2}\omega_0^2 \tau\right)}. \tag{10.161}$$

This shows that the oscillation of the charged particle no longer consists of a sharp single frequency. The intensity of the oscillations is obtained to be

$$|x(\omega)|^2 = \frac{|x(0)|^2}{4\pi^2 \left((\omega \mp \omega_0)^2 + \frac{\omega_0^4 \tau^2}{4}\right)}. \tag{10.162}$$

This shows a resonant behavior which characterizes the broadening of the line width due to radiation reaction (or the self-force).

10.7 Selected problems

1. The coordinates of a particle with charge q, moving in the $x-y$ plane, depend on time as

$$x = x_0 \, e^{-\frac{t^2}{a^2}}, \qquad y = y_0 \, e^{-\frac{t^2}{b^2}},$$

where x_0, y_0, a, b are constants. Determine $\frac{dP}{d\Omega}$ for the radiated wave.

2. A particle moves along the z-axis with a time dependence

$$z = a \cos \omega t,$$

where a, ω are constants. Show that the angular distribution of the average radiated power is given by

$$\left\langle \frac{dP}{d\Omega} \right\rangle = \frac{q^2 c \beta^4 (4 + \beta^2 \cos^2 \theta) \sin^2 \theta}{32\pi^2 a^2 (1 - \beta^2 \cos^2 \theta)^{\frac{7}{2}}}.$$

3. Show that, for a charged particle with a linear acceleration a (namely, velocity is parallel to acceleration), the "time corrected" total radiated power is given by

$$P = \frac{2q^2a^2}{3c^3} \frac{1}{(1-\beta^2)^3}.$$

(What this means is that you should correct for the retarded time and define

$$\frac{dP}{d\Omega} \rightarrow \frac{dP}{d\Omega} \frac{dt'}{dt},$$

and integrate this to obtain the total "time corrected" power radiated.)

4. For a real function, $f(t)$, show from the definition of the Fourier transformation

$$f(\omega) = \frac{1}{2\pi} \int_{-\infty}^{\infty} dt\, e^{i\omega t}\, f(t),$$

that

$$f^*(-\omega) = f(\omega).$$

Using this, derive the relation used in this chapter that

$$\int_{-\infty}^{\infty} dt\, |f(t)|^2 = 4\pi \int_{0}^{\infty} d\omega\, |f(\omega)|^2.$$

5. Consider a transparent medium with an index of refraction $n = 1.5$ in the range of visible light. Calculate the angle for the emission of Čerenkov radiation by an electron moving with a speed $0.9c$. Determine the number of photons of wavelength in the interval $4000 - 6000$ Angstroms, emitted per unit length of the trajectory.

6. A classical relativistic theory of the electron (due to Dirac) describes the electron motion by

$$\frac{dp^\mu}{d\tau} = f^\mu_{\text{ext}} + f^\mu_{\text{self}},$$

where $p^\mu = mu^\mu$ with u^μ representing the four velocity of the particle. m, τ denote respectively the rest mass and the proper time of the electron. Similarly, $f^\mu_{\text{ext}}, f^\mu_{\text{self}}$ denote respectively the relativistic generalization of the external force and the self-force.

From the fact that a relativistic force has to satisfy $u_\mu f^\mu = 0$ (discussed in chapter **12**), show that

$$f^\mu_{\text{self}} = \frac{2e^2}{3mc^3} \left[\frac{\mathrm{d}^2 p^\mu}{\mathrm{d}\tau^2} - \frac{p^\mu}{m^2 c^2} \left(\frac{\mathrm{d}p^\nu}{\mathrm{d}\tau} \frac{\mathrm{d}p_\nu}{\mathrm{d}\tau} \right) \right].$$

CHAPTER 11

Plasma

11.1 General features of a plasma

Conventionally, a partially or a fully ionized gas is called a plasma. The ionosphere in our atmosphere and the ionized gas in a discharge tube such as a diode provide examples of a plasma. However, one can generalize the definition and think of a plasma as a medium consisting of (free) positive and negative charges such that in any arbitrary volume the total charge is zero. That is, a plasma defines a charge neutral medium. If we consider a plasma of ionized gas, say for example, then while both the electrons as well as the positively charged ions are free to move, by virtue of their large mass, the positive charges (ions) do not move very much. Thus, we can think of only the negatively charged electrons in a plasma to have appreciable motion. In some sense, therefore, we can think of the plasma as consisting of a large number of electrons moving freely in a positively charged background. This seems very much like the free electron theory of metals (in condensed matter physics). However, there are essential differences. First, the number density of electrons in a plasma (either in the laboratory or in nature) is much smaller than that in a metal. As a result, a classical description of a plasma leads to quite accurate results. Second, since the electron density is so dilute in a plasma, the effects of collision can truly be neglected.

A plasma, therefore, appears to be a highly conducting medium. However, as we will see, it has very different behavior from the conductors that we have studied so far. A plasma in equilibrium develops strong restoring forces when disturbed externally. For example, let us suppose that we displace the electrons in a block of an infinite plasma (for simplicity) by an infinitesimal distance x along the x-axis. This would then lead to two charged surfaces, one positively charged on the left and the other negatively charged on the right as shown in Fig. 11.1. Each of these surfaces will have equal, but opposite charge. For

Figure 11.1: The two charged surfaces which develop when the electrons are displaced to the right along the x-axis.

example, if we assume the number density of electrons in the plasma to be N (which will also represent the number density for positively charged ions), then the surface charge density on the two surfaces will have the magnitude Nex. (We are assuming that $e > 0$.) Each of these surfaces will produce an electric field that can be calculated using the methods of electrostatics. In fact, we have already calculated the electric field produced by an infinite charged surface and taking over the results from (2.79) (for a single plane the field is derived in (1.42)), we obtain that such a displacement will generate an electric field (only) within the two surfaces of the form

$$\mathbf{E} = 4\pi Nex\, \hat{\mathbf{e}}_x. \tag{11.1}$$

The motion of the electrons (within the two charged surfaces) along the x-axis will now be subjected to a force leading to ($e > 0$)

$$m\ddot{x} = -eE = -4\pi Ne^2 x. \tag{11.2}$$

We recognize that this is the equation for a harmonic oscillator with a natural frequency given by

$$\omega_p^2 = \frac{4\pi Ne^2}{m}. \tag{11.3}$$

This is known as the plasma frequency and this analysis shows that, because of this displacement (disturbance), the plasma will begin to oscillate as (this is the complex notation and the coordinate of the particle will be given by the real part)

$$x(t) \sim e^{-i\omega_p t}. \tag{11.4}$$

Thus, we see that, in this simple case, the plasma will oscillate only with the plasma frequency. Although this is not exactly the "plasma oscillation" that one talks about in connection with a plasma (which we will discuss next), this simple example illustrates how the plasma tries to maintain its neutrality when disturbed slightly.

If the plasma is subject to a driving force with a given angular frequency, then it can oscillate with a frequency different from the plasma frequency (11.3). For example, suppose we have a harmonic electric field of the form (we are suppressing the coordinate dependence for simplicity)

$$\mathbf{E}(t) = \mathbf{E}(0) \, e^{-i\omega t}, \tag{11.5}$$

acting on the plasma (with ω representing the frequency of the harmonic field), then the equation for an electron in the plasma would become

$$m\ddot{\mathbf{x}} = -e\mathbf{E}(t) = -e\mathbf{E}(0) \, e^{-i\omega t}. \tag{11.6}$$

The solution, in this case, clearly would have the form

$$\mathbf{x}(t) = \mathbf{x}(0) \, e^{-i\omega t} = \frac{e}{m\omega^2} \, \mathbf{E}(0) e^{-i\omega t} = \frac{e}{m\omega^2} \, \mathbf{E}. \tag{11.7}$$

Namely, in this case the plasma will oscillate with the frequency of the driving force. It also follows from Eq. (11.7) that we can obtain the velocity of motion to be

$$\mathbf{v} = -\frac{ie}{m\omega} \, \mathbf{E}. \tag{11.8}$$

This is, in fact, very suggestive in that we see that the current associated with the motion of a single electron can be written as

$$\mathbf{j} = -e\mathbf{v} = \frac{ie^2}{m\omega} \, \mathbf{E} = \frac{i\omega_{\mathrm{p}}^2}{4\pi N\omega} \, \mathbf{E}, \tag{11.9}$$

where we have used the definition of the plasma frequency in (11.3). Relation (11.9) is very interesting in that it is reminiscent of the Ohm's law,

$$\mathbf{J} = N\mathbf{j} = -Ne\mathbf{v} = \frac{iNe^2}{m\omega} \, \mathbf{E} = \frac{i\omega_{\mathrm{p}}^2}{4\pi\omega} \, \mathbf{E} = \sigma \, \mathbf{E}, \tag{11.10}$$

except for the factor of "i" in the proportionality constant. This factor simply implies that the electric field and the current are not in phase. In this sense, a plasma is somewhat like a dielectric medium and, as we will see shortly, this phase difference has important consequences.

In a plasma, of course, there are always random thermal oscillations. When we can neglect thermal motions (because of their small magnitude at low temperatures or because of averaging of the random thermal motion), we can talk of a cold plasma. In such a case, the effect due to the driving force is dominant and (11.9) should hold. If the plasma is at a high temperature so that the thermal oscillations are not negligible, then one must use a statistical description of the plasma. Nonetheless, relations of the form (11.9) turn out to be fairly accurate within such derivations.

Let us next analyze a simple example within the context of a plasma at finite temperature. Let us consider a plasma in thermal equilibrium at a temperature T. Let us introduce a single static charged particle carrying a charge q into the plasma. This will generate a static electric field in the plasma and we would like to determine this field. Since the electric field is static, in order to calculate the field, we only need to determine the scalar potential Φ which will be spherically symmetric.

In the presence of the electric field generated by the static charge, the charged particles in the plasma will experience a force and, consequently, will rearrange themselves so as to attain a configuration of minimum (free) energy. We know that at a finite temperature T, the equilibrium density of particles is determined by the Boltzmann factor $e^{-\frac{E}{kT}}$, where E, k represent the energy of the particle and the Boltzmann constant respectively. Thus, in the presence of the charge, the electron and the ion densities will take the forms

$$N_{\mathrm{e}} = N\, e^{\frac{e\Phi}{kT}},$$

$$N_{\mathrm{i}} = N\, e^{-\frac{e\Phi}{kT}}, \tag{11.11}$$

where $N = e^{-\frac{E}{kT}}$ denotes the equilibrium density of the charged particles in the plasma before the introduction of the external charge. The scalar potential (and, therefore, the electric field) away from the location of the charged particle can be determined from Gauss' law (the first of the Maxwell's equations), namely,

$$\nabla^2\,\Phi = -4\pi\rho,$$

$$\text{or,}\quad \frac{1}{r^2}\frac{\mathrm{d}}{\mathrm{d}r}\left(r^2\frac{\mathrm{d}}{\mathrm{d}r}\right)\Phi = -4\pi e(N_{\mathrm{i}} - N_{\mathrm{e}}),$$

$$\text{or,}\quad \frac{\mathrm{d}^2\Phi}{\mathrm{d}r^2} + \frac{2}{r}\frac{\mathrm{d}\Phi}{\mathrm{d}r} = 8\pi Ne\,\sinh\left(\frac{e\Phi}{kT}\right). \tag{11.12}$$

Here, we have used the spherical symmetry of the scalar potential.

Equation (11.12) is difficult to solve in closed form, in general. However, let us assume that we have a plasma in equilibrium at a very high temperature $kT \gg e\Phi$. In this case, we can approximate the right hand side in (11.12) and write

$$\frac{\mathrm{d}^2\Phi}{\mathrm{d}r^2} + \frac{2}{r}\frac{\mathrm{d}\Phi}{\mathrm{d}r} - \frac{8\pi Ne^2}{kT}\,\Phi = 0. \tag{11.13}$$

We recognize this as the spherical Bessel equation of order zero and the solution that vanishes asymptotically has the form

$$\Phi(r) = \frac{C}{r}\,e^{-\frac{r}{r_\mathrm{D}}}, \tag{11.14}$$

where C is a constant and we have defined the Debye length as

$$r_\mathrm{D} = \sqrt{\frac{kT}{8\pi Ne^2}} = \sqrt{\frac{kT}{2m\omega_\mathrm{p}^2}}. \tag{11.15}$$

The electric field is now easily obtained from

$$\mathbf{E} = -\boldsymbol{\nabla}\,\Phi = \hat{\mathbf{r}}\,\frac{C(r + r_\mathrm{D})}{r^2 r_\mathrm{D}}\,e^{-\frac{r}{r_\mathrm{D}}}. \tag{11.16}$$

Thus, we see that the scalar potential as well as the electric field fall off rapidly for $r > r_\mathrm{D}$. Namely, the charged particles of the plasma will reorganize themselves so as to screen the external charge beyond the Debye length. For this reason, the Debye length is also sometimes referred to as the screening length. These examples illustrate that when a plasma is disturbed, it tries to restore its charge neutrality.

11.2 Propagation of electromagnetic waves through a plasma

Let us next consider the question of propagation of electromagnetic waves through a plasma. We assume that a harmonic electromagnetic wave of frequency ω is incident on a plasma. Let the electron density of the plasma in equilibrium be N_eq which will also be the density of positive ions in equilibrium. The incident electromagnetic wave will set up a local fluctuation in the electron density (we are assuming that the positive ions do not have appreciable motion) so that we can write

$$N_\mathrm{e}(\mathbf{x}, t) = N_\mathrm{eq} + n(\mathbf{x}, t), \tag{11.17}$$

where the fluctuation from the equilibrium value, $n(\mathbf{x}, t)$, is considered to be small. As we have already seen, there will be oscillations in the plasma leading to a current density (see (11.9))

$$\mathbf{J} = N_{\mathrm{e}}(\mathbf{x}, t)\mathbf{j} \approx \frac{iN_{\mathrm{eq}}e^2}{m\omega}\mathbf{E} = \frac{i\omega_{\mathrm{p}}^2}{4\pi\omega}\mathbf{E}, \tag{11.18}$$

where we are neglecting terms quadratic in the fluctuations, an approximation which is also known as the linearized approximation (namely, since the velocity $\mathbf{v}$ is already a fluctuation, $n(\mathbf{x}, t)\mathbf{v}$ is quadratic in the fluctuations).

The first of the Maxwell's equations, in this case, takes the form

$$\boldsymbol{\nabla} \cdot \mathbf{E} = 4\pi e(N_{\mathrm{i}} - N_{\mathrm{e}}) = -4\pi e n(\mathbf{x}, t) = 4\pi\rho(\mathbf{x}, t), \tag{11.19}$$

where we have defined $\rho(\mathbf{x}, t) = -en(\mathbf{x}, t)$ to correspond to the fluctuation in the electron charge density. The other Maxwell's equations have the forms

$$\boldsymbol{\nabla} \cdot \mathbf{B} = 0,$$

$$\boldsymbol{\nabla} \times \mathbf{E} = \frac{i\omega}{c}\mathbf{B},$$

$$\boldsymbol{\nabla} \times \mathbf{B} = \frac{4\pi}{c}\mathbf{J} - \frac{i\omega}{c}\mathbf{E} = -\frac{i\omega}{c}\left(1 - \frac{\omega_{\mathrm{p}}^2}{\omega^2}\right)\mathbf{E}, \tag{11.20}$$

where we have used (11.18). The last equation in (11.20) suggests that we can define a permittivity for the plasma of the form

$$\epsilon_{\mathrm{p}} = 1 - \frac{\omega_{\mathrm{p}}^2}{\omega^2}. \tag{11.21}$$

It is worth noting that $\epsilon_{\mathrm{p}} \leq 1$ as opposed to the case of a dielectric for which $\epsilon \geq 1$.

There are now several interesting cases to be discussed. If the frequency of the harmonic field coincides exactly with that of the plasma, namely, if $\omega = \omega_{\mathrm{p}}$, then we have

$$\epsilon_{\mathrm{p}} = 0. \tag{11.22}$$

In this case, the second and the fourth of Maxwell's equations give

$$\boldsymbol{\nabla} \cdot \mathbf{B} = 0,$$

$$\boldsymbol{\nabla} \times \mathbf{B} = 0. \tag{11.23}$$

Namely, the electron current, in this case, exactly cancels the displacement current so that the curl of the magnetic field vanishes. In the absence of sources, these two equations imply that the magnetic field vanishes identically, namely,

$$\mathbf{B} = 0. \tag{11.24}$$

The other two of Maxwell's equations take the forms

$$\boldsymbol{\nabla} \cdot \mathbf{E} = 4\pi\rho(\mathbf{x}, t),$$

$$\boldsymbol{\nabla} \times \mathbf{E} = 0. \tag{11.25}$$

These are equations of the type in electrostatics implying that the electric field can be expressed as the gradient of a scalar potential. The difference is that here the fluctuations in the electron density have a time dependence leading to a time dependence of the electric field.

This is the case normally referred to as the "plasma oscillations". This can be seen more clearly as follows. The continuity equation gives

$$\frac{\partial\rho}{\partial t} + \boldsymbol{\nabla} \cdot \mathbf{J} = 0,$$

or, $\quad \dfrac{\partial\rho}{\partial t} + i\omega_{\mathrm{p}}\,\rho = 0. \tag{11.26}$

Here, we have used Eqs. (11.18) as well as (11.19) (remember $\omega = \omega_{\mathrm{p}}$). Taking the time derivative one more time, we obtain

$$\frac{\partial^2\rho}{\partial t^2} + \omega_{\mathrm{p}}^2\,\rho = 0. \tag{11.27}$$

Namely, the density of electrons fluctuates in time with the plasma frequency. This is a cooperative phenomenon in the sense that, not one electron, but the plasma of electrons as a whole oscillates. However, there is no traveling disturbance that is generated. This should be contrasted with the transverse traveling wave solutions of Maxwell's equations for which

$$\boldsymbol{\nabla} \cdot \mathbf{E} = 0,$$

corresponding to the fact that there is no charge density. (It is worth noting here that in a conducting medium, conductivity σ is real leading to a dissipative solution $\rho(t) \sim e^{-\frac{t}{\tau}}$ as we have seen in (8.55). In contrast in the case of a plasma the conductivity σ can be thought of as purely imaginary as noted in (11.10) which is the reason for the oscillatory solution.)

When $\omega \neq \omega_p$, there can be traveling waves in the plasma. This can be seen as follows. We note that when $\omega \neq \omega_p$, then we can truly associate the permittivity ϵ_p in (11.21) with the plasma. For example, we see that for a harmonic field, the continuity equation together with (11.18) leads to

$$- i\omega\rho = -\boldsymbol{\nabla} \cdot \mathbf{J} = -\frac{i\omega_p^2}{4\pi\omega} (\boldsymbol{\nabla} \cdot \mathbf{E}),$$

or, $\quad \rho = \frac{\omega_p^2}{4\pi\omega^2} (\boldsymbol{\nabla} \cdot \mathbf{E}).$ \hfill (11.28)

Using this in (11.19) as well as the definition in (11.21), we see that we can write the Maxwell's equation, in this case, as

$$\boldsymbol{\nabla} \cdot (\epsilon_p \mathbf{E}) = 0,$$

$$\boldsymbol{\nabla} \cdot \mathbf{B} = 0,$$

$$\boldsymbol{\nabla} \times \mathbf{E} = \frac{i\omega}{c} \mathbf{B},$$

$$\boldsymbol{\nabla} \times \mathbf{B} = -\frac{i\epsilon_p\omega}{c} \mathbf{E},$$ \hfill (11.29)

where we have assumed $\mu_p = 1$. Equations (11.29) coincide exactly with (source free) Maxwell's equations in a dielectric of permittivity ϵ_p. (Plasma behaves like an isotropic dielectric medium since ϵ_p is a scalar.) It follows from (11.29) that both the electric and the magnetic fields satisfy the wave equation

$$\nabla^2 \mathbf{E} + \frac{\epsilon_p \omega^2}{c^2} \mathbf{E} = 0,$$

$$\nabla^2 \mathbf{B} + \frac{\epsilon_p \omega^2}{c^2} \mathbf{B} = 0.$$ \hfill (11.30)

These have traveling wave solutions of the form

$$\mathbf{E}(\mathbf{x}) \sim e^{i\mathbf{k}\cdot\mathbf{x}}, \qquad \mathbf{B}(\mathbf{x}) \sim e^{i\mathbf{k}\cdot\mathbf{x}},$$ \hfill (11.31)

where

$$k^2 = \mathbf{k}^2 = \frac{\epsilon_p \omega^2}{c^2},$$

or, $\quad c^2 k^2 = \omega^2 - \omega_p^2.$ \hfill (11.32)

We have used here the definition of the permittivity given in (11.21).

Relation (11.32) is interesting in the sense that it has exactly the same form as the wave number that we have already seen in the case of a wave guide. It implies that when $\omega > \omega_p$, there is propagation of the electromagnetic wave inside the plasma without any attenuation. However, there cannot be any propagation of electromagnetic waves inside the plasma when $\omega < \omega_p$. Therefore, we see that ω_p is the analogue of the cut-off frequency in a wave guide. There is a difference though. Unlike the wave guide, where there is loss of energy through dissipation when $\omega < \omega_c$, here the non-propagation does not imply dissipation. Rather, this simply implies that the incident wave is totally reflected in such a case. In fact, recalling that the index of refraction for a medium is defined as

$$n = \sqrt{\epsilon\mu},$$

we note that the index of refraction for a plasma ($\mu_p = 1$),

$$n_p = \sqrt{\epsilon_p} = \sqrt{1 - \frac{\omega_p^2}{\omega^2}}, \tag{11.33}$$

becomes purely imaginary when $\omega < \omega_p$. Taking the results of reflection from a dielectric at normal incidence (see (6.51)), for simplicity, we see that, in the case of reflection of such a harmonic wave from a plasma at normal incidence, we have

$$\frac{E_r}{E_i} = \frac{1 - n_p}{1 + n_p}. \tag{11.34}$$

Here we are assuming that the wave is traveling in vacuum before being reflected from the plasma. Note that since n_p is purely imaginary for $\omega < \omega_p$, the right hand side of (11.34) has the absolute magnitude unity, implying that the coefficient of reflection is unity. Therefore, the wave is totally reflected.

Thus, we see that an electromagnetic wave is totally reflected from a plasma when $\omega < \omega_p$. Using the definition of ω_p in (11.3), this condition is also sometimes expressed by saying that electromagnetic waves will reflect from a plasma if the number density of electrons will satisfy

$$N > \frac{m\omega^2}{4\pi e^2}, \tag{11.35}$$

or simply if the plasma is overdense. This can happen if the density of the plasma is high or the frequency of the harmonic wave is low.

This is, in fact, the basic principle used in transmitting low frequency radio waves. Let us recall that the ionosphere consists of a plasma of electrons and positive ions where the ionization is a consequence of radiations coming from the sun. As a result, low frequency radio waves cannot penetrate the ionosphere and are reflected back, leading to a transmission of such waves around the globe. The theory of propagation of electromagnetic waves in the ionosphere is, however, slightly more involved owing to the fact that the density of electrons in the ionosphere is not uniform. Rather, it changes with the height from the surface of the earth.

11.3 Motion of the positive ions

We have so far neglected the motion of the positive ions because of their heavy mass. Under certain circumstances, however, their motion becomes important as we will see later. Let us see how the motion of the positive ions can be included into our analysis and how, under the conditions that we have assumed, their contribution can be neglected in the earlier analysis.

Let us recall from (11.6) and (11.8) that the motion of electrons in a plasma driven by a harmonic electric field leads to

$$m_e \dot{\mathbf{v}}_e = -e\mathbf{E},$$

$$\mathbf{v}_e = -\frac{ie}{m_e\omega}\,\mathbf{E}. \tag{11.36}$$

Similarly, the motion of the positive ions leads to

$$m_i \dot{\mathbf{v}}_i = e\mathbf{E},$$

$$\mathbf{v}_i = \frac{ie}{m_i\omega}\,\mathbf{E}. \tag{11.37}$$

We have used the subscripts "e" and "i" to represent the respective quantities associated with electrons and ions. We note now that the total current density associated with the plasma (including the contribution due to positive ions) takes the form

$$\mathbf{J} = Ne\,(\mathbf{v}_i - \mathbf{v}_e) = \frac{iNe^2}{\omega}\left(\frac{1}{m_i} + \frac{1}{m_e}\right)\mathbf{E}$$

$$= \frac{i}{4\pi\omega}\left(\omega_{p,i}^2 + \omega_{p,e}^2\right)\mathbf{E} = \sigma\,\mathbf{E}, \tag{11.38}$$

where we have defined the plasma frequencies associated with the electrons and the ions respectively as

$$\omega_{p,e}^2 = \frac{4\pi N e^2}{m_e},$$

$$\omega_{p,i}^2 = \frac{4\pi N e^2}{m_i} \ll \omega_{p,e}^2. \tag{11.39}$$

Correspondingly, when the motion of the positive ions is taken into account, the conductivity can be written as

$$\sigma = \frac{i}{4\pi\omega} \left(\omega_{p,e}^2 + \omega_{p,i}^2 \right). \tag{11.40}$$

The fact that the conductivity σ is a scalar simply signifies that the plasma behaves like an isotropic medium. If we now substitute (11.40) into the Maxwell's equations, we can derive, in this case, that the permittivity of the plasma takes the form

$$\epsilon_p = 1 - \frac{1}{\omega^2} \left(\omega_{p,e}^2 + \omega_{p,i}^2 \right). \tag{11.41}$$

Namely, the permittivity also continues to be a scalar signifying that the refractive index of the isotropic medium has the form

$$n_p = \sqrt{\epsilon_p} = \sqrt{1 - \frac{1}{\omega^2} \left(\omega_{p,e}^2 + \omega_{p,i}^2 \right)}. \tag{11.42}$$

We note now from Eq. (11.39) that

$$\omega_{p,e}^2 \gg \omega_{p,i}^2, \tag{11.43}$$

since $m_i \gg m_e$ (recall that a proton is about 2000 times heavier than the electron and the positive ion will be at least as heavy as the proton). Consequently, in our earlier analysis we can safely neglect the motion of the positive ions. However, there may be situations where the motion of the positive ions plays an important role and we will discuss such an example later.

11.4 Effect of a background magnetic field

Let us next consider the behavior of a plasma in the background of a constant magnetic field. This is quite important in the analysis of the propagation of electromagnetic waves through the ionosphere. This is because there is a magnetic field associated with earth and, consequently, the plasma in the ionosphere is subjected to this constant background magnetic field. As we will see, this changes the nature of the plasma as a medium.

Let us concentrate on the motion of the electron. In the presence of a driving electric field which is harmonic and a constant magnetic field, the equation for the electron takes the form

$$m\dot{\mathbf{v}} = -e\left(\mathbf{E} + \frac{\mathbf{v}}{c} \times \bar{\mathbf{B}}\right),$$

$$\text{or,} \quad \mathbf{v} = -\frac{ie}{m\omega}\left(\mathbf{E} + \frac{\mathbf{v}}{c} \times \bar{\mathbf{B}}\right), \tag{11.44}$$

where we have used the fact that the velocity has to be harmonic (with the same frequency) since the driving electric field is. Here, we have denoted the background magnetic field by $\bar{\mathbf{B}}$. (As a parenthetical remark, let us note that a driving electromagnetic field can also lead to a Lorentz force. However, since the velocity is already a fluctuation as is the driving magnetic field, such a term is neglected under our approximation of linearization. This is why we did not have a Lorentz force in the earlier analysis.)

Since the velocity occurs on the right hand side of (11.44), solving for the velocity is a little more involved. To invert the relation, let us define the cyclotron frequency associated with the background magnetic field as

$$\mathbf{\Omega} = \frac{e\bar{\mathbf{B}}}{mc}. \tag{11.45}$$

In terms of this, the velocity can be expressed as

$$\left(\mathbf{v} + i\mathbf{v} \times \frac{\mathbf{\Omega}}{\omega}\right) = -\frac{ie}{m\omega}\mathbf{E}, \tag{11.46}$$

$$\text{or,} \ \mathbf{v} = -\frac{ie}{m\omega(1 - (\frac{\Omega}{\omega})^2)}\left(\mathbf{E} - \frac{1}{\omega^2}\left(\mathbf{\Omega} \cdot \mathbf{E}\right)\mathbf{\Omega} - \frac{i}{\omega}\left(\mathbf{E} \times \mathbf{\Omega}\right)\right),$$

where $\Omega^2 = \mathbf{\Omega}^2$. The inversion in (11.46) is best done in matrix form. We note that we can write (11.46) in component form as

$$P_{ij}v_j = -\frac{ie}{m\omega}E_i, \tag{11.47}$$

where

$$P_{ij} = \delta_{ij} + \frac{i}{\omega}\epsilon_{ijk}\Omega_k. \tag{11.48}$$

The inverse of this matrix can be determined in a straightforward manner and corresponds to

$$P_{ij}^{-1} = \frac{1}{1 - \left(\frac{\Omega}{\omega}\right)^2} \left(\delta_{ij} - \frac{\Omega_i \Omega_j}{\omega^2} - \frac{i}{\omega}\epsilon_{ijk}\Omega_k\right), \tag{11.49}$$

which leads to the solution of (11.47) in components to be

$$v_i = -\frac{ie}{m\omega} P_{ij}^{-1} E_j$$

$$= -\frac{ie}{m\omega(1 - (\frac{\Omega}{\omega})^2)} \left(\delta_{ij} - \frac{1}{\omega^2}\Omega_i \Omega_j - \frac{i}{\omega}\epsilon_{ijk}\Omega_k\right) E_j \tag{11.50}$$

$$= -\frac{ie}{m\omega(1 - (\frac{e\bar{B}}{mc\omega})^2)} \left(\delta_{ij} - \frac{e^2}{m^2c^2\omega^2}\bar{B}_i\bar{B}_j - \frac{ie}{mc\omega}\epsilon_{ijk}\bar{B}_k\right) E_j,$$

where we have defined $\bar{B}^2 = \bar{\mathbf{B}}^2$ and this can be compred with the solution obtained in (11.46).

Relation (11.50) is interesting because it shows that even though there is a relation between the components of the velocity and the electric field, the proportionality constant is a tensor in the presence of a background magnetic field. Consequently, in this case, the current density takes the form

$$J_i = -Nev_i = \sigma_{ij}E_j, \tag{11.51}$$

where

$$\sigma_{ij} = \frac{iNe^2}{m\omega(1 - (\frac{\Omega}{\omega})^2)} \left(\delta_{ij} - \frac{1}{\omega^2}\Omega_i\Omega_j - \frac{i}{\omega}\epsilon_{ijk}\Omega_k\right)$$

$$= \frac{i\omega_{\mathrm{p}}^2}{4\pi\omega(1 - (\frac{\Omega}{\omega})^2)} \left(\delta_{ij} - \frac{1}{\omega^2}\Omega_i\Omega_j - \frac{i}{\omega}\epsilon_{ijk}\Omega_k\right). \tag{11.52}$$

Substituting (11.51) and (11.52) into the Maxwell's equations (as well as using the continuity equation), we find that in the present case, they take the forms (repeated indices are summed)

$$\nabla_i(\epsilon_{\mathrm{p},\,ij}E_j) = 0,$$

$$\nabla_i B_i = 0,$$

$$(\nabla \times \mathbf{E})_i = \frac{i\omega}{c} B_i,$$

$$(\nabla \times \mathbf{B})_i = -\frac{i\omega}{c}\epsilon_{\mathrm{p},\,ij}E_j, \tag{11.53}$$

where the permittivity also has a tensor structure given by

$$\epsilon_{p,ij} = \delta_{ij} - \frac{4\pi\sigma_{ij}}{i\omega}$$

$$= \delta_{ij} - \frac{\omega_p^2}{(\omega^2 - \Omega^2)} \left(\delta_{ij} - \frac{\Omega_i\Omega_j}{\omega^2} - \frac{i}{\omega}\epsilon_{ijk}\Omega_k \right). \tag{11.54}$$

Thus, we see that in the presence of a magnetic field, the conductivity as well as the permittivity become complex tensors. We are already familiar with the complex nature of such quantities. The tensor structure is new and simply signifies that the plasma ceases to behave like an isotropic medium in the presence of a background magnetic field. Namely, the properties of propagation depend on the direction and the magnetic field is responsible for this anisotropy. We note from (11.53) that, as is the case with harmonic fields, we need to concentrate only on the last two equations. (The first two are consequences of the last two.) Let us choose a traveling plane wave solution of the form

$$E_i \sim e^{i\mathbf{k}\cdot\mathbf{x}}, \qquad B_i \sim e^{i\mathbf{k}\cdot\mathbf{x}}. \tag{11.55}$$

Then, taking the curl of the third equation in (11.53) (and using the fourth), we obtain

$$\left(\delta_{ij}k^2 - k_i k_j\right) E_j = \frac{\omega^2}{c^2} \epsilon_{p\,ij} E_j,$$

$$\text{or,} \quad \left(\epsilon_{p\,ij} - n^2 \left(\delta_{ij} - \frac{k_i k_j}{k^2}\right)\right) E_j = 0, \tag{11.56}$$

where we have defined the refractive index for the medium as

$$n^2 = \frac{c^2 k^2}{\omega^2}. \tag{11.57}$$

A nontrivial solution of (11.56) would exist only if the determinant of the coefficient matrix vanishes. Let us analyze this a little carefully. To simplify the analysis, let us assume that the magnetic field $\bar{\mathbf{B}}$ lies along the z-axis. In this case, the only non-vanishing component of the cyclotron frequency is given by $\Omega_z = \Omega_3 = \Omega$. Furthermore, let us define the dimensionless quantities

$$X = \frac{\omega_p^2}{\omega^2}, \qquad Y = \frac{\Omega}{\omega},$$

$$\epsilon_0 = 1 - X, \qquad \epsilon_1 = 1 - \frac{X}{1-Y^2}, \qquad \epsilon_2 = \frac{XY}{1-Y^2}, \tag{11.58}$$

so that the permittivity can be written in the simple form

$$
\epsilon_{p\,ij} = \begin{pmatrix} \epsilon_1 & i\epsilon_2 & 0 \\ -i\epsilon_2 & \epsilon_1 & 0 \\ 0 & 0 & \epsilon_0 \end{pmatrix}.
\tag{11.59}
$$

We note that without any loss of generality, we can choose our coordinate axes (x and y) such that the direction of propagation lies in the $y - z$ plane, namely,

$$
\mathbf{k} = k(0, \sin\theta, \cos\theta).
\tag{11.60}
$$

With this, we can represent

$$
n^2 \left(\delta_{ij} - \frac{k_i k_j}{k^2} \right) = \begin{pmatrix} n^2 & 0 & 0 \\ 0 & n^2 \cos^2\theta & -n^2 \sin\theta\cos\theta \\ 0 & -n^2 \sin\theta\cos\theta & n^2 \sin^2\theta \end{pmatrix}.
\tag{11.61}
$$

Consequently, we are interested in the vanishing of the determinant of the matrix

$$
\epsilon_{p\,ij} - n^2 \left(\delta_{ij} - \frac{k_i k_j}{k^2} \right)
$$

$$
= \begin{pmatrix} (\epsilon_1 - n^2) & i\epsilon_2 & 0 \\ -i\epsilon_2 & (\epsilon_1 - n^2 \cos^2\theta) & n^2 \sin\theta\cos\theta \\ 0 & n^2 \sin\theta\cos\theta & (\epsilon_0 - n^2 \sin^2\theta) \end{pmatrix}.
\tag{11.62}
$$

Let us look at the solutions for the vanishing determinant in two special cases. First, if $\theta = 0$, namely, if the direction of propagation is along the direction of the magnetic field, we see from (11.62) that the vanishing of the determinant gives

$$
\epsilon_0 \left((\epsilon_1 - n^2)^2 - \epsilon_2^2 \right) = \epsilon_0 \left(n^2 - \epsilon_1 - \epsilon_2 \right) \left(n^2 - \epsilon_1 + \epsilon_2 \right) = 0.
\tag{11.63}
$$

If $\epsilon_0 = 0$, the index of refraction is undetermined and we see from (11.58) that this happens when $\omega = \omega_{\mathrm{p}}$. Namely, in this case, the plasma oscillates unaffected by the presence of the magnetic field (since the motion is parallel to the direction of the magnetic field).

In fact, it is easy to see from (11.56) that, in this case, $E_x = 0 = E_y$ and only E_z is nonzero so that the electron motion is along the z-axis and it does not feel the presence of the magnetic field. On the other hand, if $\omega \neq \omega_\mathrm{p}$ (namely, if $\epsilon_0 \neq 0$), then we have

$$n^2 = \epsilon_1 \pm \epsilon_2 = 1 - \frac{X}{1 - Y^2} \pm \frac{XY}{1 - Y^2} = 1 - \frac{X(1 \mp Y)}{1 - Y^2}$$

$$= 1 - \frac{X}{1 \pm Y} = 1 - \frac{\omega_\mathrm{p}^2}{\omega(\omega \pm \Omega)}, \tag{11.64}$$

where we have used (11.58). Note that, in this case, it follows from (11.56) that $E_z = 0$ and since the electric field is perpendicular to the direction of propagation (and the magnetic field), the electrons feel the effect of the magnetic field.

The other special case is when the direction of propagation is perpendicular to the direction of the magnetic field, namely, $\theta = \frac{\pi}{2}$. In this case, we see from (11.62) that the vanishing of the determinant leads to

$$\left(\epsilon_0 - n^2\right)\left(\epsilon_1(\epsilon_1 - n^2) - \epsilon_2^2\right) = \left(n^2 - \epsilon_0\right)\left(n^2\epsilon_1 - \epsilon_1^2 + \epsilon_2^2\right) = 0. \tag{11.65}$$

The refractive index of the medium is now determined to be

$$n^2 = \epsilon_0 = 1 - \frac{\omega_\mathrm{p}^2}{\omega^2},$$

$$\text{or,} \quad n^2 = \frac{\epsilon_1^2 - \epsilon_2^2}{\epsilon_1} = 1 - \frac{X}{1 - Y^2} - \frac{X^2Y^2}{(1 - Y^2)(1 - X - Y^2)}$$

$$= 1 - \frac{X(1 - X)}{1 - X - Y^2} = 1 - \frac{\omega_\mathrm{p}^2\left(1 - \frac{\omega_\mathrm{p}^2}{\omega^2}\right)}{\omega^2 - \omega_\mathrm{p}^2 - \Omega^2}. \tag{11.66}$$

The first case is interesting in that the refractive index is insensitive to the magnetic field. This is understood from the fact that Eq. (11.56), in this case, leads to $E_x = 0 = E_y$. Only E_z is nonzero so that the motion of the electrons is along the direction of the magnetic field. Correspondingly, they do not feel its effect. In the second case, on the other hand, $E_z = 0$ and the electric fields are transverse to the direction of the magnetic field and, consequently, the electrons do feel its effect. Although one can solve the vanishing of the determinant for an arbitrary angle θ, these two special cases show the variation of the refractive index with the direction of propagation.

11.5 Faraday rotation

Let us go back to Eq. (11.64) and analyze the result in some more detail. We see that when the direction of propagation is along the direction of the magnetic field (which we have chosen to correspond to the z-axis), $E_z = 0$ if $\omega \neq \omega_{\mathrm{p}}$. Let us denote the two eigenvalues for the index of refraction, in this case, as

$$n_\pm^2 = \epsilon_1 \pm \epsilon_2 = 1 - \frac{\omega_{\mathrm{p}}^2}{\omega(\omega \pm \Omega)}. \qquad (11.67)$$

The meaning of the two eigenvalues becomes clear once we recognize that since $E_z = 0$, we can restrict ourselves to the study of the matrix in (11.56) (or (11.62)) to the upper left 2×2 space. In this space, for example for $n = n_+$, we have

$$
\begin{aligned}
A_{\alpha\beta} &= \epsilon_{\alpha\beta} - n_+^2 \left(\delta_{\alpha\beta} - \frac{k_\alpha k_\beta}{k^2} \right) = \begin{pmatrix} \epsilon_1 - n_+^2 & i\epsilon_2 \\ -i\epsilon_2 & \epsilon_1 - n_+^2 \end{pmatrix} \\
&= \begin{pmatrix} \epsilon_1 - (\epsilon_1 + \epsilon_2) & i\epsilon_2 \\ -i\epsilon_2 & \epsilon_1 - (\epsilon_1 + \epsilon_2) \end{pmatrix} = -\epsilon_2 \begin{pmatrix} 1 & -i \\ i & 1 \end{pmatrix},
\end{aligned}
$$
$$(11.68)$$

where $\alpha, \beta = 1, 2$. It is immediately clear from the structure of this matrix that

$$A \begin{pmatrix} 1 \\ i \end{pmatrix} = -2\epsilon_2 \begin{pmatrix} 1 \\ i \end{pmatrix}, \qquad A \begin{pmatrix} 1 \\ -i \end{pmatrix} = 0. \qquad (11.69)$$

On the other hand, for $n = n_-$, we have

$$A_{\alpha\beta} = \epsilon_{\alpha\beta} - n_-^2 \left(\delta_{\alpha\beta} - \frac{k_\alpha k_\beta}{k^2} \right) = \epsilon_2 \begin{pmatrix} 1 & i \\ -i & 1 \end{pmatrix}. \qquad (11.70)$$

In this case, we find that

$$A \begin{pmatrix} 1 \\ i \end{pmatrix} = 0, \qquad A \begin{pmatrix} 1 \\ -i \end{pmatrix} = 2\epsilon_2 \begin{pmatrix} 1 \\ -i \end{pmatrix}. \qquad (11.71)$$

The vectors $\begin{pmatrix} 1 \\ -i \end{pmatrix}$ and $\begin{pmatrix} 1 \\ i \end{pmatrix}$ denote respectively basis vectors for right circularly polarized and left circularly polarized waves.

Let us next recall that right and left circularly polarized waves traveling along the z-axis can be represented respectively as (see, for example, (6.37))

$$\mathbf{E}_\mathrm{R} = (\hat{\mathbf{x}} - i\hat{\mathbf{y}}) \, E_0 \, e^{-i\omega t + ik_\mathrm{R} z},$$

$$\mathbf{E}_\mathrm{L} = (\hat{\mathbf{x}} + i\hat{\mathbf{y}}) \, E_0 \, e^{-i\omega t + ik_\mathrm{L} z}. \tag{11.72}$$

It follows from this as well as Eqs. (11.69) and (11.71) that traveling along the direction of the magnetic field in a plasma, the right and the left circularly polarized waves will suffer different rotations of phase – the right circularly polarized wave rotating with the index of refraction n_+ while the left circularly polarized wave rotates with the index of refraction n_-. The medium responds differently to right and left circularly polarized waves. This phenomenon is commonly known as the Faraday rotation.

We note that in vacuum, a linearly polarized wave can always be written as a sum of a right and a left circularly polarized wave. If such a wave, initially linearly polarized along the x-axis, is incident along the z-axis (the direction of the magnetic field) on a plasma (at $z = 0$), then traveling through the plasma, the planes of polarization of the right and the left circularly polarized waves will rotate as

$$\mathbf{E} = \frac{1}{2} \left(\mathbf{E}_\mathrm{R} + \mathbf{E}_\mathrm{L} \right)$$

$$= \frac{E_0}{2} e^{-i\omega t} \left[(\hat{\mathbf{x}} - i\hat{\mathbf{y}}) \, e^{ik_\mathrm{R} z} + (\hat{\mathbf{x}} + i\hat{\mathbf{y}}) \, e^{ik_\mathrm{L} z} \right], \tag{11.73}$$

where

$$k_\mathrm{R} = \frac{n_+\omega}{c}, \qquad k_\mathrm{L} = \frac{n_-\omega}{c}. \tag{11.74}$$

As a result, in traveling through a certain distance z, the tilt in the polarization will be given by

$$\psi = \frac{1}{2}(k_\mathrm{R} - k_\mathrm{L})z = \frac{\omega}{2c}(n_+ - n_-)z. \tag{11.75}$$

If the harmonic frequency is high compared to the plasma frequency as well as the cyclotron frequency, $\omega \gg \omega_\mathrm{p}, \Omega$, then we obtain from Eqs. (11.67) that

$$n_\pm \approx 1 - \frac{\omega_\mathrm{p}^2}{2\omega^2} \left(1 \mp \frac{\Omega}{\omega} \right). \tag{11.76}$$

This leads to

$$\psi = \frac{\omega}{2c}(n_+ - n_-)z \approx \frac{\omega_p^2 \Omega}{2c\omega^2}\, z$$

$$= \frac{2\pi e^3 \bar{B}}{m^2 c^2 \omega^2}\, Nz. \tag{11.77}$$

This shows that the tilt in the polarization planes is proportional to the distance traveled in the plasma as well as to the number density of electrons in the plasma. This has, of course, been derived assuming that the electron density is a constant in the plasma. If the density changes with distance, as is the case in the ionosphere, then the tilt is obtained to be

$$\psi = \frac{2\pi e^3 \bar{B}}{m^2 c^2 \omega^2} \int_0^z \mathrm{d}z'\, N(z'). \tag{11.78}$$

In either case, it is clear that the number density of electrons in a plasma (or the total number of electrons in a volume of unit height) can be determined from a study of the tilt in the polarization planes.

11.6 Alfvén waves

As we have seen earlier, in the absence of a magnetic field the effect of the positive ions can be safely neglected. In the presence of a magnetic field, however, the positive ions play an important role under certain circumstances which we would like to discuss.

Including the contributions of the positive ions to our analysis of section 11.4 is quite easy. First, let us define the mass ratio between the electron and the ion as

$$\eta = \frac{m_e}{m_i} \ll 1. \tag{11.79}$$

With this, we note that we can identify

$$\Omega_i = \frac{e\bar{B}}{m_i c} = \eta \Omega_e,$$

$$\omega_{p,i}^2 = \frac{4\pi N e^2}{m_i} = \eta \omega_{p,e}^2,$$

$$X_i = \frac{\omega_{p,i}^2}{\omega^2} = \eta X_e,$$

$$Y_i = \frac{\Omega_i}{\omega} = \eta Y_e. \tag{11.80}$$

We recognize that the positive ions will also contribute to the conductivity tensor as well as to the permittivity. The qualitative structures of these tensors will be the same and, if we choose the magnetic field to be along the z-axis, we can write the permittivity tensor as in (11.59)

$$\epsilon_{p,ij} = \begin{pmatrix} \epsilon_1 & i\epsilon_2 & 0 \\ -i\epsilon_2 & \epsilon_1 & 0 \\ 0 & 0 & \epsilon_0 \end{pmatrix}, \tag{11.81}$$

where

$$\epsilon_0 = 1 - X_e - X_i = 1 - (1 + \eta)X_e$$

$$\approx 1 - X_e,$$

$$\epsilon_1 = 1 - \frac{X_e}{1 - Y_e^2} - \frac{X_i}{1 - Y_i^2}$$

$$= 1 - \frac{(1 + \eta)X_e(1 - \eta Y_e^2)}{(1 - Y_e^2)(1 - \eta^2 Y_e^2)}$$

$$\approx 1 - \frac{X_e(1 - \eta Y_e^2)}{(1 - Y_e^2)(1 - \eta^2 Y_e^2)},$$

$$\epsilon_2 = \frac{X_e Y_e}{1 - Y_e^2} - \frac{X_i Y_i}{1 - Y_i^2}$$

$$= \frac{(1 - \eta^2)X_e Y_e}{(1 - Y_e^2)(1 - \eta^2 Y_e^2)}$$

$$\approx \frac{X_e Y_e}{(1 - Y_e^2)(1 - \eta^2 Y_e^2)}. \tag{11.82}$$

We recognize that $\eta \ll 1$ and it is clear from (11.82) that when $\omega^2 \gg \eta\Omega_e^2 = \Omega_e\Omega_i$ (or equivalently, $\eta Y_e^2 \ll 1$), we can neglect the contributions from the positive ions. This is exactly like the analysis before. However, at very low frequencies, $\omega \ll \Omega_i = \eta\Omega_e$, or equivalently, when $\eta Y_e \gg 1$, the contributions from the positive ions play an important role. We see from (11.82) that in this limit, we can write

$$\epsilon_0 \approx 1 - X_e,$$

$$\epsilon_1 \approx 1 - \frac{X_e(1 - \eta Y_e^2)}{(1 - Y_e^2)(1 - \eta^2 Y_e^2)}$$

$$\approx 1 - \frac{X_e(-\eta Y_e^2)}{(-Y_e^2)(-\eta^2 Y_e^2)} = 1 + \frac{X_e}{\eta Y_e^2},$$

$$\epsilon_2 \approx \frac{X_e Y_e}{(1 - Y_e^2)(1 - \eta^2 Y_e^2)}$$

$$\approx \frac{X_e Y_e}{(-Y_e^2)(-\eta^2 Y_e^2)} = \frac{X_e}{\eta^2 Y_e^3}. \tag{11.83}$$

Since $Y_e \gg 1$, it follows that

$$\epsilon_1 \gg \epsilon_2, \tag{11.84}$$

and, therefore, ϵ_2 can be neglected in all our manipulations.

The relation (11.84) is significant in simplifying all the expressions. For example, we can now write the matrix (11.62) as

$$\epsilon_{\mathrm{p},ij} - n^2 \left(\delta_{ij} - \frac{k_i k_j}{k^2} \right)$$

$$= \begin{pmatrix} \epsilon_1 - n^2 & 0 & 0 \\ 0 & \epsilon_1 - n^2 \cos^2\theta & n^2 \sin\theta\cos\theta \\ 0 & n^2 \sin\theta\cos\theta & \epsilon_0 - n^2 \sin^2\theta \end{pmatrix}. \tag{11.85}$$

The vanishing of the determinant of this matrix is now easily seen to give

$$n^2 = \epsilon_1 \approx 1 + \frac{X_e}{\eta Y_e^2} = 1 + \frac{\omega_{\mathrm{p,e}}^2}{\eta \Omega_e^2},$$

$$n^2 = \frac{\epsilon_1 \epsilon_0}{\epsilon_0 \cos^2\theta + \epsilon_1 \sin^2\theta}$$

$$\approx \left(1 + \frac{\omega_{\mathrm{p,e}}^2}{\eta \Omega_e^2} \right) \frac{\omega_{\mathrm{p,e}}^2 - \omega^2}{\omega_{\mathrm{p,e}}^2 - (1 + \frac{\omega_{\mathrm{p,e}}^2}{\eta \Omega_e^2} \sin^2\theta)}. \tag{11.86}$$

The electric fields corresponding to the two roots can be determined from the form in (11.85). For $n^2 = \epsilon_1$, we note that $E_y = 0 = E_z$ while E_x is nontrivial. The electric field is transverse to the magnetic field in this case. On the other hand, for the second root, we see from (11.85) that $E_x = 0$ and

$$\frac{E_y}{E_z} = -\frac{n^2 \sin\theta\cos\theta}{\epsilon_1 - n^2 \cos^2\theta} = -\frac{\epsilon_0}{\epsilon_1} \cot\theta. \tag{11.87}$$

In other words, in this case, the electric field has a component parallel to the direction of the magnetic field in addition to a perpendicular component. Let us note from (11.50) that in this low frequency limit, we can write for the first root (remember that the magnetic field is along the z-axis),

$$v_{e,y} \approx -\frac{ie}{m_e\omega}\left(-\frac{\omega^2}{\Omega_e^2}\right)\left(\frac{i\Omega_e}{\omega}\right)E_x = -\frac{e}{m_e\Omega_e}E_x,$$

$$v_{i,y} \approx \frac{ie}{m_i\omega}\left(-\frac{\omega^2}{\Omega_i^2}\right)\left(-\frac{i\Omega_i}{\omega}\right)E_x = -\frac{e}{m_e\Omega_e}E_x. \tag{11.88}$$

Namely, the component of the velocities perpendicular to the magnetic field is the same for both the electrons and the ions. (Note that, for the positive ions, the magnetic interaction changes sign so that the imaginary term in the inverse relation also changes sign, which is why the velocities for the electron and the ion have the same sign.) Similarly, for the second root, we obtain

$$v_{e,x} \approx \frac{e}{m_e\Omega_e}E_y,$$

$$v_{i,x} \approx \frac{e}{m_e\Omega_e}E_y. \tag{11.89}$$

Once again, we see that the velocity of the electrons and the ions perpendicular to the direction of the magnetic field are the same. The plasma, therefore, moves as a whole along the direction perpendicular to the magnetic field.

These low frequency waves in a plasma are also known as the Alfvén waves. We note, in particular that when $\omega \ll \Omega_i$ and we have further $\omega_{p,e}^2 \gg \eta\Omega_e^2 = \Omega_e\Omega_i$ (which is quite natural), the roots in (11.86) take even simpler forms

$$n^2 \approx \frac{\omega_{p,e}^2}{\eta\Omega_e^2}, \quad \frac{\omega_{p,e}^2}{\eta\Omega_e^2\cos^2\theta}. \tag{11.90}$$

Conventionally, the quantity

$$v_A = \frac{\sqrt{\Omega_e\Omega_i}}{\omega_{p,e}}c, \tag{11.91}$$

is also known as the Alfvén speed.

11.7 Collisions

We have so far talked about an idealized plasma where there is no collision between the constituents. While this is the case for a very dilute plasma, in a realistic plasma such as in the ionosphere, there are collisions that need to be taken into account. In a partially ionized gas (or in the ionosphere), the electrons can undergo collisions with charge neutral molecules. Charged particles can also "collide" with one another, although the concept of "collision" is different in such a case. The effect of the collisions can be introduced into the dynamical equations through a friction force. For example, the equation of motion for the electron under the influence of an external electric field, when collisions are taken into account, takes the form

$$m\dot{\mathbf{v}} + m\nu\mathbf{v} = -e\mathbf{E}. \tag{11.92}$$

Here, ν denotes the frequency of collisions which can be thought of as the inverse of the average time interval of travel between collisions for an electron.

For a harmonic driving force of frequency ω, we obtain from (11.92)

$$-im\omega\mathbf{v} + m\nu\mathbf{v} = -e\mathbf{E},$$

$$\text{or,} \quad \mathbf{v} = -\frac{ie}{m(\omega + i\nu)}\mathbf{E}. \tag{11.93}$$

As a result, the electron current density can be written as

$$\mathbf{J} = -Ne\mathbf{v} = \frac{iNe^2}{m(\omega + i\nu)}\mathbf{E} = \sigma\mathbf{E}, \tag{11.94}$$

where we have identified

$$\sigma = \frac{iNe^2}{m(\omega + i\nu)} = \frac{i\omega_{\mathrm{p}}^2}{4\pi(\omega + i\nu)} = \frac{i\omega_{\mathrm{p}}^2(\omega - i\nu)}{4\pi(\omega^2 + \nu^2)}. \tag{11.95}$$

We note that this reduces to (11.10) when $\nu = 0$. In the presence of collisions, we see that the proportionality between the current density and the electric field becomes complex which is the behavior of a lossy dielectric. The real part of this proportionality constant can be identified with the conductivity of the plasma and has the form

$$\mathrm{Re}\,\sigma = \mathrm{Re}\,\frac{i\omega_{\mathrm{p}}^2(\omega - i\nu)}{4\pi(\omega^2 + \nu^2)} = \frac{\omega_{\mathrm{p}}^2\nu}{4\pi(\omega^2 + \nu^2)}. \tag{11.96}$$

Substituting the form of the current density in (11.94) into the Maxwell's equations (and using (11.95)), we can determine the permittivity of such a plasma to be

$$\epsilon_p = 1 - \frac{4\pi\sigma}{i\omega} = 1 - \frac{\omega_p^2(\omega - i\nu)}{\omega(\omega^2 + \nu^2)}$$

$$= 1 - \frac{\omega_p^2}{(\omega^2 + \nu^2)} + \frac{i\omega_p^2\nu}{\omega(\omega^2 + \nu^2)}. \tag{11.97}$$

The permittivity is complex signifying that propagation of electromagnetic waves in such a medium will be accompanied by attenuation. The imaginary part of the permittivity, as is clear from (11.96) and (11.97), is related to the conductivity. The analysis of the solutions of the Maxwell's equations can now be undertaken as we have done earlier, but we will not go into the details of this.

11.8 Selected problems

1. i) For a laboratory plasma, the number density of electrons lies between $N = 10^{12}/\text{cm}^3 - 10^{18}/\text{cm}^3$. Determine ω_p for such a plasma. Calculate the same for the plasma in the ionosphere at the F-level, where $N \simeq 10^6/\text{cm}^3$.

   ii) Determine the Debye length for a laboratory plasma at $T = 2000°\text{K}$.

2. Consider a monochromatic plane wave of frequency ω propagating along the z-axis through a plasma. Assume that the nontrivial components of the electric and the magnetic fields are along the x-axis and the y-axis respectively. Calculate the time averaged radiated power per unit area normal to the z-axis when $\omega > \omega_p$ and $\omega < \omega_p$.

3. A monochromatic plane wave in vacuum is incident normally on the plane boundary of a semi-infinite uniform plasma. Find the reflected and the transmitted waves, considering frequencies above and below the plasma frequency. For what value of the frequency is there a change of phase of $\frac{\pi}{2}$ on reflection?

CHAPTER 12

Interaction of charged particles with electromagnetic fields

12.1 Relativistic Lagrangian description

We have seen earlier that Maxwell's equations are manifestly Lorentz covariant. This is easily achieved by combining the electric and the magnetic fields into a second rank anti-symmetric field strength tensor of the form

$$F_{\mu\nu} = \partial_\mu A_\nu - \partial_\nu A_\mu = -F_{\nu\mu}, \qquad \mu, \nu = 0, 1, 2, 3, \tag{12.1}$$

where the four component vector potentials are defined as (see (6.151))

$$A_\mu = (\Phi, -\mathbf{A}). \tag{12.2}$$

It is easily seen from the definition in (12.1) that

$$F_{0i} = \partial_0 A_i - \partial_i A_0 = \frac{1}{c}\frac{\partial A_i}{\partial t} - \nabla_i \Phi = E_i,$$

$$F_{ij} = \partial_i A_j - \partial_j A_i = \nabla_i A_j - \nabla_j A_i$$

$$= -\epsilon_{ijk}(\nabla \times \mathbf{A})_k = -\epsilon_{ijk}B_k, \tag{12.3}$$

where we have used the fact that $A_i = -(\mathbf{A})_i$, $\partial_0 = \frac{1}{c}\frac{\partial}{\partial t}$ and $\partial_i = \nabla_i$.

Given the field strength tensor $F_{\mu\nu}$, we can define a Lorentz invariant action for the free Maxwell theory of the form

$$S = \int \mathrm{d}^4 x\, \mathcal{L} = \int \mathrm{d}^4 x \left(-\frac{1}{16\pi}\eta^{\mu\lambda}\eta^{\nu\rho}F_{\mu\nu}F_{\lambda\rho}\right). \tag{12.4}$$

In addition to being Lorentz invariant, this action is also invariant under a gauge transformation of the form (see (6.166))

$$A_\mu(x) \to A_\mu(x) + \partial_\mu \alpha(x), \tag{12.5}$$

where $\alpha(x)$ is an arbitrary space-time dependent parameter of gauge transformation. Under such a transformation,

$$F_{\mu\nu} \to \partial_\mu(A_\nu + \partial_\nu\alpha) - \partial_\nu(A_\mu + \partial_\mu\alpha) = \partial_\mu A_\nu - \partial_\nu A_\mu$$

$$= F_{\mu\nu}. \tag{12.6}$$

Namely, the field strengths are invariant under a gauge transformation, something that we already know. However, as a consequence of this, it follows that the action in (12.4) is also invariant under the gauge transformation (12.5).

We can derive the dynamical equations from the action in (12.4) as the Euler-Lagrange equations. In this case, since the Lagrangian density depends on the dynamical variable A_μ only through derivatives, the Euler-Lagrange equation takes the form

$$\partial_\mu \frac{\partial \mathcal{L}}{\partial \partial_\mu A_\nu} = 0,$$

or, $\partial_\mu F^{\mu\nu} = 0.$ \tag{12.7}

As we have seen earlier (see (6.157)), Eq. (12.7) gives only two of Maxwell's equations (in vacuum in the absence of sources), namely,

$$\nabla \cdot \mathbf{E} = 0,$$

$$\nabla \times \mathbf{B} = \frac{1}{c} \frac{\partial \mathbf{E}}{\partial t}.$$

The other two equations are contained in the Bianchi identity satisfied by the field strength tensor

$$\partial_\mu F_{\nu\lambda} + \partial_\nu F_{\lambda\mu} + \partial_\lambda F_{\mu\nu} = 0, \tag{12.8}$$

which follows from the definition of the field strength tensors in (12.1).

We have just described Maxwell's equations in the absence of sources. Since Maxwell's equations (with sources) are also Lorentz covariant, we can try to introduce sources in a covariant manner as well. The simplest case is, of course, to consider the interaction of a charged particle with electromagnetic fields. Thus, we first need to give a relativistic description of the motion of a free particle. This is done in a simple manner as follows. First, let us consider a free particle moving along a trajectory in the four dimensional space-time manifold as shown in Fig. 12.1. Unlike the non-relativistic case, here we cannot parameterize the trajectory with t which is not Lorentz invariant. Instead, we note that the invariant interval (length) in a Minkowski manifold is defined as (recall that $x^\mu = (ct, \mathbf{x})$)

$$ds^2 = c^2 d\tau^2 = \eta_{\mu\nu} \, dx^\mu dx^\nu = c^2 dt^2 - d\mathbf{x} \cdot d\mathbf{x}. \tag{12.9}$$

The parameter s and, therefore, the proper time τ are invariant

Figure 12.1: Trajectory of a relativistic particle parameterized by τ.

under a Lorentz transformation. The trajectory can be labeled by these parameters. We note from (12.9) that in the rest frame of the particle where $\frac{d\mathbf{x}}{dt} = 0$, the coordinate time can be identified with the proper time. This will, of course, not be the case in other Lorentz frames.

Parameterizing the trajectory of the particle as $x^\mu(\tau)$, we note that the free particle Newtonian equation can be generalized to

$$m \, \frac{d^2 x^\mu(\tau)}{d\tau^2} = 0, \tag{12.10}$$

where m denotes the rest mass of the particle. This is manifestly Lorentz covariant since m and τ are Lorentz invariant scalars and x^μ is a Lorentz vector. Introducing a relativistic four velocity associated with the particle as

$$u^\mu = \frac{dx^\mu(\tau)}{d\tau} = \gamma(c, \mathbf{v}), \tag{12.11}$$

where

$$\mathbf{v} = \frac{d\mathbf{x}}{dt}, \qquad \gamma = \frac{1}{\sqrt{1 - \frac{\mathbf{v}^2}{c^2}}} = \frac{dt}{d\tau}, \tag{12.12}$$

we see that the four velocity transforms like a vector under a Lorentz transformation and that

$$\eta^{\mu\nu} u_\mu u_\nu = \gamma^2 \left(c^2 - \mathbf{v}^2\right) = c^2, \tag{12.13}$$

which follows from (12.9) (or the definitions in (12.11) and (12.12)). Thus, the components of the four velocity are not all independent, rather they are constrained by (12.13).

The free particle equation in (12.10) can now be written as

$$m \frac{\mathrm{d}u^\mu}{\mathrm{d}\tau} = 0. \tag{12.14}$$

We see that the relation (12.13) is consistent with the equations of motion (12.14). Furthermore, the form of the equation in (12.14) allows us to define a relativistic four momentum associated with the particle as

$$p^\mu = mu^\mu = \gamma mc \left(1, \frac{\mathbf{v}}{c}\right) = \gamma mc \left(1, \boldsymbol{\beta}\right), \tag{12.15}$$

so that the equation of motion, (12.14), can also be written as

$$\frac{\mathrm{d}p^\mu}{\mathrm{d}\tau} = 0. \tag{12.16}$$

It follows now from (12.13) that

$$p^2 = \eta^{\mu\nu} p_\mu p_\nu = m^2 \eta^{\mu\nu} u_\mu u_\nu = m^2 c^2. \tag{12.17}$$

Recalling that $p^\mu = (\frac{E}{c}, \mathbf{p})$ (see (6.146)), we recognize this as the Einstein's relation for a relativistic particle. Furthermore, let us note from Eq. (12.15) that we can now identify

$$E = \gamma mc^2, \qquad \mathbf{p} = \frac{E\boldsymbol{\beta}}{c}, \tag{12.18}$$

which we have used earlier in connection with the method of virtual photons (see (10.56)). In the non-relativistic limit, $\beta \ll 1$ (or $v \ll c$) and we have

$$\gamma \approx 1,$$

$$\tau \approx t,$$

$$u^\mu \approx (c, \mathbf{v}),$$

$$p^\mu \approx m(c, \mathbf{v}), \tag{12.19}$$

so that Eq. (12.16) reduces in this limit to the familiar equation

$$\frac{\mathrm{d}\mathbf{p}}{\mathrm{d}t} = 0.$$

The constraints in (12.11) or (12.17) arise because the system has a local gauge invariance, much like the Maxwell's equations. To see this, let us note that we can write the action for a massive relativistic particle as

$$S = mc \int ds = mc \int d\lambda \frac{ds}{d\lambda}$$

$$= mc \int d\lambda \left(\eta_{\mu\nu} \frac{dx^\mu}{d\lambda} \frac{dx^\nu}{d\lambda} \right)^{\frac{1}{2}} = \int d\lambda \, L, \tag{12.20}$$

where we are assuming that the trajectory of the particle is parameterized by λ and have used Eq. (12.9). It is easy to see that this action is invariant under a local symmetry transformation. Namely, under a reparameterization of the variable λ as

$$\lambda \to \xi = \xi(\lambda), \tag{12.21}$$

we have

$$\frac{dx^\mu}{d\lambda} \to \frac{d\xi}{d\lambda} \frac{dx^\mu}{d\xi}. \tag{12.22}$$

Consequently, under such a reparameterization, the action transforms as

$$S = mc \int d\lambda \left(\eta_{\mu\nu} \frac{dx^\mu}{d\lambda} \frac{dx^\nu}{d\lambda} \right)^{\frac{1}{2}}$$

$$\to mc \int d\lambda \left(\eta_{\mu\nu} \left(\frac{d\xi}{d\lambda} \right)^2 \frac{dx^\mu}{d\xi} \frac{dx^\nu}{d\xi} \right)^{\frac{1}{2}}$$

$$= mc \int d\lambda \frac{d\xi}{d\lambda} \left(\eta_{\mu\nu} \frac{dx^\mu}{d\xi} \frac{dx^\nu}{d\xi} \right)^{\frac{1}{2}}$$

$$= mc \int d\xi \left(\eta_{\mu\nu} \frac{dx^\mu}{d\xi} \frac{dx^\nu}{d\xi} \right)^{\frac{1}{2}} = S. \tag{12.23}$$

This is an invariance under a local transformation much like the gauge transformation in the case of Maxwell's theory. Consequently, we can choose a gauge and, in particular, we can chose the trajectory to be parameterized by the proper time through the identification (which is also known as a gauge choice)

$$\lambda = \tau, \tag{12.24}$$

which leads to

$$\left(\eta_{\mu\nu}\frac{dx^\mu}{d\lambda}\frac{dx^\nu}{d\lambda}\right)^{\frac{1}{2}}\bigg|_{\lambda=\tau} = c\,\frac{d\tau}{d\lambda}\bigg|_{\lambda=\tau} = c,$$

or, $\eta_{\mu\nu}u^\mu u^\nu = c^2.$ (12.25)

With the gauge choice in (12.24), we note that the Lagrangian is a functional only of $\dot{x}^\mu = \frac{dx^\mu}{d\tau}$ so that we can write the conjugate momenta as

$$p_\mu = \frac{\partial L}{\partial \dot{x}^\mu} = \frac{mc\dot{x}_\mu}{(\dot{x}^\nu \dot{x}_\nu)^{\frac{1}{2}}} = m\dot{x}_\mu = mu_\mu,$$ (12.26)

where we have used (12.25). The Euler-Lagrange equation of motion now follows to be

$$\frac{d}{d\tau}\frac{\partial L}{\partial \dot{x}^\mu} = \frac{dp_\mu}{d\tau} = 0,$$ (12.27)

which is the equation for the free particle, as we have seen in Eq. (12.15). Here, we see that it can be derived from the principle of minimum action in the Lagrangian framework (physically implying that the path followed by a free particle between two points is the shortest path). We also note from (12.26) that

$$p^2 = \eta^{\mu\nu}p_\mu p_\nu = m^2 c^2,$$ (12.28)

which is the Einstein relation (12.18). The fact that not all the components of the momenta are independent is a consequence of the gauge invariance which the system possesses. (Let us note here parenthetically that the action in (12.20) is meaningful only for time-like trajectories. For massless particles, an alternative form of the action is more useful.)

So far, we have talked about a free relativistic particle. To describe a relativistic particle subjected to a force, we can generalize the dynamical equation (12.10) as

$$m\frac{d^2x^\mu}{d\tau^2} = f^\mu,$$

or, $\frac{dp^\mu}{d\tau} = f^\mu,$ (12.29)

where f^μ represents the force four vector. From the fact that

$$u_\mu\frac{dp^\mu}{d\tau} = mu_\mu\frac{du^\mu}{d\tau} = \frac{m}{2}\frac{d(u_\mu u^\mu)}{d\tau} = 0,$$ (12.30)

which follows from (12.9), we have

$$u_\mu f^\mu = 0. \tag{12.31}$$

Namely, much like the four velocity and the four momentum, the components of the relativistic force are not all independent. In fact, explicitly we have from (12.31) that

$$u_\mu f^\mu = \gamma(cf^0 - \mathbf{v} \cdot \mathbf{f}) = 0,$$

$$\text{or,} \quad f^0 = \frac{\mathbf{v} \cdot \mathbf{f}}{c} = \boldsymbol{\beta} \cdot \mathbf{f}. \tag{12.32}$$

We can determine the form of the relativistic force as follows. First, we note that in the rest frame of the particle $\mathbf{v} = 0$. In this frame, we have

$$f^0_{\text{rest}} = 0, \qquad \mathbf{f}_{\text{rest}} = \mathbf{F}. \tag{12.33}$$

If we know the force $\mathbf{F}$ in the rest frame of the particle, then we can obtain its relativistic form by transforming to a general Lorentz frame. The coordinate vectors in a frame moving with an instantaneous velocity $\mathbf{v}$ are related to those in the rest frame as

$$x'^0 = \gamma \left(x^0 + \boldsymbol{\beta} \cdot \mathbf{x} \right),$$

$$\mathbf{x}' = \left(\mathbf{x} - \frac{\boldsymbol{\beta}(\boldsymbol{\beta} \cdot \mathbf{x})}{\beta^2} \right) + \gamma \left(\frac{\boldsymbol{\beta}(\boldsymbol{\beta} \cdot \mathbf{x})}{\beta^2} + \boldsymbol{\beta} x^0 \right). \tag{12.34}$$

The physical meaning of Eq. (12.34) is quite clear. Only the component of the coordinates along the direction of the velocity of the frame (and, of course, the time coordinate) transforms, while the components perpendicular to the velocity of the frame do not. (Note that the first parenthesis on the right hand side of the second relation in (12.34) denotes the orthogonal component while the first term in the second parenthesis describes the longitudinal component of the coordinate with respect to $\boldsymbol{\beta}$.) Using this, then, the general form of the force in a frame with an instantaneous velocity $\mathbf{v}$ can be determined from its form in the rest frame to be

$$f^0 = \gamma \boldsymbol{\beta} \cdot \mathbf{F},$$

$$\mathbf{f} = \mathbf{F} + (\gamma - 1) \frac{\boldsymbol{\beta}(\boldsymbol{\beta} \cdot \mathbf{F})}{\beta^2}. \tag{12.35}$$

It can be checked from the form of the components of the force in (12.35) that

$$u_\mu f^\mu = u^0 f^0 - \mathbf{u} \cdot \mathbf{f}$$

$$= c\gamma^2 \boldsymbol{\beta} \cdot \mathbf{F} - c\gamma(1 + (\gamma - 1)) \boldsymbol{\beta} \cdot \mathbf{F} = 0. \tag{12.36}$$

Let us observe that in evaluating the force in a general frame, quantities on the right hand side of (12.35) should be expressed in the new frame where necessary.

Thus, we see that if we know the form of the force in the rest frame of the particle, we can determine its general covariant form through a Lorentz boost. With this, we can now describe the relativistic form of the electromagnetic force acting on a charged particle. However, in the case of electromagnetic interactions, there is a simpler way of deriving this. We note from our earlier studies that the minimal electromagnetic interaction is linear in the (electromagnetic) fields. As we have seen, the electromagnetic field strength is a second rank anti-symmetric tensor. Thus, a natural candidate for a relativistic force appears to be

$$f^\mu = \alpha \, F^{\mu\nu} u_\nu, \tag{12.37}$$

where α is a constant to be determined. Because of the anti-symmetry of the field strength tensor, this form of the force automatically satisfies the constraint (12.31), namely, $u_\mu f^\mu = 0$. Furthermore, the constant α can be fixed by going to the rest frame of the particle, where the force takes the form (in the rest frame $\gamma = 1$, $u_i = 0$)

$$f^0_{\text{rest}} = \alpha \, F^{0i} u_i = 0,$$

$$f^i_{\text{rest}} = \alpha \, F^{i\mu} u_\mu = \alpha(F^{i0} u_0 + F^{ij} u_j) = \alpha \, F^{i0} u_0,$$

$$\text{or,} \quad \mathbf{f}_{\text{rest}} = \alpha c \, \mathbf{E} = \mathbf{F}. \tag{12.38}$$

Recalling that a charged particle at rest feels only an electrostatic force of the form

$$\mathbf{F} = q\mathbf{E},$$

we determine that

$$\alpha = \frac{q}{c}, \tag{12.39}$$

where q represents the charge of the particle. As a result, the relativistic form of the (electromagnetic) force is determined to be

$$f^\mu = \frac{q}{c} \, F^{\mu\nu} u_\nu, \tag{12.40}$$

so that the dynamical equation takes the form

$$\frac{dp^\mu}{d\tau} = f^\mu = \frac{q}{c} F^{\mu\nu} u_\nu. \tag{12.41}$$

It is easy to see from (12.40) now that in a general frame, the components of the force have the forms

$$f^0 = \frac{q}{c} F^{0i} u_i = \frac{\gamma q}{c} \mathbf{v} \cdot \mathbf{E},$$

$$f^i = \frac{q}{c} F^{i\mu} u_\mu = \frac{q}{c} \left(F^{i0} u_0 + F^{ij} u_j \right),$$

$$\text{or,} \quad \mathbf{f} = \gamma q \left(\mathbf{E} + \frac{1}{c} \mathbf{v} \times \mathbf{B} \right), \tag{12.42}$$

where we have used the definitions in (12.3). That this leads to the usual Lorentz force can be seen by noting that

$$\frac{d\mathbf{p}}{d\tau} = \mathbf{f},$$

$$\text{or,} \quad \frac{dt}{d\tau} \frac{d\mathbf{p}}{dt} = \gamma q \left(\mathbf{E} + \frac{1}{c} \mathbf{v} \times \mathbf{B} \right),$$

$$\text{or,} \quad \frac{d\mathbf{p}}{dt} = q \left(\mathbf{E} + \frac{1}{c} \mathbf{v} \times \mathbf{B} \right). \tag{12.43}$$

The form of the relativistic force can also be determined by the conventional method through a Lorentz boost. For example, we know that in the rest frame of the particle, a charged particle will only feel the electrostatic force so that

$$f^0_{\text{rest}} = 0,$$

$$\mathbf{f}_{\text{rest}} = \mathbf{F} = q\mathbf{E}', \tag{12.44}$$

where we have designated the electric field in the rest frame by $\mathbf{E}'$. It follows now from Eq. (12.35) that, in a frame moving with an instantaneous velocity $\mathbf{v}$, the components of the force will have the forms

$$f^0 = \gamma\boldsymbol{\beta} \cdot \mathbf{F} = \frac{\gamma q}{c} \mathbf{v} \cdot \mathbf{E}',$$

$$\mathbf{f} = \mathbf{F} + (\gamma - 1) \frac{\boldsymbol{\beta}(\boldsymbol{\beta} \cdot \mathbf{F})}{\beta^2}$$

$$= q \left(\left(\mathbf{E}' - \frac{\boldsymbol{\beta}(\boldsymbol{\beta} \cdot \mathbf{E}')}{\beta^2} \right) + \gamma \frac{\boldsymbol{\beta}(\boldsymbol{\beta} \cdot \mathbf{E}')}{\beta^2} \right). \tag{12.45}$$

The fields, on the right hand side of (12.45), are the rest frame variables. They can be transformed to the new variables by noting that the components of the electric fields parallel to the velocity of the frame do not transform while the orthogonal components do so that

$$\mathbf{E}'_{\parallel} = \mathbf{E}_{\parallel},$$

$$\mathbf{E}'_{\perp} = \gamma \left(\mathbf{E}_{\perp} + \frac{1}{c} \mathbf{v} \times \mathbf{B} \right). \tag{12.46}$$

Using these in (12.45), we obtain

$$f^0 = \frac{\gamma q}{c} \mathbf{v} \cdot \mathbf{E},$$

$$\mathbf{f} = q \left[\gamma \left(\mathbf{E} - \frac{\boldsymbol{\beta}(\boldsymbol{\beta} \cdot \mathbf{E})}{\beta^2} + \frac{\mathbf{v}}{c} \times \mathbf{B} \right) + \gamma \frac{\boldsymbol{\beta}(\boldsymbol{\beta} \cdot \mathbf{E})}{\beta^2} \right]$$

$$= \gamma q \left(\mathbf{E} + \frac{1}{c} \mathbf{v} \times \mathbf{B} \right), \tag{12.47}$$

which is the same result as in Eq. (12.42).

We can now give a Lagrangian description of the interacting theory as

$$S = \int d\lambda \, L, \tag{12.48}$$

where we have chosen to parameterize the trajectory with λ (later identified with the proper time τ of the particle through a gauge choice) and

$$L = mc \left(\eta_{\mu\nu} \dot{x}^\mu \dot{x}^\nu \right)^{\frac{1}{2}} + \frac{q}{c} A_\mu(x(\lambda)) \dot{x}^\mu. \tag{12.49}$$

Here, the first term is, of course, the Lagrangian for a free particle that we have already studied. The second represents the minimal interaction Lagrangian obtained in the standard manner,

$$S_{\text{int}} = \frac{1}{c} \int dx \, A_\mu(x) j^\mu(x)$$

$$= \frac{q}{c} \int dx \, d\lambda \, A_\mu(x) \frac{dx^\mu(\lambda)}{d\lambda} \delta(x - x(\lambda))$$

$$= \frac{q}{c} \int d\lambda \, A_\mu(x(\lambda)) \dot{x}^\mu, \tag{12.50}$$

where we have used the standard representation for the current density associated with a charged particle, namely,

$$j^\mu(x) = q \int d\lambda \, \frac{dx^\mu(\lambda)}{d\lambda} \delta(x - x(\lambda)). \tag{12.51}$$

The action in (12.48) is invariant under reparameterization and, if we choose the gauge (12.24) (see also (12.26)), then we obtain

$$\Pi_\mu = \frac{\partial L}{\partial \dot{x}^\mu} = m\dot{x}_\mu + \frac{q}{c} A_\mu = p_\mu + \frac{q}{c} A_\mu, \tag{12.52}$$

which shows that the interaction is indeed that of minimal coupling. Here we have identified

$$p^\mu = m\dot{x}^\mu = m\frac{dx^\mu}{d\tau}, \tag{12.53}$$

which is also known as the kinematic momentum (or sometimes also as the mechanical momentum) of the particle. The Euler-Lagrange equation, in this case, leads to

$$\frac{d}{d\tau} \frac{\partial L}{\partial \dot{x}^\mu} - \frac{\partial L}{\partial x^\mu} = 0,$$

or, $\quad \dfrac{d}{d\tau}\left(p_\mu + \dfrac{q}{c} A_\mu\right) - \dfrac{q}{c}(\partial_\mu A^\nu)\dot{x}_\nu = 0.$ $\tag{12.54}$

Noting that

$$\frac{dA_\mu}{d\tau} = (\partial^\nu A_\mu)\dot{x}_\nu, \tag{12.55}$$

the Euler-Lagrange equation in (12.54) takes the form

$$\frac{dp^\mu}{d\tau} = \frac{q}{c}\left(\partial^\mu A^\nu - \partial^\nu A^\mu\right)\dot{x}_\nu = \frac{q}{c} F^{\mu\nu} u_\nu. \tag{12.56}$$

This is indeed the dynamical equation that we have discussed in (12.41).

We have, of course, neglected the dynamics of the electromagnetic fields in this discussion. If we add the Maxwell term (12.4) to the action, then, the complete set of coupled equations of motion takes the form

$$m\frac{dx^\mu}{d\tau} = p^\mu,$$

$$\frac{dp^\mu}{d\tau} = \frac{q}{c} F^{\mu\nu} u_\nu,$$

$$\partial_\mu F^{\mu\nu} = \frac{4\pi}{c} j^\nu, \tag{12.57}$$

where the current density has the form in (12.51) (with the identification $\lambda = \tau$). We note here that sometimes the second equation in (12.57) is also written as

$$m \frac{\mathrm{d}p^\mu}{\mathrm{d}\tau} = \frac{q}{c} F^{\mu\nu} p_\nu, \tag{12.58}$$

with the identification in (12.15).

12.2 Motion in a uniform electric field

Let us next work out the solutions for a few interacting systems. Let us recall that the equations of motion (12.41) can be written as (recall that $p^\mu = m u^\mu = \gamma m(c, \mathbf{v})$ and $\gamma = \frac{\mathrm{d}t}{\mathrm{d}\tau}$)

$$\frac{\mathrm{d}}{\mathrm{d}t} \left(\frac{mc^2}{\sqrt{1 - \frac{\mathbf{v}^2}{c^2}}} \right) = q\mathbf{v} \cdot \mathbf{E},$$

$$\frac{\mathrm{d}}{\mathrm{d}t} \left(\frac{m\mathbf{v}}{\sqrt{1 - \frac{\mathbf{v}^2}{c^2}}} \right) = q \left(\mathbf{E} + \frac{1}{c} \mathbf{v} \times \mathbf{B} \right), \tag{12.59}$$

where we have used the identifications in (12.42). Let us note some general characteristics following from these equations. First, for a static electric field, we note that we can write

$$\mathbf{E} = -\boldsymbol{\nabla}\Phi,$$

so that the first equation in (12.59) takes the form

$$\frac{\mathrm{d}}{\mathrm{d}t} \left(\frac{mc^2}{\sqrt{1 - \frac{\mathbf{v}^2}{c^2}}} \right) = -q \frac{\mathrm{d}\mathbf{x}}{\mathrm{d}t} \cdot \boldsymbol{\nabla}\Phi = -q \frac{\mathrm{d}\Phi}{\mathrm{d}t},$$

$$\text{or,} \quad \frac{mc^2}{\sqrt{1 - \frac{\mathbf{v}^2}{c^2}}} + q\Phi = \text{constant}. \tag{12.60}$$

This is known as the energy relation since in the non-relativistic limit, it gives the energy conservation relation. (The time component of p^μ is related to the energy.) This equation also implies that when the electric field vanishes, namely, $\mathbf{E} = 0$ (or $\Phi = 0$), the speed of the particle $v = |\mathbf{v}|$ must be a constant. This would also be the case for a static magnetic field which does not produce an electric field (only a time dependent magnetic field can generate an electric field).

With these general observations, let us now examine the solutions for the motion of a charged particle in the presence of a uniform electric field. If we assume the electric field to be along the x-axis, then we have

$$\mathbf{E} = (E, 0, 0), \qquad \mathbf{B} = 0,$$

where E is a constant and the fields can be generated by a scalar potential of the form

$$\Phi = -Ex, \qquad \mathbf{A} = 0. \tag{12.61}$$

In this case, the equations of motion in (12.59) can be trivially integrated to give

$$\frac{mc^2}{\sqrt{1 - \frac{v^2}{c^2}}} = qEx,$$

$$\frac{mv}{\sqrt{1 - \frac{v^2}{c^2}}} = qEt, \tag{12.62}$$

where we are assuming that the particle is at rest at $t = 0$ so that the velocity is along the x-axis and, correspondingly, we have set the constants of integrations to zero. The ratio of the pair of equations in (12.62) determines

$$\frac{v}{c} = \frac{ct}{x}. \tag{12.63}$$

Substituting this into the second equation in (12.62), we obtain

$$\frac{\frac{mc^2 t}{x}}{\sqrt{1 - \frac{c^2 t^2}{x^2}}} = qEt,$$

$$\text{or,} \quad \sqrt{x^2 - c^2 t^2} = \frac{mc^2}{qE}, \tag{12.64}$$

which can also be rewritten as

$$x^2 - c^2 t^2 = \frac{m^2 c^4}{q^2 E^2},$$

$$\text{or,} \quad x = c\sqrt{\frac{m^2 c^2}{q^2 E^2} + t^2}. \tag{12.65}$$

The first form of the relation in (12.65) clearly shows that the motion is hyperbolic in the presence of a constant acceleration. This has to be contrasted with the non-relativistic case where

$$x = x_0 + \frac{1}{2}at^2,$$

which describes a parabolic motion. The instantaneous velocity of the particle can be obtained from (12.65) and leads to

$$v = \frac{dx}{dt} = \frac{ct}{\sqrt{\frac{m^2c^2}{q^2E^2} + t^2}}. \tag{12.66}$$

We see from this that, as $t \to \pm\infty$, the magnitude of the velocity approaches the speed of light, $v \to c$. However, as $t \to 0$, the particle comes to rest ($v \to 0$). Therefore, in the intermediate time interval ($-\infty < t \leq 0$), the particle seems to feel a deceleration. We also recognize that the motion will become non-relativistic when the electric field is weak. In particular, if

$$\frac{qE}{mc} \ll 1, \tag{12.67}$$

then, (in a time interval $t \ll \frac{mc}{qE}$) we have from Eqs. (12.65) and (12.66),

$$v \approx \frac{qE}{m}t = at,$$

$$x \approx \frac{mc^2}{qE} + \frac{1}{2}\frac{qE}{m}t^2$$

$$= x_0 + \frac{1}{2}at^2, \tag{12.68}$$

which is what we will expect. Here, we have introduced a to represent the instantaneous acceleration.

In the derivation above, we assumed that the particle is initially at rest. However, in the general case when the particle initially has a momentum $\mathbf{p}^{(0)}$, we can solve the system of equations in the following manner. First, let us note that the initial momentum and the electric field define a plane where the particle motion is confined to. Without loss of generality, we can choose this to be the $x - y$ plane (recall that we are assuming the electric field to be along the x-axis). Thus, our equations in this case lead to

$$\frac{mc^2}{\sqrt{1 - \frac{v^2}{c^2}}} = qEx,$$

$$\frac{m\mathbf{v}}{\sqrt{1 - \frac{v^2}{c^2}}} = \left(\mathbf{p}^{(0)} + qEt\right). \tag{12.69}$$

By assumption, of course, the velocity has only x and y components. From the second equation in (12.69), we obtain

$$m^2 c^2 \left(\frac{v^2}{c^2}\right) = \left(\mathbf{p}^{(0)} + qEt\right)^2 \left(1 - \frac{v^2}{c^2}\right),$$

$$\text{or,} \quad \frac{v^2}{c^2} \left(m^2 c^2 + \left(\mathbf{p}^{(0)} + qEt\right)^2\right) = \left(\mathbf{p}^{(0)} + qEt\right)^2,$$

$$\text{or,} \quad \frac{v^2}{c^2} = \frac{\left(\mathbf{p}^{(0)} + qEt\right)^2}{m^2 c^2 + \left(\mathbf{p}^{(0)} + qEt\right)^2},$$

$$\text{or,} \quad 1 - \frac{v^2}{c^2} = \frac{m^2 c^2}{m^2 c^2 + \left(\mathbf{p}^{(0)} + qEt\right)^2}. \tag{12.70}$$

Substituting this into the first equation in (12.69), we obtain

$$x^2 = \frac{\frac{m^2 c^4}{q^2 E^2}}{1 - \frac{v^2}{c^2}} = \frac{c^2}{q^2 E^2} \left(m^2 c^2 + \left(\mathbf{p}^{(0)} + qEt\right)^2\right),$$

$$\text{or,} \quad x = \frac{c}{qE} \sqrt{m^2 c^2 + \left(\mathbf{p}^{(0)} + qEt\right)^2}. \tag{12.71}$$

This, of course, reduces to Eq. (12.65) when the particle is initially at rest, namely, $\mathbf{p}^{(0)} = 0$. We can now solve for the y coordinate of the particle from (12.69), which yields

$$\frac{dy}{dt} = \frac{p_y^{(0)}}{m} \sqrt{1 - \frac{v^2}{c^2}}$$

$$= \frac{c p_y^{(0)}}{\sqrt{m^2 c^2 + (\mathbf{p}^{(0)})^2 + 2q\mathbf{p}^{(0)} \cdot \mathbf{E}t + q^2 E^2 t^2}},$$

$$\text{or,} \quad y = \frac{c p_y^{(0)}}{qE} \sinh^{-1} \left(\frac{qE^2 t + \mathbf{p}^{(0)} \cdot \mathbf{E}}{\sqrt{E^2 (m^2 c^2 + (\mathbf{p}^{(0)})^2) - (\mathbf{p}^{(0)} \cdot \mathbf{E})^2}}\right). \tag{12.72}$$

Here, we have used the standard formula from the table of integrals (see, for example, Gradshteyn and Ryzhik, 2.261). We note from Eq. (12.70) that as $t \to \pm\infty$, the speed of the particle approaches the speed of light as before. However, in the present case for the particle moving in a plane, the motion cannot be characterized as hyperbolic.

12.3 Motion in a uniform magnetic field

Let us next consider the case of a charged particle moving in a uniform magnetic field which we can take to be along the z-axis, namely, we have

$$\mathbf{E} = 0, \qquad \mathbf{B} = B\hat{\mathbf{z}}, \tag{12.73}$$

where B is a constant. In this case, the dynamical equations, (12.59), take the forms

$$\frac{\mathrm{d}}{\mathrm{d}t}\left(\frac{mc^2}{\sqrt{1 - \frac{v^2}{c^2}}}\right) = 0,$$

$$\frac{\mathrm{d}}{\mathrm{d}t}\left(\frac{m\mathbf{v}}{\sqrt{1 - \frac{v^2}{c^2}}}\right) = \frac{qB}{c}\,\mathbf{v} \times \hat{\mathbf{z}}. \tag{12.74}$$

It is clear from the first equation in (12.74) that v^2 is a constant and we can write

$$\gamma = \frac{1}{\sqrt{1 - \frac{v^2}{c^2}}} = \frac{\mathcal{E}}{mc^2}, \tag{12.75}$$

where $\mathcal{E}$ represents the energy of the system (recall that the first equation gives the energy relation and we are denoting energy by $\mathcal{E}$ to avoid any confusion with the electric field). Furthermore, from the second equation we see that v_z is also a constant. It follows, therefore, that

$$v_\perp = \sqrt{v^2 - v_z^2}, \tag{12.76}$$

is a constant as well. Since the z component of the velocity is a constant, we note that the solution for the z coordinate of the particle is straightforward

$$z = z_0 + v_z t. \tag{12.77}$$

The second equation can now be written out explicitly for the x, y components of the velocity to give

$$\frac{dv_x}{dt} = \frac{qB}{\gamma mc} v_y,$$

$$\frac{dv_y}{dt} = -\frac{qB}{\gamma mc} v_x. \tag{12.78}$$

Recalling that $v_x^2 + v_y^2 = v_\perp^2$ is a constant, we can write the solutions of Eq. (12.78) as

$$v_x = -v_\perp \sin\left(\frac{qB}{\gamma mc} t\right),$$

$$v_y = -v_\perp \cos\left(\frac{qB}{\gamma mc} t\right). \tag{12.79}$$

These can be integrated so that, together with (12.77), we have

$$x = \frac{\gamma mcv_\perp}{qB} \cos\left(\frac{qB}{\gamma mc} t\right),$$

$$y = -\frac{\gamma mcv_\perp}{qB} \sin\left(\frac{qB}{\gamma mc} t\right),$$

$$z = z_0 + v_z t. \tag{12.80}$$

Here, we have chosen the constants of integration (as well as the phases) such that at $t = 0$, the particle has coordinates $x = \frac{\gamma mcv_\perp}{qB}, y = 0, z = z_0$ and $v_x = 0, v_y = -v_\perp, v_z = $ constant. We note that the motion in the $x - y$ plane is harmonic with a frequency

$$\omega = \frac{qB}{\gamma mc} = \frac{qBc}{\mathcal{E}}, \tag{12.81}$$

where we have used (12.75). In general, we see that the frequency of motion depends on the energy of the particle. However, in the non-relativistic limit, where $\gamma \approx 1$, the expression for the frequency becomes

$$\omega \approx \frac{qB}{mc}, \tag{12.82}$$

which is independent of energy. This frequency is commonly known as the cyclotron frequency and is important in the study of cyclotrons. For a relativistic particle moving in a uniform magnetic field, however, the relation is quite different and is of significance in the study of synchrotrons. Let us also note that, for $v_z = 0$, the motion is circular in the $x - y$ plane with the radius of the circle given by

$$R = \frac{v_\perp}{\omega} = \frac{\gamma m c v_\perp}{qB} = \frac{\mathcal{E} v_\perp}{qBc}, \tag{12.83}$$

so that we can write the planar coordinates in (12.80) also as

$$x = R \cos \omega t = R \cos \left(\frac{qB}{\gamma mc} t \right),$$

$$y = -R \sin \omega t = -R \sin \left(\frac{qB}{\gamma mc} t \right). \tag{12.84}$$

12.4 Motion in uniform crossed electric and magnetic fields

Let us next consider the case of a charged particle moving in uniform electric and magnetic fields that are orthogonal to each other. For example, we may have

$$\mathbf{E} = E\hat{\mathbf{x}}, \qquad \mathbf{B} = B\hat{\mathbf{z}}, \tag{12.85}$$

so that

$$\mathbf{E} \cdot \mathbf{B} = 0. \tag{12.86}$$

We can, of course, solve the equations (12.59) as before. However, a direct solution is much more involved. There is an alternative and simpler method for analyzing the motion in this case that we describe below.

Let us recall that the electromagnetic field strength tensor $F_{\mu\nu}$ is a second rank anti-symmetric tensor. We can construct from this two independent quadratic Lorentz invariant scalars, namely

$$F_{\mu\nu} F^{\mu\nu} = \eta^{\mu\lambda} \eta^{\nu\rho} F_{\mu\nu} F_{\lambda\rho}, \qquad F_{\mu\nu} \tilde{F}^{\mu\nu} = \frac{1}{2} \epsilon^{\mu\nu\lambda\rho} F_{\mu\nu} F_{\lambda\rho}, \tag{12.87}$$

where $\epsilon^{\mu\nu\lambda\rho}$ is the four dimensional Levi-Civita tensor. From the definitions of the field strength tensors in (12.3), it follows that

$$F_{\mu\nu} F^{\mu\nu} = 2 F_{0i} F^{0i} + F_{ij} F^{ij} = 2(\mathbf{B}^2 - \mathbf{E}^2). \tag{12.88}$$

Furthermore, since ($\tilde{F}^{\mu\nu} = \frac{1}{2}\epsilon^{\mu\nu\lambda\rho}F_{\lambda\rho}$ is also known as the dual field strength tensor, $\epsilon^{0ijk} = \epsilon_{ijk}$)

$$\tilde{F}^{0i} = \frac{1}{2}\,\epsilon^{0ijk}F_{jk} = -B_i,$$

$$\tilde{F}^{ij} = \epsilon^{ij0k}F_{0k} = \epsilon_{ijk}E_k, \tag{12.89}$$

it follows that

$$F_{\mu\nu}\tilde{F}^{\mu\nu} = 2F_{0i}\tilde{F}^{0i} + F_{ij}\tilde{F}^{ij} = -4\mathbf{E}\cdot\mathbf{B}. \tag{12.90}$$

Thus, we see that $\mathbf{E}\cdot\mathbf{B}$ as well as $(\mathbf{B}^2 - \mathbf{E}^2)$ are invariant under Lorentz transformations so that they have the same value in any Lorentz frame. As a result, when $\mathbf{E}\cdot\mathbf{B} = 0$ (namely, when $\mathbf{E}$ and $\mathbf{B}$ fields are orthogonal to each other), depending on whether

$$(\mathbf{B}^2 - \mathbf{E}^2) > 0, \quad \Rightarrow |\mathbf{B}| > |\mathbf{E}|,$$

$$\text{or,} \quad (\mathbf{B}^2 - \mathbf{E}^2) < 0, \quad \Rightarrow |\mathbf{B}| < |\mathbf{E}|, \tag{12.91}$$

we can go to a Lorentz frame where either $\mathbf{E} = 0$ or $\mathbf{B} = 0$ respectively. (Note that if $\mathbf{E}\cdot\mathbf{B} \neq 0$, we cannot go to a frame where either $\mathbf{B} = 0$ or $\mathbf{E} = 0$.) Once we are in the frame where either the electric or the magnetic field is zero, then the solution in that frame is exactly the same as the ones we have found in the earlier two sections. To come back to the solutions in the original Lorentz frame, we simply have to make the appropriate inverse Lorentz transformation. Let us recall that in going to a Lorentz frame moving with a velocity $\mathbf{u}$, the electric and the magnetic fields transform as

$$\mathbf{E}'_{\parallel} = \mathbf{E}_{\parallel},$$

$$\mathbf{E}'_{\perp} = \gamma(u)(\mathbf{E}_{\perp} + \frac{\mathbf{u}}{c}\times\mathbf{B}_{\perp}),$$

$$\mathbf{B}'_{\parallel} = \mathbf{B}_{\parallel},$$

$$\mathbf{B}'_{\perp} = \gamma(u)(\mathbf{B}_{\perp} - \frac{\mathbf{u}}{c}\times\mathbf{E}_{\perp}), \tag{12.92}$$

where the "parallel" and the "perpendicular" decompositions are with respect to the velocity $\mathbf{u}$.

Let us first consider the case when $|\mathbf{E}| = E > B = |\mathbf{B}|$ and consider a frame moving with a velocity $\mathbf{u}$ such that

$$\frac{\mathbf{u}}{c} = \frac{\mathbf{E}\times\mathbf{B}}{E^2}. \tag{12.93}$$

Such a transformation (frame) is allowed since

$$\frac{u^2}{c^2} = \frac{E^2 B^2 - (\mathbf{E} \cdot \mathbf{B})^2}{E^4} = \frac{B^2}{E^2} < 1, \tag{12.94}$$

where we have used the fact that the electric and the magnetic fields are orthogonal. In this case, therefore, we have

$$\gamma(u) = \frac{1}{\sqrt{1 - \frac{u^2}{c^2}}} = \frac{1}{\sqrt{1 - \frac{B^2}{E^2}}}. \tag{12.95}$$

We note that such a velocity is orthogonal to both the electric and the magnetic fields ($\mathbf{u} \cdot \mathbf{E} = 0 = \mathbf{u} \cdot \mathbf{B}$) and, consequently, in this case, we have

$$\mathbf{E}_{\parallel} = 0 = \mathbf{B}_{\parallel}, \qquad \Rightarrow \mathbf{E}'_{\parallel} = 0 = \mathbf{B}'_{\parallel}$$

Furthermore, under such a transformation, we have (since there is no component of the fields parallel to the velocity)

$$\mathbf{E}' = \gamma(u) \left(\mathbf{E} + \frac{\mathbf{u}}{c} \times \mathbf{B} \right) = \gamma(u) \left(\mathbf{E} + \frac{(\mathbf{E} \times \mathbf{B}) \times \mathbf{B}}{E^2} \right)$$

$$= \gamma(u) \left(\mathbf{E} + \frac{(\mathbf{E} \cdot \mathbf{B})\mathbf{B} - \mathbf{E}(\mathbf{B}^2)}{E^2} \right)$$

$$= \sqrt{1 - \frac{B^2}{E^2}} \, \mathbf{E} = \gamma^{-1}(u) \, \mathbf{E},$$

$$\mathbf{B}' = \gamma(u) \left(\mathbf{B} - \frac{\mathbf{u}}{c} \times \mathbf{E} \right) = \gamma(u) \left(\mathbf{B} - \frac{(\mathbf{E} \times \mathbf{B}) \times \mathbf{E}}{E^2} \right)$$

$$= \gamma(u) \left(\mathbf{B} - \frac{(\mathbf{E}^2)\mathbf{B} - (\mathbf{E} \cdot \mathbf{B})\mathbf{E}}{E^2} \right)$$

$$= \gamma(u)(\mathbf{B} - \mathbf{B}) = 0. \tag{12.96}$$

Here, we have used the orthogonality of the electric and the magnetic fields in the intermediate steps.

This shows that for orthogonal electric and magnetic fields, when $E > B$, we can find a Lorentz frame where the magnetic field identically vanishes. Furthermore, the electric field, in this new frame, is along the same direction as the original field, only scaled by a Lorentz factor. The solution to this problem is, as before, unbounded motion and is not very interesting.

Let us next consider the case when $E < B$ and choose a Lorentz transformation with

$$\frac{\mathbf{u}}{c} = \frac{\mathbf{E} \times \mathbf{B}}{B^2}. \tag{12.97}$$

In this case, we have

$$\frac{u^2}{c^2} = \frac{E^2 B^2 - (\mathbf{E} \cdot \mathbf{B})^2}{B^4} = \frac{E^2}{B^2} < 1, \tag{12.98}$$

so that it is an allowed transformation. For the present case,

$$\gamma(u) = \frac{1}{\sqrt{1 - \frac{u^2}{c^2}}} = \frac{1}{\sqrt{1 - \frac{E^2}{B^2}}}. \tag{12.99}$$

We note here parenthetically that when $E = B$, the frame moves with the speed of light and, therefore, is not meaningful.)

Once again the electric and the magnetic fields are orthogonal to the velocity $\mathbf{u}$ and, consequently,

$$\mathbf{E}_\| = 0 = \mathbf{B}_\|.$$

Furthermore, we have

$$\mathbf{E}' = \gamma(u) \left(\mathbf{E} + \frac{\mathbf{u}}{c} \times \mathbf{B} \right) = \gamma(u) \left(\mathbf{E} + \frac{(\mathbf{E} \times \mathbf{B}) \times \mathbf{B}}{B^2} \right)$$

$$= \gamma(u) \left(\mathbf{E} + \frac{(\mathbf{E} \cdot \mathbf{B})\mathbf{B} - (\mathbf{B}^2)\mathbf{E}}{B^2} \right)$$

$$= \gamma(u)(\mathbf{E} - \mathbf{E}) = 0,$$

$$\mathbf{B}' = \gamma(u) \left(\mathbf{B} - \frac{\mathbf{u}}{c} \times \mathbf{E} \right) = \gamma(u) \left(\mathbf{B} - \frac{(\mathbf{E} \times \mathbf{B}) \times \mathbf{E}}{B^2} \right)$$

$$= \gamma(u) \left(\mathbf{B} - \frac{(\mathbf{E}^2)\mathbf{B} - (\mathbf{E} \cdot \mathbf{B})\mathbf{E}}{B^2} \right)$$

$$= \sqrt{1 - \frac{E^2}{B^2}} \, \mathbf{B} = \gamma^{-1}(u) \, \mathbf{B}. \tag{12.100}$$

Thus, in this case, we see that the electric field vanishes in the new frame and the magnetic field is along the same direction as the original field, but scaled by a field dependent Lorentz factor. For simplicity of comparison with earlier results, let us choose $\mathbf{E} = E\,\hat{\mathbf{x}}$ and $\mathbf{B} = B\,\hat{\mathbf{z}}$. In such a case, we see that

$$\frac{\mathbf{u}}{c} = \frac{\mathbf{E} \times \mathbf{B}}{B^2} = -\frac{E}{B}\,\hat{\mathbf{y}}. \tag{12.101}$$

Namely, the frame is moving along the y axis. If we assume that the particle has no initial velocity along the z-axis, then following the analysis of the last section, we can obtain the solutions which would suggest that the particle will be moving in a circle in the plane perpendicular to the magnetic field with a radius

$$R' = \frac{\gamma(v')mcv'_\perp}{qB'} = \frac{\gamma(v')mcv'_\perp}{qB\sqrt{1 - \frac{E^2}{B^2}}}. \tag{12.102}$$

In fact, with initial conditions as before, we can write the solutions to have the form (see (12.80))

$$x' = R'\cos\left(\frac{qB'}{\gamma(v')mc}\,t'\right),$$

$$y' = -R'\sin\left(\frac{qB'}{\gamma(v')mc}\,t'\right) = -\sqrt{R'^2 - x'^2}. \tag{12.103}$$

The solution can now be transformed back into the original frame. Since the frame is moving along the y-axis, the Lorentz transformations take the forms

$$t' = \gamma(u)\left(t - \frac{\beta(u)}{c}y\right),$$

$$x' = x,$$

$$y' = \gamma(u)\,(y - \beta(u)ct),$$

$$z' = z, \tag{12.104}$$

and this leads to

$$x = x' = R\cos\left[\frac{qB\sqrt{1 - \frac{E^2}{B^2}}}{\gamma(v)mc(1 - \frac{qER}{\gamma(v)mc^2})}\left(t - \frac{E}{Bc}y\right)\right],$$

$$y' = \gamma(u)(y - \beta(u)ct) = -\sqrt{R^2 - x^2},$$

$$\text{or,} \quad y = \frac{Ec}{B}t - \sqrt{1 - \frac{E^2}{B^2}}\sqrt{R^2 - x^2}, \tag{12.105}$$

where, as before,

$$R = \frac{\gamma(v)mcv_\perp}{qB}.$$

Here we have also used the fact that, under an inverse Lorentz transformation,

$$R' \rightarrow R, \quad v' \rightarrow v, \quad v'_\perp \rightarrow v_\perp,$$

and so on. In this case, we see that the motion corresponds to an ellipsoidal motion superimposed with a constant "drift" velocity along the y axis. This is known as a trochoidal motion.

We note from the equations of motion (12.59), that when

$$\frac{\mathbf{v}}{c} = \frac{\mathbf{E} \times \mathbf{B}}{B^2} = \frac{\mathbf{u}}{c}, \tag{12.106}$$

the Lorentz force identically vanishes (see also the first relation in (12.100)), namely,

$$q\left(\mathbf{E} + \frac{(\mathbf{E} \times \mathbf{B}) \times \mathbf{B}}{B^2}\right) = q(\mathbf{E} - \mathbf{E}) = 0. \tag{12.107}$$

In this case, the particle would move along the initial trajectory completely undeflected by the presence of the fields, independent of its mass. (Namely, $\mathbf{f} = 0$ also implies that $f^0 = \frac{\mathbf{v}\cdot\mathbf{f}}{c} = 0$, see (12.32).) This is a very important feature which is utilized in creating a velocity filter. Namely, if a number of particles are incident on a region with crossed electric and magnetic fields (with $E < B$), then only those particles that have the initial velocity coinciding with (12.106) would travel undeflected. Correspondingly, one can choose different electric and magnetic fields to select the desired particles with a given velocity.

12.5 Motion in a slowly varying magnetic field

The fact that the magnetic force $\mathbf{v} \times \mathbf{B}$ introduces a curvature to the trajectory of a charged particle is exploited profitably in many physical situations. For example, this is quite useful in the study of confinement of plasma as well as in astrophysical studies of plasma. Let us recall that a charged particle in a uniform magnetic field, has a circular path with a "drift" velocity along the direction of $\mathbf{B}$ (normally taken to be along the z-axis). (Sometimes, one separates out the circular motion to talk of a uniform motion of the "guiding center".) However, in many of these physical systems of interest the magnetic field is not uniform. Although the magnetic field is static, it varies with space, particularly longitudinally. A general problem of this kind is, of course, very hard to solve. However, when the field varies slowly, we can determine the behavior of such systems quite well. Namely, if the field varies slowly within the radius of curvature of the trajectory, we can think of the particle to be still executing circular motion with a radius

$$R = \frac{\gamma m c v_\perp}{qB}, \tag{12.108}$$

where both $v_\perp, B$ will now be functions of position (although slowly varying).

If we assume that the magnetic field is everywhere parallel, then without loss of generality, we can choose this to be along the z-axis, $\mathbf{B} = B\hat{\mathbf{z}}$. However, in this case, it follows from the second equation of Maxwell

$$\boldsymbol{\nabla} \cdot \mathbf{B} = 0,$$

that $B = B(x, y)$ and that, since in general, the curl of the magnetic field will not vanish, such a configuration must have an associated steady current and we see that, in this case, the magnetic field cannot depend on the (longitudinal) z coordinate. As we noted earlier, in some physical situations, we do need a dependence of the magnetic field on the longitudinal coordinate which can arise only if the magnetic field is not parallel everywhere. For confinement of plasma to a small region in space, for example, we would expect the plasma not to extend beyond a certain vertical and horizontal dimension. Under the action of a magnetic field, as we have seen, the particle moves in an orbit whose radius is determined by the magnetic field. Thus, we see that with a suitable choice of a magnetic field, the plasma can be easily confined to a given vertical dimension. Let us note also that if we have a magnetic field that is converging along the z-axis in some region, then the magnetic force acting on the charged particles will be so as to force it into the interior of the region (see Fig. 12.2). Therefore, there will be a component of the force along the z-axis which would decelerate the "drift" velocity. As a result, at some point along the horizontal (z) axis, the "drift" velocity will vanish (the "guiding center" will come to rest) and then, the direction of the drift will change. This would lead to a containment of the charged particle along the z-axis as well, as shown in Fig. 12.3. This process is known as "mirroring" and is used to trap plasma by a

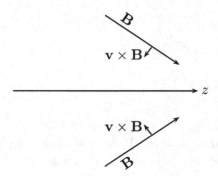

Figure 12.2: The Lorentz force directing the particle into the interior of the region.

magnetic field by forming a "bottle" with one "mirror" point at each end (namely, converging magnetic fields at both ends).

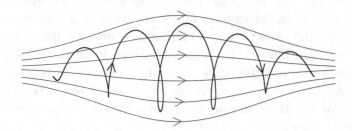

Figure 12.3: Confinement due to a magnetic field (magnetic bottle).

Keeping this qualitative picture in mind, let us consider a magnetic field with cylindrical symmetry, $\mathbf{B} = (B_\rho, 0, B_z)$. We assume that the components of the magnetic field do not depend on the angular coordinate. In this case, the second equation of Maxwell leads to

$$\boldsymbol{\nabla} \cdot \mathbf{B} = \frac{1}{\rho} \frac{\partial(\rho B_\rho)}{\partial \rho} + \frac{\partial B_z}{\partial z} = 0,$$

or, $\quad \dfrac{\partial(\rho B_\rho)}{\partial \rho} = -\rho \dfrac{\partial B_z}{\partial z}.$ \hfill (12.109)

To leading order, if we assume that $\frac{\partial B_z}{\partial z}$ is independent of ρ, this equation can be integrated and leads to

$$\rho B_\rho = -\frac{1}{2} \rho^2 \frac{\partial B_z}{\partial z},$$

$$\text{or,} \quad B_\rho = -\frac{1}{2} \rho \frac{\partial B_z}{\partial z}. \tag{12.110}$$

The constant of integration is easily seen to vanish (with the assumption that $B_\rho = 0$ at $\rho = 0$). It is clear now from (12.110), that for slowly varying fields ($|\frac{R}{B}\frac{\partial B}{\partial z}| \ll 1$), B_ρ is much smaller than B_z within a radius of the size of R. Consequently, we can approximate

$$B = |\mathbf{B}| \approx B_z, \tag{12.111}$$

to write

$$B_\rho \approx -\frac{1}{2} \rho \frac{\partial B}{\partial z}. \tag{12.112}$$

From the time component of the equations of motion, (12.59), we note that in the absence of an electric field, v is constant. However, unlike the earlier case of motion in a uniform magnetic field, here a non-vanishing B_ρ leads to a magnetic force along the z-axis. As a result, v_z and, therefore, $v_\perp$ will no longer be constant. In fact, looking at the equation for the z component of the velocity (see (12.59)) for a particle moving in an orbit of radius R, we have (γ is a constant since v is)

$$\frac{d}{dt}(\gamma m v_z) = \frac{q}{c} v_\perp B_\rho,$$

$$\text{or,} \quad \frac{dv_z}{dt} = -\frac{1}{2}\frac{qRv_\perp}{\gamma mc}\frac{\partial B}{\partial z} = -\frac{1}{2}\frac{v_\perp^2}{B}\frac{\partial B}{\partial z}$$

$$\approx -\frac{1}{2}\frac{v_\perp^2(0)}{B(0)}\frac{\partial B}{\partial z}. \tag{12.113}$$

Here, we have used the fact that the velocity of the particle, in the $x-y$ plane, is given by $\mathbf{v}_\perp = -v_\perp \hat{\phi}$ (see (12.79)), $v_\perp = \frac{qRB}{\gamma mc}$ (see (12.83)) as well as the relation in (12.112) for a particle moving in an orbit of radius R. Furthermore, since the fields are slowly varying, we have approximated the coefficient multiplying $\frac{\partial B}{\partial z}$ (which is already small) by its value at $z = 0$ which can be thought of as the leading order approximation. It now follows from (12.113) that (the assumption here is that at $t = 0$, the particle is at $z = 0$ and that at t its coordinate is z)

$$\frac{1}{2}\frac{dv_z^2}{dz} = v_z\frac{dv_z}{dz} = \frac{dz}{dt}\frac{dv_z}{dz} = \frac{dv_z}{dt} = -\frac{1}{2}\frac{v_\perp^2(0)}{B(0)}\frac{\partial B}{\partial z},$$

$$\text{or,} \quad \frac{1}{2}\left(v_z^2 - v_z^2(0)\right) = -\frac{v_\perp^2(0)}{2B(0)}\left(B(z) - B(0)\right),$$

$$\text{or,} \quad v_z^2 = v^2 - \frac{v_\perp^2(0)B(z)}{B(0)}, \tag{12.114}$$

where we have used

$$v_z^2(0) + v_\perp^2(0) = v^2(0) = v^2, \tag{12.115}$$

since v is a constant (because there is no electric field). Furthermore, using (12.115), we can rewrite the relation in (12.114) as

$$v^2 - v_z^2 = v_\perp^2 = \frac{v_\perp^2(0)B(z)}{B(0)},$$

$$\text{or,} \quad \frac{v_\perp^2}{B(z)} = \frac{v_\perp^2(0)}{B(0)}. \tag{12.116}$$

Relation (12.116) gives an adiabatic invariant associated with this problem. Namely, it says that the ratio of the square of the perpendicular velocity to the magnetic field

$$\frac{v_\perp^2}{B(z)} = \frac{qRv_\perp}{\gamma mc}, \tag{12.117}$$

is an adiabatic constant of motion. Physically, what this means is that an adiabatic invariant is not a true constant of motion because of the variation in the fields. However, when the fields change slowly (adiabatically), the change in an adiabatic invariant is even slower than that of the field. We note from (12.117) that as $B(z)$ increases, $v_\perp$ must also increase for this ratio to be constant. However, since $v = \sqrt{v_\perp^2 + v_z^2}$ is a constant, this also means that when $B(z)$ increases, v_z must decrease, something that we had seen qualitatively earlier. The magnetic flux associated with the particle motion (through a circle of radius R) can now be seen to be an adiabatic invariant for the problem as well. Namely,

$$\pi R^2 B = \pi \frac{\gamma^2 m^2 c^2 v_\perp^2}{q^2 B^2} B$$

$$= \frac{\pi \gamma^2 m^2 c^2}{q^2} \frac{v_\perp^2}{B(z)}$$

$$= \frac{\pi \gamma^2 m^2 c^2}{q^2} \frac{v_\perp^2(0)}{B(0)}, \qquad (12.118)$$

where we have used (12.108) and Eq. (12.118) follows by virtue of (12.116). Similarly, we can define a magnetic moment associated with the particle, which will also be an adiabatic invariant for the motion, namely,

$$\mu = \pi R^2 \frac{q\omega}{2\pi} = \frac{q\omega R^2}{2}$$

$$= \frac{\gamma mc}{2} \frac{v_\perp^2}{B(z)} = \frac{\gamma mc}{2} \frac{v_\perp^2(0)}{B(0)}. \qquad (12.119)$$

All the adiabatic invariants quantify the same qualitative behavior that we had discussed earlier. For example, from (12.117) we see that as $B(z)$ increases, $v_\perp$ increases and, therefore, R decreases. In this case, it follows from (12.114) that v_z decreases. This shows that by having a strong convergent magnetic field, a plasma can be confined to a narrow region in space both vertically and horizontally. The points $z = z_0$ where v_z vanishes are known as "mirror" points. The plasma is turned back at this point and, consequently, one can construct a magnetic bottle with two "mirror" points at the two ends. The standard mechanism for confining plasmas is, therefore, to have a region with stronger magnetic fields at the two ends and a weaker field in the middle. In this case, the plasma is trapped oscillating back and forth between the two "mirror" points. A similar natural phenomenon is also observed in earth's magnetic field. Because the magnetic fields are strong at the poles and weaker at the center, positive and negatively charged ions in the atmosphere are trapped and oscillate back and forth between the "Van Allen" belts.

12.6 Particles with spin and the anomalous magnetic moment

We have thus far considered the motion of particles carrying only a charge. However, a particle may also have an intrinsic structure leading to an intrinsic magnetic dipole moment $\boldsymbol{\mu}$, or an intrinsic electric dipole moment $\mathbf{d}$. In the presence of electric and magnetic fields, these intrinsic moments lead to interaction energies of the forms

$$\mathcal{E}_{\text{magnetic}} = \boldsymbol{\mu} \cdot \mathbf{B}, \qquad \mathcal{E}_{\text{electric}} = \mathbf{d} \cdot \mathbf{E}. \tag{12.120}$$

In a quantum theory, these moments are described in terms of the spin of the particle as

$$\boldsymbol{\mu} = \frac{g}{2} \frac{q}{mc} \mathbf{S}, \qquad \mathbf{d} = \frac{f}{2} \frac{q}{mc} \mathbf{S}, \tag{12.121}$$

where $g = 2$ and $f = 0$ for point charged particles as predicted by Dirac's theory and any deviation of g from this value (or a non-zero value for $(g - 2)$) is known as the anomalous magnetic moment of the particle. Quantum mechanically, spin is, of course, an operator. However, by Ehrenfest's theorem, the expectation value of this operator can be taken to correspond to the classical spin degrees of freedom of a relativistic particle.

Just as under a Lorentz transformation electric and magnetic fields get mixed, so do the electric and the magnetic dipole moments under such a transformation. In fact, the electric and the magnetic dipole moments (six in number like the components of the electromagnetic field strength tensor) can be combined into a second rank anti-symmetric tensor $M_{\mu\nu} = -M_{\nu\mu}$ with

$$M_{0i} = -d_i, \qquad M_{ij} = -\epsilon_{ijk}\mu_k, \tag{12.122}$$

so that the additional interaction energy of the particle (with intrinsic moments) in the presence of electric and magnetic fields takes the relativistic form

$$\begin{aligned} \mathcal{E}_{\text{intrinsic}} &= \frac{1}{2} M^{\mu\nu} F_{\mu\nu} = M^{0i} F_{0i} + \frac{1}{2} M^{ij} F_{ij} \\ &= -M_{0i} F_{0i} + \frac{1}{2} M_{ij} F_{ij} \\ &= -(-d_i) E_i + \frac{1}{2} (-\epsilon_{ijk}\mu_k)(-\epsilon_{ijm} B_m) \\ &= \mathbf{d} \cdot \mathbf{E} + \boldsymbol{\mu} \cdot \mathbf{B}. \end{aligned} \tag{12.123}$$

The important thing to note from (12.123) is that this interaction energy is invariant under a Lorentz transformation, much like the rest mass of a particle. (Recall that energy normally transforms like the time component of a four vector.) Therefore, we can obtain a

Lagrangian description of such an interacting system from (12.49) by replacing

$$mc \to mc \left(1 + \frac{1}{2mc^2} M^{\mu\nu} F_{\mu\nu}\right) = mc\,\Delta, \tag{12.124}$$

where we have identified

$$\Delta = \frac{1}{2mc^2} M^{\mu\nu} F_{\mu\nu}. \tag{12.125}$$

Namely, we can consider a Lagrangian of the form

$$L = mc\,\Delta(\eta_{\mu\nu}\dot{x}^\mu \dot{x}^\nu)^{\frac{1}{2}} + \frac{q}{c} A_\mu(x)\dot{x}^\mu, \tag{12.126}$$

with such a nonminimal coupling. This Lagrangian leads to the canonical momentum (in the gauge (12.24))

$$\Pi_\mu = \frac{\partial L}{\partial \dot{x}^\mu} = m\Delta \dot{x}_\mu + \frac{q}{c} A_\mu = \Delta p_\mu + \frac{q}{c} A_\mu. \tag{12.127}$$

Let us first consider the case when $M_{\mu\nu}$ is a constant tensor. In this case, the Euler-Lagrange equations following from the Lagrangian, in the gauge (12.24), lead to

$$m \frac{\mathrm{d}x^\mu}{\mathrm{d}\tau} = p^\mu = mu^\mu,$$

$$\frac{\mathrm{d}}{\mathrm{d}\tau} \frac{\partial L}{\partial \dot{x}^\mu} - \frac{\partial L}{\partial x^\mu} = 0,$$

$$\text{or,} \quad \frac{\mathrm{d}p^\mu}{\mathrm{d}\tau} = \frac{1}{\Delta}\left[\frac{q}{c} F^{\mu\nu} u_\nu + m(\partial_\nu \Delta)\left(c^2 \eta^{\mu\nu} - u^\mu u^\nu\right)\right]$$

$$= \frac{1}{\Delta}\left[\frac{q}{c} F^{\mu\nu} u_\nu + \frac{1}{2c^2} M^{\lambda\rho} F_{\lambda\rho,\nu}\left(c^2 \eta^{\mu\nu} - u^\mu u^\nu\right)\right]. \tag{12.128}$$

Here "comma" denotes a derivative. We note that the first term, inside the square bracket, represents the usual minimal interaction of a charged particle (although the coefficient is now modified by the Δ term). The second term which represents a new force involving derivatives of the field strength tensor, is manifestly transverse to u_μ as any relativistic force should be (see Eq. (12.31)). Second, we note that for

$$\frac{1}{2mc^2} M^{\mu\nu} F_{\mu\nu} \ll 1, \tag{12.129}$$

which would normally be the case, we can approximate

$$\Delta \approx 1, \tag{12.130}$$

in which case, the dynamical equation for the momentum takes the form

$$\frac{\mathrm{d}p^\mu}{\mathrm{d}\tau} \approx \left[\frac{q}{c} F^{\mu\nu} u_\nu + \frac{1}{2c^2} M^{\lambda\rho} F_{\lambda\rho,\nu} \left(c^2 \eta^{\mu\nu} - u^\mu u^\nu \right) \right]. \tag{12.131}$$

The set of equations in (12.128), however, is incomplete in the sense that there is no dynamics for the spin variable. We do know that a spin angular momentum precesses in the presence of a magnetic field. However, the reason that we have no dynamics for the spin variable is because our derivation of the equations has been from the Lagrangian and in order to have the dynamical equations for the spin, we should also have a dynamical Lagrangian for the new variables representing the spin degrees of freedom. This can, of course, be done. However, it is much simpler to obtain the equations for the spin in an alternative manner. Let us note that since both the electric and the magnetic dipole moments are proportional to spin, the tensor $M_{\mu\nu}$ describing spin must be highly constrained. Let us define a spin four vector from the dual of $M_{\mu\nu}$ as

$$s^\mu = \frac{1}{2q} \epsilon^{\mu\nu\lambda\rho} p_\nu M_{\lambda\rho}. \tag{12.132}$$

This is known as the Pauli-Lubanski spin variable. In the rest frame of the particle $p^\mu = (mc, 0, 0, 0)$ and then, it is easy to see that the space component of s^μ is proportional to the magnetic dipole moment and, therefore, to spin. It is also clear from the definition in (12.132) that

$$u_\mu s^\mu = \frac{1}{m} p_\mu s^\mu = 0. \tag{12.133}$$

Namely, much like the relativistic force, not all the components of s^μ are independent. Rather, we have

$$s^0 = \frac{\mathbf{v} \cdot \mathbf{s}}{c}. \tag{12.134}$$

In the rest frame of the particle, the spin four vector takes the form

$$s^\mu_{\text{rest}} = (0, \mathbf{S}). \tag{12.135}$$

For completeness, let us also note that we can invert the relation (12.132) to write

$$M_{\mu\nu} = \frac{q}{2}\epsilon_{\mu\nu\lambda\rho}p^{\lambda}s^{\rho}. \tag{12.136}$$

It follows from this that

$$p^{\mu}M_{\mu\nu} = mu^{\mu}M_{\mu\nu} = 0 = s^{\mu}M_{\mu\nu}. \tag{12.137}$$

Let us note that in the rest frame of the particle, we can identify the space components of the spin variable as the spin of the particle (up to multiplicative factors). Furthermore, in the rest frame of the particle, we know the equation for the spin in the presence of external electromagnetic fields to correspond to

$$\frac{d\mathbf{S}}{dt} = \frac{q}{mc}\left(\frac{g}{2}\mathbf{S}\times\mathbf{B} + \frac{f}{2}\mathbf{S}\times\mathbf{E}\right). \tag{12.138}$$

From this, we can transform to any other Lorentz frame in a simple manner. First, let us note that we can write

$$(\mathbf{S}\times\mathbf{B})_i = -F_{ij}S_j = F_i^{\ j}S_j = F_i^{\ \nu}s_{\nu}^{\text{rest}},$$

$$(\mathbf{S}\times\mathbf{E})_i = \epsilon_{ijk}S_jF_{0k} = \tilde{F}_i^{\ \nu}s_{\nu}^{\text{rest}}, \tag{12.139}$$

where we have defined the dual of the field strength tensor as

$$\tilde{F}^{\mu\nu} = \frac{1}{2}\epsilon^{\mu\nu\lambda\rho}F_{\lambda\rho}. \tag{12.140}$$

With this, it is easy to see that the generalization of the equation for the spin to any frame can be given by

$$\frac{ds^{\mu}}{d\tau} = \frac{q}{mc}\left[\frac{g}{2}s^{\rho}F_{\lambda\rho}\left(\eta^{\lambda\mu} - \frac{u^{\lambda}u^{\mu}}{c^2}\right) + \frac{f}{2}s^{\rho}\tilde{F}_{\lambda\rho}\left(\eta^{\lambda\mu} - \frac{u^{\lambda}u^{\mu}}{c^2}\right)\right]$$

$$- \frac{1}{c^2}s_{\lambda}\frac{du^{\lambda}}{d\tau}u^{\mu}. \tag{12.141}$$

It is easy to check that this equation reduces to (12.138) in the rest frame of the particle. The last term in (12.141) vanishes in the rest frame of the particle. However, it is essential in a relativistic generalization in order to satisfy the constraint in (12.133). For example, let us note from (12.141) that the terms in the square bracket are orthogonal to u_{μ}. Consequently, it follows that

$$u_{\mu}\frac{ds^{\mu}}{d\tau} = -s_{\lambda}\frac{du^{\lambda}}{d\tau},$$

or, $$\frac{d(u_{\mu}s^{\mu})}{d\tau} = 0, \tag{12.142}$$

which is consistent with (12.133) and holds only because of the presence of the last term in (12.141).

Equation (12.141) is still not in a simple form. It can be further simplified using the dynamical equations for the particle in (12.128). In fact, for a uniform field (and for $\Delta \approx 1$), we can use (12.128) to write the equation for the spin in any Lorentz frame as

$$
\frac{ds^\mu}{d\tau} = \frac{q}{mc} \left[\frac{g}{2} F^{\mu\nu} s_\nu + \frac{1}{c^2} \left(\frac{g}{2} - 1 \right) s_\lambda F^{\lambda\rho} u_\rho u^\mu \right.
$$
$$
\left. + \frac{f}{2} \left(\tilde{F}^{\mu\nu} s_\nu + \frac{1}{c^2} s_\lambda \tilde{F}^{\lambda\rho} u_\rho u^\mu \right) \right]. \tag{12.143}
$$

There are several things to observe from this equation. We note that for a particle without an electric dipole moment ($f = 0$) moving in a uniform magnetic field ($F_{0i} = 0$), the equation for the spin takes the form

$$
\frac{ds^\mu}{d\tau} = \frac{q}{2mc} \left(g F^{\mu\nu} s_\nu + \frac{(g-2)}{c^2} s_\lambda F^{\lambda\rho} u_\rho u^\mu \right). \tag{12.144}
$$

Writing out the equations explicitly, we obtain,

$$
\frac{d\mathbf{s}}{d\tau} = \frac{gq}{2mc} \mathbf{s} \times \mathbf{B} + \frac{\gamma^2 q(g-2)}{2mc^3} \left(\mathbf{v} \cdot (\mathbf{s} \times \mathbf{B}) \right) \mathbf{v},
$$
$$
\frac{ds^0}{d\tau} = \frac{1}{c} \frac{d(\mathbf{v} \cdot \mathbf{s})}{d\tau} = \frac{\gamma q(g-2)}{2mc^2} \mathbf{v} \cdot (\mathbf{s} \times \mathbf{B}), \tag{12.145}
$$

where we have made the identification in (12.134). The time rate of variation of the longitudinal component of the spin can now be calculated using (12.128) and the second relation in (12.145).

$$
\frac{d(\hat{\mathbf{v}} \cdot \mathbf{s})}{d\tau} = \frac{1}{v} \frac{d(\mathbf{v} \cdot \mathbf{s})}{d\tau} - \frac{(\mathbf{v} \cdot \mathbf{s})}{v^3} \mathbf{v} \cdot \frac{d\mathbf{v}}{d\tau}
$$
$$
= \frac{1}{v} \frac{d(\mathbf{v} \cdot \mathbf{s})}{d\tau}
$$
$$
= \frac{\gamma q(g-2)}{2mc^2} \hat{\mathbf{v}} \cdot (\mathbf{s} \times \mathbf{B}),
$$

$$
\text{or,} \quad \frac{d(\hat{\mathbf{v}} \cdot \mathbf{s})}{dt} = \frac{q(g-2)}{2mc^2} \hat{\mathbf{v}} \cdot (\mathbf{s} \times \mathbf{B}). \tag{12.146}
$$

Here the second term on the right hand side in the first line vanishes because $\mathbf{v} \cdot \frac{d\mathbf{v}}{d\tau} = 0$ which follows from the second equation in (12.128) for a uniform (constant) magnetic field. Equations (12.145) and (12.146) show that when $g = 2$, the longitudinal component of the spin (also known as the helicity) does not change with time, while the spin precesses with the same frequency as the cyclotron frequency

$$\omega = \frac{qB}{\gamma mc}.$$

However, when the particle has an anomalous magnetic moment and $(g - 2) \neq 0$, then the longitudinal component of the spin changes with time and, therefore, along the trajectory of the particle. Consequently, the anomalous magnetic moment associated with a charged particle can be experimentally determined and the current measurements have it vanishing for a point particle up to an accuracy of about 10^{-10}. Furthermore, the experimental measurements are in excellent agreement with the predictions of quantum electrodynamics.

CHAPTER 13

Scattering and diffraction of electromagnetic waves

In classical mechanics as well as in quantum mechanics, we are familiar with the phenomenon of scattering where a particle or a number of particles incident on some center of force suffer deflection in their trajectories because of the action of the force. Similarly, electromagnetic waves can also undergo scattering. For example, we can imagine a plane wave incident on a perfectly conducting sphere. The incident fields would induce a surface current on the sphere which, in turn, would generate electromagnetic fields and would lead to total fields (incident plus scattered) which will be different from those associated with the incident wave. So, conceptually the scattering set up in electromagnetism is quite similar to what we are used to. The meaningful concept in such an experiment is again the cross section. However, in the present context, the differential cross section is defined as (classically we have only electromagnetic waves and no particles) the ratio of the average energy scattered per unit time per unit solid angle along the direction (θ, ϕ) to the average incident energy per unit time per unit area.

$$\sigma(\theta, \phi) = \frac{\text{Average scattered energy/second/solid angle}}{\text{Average incident energy/second/unit area}}. \quad (13.1)$$

We note that in a scattering experiment involving waves, there are three length scales involved, namely, the size of the scattering source d, the wave length λ of the incident wave and the distance R from the scattering source where observations are made. The observations are, of course, made far away from the scattering source so that $R \gg \lambda, d$. When $\lambda \gg d$, the phenomenon is commonly characterized as scattering while for $\lambda \ll d$, it is called diffraction. As we will see, the techniques involved in the study of the two phenomena are different.

13.1 Scattering from a perfectly conducting sphere

It is clear that the problem of scattering corresponds to solving a boundary value problem, much like what we have done in electrostatics or magnetostatics. However, in the time dependent case, there are very few problems that can be solved exactly. Scattering from a perfectly conducting sphere is one such example. In general, therefore, we have to develop some approximate methods for dealing with such problems. In this section, however, let us see how the boundary value problem can be solved exactly in this case.

Let us assume that the scattering sphere has a radius d and that a monochromatic plane wave of frequency ω is incident on it along the z-axis. We assume that the center of the sphere coincides with the origin of our coordinate system. First, let us recall some facts. We know from the study of scattering (in quantum mechanics) that a scalar plane wave, which is a solution of the Helmholtz equation, can be expanded in terms of spherical harmonics as

$$e^{i\mathbf{k}\cdot\mathbf{x}} = e^{ikz} = e^{ikr\cos\theta}$$

$$= \sum_{\ell=0}^{\infty} i^\ell \, (2\ell+1) \, j_\ell(kr) P_\ell(\cos\theta)$$

$$= \sum_{\ell=0}^{\infty} i^\ell \, \sqrt{4\pi(2\ell+1)} \, j_\ell(kr) Y_{\ell,0}(\theta), \tag{13.2}$$

where $j_\ell(kr)$ denote spherical Bessel functions and we have used the definition

$$Y_{\ell,0}(\theta) = \sqrt{\frac{(2\ell+1)}{4\pi}} \, P_\ell(\cos\theta).$$

This shows that a scalar plane wave contains only waves with angular momentum projection $m = 0$.

Let us next recall that a circularly polarized harmonic electromagnetic wave traveling along the z-axis in vacuum will have electric and magnetic fields of the forms (factoring out the time dependence, see also (11.72))

$$\mathbf{E}_{\mathrm{R,L}}(\mathbf{x}) = \frac{1}{\sqrt{2}} \, (\hat{\mathbf{x}} \mp i\hat{\mathbf{y}}) \, e^{ikz},$$

$$\mathbf{B}_{\mathrm{R,L}}(\mathbf{x}) = \hat{\mathbf{z}} \times \mathbf{E}_{\mathrm{R,L}}(\mathbf{x}) = \pm i\,\mathbf{E}_{\mathrm{R,L}}(\mathbf{x}), \tag{13.3}$$

where the first sign corresponds to a right circularly polarized wave and the second to a left circularly polarized wave. (For simplicity, we are assuming fields of unit intensity.) Of course, the right and left circularly polarized fields in (13.3) can be expressed as superpositions of electric and magnetic multipole fields which we have discussed earlier. Therefore, for the fields in (13.3), we can write in general (see the discussion in sections 9.4 and 9.5 on multipole expansions, in particular, see (9.116) and (9.117))

$$
\mathbf{E}_{R,L}(\mathbf{x}) = \sum_{\ell,m} \left[a_\mp(\ell,m) j_\ell(kr) \mathbf{Y}_{\ell,m} \right.
$$

$$
\left. + \frac{i}{k} b_\mp(\ell,m) \boldsymbol{\nabla} \times (j_\ell(kr) \mathbf{Y}_{\ell,m}) \right],
$$

$$
\mathbf{B}_{R,L}(\mathbf{x}) = \sum_{\ell,m} \left[-\frac{i}{k} a_\mp(\ell,m) \boldsymbol{\nabla} \times (j_\ell(kr) \mathbf{Y}_{\ell,m}) \right.
$$

$$
\left. + b_\mp(\ell,m) j_\ell(kr) \mathbf{Y}_{\ell,m} \right]. \tag{13.4}
$$

Here, we have used the well behaved spherical Bessel functions for the expansion of the plane waves (and not the spherical Neumann function) since the plane wave is well behaved everywhere. The constants $a_\mp(\ell,m)$ and $b_\mp(\ell,m)$ specify respectively the amounts of magnetic multipole terms and the electric multipole terms present in the wave corresponding to a given (ℓ,m). Furthermore, $\mathbf{Y}_{\ell,m}(\theta,\phi)$ represent the vector spherical harmonics defined in (9.114).

To determine the expansion coefficients $a_\mp(\ell,m), b_\mp(\ell,m)$, we need to understand some of the properties of the vector spherical harmonics. We have already seen in (9.115) that

$$
\int d\Omega \, \mathbf{Y}^*_{\ell',m'}(\theta,\phi) \cdot \mathbf{Y}_{\ell,m}(\theta,\phi) = \delta_{\ell\ell'}\delta_{mm'}. \tag{13.5}
$$

Let us next note that for any arbitrary radial function $f_\ell(r)$, we have

$$
\boldsymbol{\nabla} \times (f_\ell(r) \mathbf{Y}_{\ell,m}(\theta,\phi)) = (\boldsymbol{\nabla} f_\ell(r)) \times \mathbf{Y}_{\ell,m} + f_\ell(r) \boldsymbol{\nabla} \times \mathbf{Y}_{\ell,m}
$$

$$
= \frac{d f_\ell(r)}{dr} \hat{\mathbf{r}} \times \mathbf{Y}_{\ell,m} + \frac{f_\ell(r)}{\sqrt{\ell(\ell+1)}} \boldsymbol{\nabla} \times \mathbf{L} Y_{\ell,m}, \tag{13.6}
$$

where we have used the definition of the vector spherical harmonics given in (9.114). Furthermore, using the relation (9.122) we obtain

$$
\boldsymbol{\nabla} \times \mathbf{L} Y_{\ell,m} = -i \left(\mathbf{r} \nabla^2 - \boldsymbol{\nabla} \left(1 + r \frac{\partial}{\partial r} \right) \right) Y_{\ell,m}
$$

$$= -i r (\nabla^2 Y_{\ell,m}) + i (\nabla Y_{\ell,m})$$

$$= \frac{i r}{r^2} \ell(\ell+1) Y_{\ell,m} + \frac{1}{r} \hat{\mathbf{r}} \times \mathbf{L} Y_{\ell,m}. \tag{13.7}$$

Here, we have used the fact that the spherical harmonics depend only on the angular coordinates and are eigenstates of $\mathbf{L}^2$ (see Eq. (9.97)) and that

$$\nabla^2 = \frac{1}{r^2} \frac{\partial}{\partial r} \left(r^2 \frac{\partial}{\partial r} \right) - \frac{1}{r^2} \mathbf{L}^2. \tag{13.8}$$

We have also used the decomposition of the gradient given in (9.108). Substituting all these relations into (13.6), we obtain

$$\nabla \times (f_\ell(r) \mathbf{Y}_{\ell,m}) = i \sqrt{\ell(\ell+1)} \frac{\hat{\mathbf{r}}}{r} f_\ell(r) Y_{\ell,m}$$

$$+ \left(\frac{\mathrm{d} f_\ell(r)}{\mathrm{d} r} + \frac{f_\ell(r)}{r} \right) \hat{\mathbf{r}} \times \mathbf{Y}_{\ell,m}. \tag{13.9}$$

It follows now that

$$\int \mathrm{d}\Omega \, \mathbf{Y}_{\ell',m'}^* \cdot (\nabla \times (f_\ell(r) \mathbf{Y}_{\ell,m}))$$

$$= \int \mathrm{d}\Omega \, Y_{\ell',m'}^* \mathbf{L} \cdot (\nabla \times (f_\ell(r) \mathbf{Y}_{\ell,m}))$$

$$= \frac{1}{r} \left(\frac{\mathrm{d} f_\ell(r)}{\mathrm{d} r} + \frac{f_\ell(r)}{r} \right) \int \mathrm{d}\Omega \, Y_{\ell',m'}^* \mathbf{L} \cdot (\mathbf{r} \times \mathbf{Y}_{\ell,m})$$

$$= \frac{1}{r \sqrt{\ell(\ell+1)}} \left(\frac{\mathrm{d} f_\ell}{\mathrm{d} r} + \frac{f_\ell}{r} \right) \int \mathrm{d}\Omega \, Y_{\ell',m'} \mathbf{L} \cdot (\mathbf{r} \times \mathbf{L}) Y_{\ell,m}, \tag{13.10}$$

where we have used (9.98). Furthermore, from the definition of the angular momentum operator in (9.95), it follows that

$$\mathbf{L} \cdot (\mathbf{r} \times \mathbf{L}) = \epsilon_{ijk} L_i r_j L_k$$

$$= -i \epsilon_{ijk} \epsilon_{ipq} r_p \nabla_q r_j L_k$$

$$= -i \left(r_j \nabla_k r_j L_k - r_k \nabla_j r_j L_k \right)$$

$$= -i \left(\mathbf{r} \cdot \mathbf{L} + r^2 \nabla \cdot \mathbf{L} + \mathbf{r} \cdot \mathbf{L} - \nabla_j (r_j \mathbf{r} \cdot \mathbf{L}) \right)$$

$$= 0. \tag{13.11}$$

Here we have used (9.98) as well as the fact that the defining relation in (9.95) leads to

$$\nabla \cdot \mathbf{L} = 0. \tag{13.12}$$

As a consequence of (13.12), it follows now that

$$\int d\Omega\, \mathbf{Y}^*_{\ell',m'} \cdot (\boldsymbol{\nabla} \times (f_\ell(r)\mathbf{Y}_{\ell,m})) = 0. \tag{13.13}$$

With these identities, we are now ready to determine the coefficients of expansion in (13.4). Using (13.5) and (13.13), we obtain from Eq. (13.4) that

$$a_\mp(\ell,m)j_\ell(kr) = \int d\Omega\, \mathbf{Y}^*_{\ell,m} \cdot \mathbf{E}_{\mathrm{R,L}}$$

$$= \frac{1}{\sqrt{2\ell(\ell+1)}} \int d\Omega\, (\mathbf{L}Y_{\ell,m})^* \cdot (\hat{\mathbf{x}} \mp i\hat{\mathbf{y}})\, e^{ikz}$$

$$= \frac{1}{\sqrt{2\ell(\ell+1)}} \int d\Omega\, (L_\pm Y_{\ell,m})^* \, e^{ikz}$$

$$= \sqrt{\frac{(\ell\mp m)(\ell\pm m+1)}{2\ell(\ell+1)}} \int d\Omega\, Y^*_{\ell,m\pm1}\, e^{ikz}, \tag{13.14}$$

where we have used the standard definition

$$L_\pm = L_x \pm iL_y = \mathbf{L}\cdot(\hat{\mathbf{x}}\pm i\hat{\mathbf{y}}), \tag{13.15}$$

as well as the action of $L_\pm$ on spherical harmonics. If we now use the expansion of the plane wave in (13.2) as well as the orthonormality of the spherical harmonics, we obtain

$$a_\mp(\ell,m)j_\ell(kr) = \sqrt{\frac{(\ell\mp m)(\ell\pm m+1)}{2\ell(\ell+1)}}$$

$$\times \sum_{\ell'} i^{\ell'} \sqrt{4\pi(2\ell'+1)}\, j_{\ell'}(kr) \int d\Omega\, Y^*_{\ell,m\pm1} Y_{\ell',0}$$

$$= \sqrt{\frac{(\ell\mp m)(\ell\pm m+1)}{\ell(\ell+1)}}$$

$$\times \sum_{\ell'} i^{\ell'} \sqrt{2\pi(2\ell'+1)}\, j_{\ell'}(kr)\, \delta_{\ell,\ell'}\delta_{m,\mp1}$$

$$= i^\ell \sqrt{2\pi(2\ell+1)}\, j_\ell(kr)\delta_{m,\mp1},$$

$$\text{or,}\quad a_\mp(\ell,m) = i^\ell \sqrt{2\pi(2\ell+1)}\, \delta_{m,\mp1}. \tag{13.16}$$

Similarly, using (13.3) and (13.16), we can determine

$$b_\mp(\ell, m)j_\ell(kr) = \int d\Omega\, \mathbf{Y}_{\ell,m}^* \cdot \mathbf{B}_{R,L}$$

$$= \pm i \int d\Omega\, \mathbf{Y}_{\ell,m}^* \cdot \mathbf{E}_{R,L}$$

$$= \pm i a_\mp(\ell, m)j_\ell(kr),$$

$$\text{or,} \quad b_\mp(\ell, m) = \pm i a_\mp(\ell, m)$$

$$= \pm i^{\ell+1}\sqrt{2\pi(2\ell + 1)}\,\delta_{m,\mp 1}. \tag{13.17}$$

Substituting Eqs. (13.16) and (13.17) into (13.4), we obtain

$$\mathbf{E}_{R,L} = \sum_{\ell=1}^{\infty} i^\ell \sqrt{2\pi(2\ell + 1)}$$

$$\times \left[j_\ell(kr)\mathbf{Y}_{\ell,\mp 1} \mp \frac{1}{k}\boldsymbol{\nabla} \times (j_\ell(kr)\mathbf{Y}_{\ell,\mp 1}) \right],$$

$$\mathbf{B}_{R,L} = \sum_{\ell=1}^{\infty} i^\ell \sqrt{2\pi(2\ell + 1)}$$

$$\times \left[-\frac{i}{k}\boldsymbol{\nabla} \times (j_\ell(kr)\mathbf{Y}_{\ell,\mp 1}) \pm i j_\ell(kr)\mathbf{Y}_{\ell,\mp 1} \right]. \tag{13.18}$$

We see that since the projection of the angular momentum along the z axis is unity for circularly polarized waves, only the $m = \pm 1$ azimuthal quantum numbers enter into the multipole expansion of the electric and the magnetic fields (which is the reason $\ell = 0$ is excluded from the sum). This has to be contrasted with the expansion of the scalar plane wave which involves only $m = 0$.

With these basics, let us now discuss the scattering of a plane electromagnetic wave from a perfectly conducting sphere of radius d. The incident wave can be right or left circularly polarized (or linearly polarized which is a superposition of these two) and would scatter due to the presence of the sphere. Therefore, at large distances away from the scattering source, we expect outgoing spherical scattered waves. We can, therefore, write the total electric and magnetic fields to have the forms

$$\mathbf{E}_{R,L} = \mathbf{E}_{R,L}^{(\text{inc})} + \mathbf{E}_{R,L}^{(\text{sc})}, \qquad \mathbf{B}_{R,L} = \mathbf{B}_{R,L}^{(\text{inc})} + \mathbf{B}_{R,L}^{(\text{sc})}, \tag{13.19}$$

where we can assume the incident plane waves $\mathbf{E}_{R,L}^{(\text{inc})}, \mathbf{B}_{R,L}^{(\text{inc})}$ of unit intensity to be given by (13.3) or (13.18). On the other hand, we expect the scattered waves to have outgoing spherical wave forms at large distances. Therefore, following our earlier discussions, we can write

$$
\mathbf{E}_{R,L}^{(sc)} = \frac{1}{2} \sum_{\ell=1}^{\infty} i^{\ell} \sqrt{2\pi(2\ell+1)}
$$

$$
\times \left[c_{\mp}(\ell) h_{\ell}^{(1)}(kr) \mathbf{Y}_{\ell,\mp 1} \mp \frac{d_{\mp}(\ell)}{k} \boldsymbol{\nabla} \times \left(h_{\ell}^{(1)}(kr) \mathbf{Y}_{\ell,\mp 1} \right) \right],
$$

$$
\mathbf{B}_{R,L}^{(sc)} = \frac{1}{2} \sum_{\ell=1}^{\infty} i^{\ell} \sqrt{2\pi(2\ell+1)}
$$

$$
\times \left[-\frac{ic_{\mp}(\ell)}{k} \boldsymbol{\nabla} \times \left(h_{\ell}^{(1)}(kr) \mathbf{Y}_{\ell,\mp 1} \right) \pm id_{\mp}(\ell) h_{\ell}^{(1)}(kr) \mathbf{Y}_{\ell,\mp 1} \right],
$$

$$
(13.20)
$$

where $h_{\ell}^{(1)}(kr)$ is the spherical Hankel function of the first kind defined in (9.91) which is spherically outgoing at large distances. The fields in (13.20) have exactly the same forms as in (13.18) except for the constant coefficients $c_{\mp}(\ell), d_{\mp}(\ell)$ which will be determined from the boundary conditions.

As we have already seen, there are no electric and magnetic fields inside a perfect conductor and the conditions satisfied by the electric and the magnetic fields on the boundary surface are given by (see (7.4))

$$
\hat{\mathbf{n}} \times \mathbf{E}| = 0, \qquad \hat{\mathbf{n}} \cdot \mathbf{B}| = 0. \tag{13.21}
$$

For a perfectly conducting sphere of radius d, the boundary is at $r = d$ (we are assuming that the origin of the coordinate system lies at the center of the sphere) and the unit vector normal to the boundary is the radial unit vector $\hat{\mathbf{r}}$. Thus, imposing the boundary condition on the electric field, we obtain

$$
\hat{\mathbf{r}} \times \left(\mathbf{E}_{R,L}^{(inc)} + \mathbf{E}_{R,L}^{(sc)} \right) \Big|_{r=d} = 0. \tag{13.22}
$$

Let us recall from Eq. (13.9) that

$$
\hat{\mathbf{r}} \times \left(\boldsymbol{\nabla} \times (f_{\ell}(r) \mathbf{Y}_{\ell,m}) \right) = \left(\frac{df_{\ell}(r)}{dr} + \frac{f_{\ell}(r)}{r} \right) \hat{\mathbf{r}} \times (\hat{\mathbf{r}} \times \mathbf{Y}_{\ell,m})
$$

$$
= -\left(\frac{df_{\ell}(r)}{dr} + \frac{f_{\ell}(r)}{r} \right) \mathbf{Y}_{\ell,m}
$$

$$
= -\frac{1}{r} \frac{d}{dr} \left(r f_{\ell}(r) \right) \mathbf{Y}_{\ell,m}, \tag{13.23}
$$

where we have used the definition of the vector spherical harmonics in (9.114) as well as the orthogonality relation (9.98). Furthermore, we note that

$$\int d\Omega\, \mathbf{Y}^*_{\ell',m'} \cdot (\hat{\mathbf{r}} \times \mathbf{Y}_{\ell,m}) = 0, \tag{13.24}$$

which follows from (13.11).

Using these identities, we obtain from (13.22)

$$\int d\Omega\, \mathbf{Y}^*_{\ell,m} \cdot (\hat{\mathbf{r}} \times (\mathbf{E}^{(\mathrm{inc})}_{\mathrm{R,L}} + \mathbf{E}^{(\mathrm{sc})}_{\mathrm{R,L}}))\Big|_{r=d} = 0,$$

or,
$$\left[\frac{\partial}{\partial r}(r j_\ell(kr)) + \frac{d_{\mp}(\ell)}{2}\frac{\partial}{\partial r}\left(r h^{(1)}_\ell(kr)\right)\right]_{r=d} = 0, \tag{13.25}$$

where we have used the orthonormality of the vector spherical harmonics in (9.115). Similarly, the boundary condition on the magnetic field (see (13.21)) leads to

$$j_\ell(kd) + \frac{c_{\mp}(\ell)}{2} h^{(1)}_\ell(kd) = 0. \tag{13.26}$$

Relations (13.25) and (13.26) determine the arbitrary coefficients present in the definitions of the scattered waves, namely,

$$c_{\mp}(\ell) = -\frac{2 j_\ell(kd)}{h^{(1)}_\ell(kd)} = \left(-\frac{h^{(2)}_\ell(kd)}{h^{(1)}_\ell(kd)} - 1\right),$$

$$d_{\mp}(\ell) = -\frac{2\frac{\partial}{\partial r}(r j_\ell(kr))}{\frac{\partial}{\partial r}(r h^{(1)}_\ell(kr))}\Bigg|_{r=d}$$

$$= \left[-\frac{\frac{\partial}{\partial r}(r h^{(2)}_\ell(kr))}{\frac{\partial}{\partial r}(r h^{(1)}_\ell(kr))} - 1\right]_{r=d}, \tag{13.27}$$

where we have used the definition of the spherical Bessel functions in (9.91).

Let us recall that $h^{(2)}_\ell(kr) = (h^{(1)}_\ell(kr))^*$. Therefore, we see that the first term on the right hand side of each of the defining relations in (13.27) has unit modulus. This allows us to define phase angles of the forms

$$e^{2i\delta_\ell} = -\frac{h^{(2)}_\ell(kd)}{h^{(1)}_\ell(kd)},$$

$$e^{2i\delta'_\ell} = -\frac{\frac{\partial}{\partial r}(r h^{(2)}_\ell(kr))}{\frac{\partial}{\partial r}(r h^{(1)}_\ell(kr))}\Bigg|_{r=d}. \tag{13.28}$$

We can think of $\delta_\ell, \delta'_\ell$ respectively as the magnetic and the electric phase shifts in the ℓth multipole fields. This is completely parallel to the analysis of scattering in quantum mechanics. In fact, in terms of these phase shifts, we can write

$$c_\mp(\ell) = (e^{2i\delta_\ell} - 1) = 2ie^{i\delta_\ell} \sin \delta_\ell,$$

$$d_\mp(\ell) = (e^{2i\delta'_\ell} - 1) = 2ie^{i\delta'_\ell} \sin \delta'_\ell. \tag{13.29}$$

Correspondingly, the scattered electric and magnetic fields can be written as

$$\mathbf{E}_{R,L}^{(sc)} = \frac{1}{2} \sum_{\ell=1}^{\infty} i^\ell \sqrt{2\pi(2\ell+1)} \Big[(e^{2i\delta_\ell} - 1)h_\ell^{(1)}(kr)\mathbf{Y}_{\ell,\mp 1}$$

$$\mp \frac{(e^{2i\delta'_\ell} - 1)}{k} \boldsymbol{\nabla} \times (h_\ell^{(1)}(kr)\mathbf{Y}_{\ell,\mp 1}) \Big],$$

$$\mathbf{B}_{R,L}^{(sc)} = \frac{1}{2} \sum_{\ell=1}^{\infty} i^\ell \sqrt{2\pi(2\ell+1)} \Big[-\frac{i(e^{2i\delta_\ell} - 1)}{k} \boldsymbol{\nabla} \times (h_\ell^{(1)}(kr)\mathbf{Y}_{\ell,\mp 1})$$

$$\pm i(e^{2i\delta'_\ell} - 1)h_\ell^{(1)}(kr)\mathbf{Y}_{\ell,\mp 1} \Big]. \tag{13.30}$$

Let us next consider the asymptotic forms of these fields for large r, far away from the scattering source. In this case, we know that

$$h_\ell^{(1)}(kr) \rightarrow (-i)^{\ell+1} \frac{e^{ikr}}{kr}. \tag{13.31}$$

It follows, therefore, from Eq. (13.9) as well as (13.31) that for large r

$$\boldsymbol{\nabla} \times (h_\ell^{(1)}(kr)\mathbf{Y}_{\ell,\mp 1}) \rightarrow \frac{(-i)^\ell e^{ikr}}{r} \hat{\mathbf{r}} \times \mathbf{Y}_{\ell,\mp 1}. \tag{13.32}$$

Substituting these into the expression for the scattered electric and magnetic fields, we obtain that, for large r,

$$\mathbf{E}_{R,L}^{(sc)} \rightarrow \frac{e^{ikr}}{r} \mathbf{f}_\mp(\theta, \phi),$$

$$\mathbf{B}_{R,L}^{(sc)} \rightarrow \hat{\mathbf{r}} \times \mathbf{E}_{R,L}^{(sc)}, \tag{13.33}$$

where we have defined

$$\mathbf{f}_\mp(\theta, \phi) = \sum_{\ell=1}^{\infty} \frac{\sqrt{2\pi(2\ell+1)}}{k}$$

$$\times \Big[e^{i\delta_\ell} \sin \delta_\ell \mathbf{Y}_{\ell,\mp 1} \mp ie^{i\delta'_\ell} \sin \delta'_\ell \hat{\mathbf{r}} \times \mathbf{Y}_{\ell,\mp 1} \Big]. \tag{13.34}$$

In analogy with the discussion of scattering in quantum mechanics, we can think of $\mathbf{f}_{\mp}(\theta, \phi)$ as the vector scattering amplitude. We note that the second relation in (13.33) is what we would expect in the far (radiation) zone.

Let us note now that the average power radiated by the scattered wave per unit solid angle through a large spherical surface of radius R is given by

$$
\frac{\mathrm{d}P_{\mathrm{R,L}}^{(\mathrm{sc})}}{\mathrm{d}\Omega} = \frac{cR^2}{8\pi} \operatorname{Re} \hat{\mathbf{r}} \cdot \left(\mathbf{E}_{\mathrm{R,L}}^{(\mathrm{sc})} \times (\mathbf{B}_{\mathrm{R,L}}^{(\mathrm{sc})})^* \right)
$$

$$
= \frac{cR^2}{8\pi} |\mathbf{E}_{\mathrm{R,L}}^{(\mathrm{sc})}|^2
$$

$$
= \frac{c}{8\pi} |\mathbf{f}_{\mp}(\theta, \phi)|^2. \tag{13.35}
$$

We also note from the form of the incident wave (13.3) that it is a plane wave traveling along the z-axis. Therefore, the average incident power per unit area is given by

$$
I_{\mathrm{R,L}} = \frac{c}{8\pi} \operatorname{Re} \hat{\mathbf{z}} \cdot \left(\mathbf{E}_{\mathrm{R,L}}^{(\mathrm{inc})} \times (\mathbf{B}_{\mathrm{R,L}}^{(\mathrm{inc})})^* \right)
$$

$$
= \frac{c}{8\pi} |\mathbf{E}_{\mathrm{R,L}}^{(\mathrm{inc})}|^2 = \frac{c}{8\pi}. \tag{13.36}
$$

In this case, we obtain the differential cross section for scattering, (13.1), to be

$$
\sigma_{\mathrm{R,L}}(\theta, \phi) = |\mathbf{f}_{\mp}(\theta, \phi)|^2. \tag{13.37}
$$

This clarifies the meaning of $\mathbf{f}_{\mp}(\theta, \phi)$ as the (vector) scattering amplitude. The total scattering cross section can now be obtained by integrating the differential cross section and, using the orthonormality relations for the spherical harmonics, we obtain

$$
\sigma_{\mathrm{R,L}}^{\mathrm{total}} = \int \mathrm{d}\Omega \, |\mathbf{f}_{\mp}(\theta, \phi)|^2
$$

$$
= \frac{2\pi}{k^2} \sum_{\ell=1}^{\infty} (2\ell + 1) \left[\sin^2 \delta_\ell + \sin^2 \delta_\ell' \right]. \tag{13.38}
$$

This shows that the magnetic and the electric multipole fields contribute incoherently to the total scattering cross section (there is no cross term because of (13.24)) although, as we will see, there can be interference terms present in the differential cross section.

With these general observations, let us calculate the differential scattering cross section in the long wave length limit $kd \ll 1$. This would particularly be true for scattering by small particles. In this case, from the definitions of the spherical Bessel functions we obtain that

$$\tan \delta_\ell = \frac{j_\ell(kd)}{\eta_\ell(kd)} \to -\frac{(kd)^{2\ell+1}}{(2\ell+1)[(2\ell-1)!!]^2}$$

or, $\quad \delta_\ell \approx -\frac{(kd)^{2\ell+1}}{(2\ell+1)[(2\ell-1)!!]^2},$

$$\tan \delta'_\ell = \left.\frac{\frac{\partial}{\partial r}(r j_\ell(kr))}{\frac{\partial}{\partial r}(r \eta_\ell(kr))}\right|_{r=d} \to \frac{\ell+1}{\ell}\frac{(kd)^{2\ell+1}}{(2\ell+1)[(2\ell-1)!!]^2}$$

or, $\quad \delta'_\ell \approx -\frac{\ell+1}{\ell}\delta_\ell.$ $\hfill (13.39)$

This shows that the higher angular momentum components of the phase shifts fall off rapidly with ℓ in the long wave length limit. Therefore, we can approximate the expression for the scattering amplitude by keeping only the lowest order term corresponding to $\ell = 1$. We note from (13.39) that

$$\delta_1 \approx -\frac{(kd)^3}{3}, \qquad \delta'_1 \approx -2\delta_1. \hfill (13.40)$$

Using this, the scattering amplitude in this approximation becomes

$$\mathbf{f}_\mp(\theta, \phi) \approx \frac{\sqrt{6\pi}}{k}\delta_1 \left(\mathbf{Y}_{1,\mp 1} \pm 2i\,\hat{\mathbf{r}} \times \mathbf{Y}_{1,\mp 1}\right). \hfill (13.41)$$

Furthermore, using the properties of the angular momentum operators and the spherical harmonics, we obtain

$$\sigma_{\mathrm{R,L}}(\theta, \phi) = |\mathbf{f}_\mp(\theta, \phi)|^2$$

$$\approx \frac{6\pi}{k^2}|\delta_1|^2 \left(|\mathbf{Y}_{1,\mp 1}|^2 + 4|\mathbf{Y}_{1,\mp 1}|^2\right.$$

$$\left. \pm 4\mathrm{Im}\left(\mathbf{Y}^*_{1,\mp 1} \cdot (\hat{\mathbf{r}} \times \mathbf{Y}_{1,\mp 1})\right)\right)$$

$$= d^2(kd)^4\left[\frac{5}{8}(1 + \cos^2\theta) - \cos\theta\right]. \hfill (13.42)$$

Here, we have used the definitions of the spherical harmonics in (9.138) as well as $\mathbf{L}$ in (9.95). The $\cos\theta$ term represents the interference between the magnetic and the electric dipole terms ($\ell = 1$).

We note that the differential cross section is the same for both the right and the left circularly polarized waves. Furthermore, it is proportional to the fourth power of the frequency and has a peak in the backward direction $\theta = \pi$. Integrating this, we obtain

$$\sigma_{\text{R,L}}^{\text{total}} = \int d\Omega \, |\mathbf{f}_{\mp}(\theta, \phi)|^2 = \frac{10\pi d^2}{3} \, (kd)^4. \tag{13.43}$$

The interference term does not contribute to the total cross section. The dependence of the scattering cross section on the fourth power of the frequency is a characteristic of dipole fields and is known as Rayleigh's law.

13.2 Kirchhoff's approximation

We have already discussed briefly about the Kirchhoff's representation in section 6.9. However, the discussion there was in terms of the scalar potential. The basic observable fields in the Maxwell theory are, on the other hand, vector fields, namely, the electric and the magnetic fields. Thus, in this section we will generalize the discussion of section 6.9 to vector fields. Let us, however, emphasize that the components of the electric and magnetic fields can be thought of as scalar functions (as far as the discussion of the Kirchhoff's representation is concerned) and, as a result, we can, in principle, carry out the discussion from the results already derived earlier. But, for completeness as well as continuity with the earlier section on diffraction, we will discuss here Kirchhoff's representation for vector fields.

Let us recall that if we Fourier transform the time variable (see discussion in section 9.4), Maxwell's equations in a dielectric medium take the forms

$$\boldsymbol{\nabla} \cdot \mathbf{E} = \frac{4\pi}{\epsilon} \, \rho,$$

$$\boldsymbol{\nabla} \cdot \mathbf{B} = 0,$$

$$\boldsymbol{\nabla} \times \mathbf{E} = \frac{i\omega}{c} \, \mathbf{B},$$

$$\boldsymbol{\nabla} \times \mathbf{B} = \frac{4\pi\mu}{c} \, \mathbf{J} - \frac{i\epsilon\mu\omega}{c} \, \mathbf{E}. \tag{13.44}$$

Defining

$$k = \frac{\sqrt{\epsilon\mu}\,\omega}{c},$$

we note that (13.44) leads to

$$\nabla \times (\nabla \times \mathbf{E}) - k^2 \mathbf{E} = \frac{4i\pi\mu\omega}{c^2} \mathbf{J},$$

$$\nabla \times (\nabla \times \mathbf{B}) - k^2 \mathbf{B} = \frac{4\pi\mu}{c} \nabla \times \mathbf{J}. \tag{13.45}$$

In the absence of sources, both the electric and the magnetic fields are transverse and, consequently, (13.45) reduces to the Helmholtz equations

$$\left(\nabla^2 + k^2\right) \mathbf{E}(\mathbf{x}, \omega) = 0,$$

$$\left(\nabla^2 + k^2\right) \mathbf{B}(\mathbf{x}, \omega) = 0, \tag{13.46}$$

the solutions of which we have discussed in detail in section 9.4.

In the presence of sources, on the other hand, we see from (13.45) that both the electric and the magnetic fields satisfy an equation of the form

$$\nabla \times (\nabla \times \mathbf{V}) - k^2 \mathbf{V} = 4\pi \mathbf{f},$$

$$\text{or,} \quad \left(\partial_i \partial_j - \delta_{ij} \left(\nabla^2 + k^2\right)\right) V_j = 4\pi f_i, \tag{13.47}$$

where V_i stands for the three components of either the electric or the magnetic fields and f_i the sources. (Repeated indices are assumed to be summed.) To solve such an inhomogeneous equation, we will make use of the method of Green's functions. In this case, we note that the Green's function will be a tensor and from (13.47) it follows that the equation satisfied by the Green's function would have the form

$$\left(\partial_i \partial_m - \delta_{im} \left(\nabla^2 + k^2\right)\right) G_{mj}(\mathbf{x} - \mathbf{x}') = 4\pi \delta_{ij} \delta^3(x - x'). \tag{13.48}$$

From the definition in (13.48) we see that the Green's function is a second rank symmetric tensor, namely,

$$G_{ij}(\mathbf{x} - \mathbf{x}') = G_{ji}(\mathbf{x}' - \mathbf{x}). \tag{13.49}$$

Furthermore, the particular solution of an inhomogeneous Helmholtz equation of the form (13.47) can now be written as

$$V_i(\mathbf{x}, \omega) = \int d^3 x' \, G_{ij}(\mathbf{x} - \mathbf{x}') f_j(\mathbf{x}'). \tag{13.50}$$

The form of the Green's function satisfying (13.48) can now be determined along the lines discussed in section 6.8. First, we note that the Green's function for the scalar Helmholtz equation (9.87) satisfying the boundary conditions that at large distances it represents an outgoing wave has the form

$$g(\mathbf{x}) = \frac{e^{ikx}}{x}, \tag{13.51}$$

where $x = |\mathbf{x}|$, namely,

$$\left(\boldsymbol{\nabla}^2 + k^2\right) g(\mathbf{x}) = -4\pi\delta^3(x).$$

It can now be easily checked that the tensor Green's function in (13.48) is related to the scalar Green's function as

$$G_{ij}(\mathbf{x} - \mathbf{x}') = \left(\frac{1}{k^2}\,\partial_i\partial_j + \delta_{ij}\right) g(\mathbf{x} - \mathbf{x}'). \tag{13.52}$$

The symmetry (13.49) of the Green's function is manifest in the form (13.52).

The Green's identity (3.67) or (6.197), in this case, can be derived as follows. Let us note that for an arbitrary vector V_i and a second rank tensor A_{ij} (not necessarily symmetric), we can write

$$\int d^3x' \left[V_i \left(\partial_i'\partial_j' - \delta_{ij}\boldsymbol{\nabla}'^2 \right) A_{jk} - A_{ik} \left(\partial_i'\partial_j' - \delta_{ij}\boldsymbol{\nabla}'^2 \right) V_j \right]$$

$$= \int d^3x' \, \partial_i' \left[V_j \left(\partial_j' A_{ik} - \partial_i' A_{jk} \right) + A_{jk} \left(\partial_i' V_j - \partial_j' V_i \right) \right]$$

$$= \int ds' \, n_i' \left[V_j \left(\partial_j' A_{ik} - \partial_i' A_{jk} \right) + A_{jk} \left(\partial_i' V_j - \partial_j' V_i \right) \right]$$

$$= \int ds' \left[- \left(n_i' V_j - n_j' V_i \right) \partial_i' A_{jk} + A_{jk} n_i' \left(\partial_i' V_j - \partial_j' V_i \right) \right], \tag{13.53}$$

where we have used Gauss' theorem. Furthermore, $\mathbf{n}'$ denotes the unit vector normal to the surface S'. There are now two cases to consider. If we identify

$$V_i(\mathbf{x}') = E_i(\mathbf{x}', \omega), \qquad A_{ij}(\mathbf{x}, \mathbf{x}') = G_{ij}(\mathbf{x} - \mathbf{x}'), \tag{13.54}$$

then, in a region within the bounding surface (free of sources), (13.53) leads to

$$E_i(\mathbf{x},\omega) = -\frac{1}{4\pi}\int ds' \left[\left(n'_j E_m - n'_m E_j\right) \partial'_j G_{mi} \right.$$

$$\left. -\frac{ik}{\sqrt{\epsilon\mu}}\,\epsilon_{jm\ell}G_{mi}n'_j B_\ell \right]$$

$$= -\frac{1}{4\pi}\int ds' \left[\epsilon_{j\ell m}\left(\mathbf{n}'\times\mathbf{E}\right)_m \partial'_j G_{\ell i} - \frac{ik}{\sqrt{\epsilon\mu}}\,B_m\epsilon_{mj\ell}n'_j G_{\ell i} \right].$$

$$(13.55)$$

Here we have used (13.44) in a source free region as well as (13.48). It is clear that (13.55) can determine uniquely the electric field when the values of the tangential components (of the electric field) are specified on a given surface provided the Green's function satisfies the boundary condition

$$\epsilon_{j\ell m}n'_\ell G_{mp}(\mathbf{x}-\mathbf{x}') = 0. \qquad (13.56)$$

The Green's function satisfying such a boundary condition is known as the electric tensor Green's function and is denoted by $G^{(\mathrm{e})}_{ij}$. Choosing such a boundary condition, we obtain from (13.55)

$$E_i(\mathbf{x},\omega) = -\frac{1}{4\pi}\int ds'\, \left(\mathbf{n}'\times\mathbf{E}(\mathbf{x}')\right)_m \epsilon_{mj\ell}\partial'_j G^{(\mathrm{e})}_{\ell i}(\mathbf{x}-\mathbf{x}'). \quad (13.57)$$

The magnetic field can now be determined from (13.44) to be

$$B_i(\mathbf{x},\omega) = -\frac{i\sqrt{\epsilon\mu}}{k}\left(\boldsymbol{\nabla}\times\mathbf{E}(\mathbf{x},\omega)\right)_i$$

$$= -\frac{i\sqrt{\epsilon\mu}}{4\pi k}\int ds'\, \left(\mathbf{n}'\times\mathbf{E}(\mathbf{x}')\right)_j \epsilon_{j\ell m}\epsilon_{ipq}\partial'_\ell\partial'_p G^{(\mathrm{e})}_{mq}(\mathbf{x}-\mathbf{x}').$$

$$(13.58)$$

The combination

$$\epsilon_{j\ell m}\epsilon_{ipq}\partial'_\ell\partial'_p G^{(\mathrm{e})}_{mq}(\mathbf{x}-\mathbf{x}'), \qquad (13.59)$$

in the earlier equation is sometimes also referred to as the magnetic tensor Green's function (up to a multiplicative constant). Thus, we see that the Green's identity allows us to solve for the electric and the magnetic fields in terms of the boundary values of the tangential components of the electric field on a given surface and the electric tensor Green's function.

Similarly, identifying

$$V_i(\mathbf{x}') = B_i(\mathbf{x}'), \qquad A_{ij}(\mathbf{x}, \mathbf{x}') = G_{ij}(\mathbf{x} - \mathbf{x}'), \qquad (13.60)$$

in a source free region (within the bounding surface), we obtain from (13.53),

$$
\begin{aligned}
B_i(\mathbf{x}, \omega) &= -\frac{1}{4\pi} \int ds' \left[\left(n_j' B_m - n_m' B_j \right) \partial_j' G_{mi} \right. \\
&\qquad\qquad \left. + i\sqrt{\epsilon\mu} k \, \epsilon_{jm\ell} G_{mi} n_j' E_\ell \right] \\
&= -\frac{1}{4\pi} \int ds' \left[\epsilon_{j\ell m} \left(\mathbf{n}' \times \mathbf{B} \right)_m \partial_j' G_{\ell i} + i\sqrt{\epsilon\mu} k \, E_m \epsilon_{mj\ell} n_j' G_{\ell i} \right].
\end{aligned}
$$

$$(13.61)$$

We see once again that if we have the electric tensor Green's function, then the magnetic field can be solved uniquely in terms of the boundary values of its tangential components specified on a given surface as

$$B_i(\mathbf{x}, \omega) = -\frac{1}{4\pi} \int ds' \left(\mathbf{n}' \times \mathbf{B}(\mathbf{x}') \right)_m \epsilon_{mj\ell} \partial_j' G_{\ell i}^{(e)} (\mathbf{x} - \mathbf{x}').$$

$$(13.62)$$

The electric field then follows from (13.44) to be

$$
\begin{aligned}
E_i(\mathbf{x}, \omega) &= \frac{i}{\sqrt{\epsilon\mu} k} \left(\mathbf{\nabla} \times \mathbf{B}(\mathbf{x}, \omega) \right)_i \\
&= \frac{i}{4\pi\sqrt{\epsilon\mu} k} \int ds' \left(\mathbf{n}' \times \mathbf{B}(\mathbf{x}') \right)_j \epsilon_{j\ell m} \epsilon_{ipq} \partial_\ell' \partial_p' G_{mq}^{(e)} (\mathbf{x} - \mathbf{x}').
\end{aligned}
$$

$$(13.63)$$

This analysis makes it clear that the electric and the magnetic fields can be determined uniquely if the values of the tangential components of either the electric or the magnetic field are given on a given boundary surface. However, specifying the tangential components of both the electric as well as the magnetic fields on a boundary over-specifies the system unless the tangent components of the electric and the magnetic fields on the boundary are consistent. In such a case, the electric field at any point within the region bounded by the surface is given by (13.55), namely,

$$
\begin{aligned}
E_i(\mathbf{x}, \omega) &= -\frac{1}{4\pi} \int ds' \left[\epsilon_{j\ell m} \left(\mathbf{n}' \times \mathbf{E} \right)_m \partial_j' G_{\ell i} \right. \\
&\qquad\qquad \left. - \frac{ik}{\sqrt{\epsilon\mu}} B_m \epsilon_{mj\ell} n_j' G_{\ell i} \right].
\end{aligned}
$$

$$(13.64)$$

The relations obtained above are exact in the sense that there has been no approximation used so far. However, determining the boundary conditions in a given problem, (namely, determining the tangential components of the electric and the magnetic fields on the boundary) is in general difficult and it is here that approximations creep in. In particular, Kirchhoff's approximation uses the notions from geometrical optics in estimating the tangential components of the fields on a boundary surface. For example, let us assume that we are considering diffraction of electromagnetic waves from a spherical conducting surface. In this case, the surface of the sphere can be divided into two parts, one that is illuminated by the incident wave and the other that is in the shadow. Kirchhoff's approximation consists of assuming that in the illuminated part of the spherical surface, we have already seen from our studies on reflection from a perfectly conducting surface (see the discussion in sections 8.2 and 8.3) that

$$\mathbf{n} \times \mathbf{E} = \mathbf{n} \times \left(\mathbf{E}^{(\text{inc})} + \mathbf{E}^{(\text{refl})} \right) = 0,$$

$$\mathbf{n} \times \mathbf{B} = \mathbf{n} \times \left(\mathbf{B}^{(\text{inc})} + \mathbf{B}^{(\text{refl})} \right) = 2\mathbf{n} \times \mathbf{B}^{(\text{inc})}. \tag{13.65}$$

Using these we conclude that, in the illuminated part of the sphere, we can write

$$\mathbf{n} \times \mathbf{E}^{(\text{inc})} = -\mathbf{n} \times \mathbf{E}_{\text{I}}^{(\text{sc})}, \qquad \mathbf{n} \times \mathbf{B}^{(\text{inc})} = \mathbf{n} \times \mathbf{B}_{\text{I}}^{(\text{sc})}. \tag{13.66}$$

On the other hand, in the shadow part of the spherical surface, Kirchhoff's approximation assumes that there is no total field so that

$$\mathbf{n} \times \mathbf{E}^{(\text{inc})} = -\mathbf{n} \times \mathbf{E}_{\text{II}}^{(\text{sc})}, \qquad \mathbf{n} \times \mathbf{B}^{(\text{inc})} = -\mathbf{n} \times \mathbf{B}_{\text{II}}^{(\text{sc})}. \tag{13.67}$$

Here the subscripts I and II refer to the two regions of the surface of the sphere.

Without going into too much technical details, let us indicate how Kirchhoff's approximation can be used to calculate the diffraction of a plane wave from a conducting sphere. Let us assume that the plane wave is incident in vacuum ($\epsilon = 1 = \mu$) along the z-axis on a perfectly conducting sphere of radius a. Thus, we identify $\hat{\mathbf{k}} = \hat{\mathbf{z}} = \mathbf{n}'$. We assume that the origin of the coordinate system coincides with the center of the sphere. Then, we can consider a point outside the sphere to be contained in a region bounded by the surface of the conducting sphere as well as the large spherical surface at infinity. With a little bit of analysis, it can be shown that the surface integral over the large sphere at infinity vanishes. Thus, we can write the scattered

electric field in terms of the fields on the surface of the conducting
sphere as

$$
E_i^{(\text{sc})}(\mathbf{x}) = -\frac{1}{4\pi} \int ds' \left[\epsilon_{j\ell m} \left(\mathbf{n}' \times \mathbf{E}^{(\text{sc})} \right)_m \partial'_j G_{\ell i} \right.
$$
$$
\left. - ik\, B_m^{(\text{sc})} \epsilon_{mj\ell} n'_j G_{\ell i} \right]
$$
$$
= -\frac{1}{4\pi} \int ds' \left[\epsilon_{j\ell m} \left(\hat{\mathbf{k}} \times \mathbf{E}^{(\text{sc})} \right)_m \partial'_j G_{\ell i} - i\, B_m^{(\text{sc})} \epsilon_{mj\ell} k_j G_{\ell i} \right],
$$
$$
(13.68)
$$

where the surface integral is over the conducting surface. As we have
noted earlier, the surface integral can be divided into two parts, one
over the illuminated region and the other over the shadow region and
the boundary conditions are different for the two regions in Kirchhoff's approximation. We see from (13.67) that we can write the
contribution from the shadow region as

$$
E_{\text{II},i}^{(\text{sc})}(\mathbf{x}) = \frac{1}{4\pi} \int ds' \left[\epsilon_{j\ell m} \left(\hat{\mathbf{k}} \times \mathbf{E}^{(\text{inc})} \right)_m \partial'_j G_{\ell i} \right.
$$
$$
\left. - i B_m^{(\text{inc})} \epsilon_{mj\ell} k_j G_{\ell i} \right].
$$
$$
(13.69)
$$

The shadow region has an interesting property that as long as
the observation point $\mathbf{x}$ does not lie inside the surface of the sphere,
the shadow integral can be evaluated over any surface bounded by
the diametric vertical plane (that separates the illuminated and the
shadow regions). This property can be easily proved using Gauss'
theorem, but the consequence of this is that the shadow integral can
be simplified and evaluated over the diametric plane. Here $z = 0$
and we can parameterize $\mathbf{x}' = \boldsymbol{\rho} = x\hat{\mathbf{e}}_x + y\hat{\mathbf{e}}_y$. Furthermore, if the
observation point is very far away from the surface of the sphere,
namely, $x \gg x'$, then we can approximate

$$
g(\mathbf{x} - \mathbf{x}') \approx \frac{e^{ikx}}{x} e^{-i\mathbf{k}\cdot\mathbf{x}'} = g(x)\, e^{-i\mathbf{k}\cdot\mathbf{x}'},
$$
$$
G_{ij}(\mathbf{x} - \mathbf{x}') \approx \left(-\frac{k_i k_j}{k^2} + \delta_{ij} \right) g(x)\, e^{-i\mathbf{k}\cdot\mathbf{x}'},
$$
$$
(13.70)
$$

where we have used the definition of the tensor Green's function in
(13.52). Using this in (13.69), we obtain that at large distances away
from the conducting sphere, the contribution coming from the shadow

region has the form

$$E_{\mathrm{II},i}^{(\mathrm{sc})}(\mathbf{x}) \approx -\frac{ik}{4\pi}\,g(x)\int ds'\left[-E_i^{(\mathrm{inc})} + \left(\hat{\mathbf{k}}\times\mathbf{B}^{(\mathrm{inc})}\right)_i\right]e^{-i\mathbf{k}\cdot\mathbf{x}'}$$

$$= \frac{ik}{2\pi}\,g(x)E_i^{(\mathrm{inc})}\int ds'\,e^{-i\mathbf{k}\cdot\mathbf{x}'}, \qquad (13.71)$$

where $E_i^{(\mathrm{inc})}$ represents the amplitude of the incident wave and we have used the relation

$$\hat{\mathbf{k}}\times\mathbf{B}^{(\mathrm{inc})} = -\mathbf{E}^{(\mathrm{inc})}. \qquad (13.72)$$

The surface integral can now be done in a straightforward manner

$$\int ds'\,e^{-i\mathbf{k}\cdot\mathbf{x}'} = \int \rho d\rho d\phi'\,e^{-ik\rho\sin\theta\cos(\phi-\phi')}$$

$$= \int_0^a \rho d\rho\,2\pi J_0\left(k\rho\sin\theta\right)$$

$$= 2\pi\,\frac{aJ_1(ka\sin\theta)}{k\sin\theta}, \qquad (13.73)$$

where we have used the formulae from the standard tables (see, for example, Gradshteyn and Ryzhik, 6.5615 and 8.4111). Putting together all these, the contribution from the shadow region at large distances can be written as

$$E_{\mathrm{II},i}^{(\mathrm{sc})}(\mathbf{x}) \approx \frac{iaJ_1(ka\sin\theta)}{\sin\theta}\,\frac{e^{ikx}}{x}\,E_i^{(\mathrm{inc})}. \qquad (13.74)$$

The contribution from the illuminated region can also be calculated in a similar manner. However, the integral, in this case, is slightly involved (simply because it has to be carried out over the hemisphere) and we will not go into the details of that.

Index